Ferroelectric Crystals for Laser Radiation Control

The Adam Hilger Series on Optics and Optoelectronics

Series Editors: **E R Pike** FRS and **W T Welford** FRS

The Adam Hilger Series on Optics and Optoelectronics

Ferroelectric Crystals for Laser Radiation Control

A M Prokhorov and Yu S Kuz'minov

General Physics Institute, USSR Academy of Sciences, Moscow

Translated from the Russian by
Marianna Tsaplina

Adam Hilger
Bristol, Philadelphia and New York

British Library Cataloguing in Publication Data

Prokhorov, A. M.
   Ferroelectric crystals for laser radiation control.
   1. Crystalline solids. Effects of ionising radiation
   I. Title   II. Kuz'minov, Yu S.
   530.41

   ISBN 0-7503-0047-7

Library of Congress Cataloging-in-Publication Data are available

Published under the Adam Hilger imprint by IOP Publishing Ltd
Techno House, Redcliffe Way, Bristol BS1 6NX, England
335 East 45th Street, New York, NY 10017-3483, USA

US Editorial Office: 1411 Walnut Street, Philadelphia, PA 19102

Typeset by P&R Typesetters Ltd, Salisbury, Wilts.
Printed in Great Britain by Galliard (Printers) Ltd, Norfolk

Contents

Series Editors' Preface

Optics has been a major field of pure and applied physics since the mid 1960s. Lasers have transformed the work of, for example, spectroscopists, metrologists, communication engineers and instrument designers in addition to leading to many detailed developments in the quantum theory of light. Computers have revolutionized the subject of optical design and at the same time new requirements such as laser scanners, very large telescopes and diffractive optical systems have stimulated developments in aberration theory. The increasing use of what were previously not very familiar regions of the spectrum, e.g. the thermal infrared band, has led to the development of new optical materials as well as new optical designs. New detectors have led to better methods of extracting the information from the available signals. These are only some of the reasons for having an *Adam Hilger Series on Optics and Optoelectronics*.

The name Adam Hilger, in fact, is that of one of the most famous precision optical instrument companies in the UK; the company existed as a separate entity until the mid 1940s. As an optical instrument firm Adam Hilger had always published books on optics, perhaps the most notable being Frank Twyman's *Prism and Lens Making*.

Since the purchase of the book publishing company by The Institute of Physics in 1976 their list has been expanded into all areas of physics and related subjects. Books on optics and quantum optics have continued to comprise a significant part of Adam Hilger's output, however, and the present series has some twenty titles in print or to be published shortly. These constitute an essential library for all who work in the optical field.

Preface

In the practical application of the achievements of quantum electronics an important role is played by crystals permitting laser radiation control (frequency modulation, deviation and conversion). Unfortunately, there are only a few such crystals known at the present time. There are no universal crystals: each of them has applicability limits. The present book, along with *Physics and Chemistry of Crystalline Lithium Niobate* by the same authors (published by Adam Hilger), embraces practically all the crystals of alkaline and alkaline earth niobates and tantalates which are already widely used or will be used in the near future to control laser radiation in the visible and near IR ranges.

The book provides a description of the ferroelectric, electro-optic and nonlinear optical properties of crystals and touches upon physico-chemical aspects of the technology of growth of single crystals of concrete compounds. It is the first attempt to obtain and employ materials for nonlinear optics.

This book will promote a wider introduction, use and production of single crystals of niobates and tantalates of alkaline earth metals and will advance the development of laser engineering.

A M Prokhorov
Yu S Kuz'minov
Moscow 1990

List of Symbols

B_0	natural birefringence of the crystal
c	speed of electromagnetic radiation in free space
c	concentration
C	sample capacitance
C_0	Curie–Weiss constant
d_{ij}	nonlinear optic coefficient
d	sample thickness
$\bar{D}$	dielectric displacement vector
D	diffuse coefficient
$\bar{E}$	electric field vector
E	activation energy
E_c	coercive field
e	charge of electron
F	Fermi level
F_0	free energy for zero polarization
f_i	oscillator strength
G	Poynting's vector
$g_{ijkl}(P)$	quadratic electro-optic coefficient
h	Planck's constant
I_0	intensity of the incident beam
I	molar fraction
$J_{3/2}$	Bessel function
$\bar{k}$	wavenumber vector
k	Boltzmann constant
K_G	Glass constant
K_L, K_S	heat conductivities of liquid and solid phase
K_t	electromechanical coupling factor
l_{ij}	coherent wavelength
l	optical path length
L	domain size
L	period diffraction grating

m	mass of electron
m	interference pattern contrast
N	Fresnel number for deflector
n_t	concentration of filled traps
n_A, n_B	valences of ions
n_o, n_e	refractive index of medium
Δn	change of refractive index
Δn_s	spontaneous birefringence
$\bar{P}_s$	vector of spontaneous polarization
$\bar{P}$	vector of electrical polarization
R_s	reflectance of s-polarized radiation
$R_{ij}(E)$	quadratic electro-optic coefficient
r_{ij}	linear electro-optic coefficient
R_A, R_B, R_O	ions' radii
S	sample area
s	electrode area
S_0	Sellmeier's constant
s_{ij}	oscillator strength
SH	second harmonic
T	temperature
T_{phm}	phase matching temperature
ΔT	half-widths of SH peak
T_C	Curie temperature
t	tolerance factor
ΔT	temperature fluctuation
$V_{\lambda/2}$	half-wave voltage
v	rate pulled
$2V$	angle between the optical axes
W	light energy
α	polarizability
α	linear absorption coefficient
β	density of free charges
β	angle of refraction
γ	pyroelectrical coefficient
$\Delta\Gamma$	phase difference between s and d components of light
Δ_{ijkl}	nonlinear optical coefficient quadratic effect
δ_{ijk}	Miller's coefficient
Δn	variation of birefringence
$\mathscr{E}$	mean oscillator state
ε_0	permittivity of free space
ε_{ij}	relative dielectric permittivity constant
η	packing density
η	diffraction efficiency
θ_{phm}	phase matching angle

θ_i	angle of diffraction
θ	angle of refraction
λ_o	oscillator state
λ	wavelength
λ_{rec}	wavelength of the recording light
λ_{read}	wavelength of the reading light
μ	electrical mobility
μ	effective electrical mobility
ν	frequency
π_{ij}	piezo-optic coefficient
ρ	density of free charges
ρ_{ph}	photoconduction
ρ	resistivity ($\Omega\ cm^{-1}$)
σ	electrical conductivity
σ	mechanical stress
τ	collision relaxation time of electron
τ	time constant
φ	phase change of transmission
$\Delta\phi$	phase difference between ordinary and extraordinary beams
ω	angular frequency
ω	rotation rate

Introduction

Lasers, discovered by Soviet physicists H G Basov and A M Prokhorov, as well as by the American scientist C Townes, find various applications in physical experiments, technology and medicine. Laser radiation is used to stimulate thermonuclear reactions, to weld the elements of micro-circuits, to cut glass, to grow crystals, to make surgical operations and to obtain three-dimensional images of objects. In the near future laser radiation will be applied in computers, which will speed up their operation.

To make use of laser radiation, it is of importance to control its operation, i.e. to rotate beams, to modulate in intensity, phase and polarization, to transform IR radiation into visible and visible into UV. Crystals possessing the electro-optic effect and nonlinear optical properties are widely used for laser control. It should be noted that linear and nonlinear optical effects are inherent in crystals whose lattice has no centre of symmetry.

Over 2300 known compounds belonging to the acentric class are known at present, of which about 60% possess piezoelectric and about 15% ferro-, antiferro- and ferroelectric properties. Electro-optic coefficients have been measured for nearly 100 compounds. The electro-optic effect was discovered in a comparatively small number of substances out of the large number of potential candidates because it is difficult to obtain single crystals suitable for electro-optic studies.

The first crystal applied for laser radiation control was potassium dihydrogen phosphate KH_2PO_4 (KDP) and its analogues in which the potassium was replaced by ammonium and the hydrogen by its heavy isotope deuterium. KDP crystals are grown from aqueous solutions; the technology of their growth has already reached a high level and is now being improved. The KDP crystals are of high optical quality and find wide application, but these crystals have some shortcomings, for example comparatively low electro-optic and nonlinear coefficients responsible for the electro-optic effect and for second harmonic generation (SHG).

In recent years, ferroelectric single crystals of niobates and tantalates of alkaline earth metals have been synthesized and studied in detail. They appeared to possess high electro-optic, piezoelectric, pyroelectric and nonlinear properties. The physical properties of these crystals enable them to be widely used in devices for modulation, beam rotation and conversion of laser radiation frequency, as well as in parametric generators of light. Crystals of this class of compounds have nonlinear and electro-optic coefficients exceeding greatly those of other crystals. Suffice it to say that barium sodium niobate crystals yield 100% conversion of radiation with wavelength $\lambda = 1.06$ μm into that with $\lambda = 0.53$ μm, and crystals of a solid solution of barium strontium niobate have a half-wave voltage of 80 V, which is 40 times less than in lithium niobate and lithium tantalate and 100 times less than in the widely used crystals of potassium dihydrogen phosphate.

The potential of these crystals lies in the specificity of the crystal structure of this class of compounds. The study of the physics of the ferroelectric state, phase transitions, domain structure, as well as the interaction between optical radiation and these crystals (SHG and electro-optic effect), is of considerable scientific and practical interest.

Niobates and tantalates possess a characteristic feature — the violation of stoichiometry in the process of crystal growth. The deviation of the composition from stoichiometry leads to the appearance of imperfections in the crystal lattice which have a substantial effect upon the properties of these compounds. Therefore the ferroelectric, electro-optic and nonlinear properties of these crystals should be considered to be dependent on the chemical composition of the grown crystal rather than of the initial melt.

In view of the prospects for wide application of these crystals, it appears necessary to master the growth technology which should make allowance for the physical and chemical specificities of the corresponding compounds.

This book presents the basic physical and chemical properties of single crystals of these compounds, and gives a brief description of their growth technique and the various aspects of their application in quantum electronics and nonlinear optics. It also contains some reference material on these crystals.

Subject to the nature of the alkaline earth metals and proportions of the components, the compounds have different crystal structures, of which the most important from the viewpoint of the application in quantum electronics are perovskite, pseudo-ilmenite and potassium tungsten bronze-type structures.

Chapters 1, 2 and 3 deal with perovskites which include potassium niobate, potassium niobate tantalate, lead magnesium and zinc niobates. The latter compounds have found their application in industry. The physical and chemical characteristics of these compounds, phase diagrams, crystal structure and phase transitions are considered. The growth of single crystals of these compounds is discussed. The most important physical, optical and nonlinear characteristics of these crystals necessary for their device use are presented.

The subject of chapter 4 is the solid solution of barium strontium niobate. Single crystals of these compounds possess the highest electro-optic coefficients.

The growth of these crystals is also connected with the specificities of the components and the crystal structure. The physico-chemical data, phase diagrams, high-temperature phase transitions, polarization and methods of growth are examined. The dependence of optical and ferroelectric properties on the crystal or melt composition is described.

Chapter 5 deals with barium sodium niobate (BNN) crystals which give 100% conversion of laser radiation with $\lambda = 1.06$ μm into the second harmonic. The physico-chemical characteristics and phase diagrams of this compound are presented, the stoichiometry violations occurring are pointed out, and the compositions recommended as congruent are listed. The optical and electro-optic properties and the efficiency of second harmonic generation, depending on the composition, growth technique and thermoelectric treatment in the course of polarization and detwinning of these crystals, are discussed. The methods of growing BNN crystals, their polarization and detwinning are outlined.

Chapter 6 is devoted to the analogues of barium sodium and barium strontium niobates obtained by partial or complete replacement of alkaline earth elements by alkaline or rare earth ones, as well as the replacement of niobium by tantalum.

Chapter 7 presents a consideration of two crystals which were discovered not long ago but have already found wide application in nonlinear optics. These are potassium titanate phosphate (KTP) and a low-temperature modification of barium borate (BBO). They are highly resistant to laser radiation and have relatively high nonlinear optical coefficients and transmission in the ultraviolet, which makes them promising for obtaining higher harmonics.

Chapter 8 is concerned with the interaction of laser radiation with crystals, which leads to the formation of optical inhomogeneity. The mechanisms of laser-induced variations of the refractive indices are established, and the factors affecting the sensitivity of crystals to laser radiation and the possibilities of using this phenomenon for holographic information recording on crystals are considered.

The interaction of ferroelectric, electro-optic and nonlinear properties with structural specificities of crystals is treated in chapter 9.

This book is a logical continuation of the authors' 1975 book *Lithium Niobate and Tantalate — Materials for Nonlinear Optics* (Moscow: Nauka) and covers the wider class of compounds of the niobates and tantalates of alkaline earth metals.

PART I

PEROVSKITE-TYPE FERROELECTRICS

In recent years it has been revealed that ferroelectric single crystals may be used for laser beam modulation and control over the Q factor of laser sources, and also as nonlinear elements [1, 2]. This new field of application has been rapidly developed.

The number of ferroelectrics known at present has reached several hundred, and goes on increasing. Especially fruitful was the search for new ferroelectrics in the structural type of perovskite [3–10]. The group of perovskite-type ferroelectrics has now been extended to 500, mainly at the expense of complex compounds synthesized in the solid phase. A large contribution to this field has been made by Soviet researchers.

Ceramic technology makes it possible to obtain rather simply an extensive class of complex compounds, and to find among them substances with interesting physical properties. But a more complete and thorough analysis requires single-crystal compounds. The determination of physical parameters in different crystallographic directions helps establish crystal suitability for device use. With the examples of $BaTiO_3$, $LiNbO_3$, $K(NbTa)O_3$ and other compounds one can see the wide range of investigations and practical applications of single crystals as compared to polycrystals. In view of the fact that single crystals were obtained only in a few out of all the abundance of complex perovskites, and that some of them have already exhibited unique properties [11, 12], it becomes obvious how important and necessary it is to obtain single crystals of complex perovskites.

In the search for new perovskite-type ferroelectrics one should consider two questions. (i) What requirements should be met by ions in order that crystals possess a perovskite-type structure? (ii) What factors determine the appearance of ferroelectricity in perovskite-type crystals?

The perovskite structure (figure 1.1, chapter 1) is a very dense cubic packing of oxygen ions and A and B cations (the general formula of the compounds is ABO_3), where B cations occupy octahedra formed by O ions. It may also be specified as a structure, where BO_6 octahedra are joined by their vertices so that infinite rectilinear chains –O–B–O–B–O–B–O– are formed in three mutually perpendicular directions. Such a way of linking regular octahedra leads to a cubic lattice with the space symmetry group Pm3m, where A cations are surrounded by 12 anions positioned in the vertices of an octahedron. If we note that in this structure cations with cubic and octahedral surroundings may be different, and also that some of the cation positions may remain vacant (defect structures), it becomes clear in what manner the family may be extended at the expense of complex composition.

The decisive role in the appearance of perovskite-type structure is played by the relationship of ion sizes, which is formally expressed in terms of the 'tolerance factor' t (provided that the crystal as a whole is electrically neutral).

G A Smolensky and co-workers [13, 14] defined the geometrical conditions for the existence of complex compositions with perovskite-type structure using the factor t calculated from the mean ion radii $\bar{R}_A$, $\bar{R}_B$ and R_O:

$$\bar{R}_A + R_O = t\sqrt{2}(\bar{R}_B + R_O)$$

where $0.8 \leqslant t \leqslant 1.05$.

Here

$$\bar{R}_A = \sum_{i=1}^{q} R_{A_i} y_{A_i} \qquad \bar{R}_B = \sum_{m=1}^{r} R_{B_m} y_{B_m}$$

y_{A_i} and y_{B_m} being ion concentrations in the A and B positions respectively. It should be noted that the limits on the condition $0.8 \leqslant t \leqslant 1.05$ are established only approximately.

An essential shortcoming of the t factor is that it does not allow for qualitative distinctions between simple and complex perovskites. In complex perovskites, one and the same positions (e.g. B positions) may be occupied by ions of quite different radii, which leads to a subsequent expansion in the geometrical limits of the existence of perovskite-type structures.

An important factor in crystal formation is meeting the condition of electrical neutrality, that is, the requirement of equality between positive and negative ion valencies in any crystal.

The condition of electrical neutrality on complex perovskites without vacancies was reported in [13, 14], where it was shown that, if the formula of an oxygen-containing perovskite is

$$(A_1 y_{A_1} \ldots A_i y_{A_i})(B_1 y_{B_1} \ldots B_m y_{B_m})O_3$$

where y_{A_i} and y_{B_m} are ion concentrations in the A and B positions, then the condition of electrical neutrality will be

$$\sum_{i=1}^{q} n_{A_i} y_{A_i} + \sum_{m=1}^{r} n_{B_m} y_{B_m} = \bar{n}_A + \bar{n}_B = 6$$

where n_{A_i} and n_{B_m} are the valencies of A_i and B_m cations, and $\bar{n}_A$ and $\bar{n}_B$ are the mean ion valencies in the A and B positions.

A necessary condition for the formation of complex perovskites is the observance of the following rule: a complex chemical compound with a perovskite-type structure may exist, provided that from the formula composition one cannot separate simpler perovskites with the same cations in A and B positions. If such simple compositions exist then the complex composition is either a solid solution or a mixture, depending on whether or not the conditions of the formation of solid solutions are satisfied.

The general formula of complex compositions with perovskite-type structure admits three and more cations in A and B positions. It may be shown, however, that as distinct from perovskites with doubled cations, such cations are formally solid solutions of simpler ones [15]. So perovskites with cations tripled in B positions split into simpler ones in the following cases.

(i) Valencies of two different ions are identical. In this case

$$A(B'y_1, B''y_2, B'''y_3)O_3 = y[A(B'y_1, B''(y_2 + y_3))O_3]$$
$$+ (1 - y)[A(B'y_1 B'''(y_2 + y_3))O_3].$$

(ii) Valency of one of the ions is equal to the mean, for example, $y_1 n_1 + y_3 n_3 = \bar{n}(y_1 + y_3)$ and $n_2 = \bar{n}$. In this case

$$A(B'y_1 B''y_2 B'''y_3)O_3 = y_2[AB''O_3] + (1 - y_2)[A(B'y'B'''y''')O_3]$$

where $y' = y_1/(y_1 + y_3)$ and $y''' = 1 - y'$.

The consideration presented in [15] is entirely formal. The question of whether the perovskites with tripled cations are chemical compounds or solid solutions may be answered in each particular case only after a corresponding examination.

The ferroelectric properties in perovskites were originally explained by B ion displacement in oxygen octahedra in the lattice [1]. Soon it became clear, however, that ion displacement alone cannot account for the ferroelectricity, and that one should also take into account other factors, the most important of which are polarizability and the character of the chemical bond (the degree of covalence). It was established that the presence of partially covalent bonds, along with ion bonds, plays a decisive role in the appearance of mutually uncompensated dipoles (spontaneous polarization).

The formation of non-equivalent, mutually uncompensated chemical bonds is possible only in ion–covalent lattices due to saturation and directivity of covalent B–O bonds. The character of the B–O bond is affected by the polarization characteristics of the A ions.

The establishment of polarity over the whole crystal is promoted by the crystal lattice which contains an infinite frame of oxygen octahedra linked by rigid ion–covalent bonds which prevent non-equilibrium bonds from mutual compensation [16]. There exists mathematical evidence that such structures possess strong internal electric fields [17, 18].

According to the crystallochemical criterion formulated in [19] and [20], ferroelectricity may be inherent in crystals containing oxygen octahedra linked with their vertices and filled completely or partially by cations with a vacant penultimate shell (e.g. a noble gas) and with a large charge and a small ion radius. Later, G A Smolensky replaced this criterion by the following proposition: 'The appearance of spontaneous polarization may be expected in crystals containing lead ions in which central ions have no electron structure of inert gas atoms and do not occur from atoms with unfilled penultimate shells'.

Some time later it was pointed out that, alongside Pb^{2+} ions, Bi^{3+} and Ti^{4+} ions may also favour the appearance of the ferroelectric or antiferroelectric state. This criterion permitted the discovery of ferroelectric and antiferroelectric properties in many complex perovskites, such as $PbMg_{1/3}Nb_{2/3}O_3$, $PbSc_{1/2}Nb_{1/2}O_3$, $PbMg_{1/2}W_{1/2}O_3$ and $Na_{1/2}Bi_{1/2}TiO_3$. But the discovery of a whole number of new ferroelectrics showed that the crystallochemical criterion should be further developed.

The study of the dielectric properties of ferroelectrics with complex composition made it clear that the maxima of ε and $\tan \delta$ corresponding to phase transitions are as a rule extended along the temperature scale, and shift towards high temperatures with increasing frequency of the applied field, and that the dielectric polarization has a relaxation character [21, 22]. Such ferroelectrics are called ferroelectrics with a smeared phase transition.

In these compounds, several types of ions are present in the same crystallographic positions. Due to inhomogeneity of the composition over the crystal volume, regions with different composition pass over to the ferroelectric state at different temperatures, which smears the phase transition. Inhomogeneity of the composition may be caused by an incomplete diffusion in the course of crystal formation or by fluctuations of the composition in the static distribution of different types of ions in the same crystallographic positions.

1 Single Crystals of Potassium Niobate

1.1 Structural characteristics and phase transitions

Ferroelectricity in potassium niobate was first discovered by Matthias, Wood and Holden in 1949 [23]. The Curie point of $KNbO_3$ crystals lies at 435 °C. The structure of $KNbO_3$ above the Curie temperature is cubic with the symmetry group $Pm3m(O_h)$. Each Nb ion is surrounded by six oxygen atoms located in the corners of a regular octahedron (figure 1.1). The octahedra are linked by their corners into a framework enclosing large octahedral holes which are occupied by the K atoms. Below the Curie temperature, the crystal becomes a tetragonal ferroelectric which belongs to the point symmetry group $4mm(C_{4v})$. With further cooling through ≈ 225 °C, the crystal becomes rhombohedral with the point symmetry group $mm2(C_2)$ and then, at a temperature of -10 °C, it changes to the point group $3m(C_3)$.

At room temperature the $KNbO_3$ crystal has an orthorhombic structure, with two formula units per unit cell. The parameters now available for the $KNbO_3$ unit cell in this modification are listed in table 1.1.

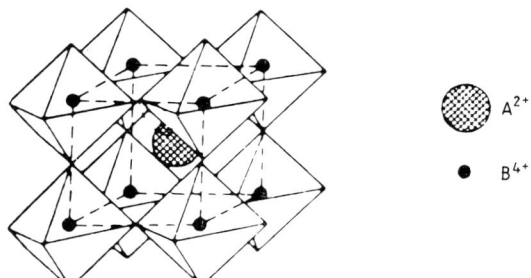

A^{2+}

B^{4+}

Figure 1.1 Perovskite structure of ABO_3 type compounds.

Table 1.1   Parameters of the unit cell of $KNbO_3$ in the rhombohedral modification (Å).

a	b	c	c/a	References
5.695	3.973	5.721	1.0046	[24]
5.692 ± 0.002	3.972 ± 0.002	5.720 ± 0.002	1.0049	[25]
5.704	3.984	5.736	1.0056	[26]
5.702	3.984	5.739	1.0065	[27]
5.697	3.971	5.720	1.0040	[28]
5.6946	3.9714	5.7203	1.0045	[29]
5.703	3.971	5.722	1.0033	[30]

In the cubic modification at a temperature of 510°C, the cell edge $a = 4.0252$ Å, in the tetragonal phase at 270°C $a = 3.9992$ Å, $c = 4.0647$ Å and $c/a = 1.0164$. No detailed studies of the temperature dependence of lattice parameters in the rhombohedral modification have yet been attempted. In [24] it is only reported that at -140°C $a = 4.016 \pm 0.002$ Å and $\alpha = 89°50' \pm 1'$.

The unit cell of the rhombic $KNbO_3$ can be represented as an ideal perovskite cell extended along the face diagonal (figure 1.2). If a crystallographic axis z is sent along this extended diagonal then, according to [28], the atoms

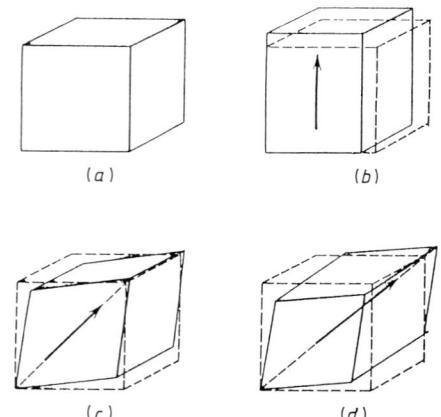

(a)   (b)

(c)   (d)

Figure 1.2   Unit cells of a $KNbO_3$ crystal in the cubic (a), tetragonal (b), orthorhombic (c) and rhombohedral (d) phases. The original cubic cell is shown by broken lines. The arrow indicates the direction of spontaneous polarization P_s in each phase.

of the orthorhombic unit cell will have the coordinates

$$\text{Nb}: (0; 0; 0) \qquad\qquad \text{O}_\text{I}: (0; \tfrac{1}{2}; z_{\text{O}_\text{I}})$$

$$\text{K}: (0; \tfrac{1}{2}; \tfrac{\overline{1}}{2} + z_\text{K}) \qquad \text{O}_\text{II}: (\tfrac{1}{4} + x_{\text{O}_\text{II}}; 0; \tfrac{1}{4} + z_{\text{O}_\text{II}})$$

where $z_\text{K} = 0.017 \pm 0.001$, $z_{\text{O}_\text{I}} = 0.021 \pm 0.002$, $z_{\text{O}_\text{II}} = 0.035 \pm 0.02$, $x_{\text{O}_\text{II}} = 0.004 \pm 0.002$.

It can be shown that oxygen octahedra remain almost regular, but the Nb are displaced from their centres by $0.030c$, i.e. by $0.17\,\text{Å}$. In the analysis of ion displacements in the ABO_3-type structure, considerable attention is given to packing and to the parameter t introduced by Goldschmidt:

$$t = \frac{(R_\text{A} + R_\text{O})1.06}{0.95(R_\text{B} - R_\text{O})\sqrt{2}}$$

where R_A, R_B and R_O are the radii of A, B and O ions in the ABO_3-type structure. The coefficients 1.06 and 0.95 were introduced by Wainer [31] as corrections due to the violation of valence, coordination number and the ratio of ion radii. The parameter t is responsible for the character of A–O and B–O ion contacts. So $t = 1$ corresponds to close contacts of the A and B ions, $t > 1$ to a loose arrangement of the B cations and $t < 1$ to a loose arrangement of the A cations.

An analysis using the above formula shows that compounds with $t > 1$ and $t \approx 1$ often prove to be ferroelectrics, while those with $t < 1$ are antiferroelectrics. For $KNbO_3$, $R_\text{K} = 1.33\,\text{Å}$, $R_\text{Nb} = 0.66\,\text{Å}$ and $R_\text{O} = 1.36\,\text{Å}$, which yields the value $t = 1.051$, typical of the ferroelectric phase.

The various model theories of perovskite-type ABO_3 ferroelectrics suggest one or another degree of covalence of ion bonds in the crystal lattice. Matthias [32], for example, believes that purely ion bonds are realized in ferroelectric crystals. Venevtsev and Zhdanov [33] express it more carefully, saying that these bonds are predominantly of ion character. Belyaev [16] suggests that the character of the bonds is close to that in corresponding oxides. According to Smolensky [34], Megaw [35] and Vousden [36], ferroelectrics are ion compounds with a certain degree of covalent bonds. The latter opinion is confirmed by Blokhin in [37]. Megaw also assumes that spontaneous polarization is associated with a sharp increase of covalent character of the bonds at the Curie point. Granovsky [38] attempted to estimate the effective ion charges in ferroelectric crystals of $KNbO_3$, and used these estimates to calculate the covalent ion radii. The results are listed in table 1.2. Granovsky determined the degree of covalent bonds in ferroelectric crystals of $KNbO_3$ as quite noticeable.

Highly symmetric $BaTiO_3$ and $KNbO_3$ phases can be described as parallel (for the tetragonal ferroelectric phase) or antiparallel (for the cubic paraelectric

Table 1.2 Effective ion charges q and ion radii r (Å) for $KNbO_3$ crystals in different phases [38]. Electron charge is taken as unit charge.

Ions Phase	K	Nb Cubic	O	K	Nb Tetragonal	O_I	O_{II}
q	+0.96	+1.92	−0.96	0.92	+1.79	−0.67	−1.02
r	1.35	0.81	1.03	1.37	0.90	0.98	1.04

Phase		Rhombohedral			Orthorhombic		
q	+0.96	+1.44	−0.80	+0.96	+1.48	−0.69	−1.06
r	1.35	0.95	0.99	1.35	0.94	0.97	1.05

phase) chains consisting of 10–25 unit cells extended along the [001] axis [39]. This model yields close agreement between experimental and theoretical values of the changes in the entropy and in the Curie constant for transitions in $BaTiO_3$ at 120°C and in $KNbO_3$ at 425°C.

Neutron diffraction analysis was made of the tetragonal phase of $KNbO_3$ at a temperature of 270°C in [40], where the phase transition in $KNbO_3$ is interpreted as the soft mode condensation under which the rigid oxygen octahedra oscillate relative to the K and Nb atoms. From the standpoint of the dynamical theory of ferroelectricity, we have analysed the available data on the structure and properties of $KNbO_3$, estimated the effective ion charges, and calculated (in satisfactory agreement with experiment) the values of spontaneous polarization, Curie constants and entropy variations in phase transitions.

Hewat [41] was the first to try and decode the structure of $KNbO_3$ crystals in the orthorhombic and rhombohedral modifications, using the neutron diffraction technique. Atomic displacements in the orthorhombic modification at 220°C (along the ⟨011⟩ direction) and in the rhombohedral phase at 230 K (along the ⟨111⟩ direction) are listed in table 1.3. Within the temperature range 435–225°C, a spontaneous polarization vector is aligned in the [001]

Table 1.3 Atomic displacement in the orthorhombic and rhombohedral phases of $KNbO_3$ (Å) [41].

Ions	Phase, direction of displacement	
	Orthorhombic, ⟨011⟩	Rhombohedral, ⟨111⟩
K	−0.006 ± 0.009	−0.013 ± 0.0032
Nb	0.070	0.078
O_I	−0.138 ± 0.006	—
O_{II}, O_{III}	−0.125 ± 0.004	−0.140 ± 0.004

direction (in the tetragonal position). The onset of spontaneous polarization is associated with a weak mechanical strain whose value is proportional to the square of polarization. At temperatures from 225 to $-10\,°C$, the polarization vector is aligned along the [110] direction, and the crystal symmetry becomes orthorhombic, and below $-10\,°C$ it becomes rhombohedral, the polarization now being directed along the [111] axis. All these are first-order transitions and are accompanied by release (or absorption) of the latent heat. The strain in the perovskite unit cell in the three phases of $KNbO_3$ crystal is schematically shown in figure 1.2. Table 1.4 illustrates phase transition temperatures upon heating and cooling, according to different authors.

It is worth mentioning that potassium niobate is the only ferroelectric crystal that exhibits the same sequence of phase transitions as that observed in $BaTiO_3$: cubic, tetragonal, orthorhombic and rhombohedral. The phase transition point of $KNbO_3$ is, however, higher than that of $BaTiO_3$. The former shows a higher latent heat of all the transitions than the latter. This is explained by a greater lattice distortion in potassium niobate as compared to barium titanate. So the Ti ion displacement from the centre of the oxygen octahedron in $BaTiO_3$ makes up 0.125 Å, whereas the Nb ion displacement in $KNbO_3$ is 0.17 Å. The entropy variation at the Curie point in both these crystals is approximately proportional to $(c/a - 1)$ in the tetragonal phase.

Shirane *et al* [24, 30] measured the latent heat ΔQ of transitions in polycrystalline samples of $KNbO_3$ and calculated the corresponding entropy variation ΔS. The results obtained by these authors, along with the corresponding data for $BaTiO_3$ taken from [46–49], are listed in table 1.5.

According to [50, 51], the latent heat of the ferroelectric–paraelectric transition in polycrystalline $KNbO_3$ makes up 134 ± 5 cal mol^{-1}, whereas some later measurements [52] gave a still smaller value of 110 ± 15 cal mol^{-1}.

Table 1.4 Phase transition temperatures in potassium niobates ($°C$).

Cubic–tetragonal		Tetragonal–orthorhombic		Orthorhombic–rhombohedral		
Heating	Cooling	Heating	Cooling	Heating	Cooling	References
420	410	220	200	-10	-55	24
435 ± 5	420 ± 5	225 ± 5	200 ± 5	†	†	27
434	—	224	—	‡	‡	42
425 ± 2	—	214 ± 2	—	—	—	43
430	—	197	—	43	—	44
431	426	222	207	-27	-52	45

† The transition remained unnoticed in the study of optical properties.
‡ The transition remained unnoticed in the study of dielectric properties up to $-190\,°C$.

Table 1.5 Latent heat ΔQ (cal mol^{-1}) and entropy variation ΔS (cal mol^{-1} K^{-1}) of phase transitions in KNbO and BaTiO$_3$.

Crystal	Thermodynamic functions	Cubic–tetragonal	Tetragonal–orthorhombic	Orthorhombic–rhombohedral	Maximal tetragonal distortion $(c/a - 1)$
KNbO$_3$	ΔQ	190 ± 15	85 ± 10	32 ± 5	0.017
	ΔS	0.28	0.17	0.12	
BaTiO$_3$	ΔQ	47–50	16–26	8–44	0.010
	ΔS	0.12–0.13	0.06–0.09	0.04–0.07	

1.2 Ferroelectric properties of potassium niobate

The discovery of ferroelectricity in $KNbO_3$ stimulated an extensive study of its dielectric properties. The temperature dependence of the dielectric constant of polydomain crystals of $KNbO_3$, which has a peak in each of the three phase transitions, proved to be very much the same as the curve obtained by Merz [53] for $BaTiO_3$. In $KNbO_3$, the temperature dependence of the dielectric constant above the Curie point can be described fairly accurately by the Curie–Weiss law with $C = 2.68 \times 10^5 °C$ and $T_0 = 350 °C$ [50, 51]. The spontaneous polarization approximately equal to 26 μC cm^{-2} [50, 51] was estimated from the dielectric hysteresis loops obtained for tetragonal single crystals in the vicinity of the Curie temperature.

Fukuda *et al* [54] investigated the dielectric properties of the orthorhombic single-domain crystal of $KNbO_3$ from room temperature to 250 °C at a frequency of 100 kHz. The rate of heating and cooling was about 3 °C min^{-1}. The dielectric constants along principal crystallographic axes at room temperature were $\varepsilon_a \approx 140$, $\varepsilon_b \approx 1200$ and $\varepsilon_c \approx 40$. These values differ slightly from those reported earlier in [55] and from $\varepsilon_c \approx 80$ given in [56]. The discrepancy is probably due to the fact that the $KNbO_3$ crystals used earlier were not completely single-domain. The authors of [54] point out that crystals with distinct dielectric constants have a constant phase matching temperature upon second harmonic generation with a wavelength of 1.06 μm. The dielectric constants of single-domain crystals of $KNbO_3$ are plotted in figure 1.3 as a function of temperature. In the range from room temperature to $\sim 200 °C$, the quantity ε_b decreases with increasing temperature, whereas ε_a and ε_c slightly increase, the former faster than the latter. A sharp change in the dielectric constant along the three principal directions occurs at 222 °C upon heating and at 200 °C upon cooling. Upon heating through 222 °C, the crystal becomes tetragonal. The absolute values of ε were determined above 222 °C because the domain structure is formed in the course of the phase transition.

Triebwasser [51] calculated the coefficients in the expression for the free energy:

$$F = F_0(T) + A(T - T_0)P^2 + BP^4 + DP^6 + \ldots$$

where F_0 is the free energy for zero polarization, $A = 2.60 \times 10^{-5} (°C)^{-1}$, $B = 5.0 \times 10^{-13} erg^{-1} cm^3$, $D = 4.10 \times 10^{-23} erg^{-2} cm^6$ and $T_0 = 360.4 \pm 5.9 °C$. The temperature dependence of P_s calculated with these values of the constants is in close agreement with experiment.

The majority of oxygen octahedral ferroelectrics are good insulators. Their electrical conductivity does not exceed 10^{-12} Ω^{-1} cm^{-1}. Several other oxygen octahedral ferroelectrics (e.g. bismuth titanate, bismuth ferrite) have a somewhat higher electrical conductivity of about 10^{-8} Ω^{-1} cm^{-1}, and can be referred to as semiconductors. Studies of polydomain single crystals of

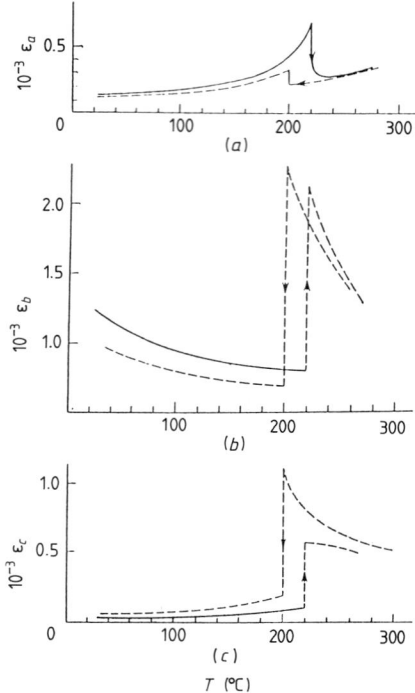

Figure 1.3   Temperature dependence of ε_a (*a*), ε_b (*b*) and ε_c (*c*) in a single-domain $KNbO_3$ crystal [55].

$KNbO_3$ revealed that this ferroelectric is a semiconductor with the forbidden bandwidth above 3 eV at room temperature and with electron-type conductivity. The forbidden bandwidth ranges from 3.6 eV at room temperature to 2.7 eV at 800 K, with the coefficient equal to 10^{-4} eV K^{-1} [44]. The values of the forbidden bandwidths undergo jump-like changes in phase transitions.

The resistance of $KNbO_3$ crystals changes drastically depending on the conditions of the growth process. For black crystals $\rho = 0.7 \times 10^4$ Ω cm, whereas for yellow samples $\rho = 4.8 \times 10^9$ Ω cm [56]. The electrical conductivity and the dielectric constant of thin lamellar single crystals of $KNbO_3$ in weak fields were measured in [57] as a function of temperature. The dependence of $\ln \sigma$ on $1/T$ in the orthorhombic phase is linear, whereas in the tetragonal phase the linearity is violated. The activation energy, calculated from the slope of the curve $\ln \sigma = f(1/T)$, in the orthorhombic modification is close to 0.034 eV, and in the tetragonal phase makes up about 0.75 eV. The character of lattice vibrations for $KNbO_3$ and their change with temperature were established by Last [58] in the absorption spectra measurements in the IR and visible regions. At room temperature, the lattice vibration spectrum of $KNbO_3$ was found to be absolutely identical with the spectrum of $BaTiO_3$

in the orthorhombic phase. These results lead to the conclusion that the IR absorption spectrum of $KNbO_3$ can be explained on the grounds of crystal symmetry and atomic positions in the cubic phase. The absorption band v_1 observed at 660 cm^{-1} is attributed to the stretching vibration (valence vibration) of the bond $Nb-O_I$ (figure 1.4(a)) while the band v_2 near 375 cm^{-1} is attributed to the bending vibration of $Nb-O_{II}$ (figure 1.4(b)). In the orthorhombic phase, the stretching vibration band must be a triplet. However, two of the lattice parameters are very close to each other, and the vibrational frequencies cannot be resolved. Last did not discover a single effect whose existence might be directly associated with the spontaneous polarization and ferroelectricity.

1.2.1 Domain structure

The domain structure of $KNbO_3$ crystals resembles that of $BaTiO_3$ in all modifications, although potassium niobate has some specific features concerning the predominant type of domain walls and the character of temperature-dependent variation of their motions, because this crystal exhibits a greater deviation from the cubic structure than $BaTiO_3$.

The static characteristics of the domains in $KNbO_3$ have not been thoroughly investigated to date, since top-quality single crystals have not been obtained. In the orthorhombic modification of $KNbO_3$ (at room temperature), the angles between the directions of spontaneous polarization of adjacent domains can be 60, 90, 120 and 180°; in the rhombohedral phase these angles are 70, 109 and 180°; in the tetragonal modification 90 and 180°.

The domain configurations in the orthorhombic phase of $KNbO_3$ are usually extremely complicated, especially if they occur from the complicated

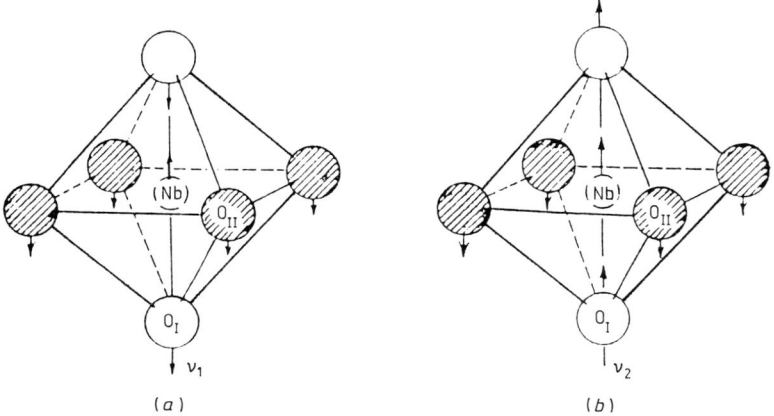

(a)   (b)

Figure 1.4 The model of high-frequency oscillations of the extension v_1 (a) and low-frequency deformation vibrations v_2 (b) of the NbO_6 octahedron in $KNbO_3$ [58].

domain pattern in the tetragonal phase. This is explained by the fact that spontaneous polarization can be directed along several equivalent crystallographic axes.

The most extensive studies of domain configurations in the orthorhombic phase have been carried out by Wiesendanger, Deshmukh, Tanaka, Kulkarni and others [59–66]. They have ascertained that 60° domains are formed, upon twinning, along pseudo-cubic {011} (orthorhombic {111}) planes which make an angle of 45° with the pseudo-cubic {100} plane. 90° domains are formed along pseudo-cubic {100} (orthorhombic {110}) planes [64]. Since the axial ratio c/a is other than unity, the angle between the orthorhombic axes of two adjacent 60° domains differs from 60° by about 57' and in 90° domains by 30' [61].

For a 90° wall to be stable in a perfectly insulating crystal, no charges should accumulate on this wall. This requires a 'head-to-tail' arrangement of the polar axes of adjacent domains. In real $KNbO_3$ crystals, 'head-to-head' and 'tail-to-tail' arrangements have been observed [60]. In this case, the charge on the wall is compensated at the expense of crystal conduction, and the wall is stable.

Applying the etching technique to $KNbO_3$ crystals in a mixture of KHF_2 and strong HNO_3 acids at a temperature of 100°C within 20 min, Wiesendanger [59] found that the crystal surface on the side of the negative end of the polar axis etches faster than on the side of the positive end.

The most effective method for studying 90° and 60° (120°) domains is the polarization optical method. The arrangement of possible domains in orthorhombic phase walls is depicted in figure 1.5 [62]. Boundaries between 90° domains, i.e. 90° walls (figure 1.5(a)) are parallel to pseudo-cubic {100} planes, which are equivalent to {110} planes in the orthorhombic phase; 60°

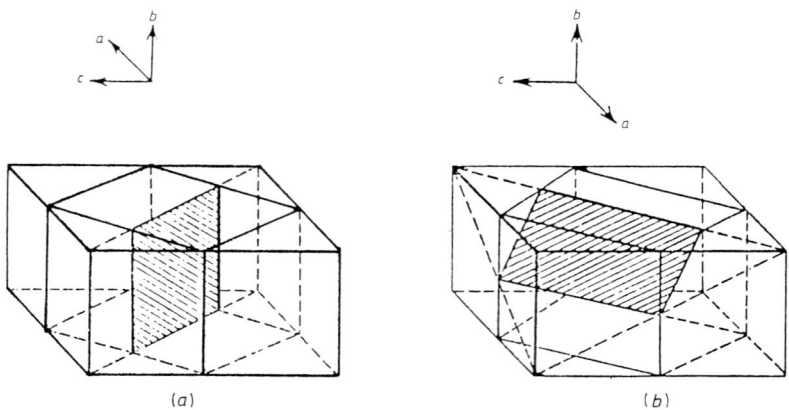

(a)   (b)

Figure 1.5   Positions of 90° (a) and 60° or 120° (b) domain walls in the orthorhombic phase of $KNbO_3$ [61].

or 120° walls (figure 1.5(*b*)) are parallel to pseudo-cubic {110} or orthorhombic {111} planes.

The main types of domains in real crystals of $KNbO_3$ are extended 90° domains with thickness from fractions of a micrometre to tens of micrometres. Such domains can be observed only with sufficiently large magnifications and only in polarized light. A more rare case is extended 60° (120°) domain walls subdividing the crystal into a net of twins with thickness from 0.1 to several millimetres. A good reflection of light makes such walls visible even with the naked eye in transparent polished plates or in rough high-quality crystals. In crystals containing a noticeable amount of solvent inclusions and other imperfections, a dense net of small (up to 10 μm) 60° or 120° domain walls can be observed. The presumable cause of their appearance may be the local mechanical stresses occurring near sub-microscopic inhomogeneities [67].

The optical technique cannot reveal the presence of 180° domains because optical indicatrices in these domains have one and the same orientation. In this case the etching technique proves efficient. As has already been mentioned [59], the negative end of the polar axis in the crystal etches faster than the positive end. Some pictures obtained for different crystallographic cuts of a single-domain $KNbO_3$ crystal using such etching techniques are presented in figure 1.6.

The domain structure of $KNbO_3$ crystals is closely connected with their general perfection. In top-quality crystals, grown with suitably chosen crystallization regimes and melt composition, only 90° and 180° domains are observed. The former have, as a rule, a comparatively large thickness, up to tens of micrometres, and run throughout the crystal. Their walls are usually parallel to the crystal growth direction. 60° or 120° domains or the twins described above are completely missing. In the tetragonal phase, such crystals are single-domain ones, as shown in [63]. Imperfect crystals contain a net of comparatively small-sized twins, and a large amount of local 60° (120°) domain walls occur [67].

1.2.2 *Formation of single-domain crystals of potassium niobate*

As mentioned above, at room temperature the crystal of $KNbO_3$ is divided into ferroelectric domains with twelve admissible directions of spontaneous polarization [69]. It was established experimentally that the most effective way to make a single-domain crystal is to apply a DC field parallel to the crystal polarization direction. Figure 1.7 shows a schematic drawing of an electric field applied to a $KNbO_3$ crystal. To achieve the largest possible size of a single-domain crystal, electrodes are deposited on natural (100) faces of the crystal. The areas covered with electrodes are shaded in the figure. Such geometry makes it possible to apply the electric field along the polar axis c_0 in the orthorhombic position. Crystals are made single domain by way of applying a DC field of 1 kV cm^{-1} at a temperature of 180°C in a silicone

(a)   (c)

(b)   (d)

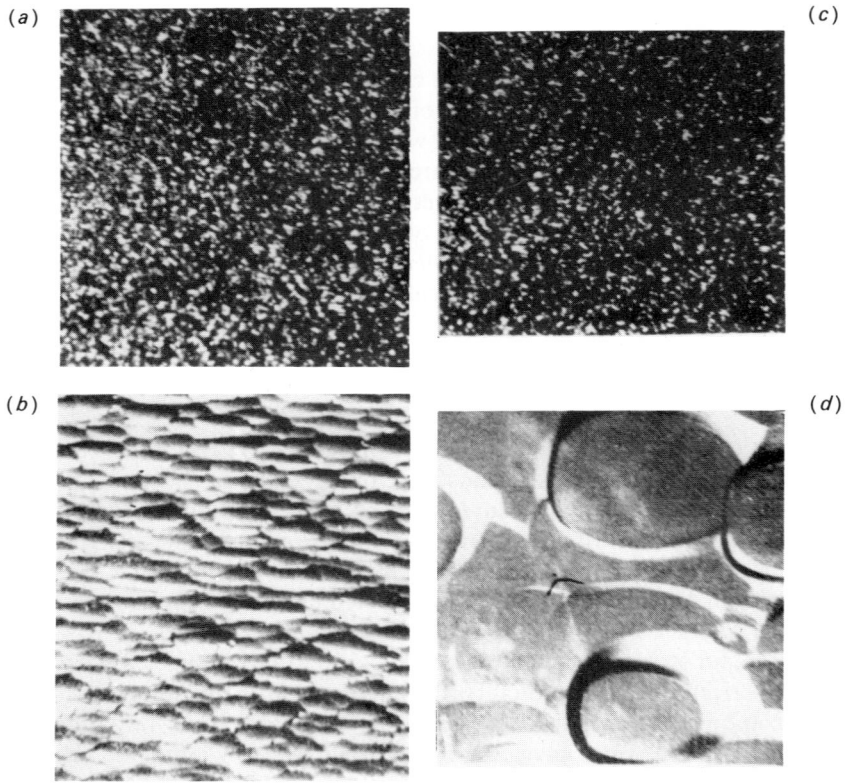

Figure 1.6 Microphotographs of etch patterns of a single-domain KNbO$_3$ crystal for different crystallographic cuts [68]: c cut $(+)$ (a); c cut $(-)$ (b); ac cut $(+)$ (c); ac cut $(-)$ (d).

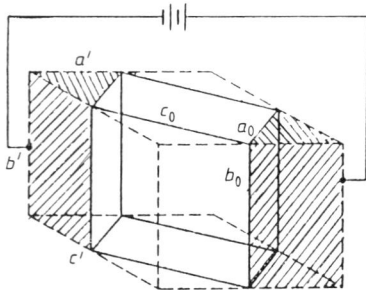

Figure 1.7 The scheme of application of an electric field to a KNbO$_3$ crystal for its polarization [68]. Broken lines indicate the natural faces to the crystal; a_0, b_0, c_0 are pseudo-cubic axes; a', b', c' are orthorhombic axes of the crystal.

bath. The process is controlled by etching techniques as well as by the output of the second harmonic of laser radiation (YAG: Nd laser).

1.3 Optical and electro-optic properties

The optical properties of $KNbO_3$ crystals, as distinguished from their dielectric properties, have not been investigated in detail, in spite of the fact that as far back as 1968 potassium niobate was found [70] to belong to the most suitable materials for nonlinear optics. This lack of knowledge concerning the optical properties is mostly due to the difficulties encountered in growing top-quality single crystals.

The most extensive study of the optical properties of $KNbO_3$ was carried out by Uematsu [71]. Prisms of 7×9 mm^2 in the base and a vertical angle of $30 \pm 1°$ were used to measure the refractive indices of the crystals. The refractive indices were measured using the minimum deviation technique for different laser wavelengths: 4880 and 5145 Å of an argon laser, 1.0642 μm of the fundamental wave and 0.5321 μm of the second harmonic of a YAG: Nd laser and 6328 Å of a He−Ne laser. The principal refractive indices of $KNbO_3$ measured at room temperature, to an accuracy of ± 0.0003, are listed in table 1.6. It should be noted that the tabulated values differ substantially from those reported earlier [72].

The dispersion curves of the refractive indices n_a, n_b and n_c are depicted in figure 1.8. The experimental values are marked by points, while full curves are calculated by choosing the parameters of equation (9.8), with use made of the least-squares technique. Thus the refractive indices measured in [71] satisfy Sellmeier's dispersion formula (9.8).

Sellmeier's constant S_0 and the oscillator states λ_0 for the three principal refractive indices are listed in table 9.3. The table also gives the mean oscillator states $\mathscr{E}_0 = hc/e\lambda_0$ and the dispersion parameters of the refractive indices $\mathscr{E}_0/S_0$. The values of the dispersion parameters 6.44, 6.67 and 6.81 $\times$ 10^{-14} eV m^2 for the three principal refractive indices are consistent with the dispersion parameters of the majority of oxide perovskites.

Uematsu established that at room temperature $KNbO_3$ belongs to negative biaxial crystals with the angle between the axes $2V = 66°47'$ for $\lambda = 5321$ Å.

Table 1.6 Refractive indices of $KNbO_3$ along a, b and c axes at room temperature [71].

λ (Å)	n_a	n_b	n_c
4 880	2.3526	2.4190	2.2275
5 145	2.3329	2.3941	2.2116
5 321	2.3224	2.3807	2.2029
6 328	2.2799	2.3291	2.1685
10 642	2.2200	2.2574	2.1196

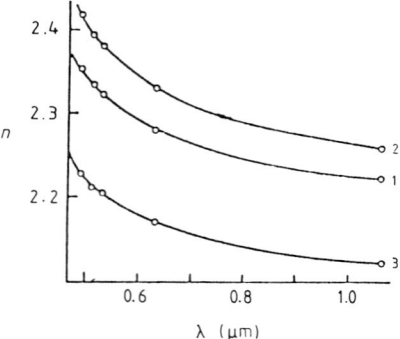

Figure 1.8 Dispersion of the principal refractive indices of $KNbO_3$ at room temperature: 1, n_a; 2, n_b; 3, n_c.

Temperature variations of the refractive indices of $KNbO_3$ for laser wavelengths of 1.0642 and 0.5321 μm are depicted in figure 1.9. The refractive indices n_a and n_c increase, whereas n_b decreases as temperature increases from room temperature to $\sim 150\,^\circ$C.

The birefringence of $KNbO_3$ crystals measured in [72] as a function of temperature is given in figure 1.10. Measurements were performed on three single-domain V-shaped plates cleaved perpendicularly to the principal crystallographic axes. At room temperature, the birefringence for $\lambda = 0.564$ μm is equal to 0.174, and gradually decreases down to 0.142 at the point of transition to the tetragonal phase. The figure represents two values of

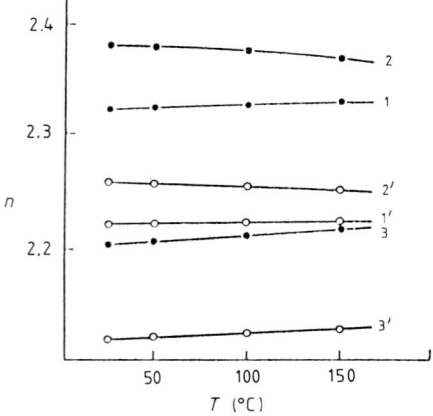

Figure 1.9 Temperature dependence of refractive indices of $KNbO_3$: 1, 1′, n_a; 2, 2′, n_b; 3, 3′, n_c for laser wavelengths. Curves 1, 2, 3 are for $\lambda = 0.532$ μm; curves 1′, 2′, 3′ are for $\lambda = 1.0642$ μm [71].

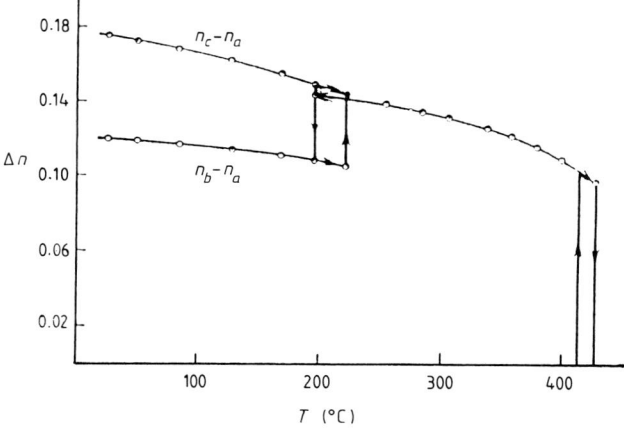

Figure 1.10 Temperature dependence of birefringence of $KNbO_3$ crystals $\lambda = 0.546 \ \mu m$ [72].

birefringence in the orthorhombic phase, the upper corresponding to $(n_c - n_a)$ and the lower to $(n_b - n_a)$. The behaviour of birefringence in the tetragonal phase is similar to that in the orthorhombic phase. Near the Curie point $\Delta n \approx 0.1$, and at the transition point the birefringence vanishes sharply.

Potassium niobate, as with other oxygen octahedral ferroelectrics, exhibits high linear electro-optic coefficients [73]. In the orthorhombic phase, the linear electro-optic matrix of $KNbO_3$ has five independent non-zero coefficients: r_{13}, r_{23}, r_{33}, r_{42} and r_{51}. In the tetragonal phase, only three of the five linear coefficients are independent since the point symmetry group 4mm requires that $r_{13} = r_{23}$ and $r_{42} = r_{51}$.

Electro-optic measurements were carried out at room temperature on single-domain samples of $KNbO_3$ at $\lambda = 0.633 \ \mu m$ and a frequency of 1 kHz [74]. The values of the coefficients obtained in these measurements are as follows (in m V^{-1}): $r_{13} = (28 \pm 2) \times 10^{-12}$, $r_{23} = (1.3 \pm 0.5) \times 10^{-12}$, $r_{33} = (64 \pm 5) \times 10^{-12}$, $r_{42} = (380 \pm 50) \times 10^{-12}$ and $r_{51} = (105 \pm 13) \times 10^{-12}$.

Crystals of $KNbO_3$ are transparent in a broad spectrum region, from 0.40 to 5.5 μm [75], but the majority of crystals usually exhibit a strong colouring due to partial niobium reduction to Nb^{4+}. Such crystals are no good for optical applications. Specimens can be partially decoloured by long oxygen annealing.

1.4 Nonlinear optical properties of potassium niobate

In investigating polycrystalline $KNbO_3$, Kurtz and Perry [70] ascertained that this material has sufficiently good nonlinear optical properties and, as distinguished from $BaTiO_3$, satisfies the synchronism conditions for Nd laser

Table 1.7 Nonlinear optical coefficients d_{ij} and coherence lengths l_{ij} of $KNbO_3$ at room temperature [71].

d_{ij}	$\dfrac{d_{ij}(KNbO_3)}{d_{33}(LiNbO_3)}$	l_{ij} (μm)	d_{ij}	$\dfrac{d_{ij}(KNbO_3)}{d_{33}(LiNbO_3)}$	l_{ij} (μm)
d_{15}	$+2.3 \pm 0.4$	2.17	d_{32}	-2.5 ± 0.2	4.88
d_{24}	-2.4 ± 0.4	1.38	d_{33}	-3.8 ± 0.2	3.19
d_{31}	$+2.2 \pm 0.2$	15.6	$d_{33}(LiNbO_3)$	1	8.59

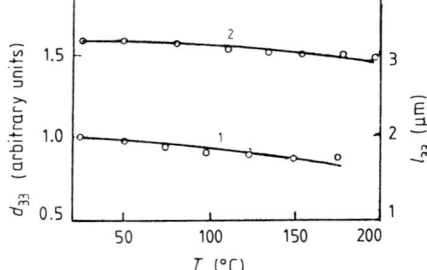

Figure 1.11 Temperature dependences of the coefficient d_{33} and of the coherence length l_{33} (1 and 2 respectively) for $KNbO_3$ [71].

transformation into the second harmonic. A more detailed nonlinear optical study of $KNbO_3$ was attempted by a group of Japanese physicists [71] who obtained high-quality optical crystals of $KNbO_3$. At room temperature, $KNbO_3$ belongs to the orthorhombic syngony (the point symmetry group 2mm) with the following independent nonlinear optic coefficients: d_{15}, d_{24}, d_{31}, d_{32} and d_{33}. The measured values of these coefficients versus d_{33} of $LiNbO_3$ and the coherent wavelengths l_{ij} calculated from these values are listed in table 1.7. The values of the coefficients d_{ij} were obtained from the Maker fringes analysis described in [73]. The relative signs of the coefficients d_{ij} for $KNbO_3$ were determined using Miller's technique [76]. The coefficients d_{33} and l_{33} as functions of temperature (figure 1.11) were estimated in [71] from separated minima of the Maker fringes, from the temperature-dependent variations of the coherent wavelength and from the second harmonic intensity. These dependences show that a non-critical phase matching is achieved at 181 °C. The temperature coefficient of $KNbO_3$, $d(n_a^\omega - n_c^{2\omega})/dT$, was estimated to be equal to 8.0×10^{-5} between 25 and 130 °C and to 1.6×10^{-4} at 180 °C.

1.5 Phase matched second harmonic generation in potassium niobate

Phase matched second harmonic generation (SHG) was attained using the YAG : Nd laser with $\lambda = 1.064$ μm. From the analysis carried out by Hobden

[77] and the refractive index measurements on $KNbO_3$, critical phase matched SHG was found to be possible for the coefficients d_{31} and d_{32}. The phase matching angles θ_{phm} for these coefficients are

$$\sin^2\theta_{phm}^{31} = \frac{(n_a^\omega)^{-2} - (n_b^{2\omega})^{-2}}{(n_c^{2\omega})^{-2} - (n_b^{2\omega})^{-2}} \qquad \sin^2\theta_{phm}^{32} = \frac{(n_b^\omega)^{-2} - (n_a^{2\omega})^{-2}}{(n_c^{2\omega})^{-2} - (n_a^{2\omega})^{-2}}.$$

Substitution into these expressions of the refractive indices of $KNbO_3$ measured at room temperature for the refractive wavelengths of 1.064 and 0.532 μm gave the phase matching angles $\theta_{phm}^{31} = 70°56'$ and $\theta_{phm}^{32} = 46°23'$. The experimental values of these angles vary in the ranges $70-72°$ and $46-48°$ respectively. The specificity of the critical phase matching lies in the fact that the Poynting vector of the fundamental wave in the crystal deviates from the second harmonic wavevector due to birefringence. For a maximum exploitation of nonlinear optical properties it is desirable that the fundamental wave and the second harmonic wave interact effectively, i.e. that $\theta_{phm} = 90°$. In this case, both beams propagate normally to the c axis without birefringence (non-critical synchronism). Figure 1.12 is a plot of the exterior and interior angles of phase matching, θ_{phm}^{31}, as a function of temperature. For the nonlinear coefficient d_{31}, the phase matching angle with the c axis becomes equal to $90°$ at a temperature of $181°C$.

The cuts of a $KNbO_3$ crystal for which the phase matching conditions are approximately satisfied in the case of normal incident light are depicted in figure 1.13. In an element with temperature tuning of synchronism, the cut perpendicular to the b axis is the working cut. For an element with angular tuning of synchronism and an acting coefficient d_{31}, the normal to the element surface makes up an angle of $20°$ with the c axis in the (bc) plane; such a cut is called a $20° bc$ cut. For the coefficient d_{32} the corresponding angle in the (ac) plane is $45°$, and such a cut is called a $45° ac$ cut. The temperature dependence of the exterior angle of synchronism for these two cuts made it possible to determine the temperature coefficients $d\theta_{phm}/dT$. They were

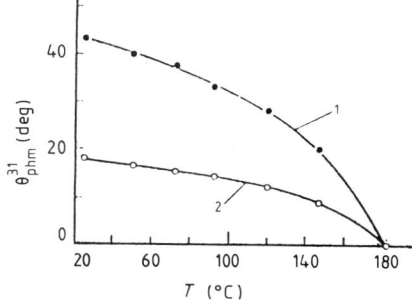

Figure 1.12 Exterior (1) and interior (2) phase matching angles θ_{phm}^{31} measured along the b axis as a function of temperature [71].

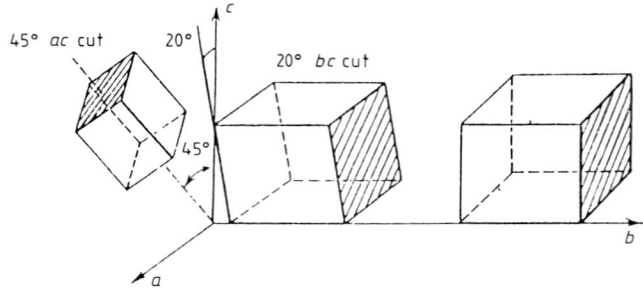

Figure 1.13 Cuts of a KNbO$_3$ crystal in SHG elements, which satisfy the phase matching conditions [68].

equal to 4.6' K^{-1} (20° cut) and 9.83' K^{-1} (45° cut). The corresponding value for LiNbO$_3$ is 14' K^{-1}.

Experiments on intra-cavity SHG with angular tuning were performed on a 20° cut of a $5 \times 5 \times 5$ mm^3 crystal of KNbO$_3$ in which the internal losses made up 0.8%. In elements with temperature tuning, the losses made up 0.25% for the optical path $l = 6.7$ mm and 12% for $l = 2.4$ mm. The authors of [68] suggest that the large losses in the former case were due to a poor quality of anti-reflection coating. The laser radiation power was 500 mW. Figure 1.14 shows the second harmonic power versus the pumping power. These data show that a SHG power equal to 150 mW was obtained for a pumping power of 1.47 kW. The transformation efficiency was calculated to be about 60%. In temperature tuning experiments, SH power equal to 360 mW was obtained for a pumping power of 1.48 kW, and the transformation coefficient was equal to 90%.

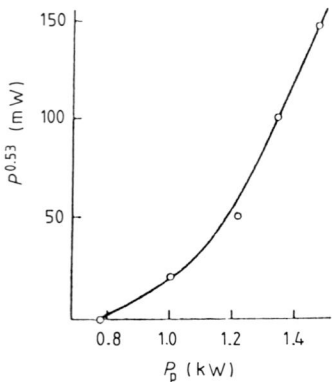

Figure 1.14 Dependence of the second harmonic power $P^{0.53}$ on the pumping power P_p [68].

We shall now give a concrete example of employing single crystals of $KNbO_3$ for intra-cavity generation of the second harmonic of YAG: Nd laser radiation [78]. Parallelepiped-shaped crystal samples were cleaved whose edges coincided with the crystallographic *a*, *b* and *c* axes. The lengths of the edges along the *a*, *b* and *c* axes were respectively $6 \times 7 \times 6$ (crystal I) and $6 \times 2 \times 6$ mm^3 (crystal II).

The surfaces normal to the *b* axis, which is parallel to the direction of propagation of light, were polished. The roughness of these surfaces did not exceed $\lambda/10$, and the deviation from plane parallelism was less than 5″. Reflective coverings for radiation of 1.06 μm were deposited on the polished surfaces. A $4.5 \times 4.5 \times 4$ mm^3 crystal of $Ba_2NaNb_5O_{15}$ was used for comparison.

In a hemispherical cavity, the power density is known to be enhanced as r_1/r_2, where r_1 and r_2 are the beam radii in a laser and nonlinear crystal respectively [79]. This enhancement is particularly important for the high transformation efficiency. The laser device is schematically shown in figure 1.15. The YAG: Nd laser cavity had two internal mirrors R_1 and R_2 with curvature radii equal to 50 and 30 cm respectively. The reflectivity of each mirror was 99.8% for $\lambda = 1.06$ μm. The transmittance of the mirror R_2 for $\lambda = 0.53$ μm was equal to 90%. The mirrors R_1 and R_2 were so arranged as to achieve single-mode TEM$_{00}$ operation of the generator. Inside the cavity, a nonlinear crystal was located at the minimum of the light spot size. The beam radii of the fundamental wave in the lasing filament and in the nonlinear crystal were $r_1 = 0.7$ mm and $r_2 = 0.16$ mm respectively. The fundamental wave and harmonic powers were measured by a calibrated thermoelectric stub. When a crystal of $KNbO_3$ was used, the output SH power was 165 mW, the pumping power being 1.48 kW. For a $Ba_2NaNb_5O_{15}$ crystal, this quantity under the same conditions was equal to 110 mW. In the case of the $KNbO_3$ crystal, no degradation of SH output was observed within many hours of laser operation. This means that SH does not affect the refractive index of $KNbO_3$, as opposed to $LiNbO_3$. With account taken of the value of the transmittance of the mirror R_2 for the second harmonic and the symmetric

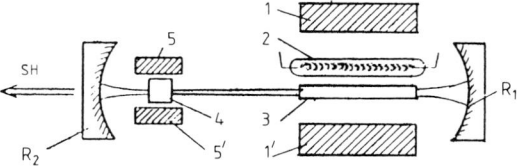

Figure 1.15 The scheme of a set for intercavity generation of the second harmonic of laser radiation [78]: 1, 1′ are elliptic reflectors; 2 is a pumping lamp; 3 is a YAG:Nd crystal; 4 is a nonlinear crystal; 5, 5′ show a furnace for heating nonlinear crystals; R_1 and R_2 are mirrors of the spherical cavity.

picture of the radiation output from the nonlinear crystal, the total SH power generated by the nonlinear crystal can be readily shown to be at least twice the indicated values.

Table 1.8 represents the values of power losses in cavity-immersed crystals (insertion losses) as well as the proper linear absorption coefficients of these crystals. The insertion losses were determined by comparison between the threshold of pumping with a nonlinear crystal in the cavity and the threshold established upon the replacement of the mirrors with a definite transmission.

The generation of an optical second harmonic inside a laser cavity was theoretically analysed by Smith [80]. He showed that for all levels of laser power, the output of the second harmonic is determined by the magnitude of nonlinearity K. The K value required for optimal matching is proportional to the linear losses L at a fundamental frequency, i.e. to the basic laser losses, and inversely proportional to the saturation S for laser transition:

$$K = 2L/S.$$

The optimal nonlinearity K of the laser system used in [78] was theoretically estimated as 3×10^{-5} cm W^{-1}, which exceeds the experimental value. Thus the system (a nonlinear crystal in a laser cavity) described in [78] is sufficiently consistent. Indeed, it has appreciable basic losses ($L = 0.085$) and a low power amplification factor as compared with the system employed in [86], in which a 100% transformation was reached. In the latter system, the losses were as low as 0.015 and the amplification factor was equal to 50.

According to Uematsu and Fukuda [78], the furnace temperature T_0, at which the output of the second harmonic is maximum, is somewhat lower than the phase matching temperature T_{phm}. This can be explained by the fact that inside a cavity, where the power density of the fundamental wave reaches several tens of watts, a nonlinear crystal even with a very low absorption coefficient is heated upon absorption of the fundamental wave. The difference between the heater temperature, which provides the maximum output of the second harmonic, and the phase matching temperature for $KNbO_3$ and barium sodium niobate (BNN) crystals, is shown in figure 1.16. It is readily seen that, irrespective of the optical length L, the temperature gradient of

Table 1.8   Insertion laser losses and absorption coefficients of crystals [78].

Crystal	Element size along the b axis (mm)	Losses (%)	Absorption coefficient (% cm^{-1})
$KNbO_3$(I)	6.7	0.24	0.18
$KNbO_3$(II)	2.4	0.12	0.25
$Ba_2NaNb_5O_{15}$	4.5	0.18	0.20

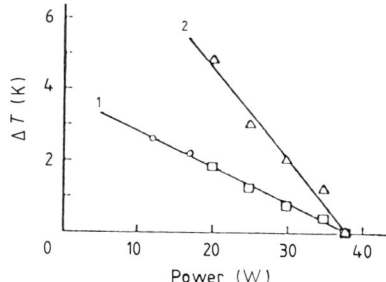

Figure 1.16   The difference of the phase matching temperature and the furnace temperature as a function of the laser radiation power inside the cavity: 1 for $KNbO_3$, 2 for BNN [78]. Optical lengths of $KNbO_3$ crystals are 0.67 and 0.24 cm (circles and squares respectively), and those of BNN crystals are 0.45 cm.

$KNbO_3$ is smaller than that of BNN, because the thermal conductivity of the latter is smaller than that of the former.

The temperature dependence of second harmonic intensity during the propagation of light in the direction of the b axis was used in [75] to estimate the quality of $KNbO_3$ crystals. As shown earlier [82], the product of the half-width ΔT of the peak in the temperature dependence of the second harmonic intensity by the optical crystal length L, as well as the symmetric position of the peak, may indicate the presence or absence of local inhomogeneities in the refractive index, i.e. may characterize the optical perfection of the crystal. The product $L \Delta T$ for potassium niobate was estimated to be 0.28–0.32 cm K. This is two to three times as low as the corresponding value for lithium niobate.

1.6   Growth of potassium niobate single crystals

The phase diagram of the system $K_2CO_3(K_2O)–Nb_2O_5$ was investigated in 1955 [83]. In this system there exist five different compounds: $3K_2O \cdot Nb_2O_5$ (I); $K_2O \cdot Nb_2O_5$ (II); $2K_2O \cdot 3Nb_2O_5$ (III); $K_2O \cdot 3Nb_2O_5$ (IV) and $3K_2O \cdot 22Nb_2O_5$ (V). Compounds II, IV and V melt incongruently at 1039, 1234 and 1279 °C respectively. Compounds I and III melt congruently at 950 and 1163 °C. As follows from the diagram (figure 1.17), due to incongruent melting of $KNbO_3$, crystals of this compound cannot be obtained from a stoichiometric melt by the Czochralski or any other frequently used method, and one has to employ the method of crystallization from the melt or the hydrothermal method.

As is known, the first attempt to grow potassium niobate crystals was made by Joly in 1877 [84]. The crystal growing technique was based on

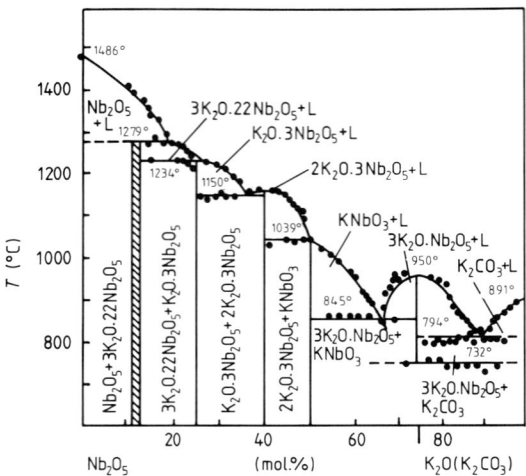

Figure 1.17   Phase diagram of the system $K_2CO_3(K_2O)-Nb_2O_5$ [83].

melting a stoichiometric amount of K_2CO_3 and Nb_2O_5 with a certain amount of CaF_2 as a flux. When reproducing this technique, Matthias and Remeika [42] did not succeed in obtaining $KNbO_3$ crystals. But using KF and KCl salts as a flux, they obtained crystals with edges measuring 1–3 mm in length. The samples were dark blue because of niobium reduction.

Later on, for growing single crystals of $KNbO_3$ the two-component system $K_2CO_3-Nb_2O_5$ with excess K_2CO_3 as solvent was usually employed [27, 56, 85]. Crystals obtained in this way were as a rule very small and possessed significant electrical conductivity. But after a 30 h melting of the $K_2CO_3-Nb_2O_5$ mixture in a platinum crucible at 1000 °C and a rapid cooling (100 °C h^{-1}), Wood [27] obtained 1–2 mm pure cubic crystals of potassium niobate. The growth conditions of single crystals of potassium niobate were examined in [56] using a differential thermal analysis of the system $K_2O-Nb_2O_5$ in the temperature range from 900 to 1400 °C. The thermograms obtained corresponded to the formation of two compounds with melting temperatures equal to 950 and 1050 °C. According to the data reported in the literature, this corresponded to the compounds $3K_2O \cdot Nb_2O_5$ (I) and $K_2O \cdot Nb_2O_5$ (II).

Next, the authors of [56] made attempts to grow single crystals of $KNbO_3$ by the high-temperature flux seeding method. The experiments suggested that overheating of the system, with the melting of the crystalline nuclei in it, leads to its overcooling as the temperature is decreased. The authors concluded that the basic difficulty in growing $KNbO_3$ crystals is the overcooling in the system. In their opinion, this can be eliminated in two ways: either by preventing the system from complete homogenization during heating or, as soon as complete homogenization is reached, by using special vessels, as low and as flat as possible. The authors managed to obtain 1 cm³ cubic crystals coloured from pale yellow to black. The appearance of a whole range of colours is explained by a partial oxygen loss by the crystals. The longer the growth period, the darker the colour.

The growth of $KNbO_3$ crystals from seeds was investigated in [86]. As a result of the change in the K_2CO_3 to Nb_2O_5 ratio from stoichiometric to $1:2$ (in weight fractions), an increase of the Nb_2O_5 concentration in the starting material was found to lead to the formation of a new non-ferroelectric compound $3Nb_2O_5.2K_2O$ melting at $1163°C$, whereas an increase of the K_2CO_3 concentration speeds up the growth, which causes imperfections such as solvent inclusions. But with a substantial increase of the K_2CO_3 concentration, the growth rate fell sharply, which led to the appearance of nodes of fine crystals, with the space between them filled with solvent.

An important contribution, which developed the methods proposed in [86], was the paper by Miller [87] on growing single crystals of potassium niobate on seeds from the melt (1.2 mol of K_2CO_3 and 1.0 mol of Nb_2O_5). He obtained crystals larger than 1 cm³. But in spite of the low cooling rate, the crystals were prone to cracking.

Clear and uncoloured $KNbO_3$ crystals were obtained in 1972 [75], following Miller, from a melted mixture with 52 mol.% of K_2CO_3. The quality of the crystals obtained was found to depend on the melt-holding temperature prior to crystallization, as well as on the temperature gradient in the furnace. Later it was established [88] that $KNbO_3$ crystals grown in oxygen flux are of higher quality. We shall discuss this paper in more detail.

Single crystals of $KNbO_3$ were obtained by the Kyropoulos method from melts composed of a mixture of K_2CO_3 and Nb_2O_5. The starting material was placed into a 250 ml platinum crucible. The growth conditions were as follows: melt composition, K_2CO_3 (52.5 mol.%) + Nb_2O_5 (47.5 mol.%); seed orientation, [001]; seed rotation, 30 rpm; cooling rate, $0.5°C\ h^{-1}$ (melt temperature control $\pm 0.1°C$); atmosphere, oxygen (a flux of 1.5 l min^{-1}). Grown under these conditions, the crystals were colourless and well faceted, reaching $40 \times 40 \times 15$ mm³ in size. Typical crystals (figure 1.18(a)) were perfect parallelepipeds faceted by $\{100\}$ planes. The crystal faces were macroscopically perfect, but the central regions of the surfaces facing the melt were rough (region A). The upper faces of the crystals (region B) exhibited striations in the [110] direction. The grown crystals were clear along the growth axis. They were relatively easily polarized by the method described above, except for region A which exhibited cracking in the course of polarization. This may supposedly imply that this region contains structural defects that induce great inner stresses. Single-domain samples cut out from regions B and C of the initial crystal are illustrated in figure 1.18(b). They do not show striations. But the sample cut out of region B possesses optical inhomogeneities which become visible if the sample is placed between crossed polaroids and exposed to a He–Ne laser beam. These inhomogeneities

(*a*)

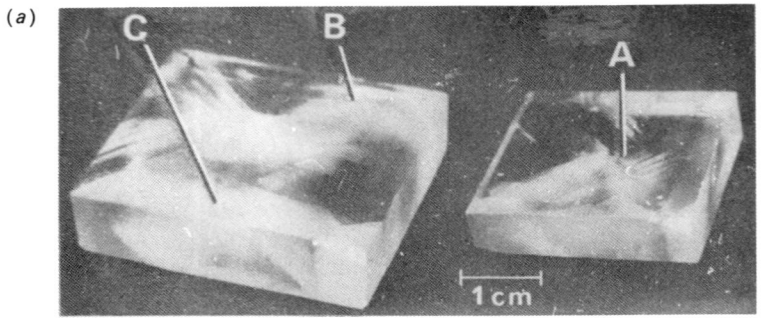

(*b*)

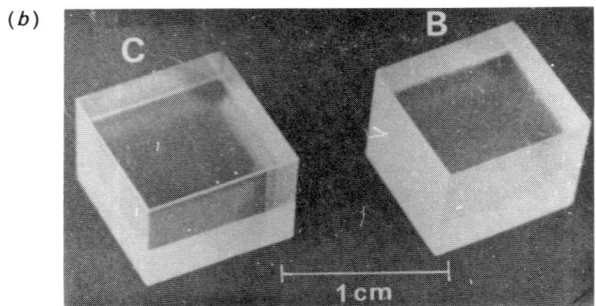

Figure 1.18 Typical Kyropoulos-grown $KNbO_3$ crystals (a) and single-domain samples (*b*) cut out of regions B and C of the original crystal [88].

correspond to growth ridges on the crystal surface. The sample obtained from region C shows neither striations nor optical inhomogeneities. Optical homogeneity of single crystals was investigated using a Twyman–Green interferometer. Small changes in the refractive index were observed only on sample edges, which are due to the stresses occurring in the course of preparation of the optical elements. The changes in the refractive index per centimetre were less than 10^{-5}, which satisfies the conditions of applicability of these elements for SHG.

To determine the mechanism of Kyropoulos crystal growth, the authors of [88] grew $KNbO_3$ crystals in unsatisfactory conditions in which the temperature gradient was $\pm 10°C$ cm^{-1}, the melt temperature was $1050°C$ and the melt hold time was 1 h. A crystal grown under such conditions had cyclic colour variations from blue to clear. The blue colour perhaps appears due to the formation of oxygen vacancies in the crystal.

Obtaining large-sized perfect perovskite-type single crystals by the Czochralski method was attempted in [89]. The authors managed to grow $LiTaO_3$, $LiNbO_3$ and $NaNbO_3$ crystals, but the attempts to grow $KNbO_3$ crystals failed. The reason was that the compound melts incongruently.

An attempt was also made to grow $KNbO_3$ crystals by the floating crystal method. This method relies on the fact that the seed remaining on the melt surface due to the surface tension force may increase in size with decreasing temperature if the melt shows a large vertical temperature gradient [90]. The crystal plates of $KNbO_3$ thus obtained had a curved crystallization front and were of poor quality.

The growth of $KNbO_3$ single crystals under hydrothermal conditions is described in [91]. Experiments were carried out in steel autoclaves lined with platinum or silver shells. KOH solution served as solvent. The temperature was varied from 400 to 600 °C; the temperature difference in the autoclave was 30–80 °C. The growth of potassium niobate crystals was observed in all cases in the upper zone of the autoclave on the shell walls. At 600 °C, the crystal formation was very energetic; it then decreased sharply through 500 °C until it ceased completely at 400 °C. The maximum size of crystals obtained using this method reached 2–3 mm. The crystals appeared blue with different colour intensity. The crystal shape depended on the temperature difference between the growth and solution zones: if it was less than 50 °C, the crystals were dendrite shaped; if more than 50 °C then isometric crystals were produced.

2 Solid Solutions of Potassium Tantalate Niobate

Electro-optic studies of perovskite-type structures have led to the discovery of a strong electro-optic effect in solid solutions of $KNbO_3$ and $KTaO_3$ (KTN), of which the solid solution of composition $KTa_{0.65}Nb_{0.35}O_3$ is of greatest interest.

2.1 Dielectric properties

The dielectric measurements of perovskite-type ferroelectrics, which include $KNbO_3$, $KNbO_3$–$KTaO_3$ solid solutions as well as $PbZn_{1/3}Nb_{2/3}O_3$ and others, can be described within the phenomenological theory due to Devonshire [1]. The free energy of an unclamped crystal can be written in the form

$$F(P, T) = F_0(T) + AP^2 + BP^4 + DP^6 + \ldots \tag{2.1}$$

where P is polarization, T is temperature, $F_0(T)$ is the part of the free energy essentially independent of polarization, and A, B and D are the temperature-dependent coefficients.

From thermodynamics it is known that

$$\partial F/\partial P = E \qquad \partial^2 F/\partial P^2 = 4\pi/(\varepsilon - 1) \tag{2.2}$$

where E is the electric field and ε is the dielectric permittivity in the paraelectric state ($P = 0$). From equations (2.1) and (2.2) it follows that

$$\varepsilon - 1 = 2\pi/A. \tag{2.3}$$

For ferroelectrics it was experimentally found that the temperature dependence of A can be explicitly written as

$$A = A'(T - T_0) \tag{2.4}$$

where A' is a temperature-independent constant and T_0 is the paraelectric Curie–Weiss temperature. Practically, for ferroelectrics $\varepsilon > 10^3$, so that one can use the approximation $\varepsilon - 1 \approx \varepsilon$. Then for the dielectric permittivity one can write

$$\varepsilon = \frac{2\pi}{A'} \frac{1}{(T - T_0)} = \frac{C}{T - T_0} \qquad (2.5)$$

where C is the Curie constant. The values of T_0 and C can be determined from a plot of $1/\varepsilon$ versus temperature in the paraelectric state of the ferroelectric. The maximum ε value is reached at a phase transition temperature T_C and is defined by the expression

$$\varepsilon_{max} = 2\pi / [A'(T_C - T_0)]. \qquad (2.6)$$

For positive B, for which $T_C = T_0$, the value ε_{max} becomes uncertain, and a second-order ferroelectric transition takes place. For negative B, for which $T_C > T_0$, the quantities A', T_0, B and D are related as

$$A'(T_C - T_0) = B^2/4D. \qquad (2.7)$$

In this case we deal with a first-order phase transition [1].

For several solid solutions of $KNb_xTa_{1-x}O_3$, the ratio $1/\varepsilon$ was plotted as a function of temperature T, and the values of A' and T_0 were determined for various $KTaO_3$ concentrations. Experimental results implied that the constant A' increases and the quantities T_0 and $(T_C - T_0)$ decrease with increasing $KTaO_3$ concentration. Then, according to equation (2.6), the peak of the dielectric permittivity ε must increase, which was shown by experiment.

From the phenomenological description of KTN it follows that the replacement of Nb ions by Ta ions affects the character of a ferroelectric first-order transition, so that it becomes a second-order transition. This is a consequence of the sign reversal of B (from negative to positive) upon which $(T_C - T_0) = 0$. In this case, according to equation (2.6) the quantity ε_{max} becomes infinitely large, but its experimental value is finite due to sample inhomogeneity, insufficiently high quality of electrodes and a finite amplitude of the signal measured. The dielectric permittivity measured in the direction parallel [1] and normal [2, 3] to the crystal growth direction is plotted in figure 2.1 as a function of temperature [2]. The peaks appearing with decreasing temperature correspond to a change of the cubic lattice into tetragonal, then through orthorhombic ultimately to rhombohedral. It should be noted that for the first transition the height and the position of peaks depend fairly little on the direction of measurement, while for the second and third transitions this difference is rather substantial. The observed temperature hysteresis decreases in the order of decreasing transition temperature and is equal to 0, $\sim$0 and 3 °C respectively. In pure $KNbO_3$ the temperature hysteresis for the same transitions equals 5, 15 and 25 °C.

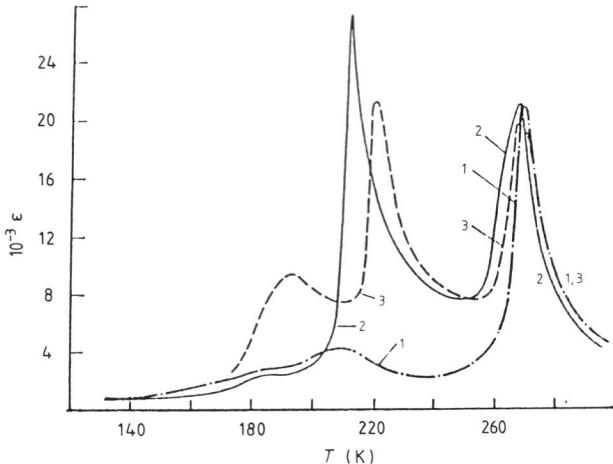

Figure 2.1 Temperature dependence of the dielectric permittivity of $KTa_{0.63}Nb_{0.37}O_3$ in the directions: 1, [001]; 2, [100]; 3, [010] [2].

The melt-grown KTN crystals exhibit optical inhomogeneity. Cubic specimens with {100} faces look homogeneous along the growth axis, while in the other two directions they show ridges (striae) perpendicular to the direction of growth. Since perovskite-type crystals should be optically isotropic, these ridges can be put down to variations of the Nb/Ta ratio in the crystal.

The dielectric permittivities of three KTN crystals were investigated in [3] as a function of temperature. The Curie–Weiss law was found to be obeyed up to temperatures 5–15 °C above the critical temperature T_C, which depends on the crystal composition. For measurements, the electric field was applied across the growth striae and parallel to the growth direction. It should be noted that the technique proposed in [3] can be employed only for measuring the ε value averaged over the entire crystal volume. For the three crystals examined (figure 2.2), the temperature dependences of $\beta = 1/\varepsilon$, in the region where the Curie–Weiss law is followed, were of the form $\beta = 0.77 \times 10^6$ $(T - 0.6)$; $\beta = 0.89 \times 10^6$ $(T - 9.9)$; and $\beta = 0.77 \times 10^6$ $(T - 20.7)$ (curves 1, 2 and 3 respectively). If the Curie–Weiss law is assumed to be followed for each individual region of the crystal, then the averaged value of β must also change linearly with temperature. However, the difference in the critical temperatures of these regions may lead to a smeared minimum of β (maximum of ε). In this respect, the crystal corresponding to curve 3 is apparently more homogeneous. The peak of the dielectric permittivity for this crystal is at 38 000, whereas for the other crystals it is only 18 000.

Ferroelectric phase transitions in the system $KNbO_3$–$KTaO_3$ were examined by Triebwasser [4] from the temperature dependence of the dielectric permittivity. Examinations were carried out on single crystals of

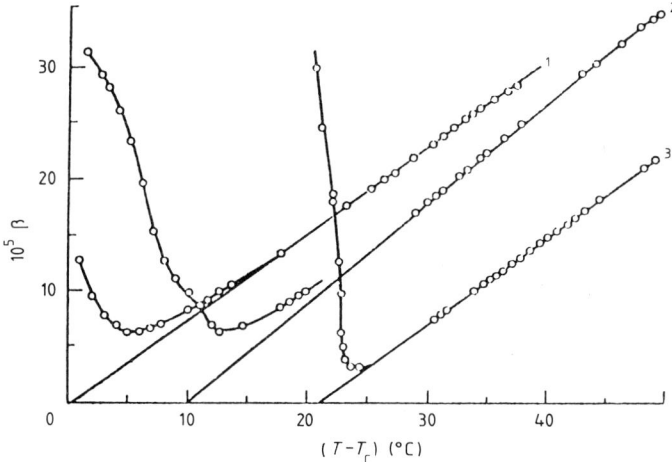

Figure 2.2 The Curie–Weiss law for KTN crystals [3].

compositions from 0 to 80 mol. % of potassium tantalate in the temperature range from −180 to +450°C. Based on these data, the schematic diagram of the KNbO$_3$–KTaO$_3$ system equilibrium in the cubic, tetragonal, orthorhombic and rhombohedral phases is depicted in figure 2.3.

In KTN crystals, the Curie temperature, the magnitude of the temperature hysteresis decrease, and the dielectric permittivity all increase with increasing percentage of potassium tantalate. The ferroelectric transition, which is first

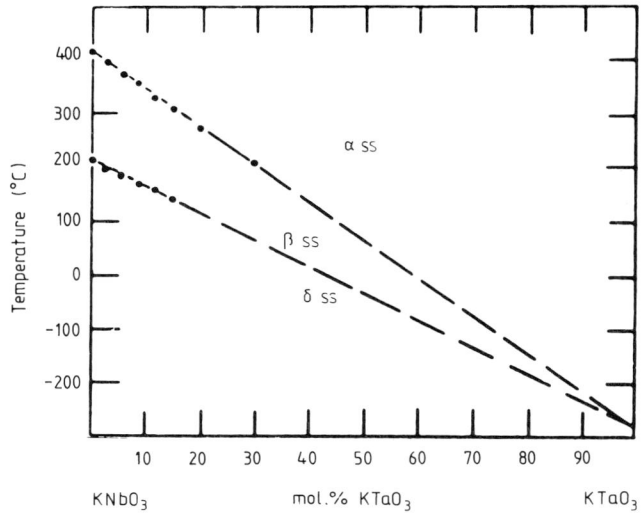

Figure 2.3 Phase diagram of the system KNbO$_3$–KTaO$_3$ [4].

order in pure potassium niobate, becomes second order when the $KTaO_3$ concentration in a solid solution exceeds 55%.

2.2 Electro-optic properties

The crystals of KTN may in principle be used as high-speed electro-optic elements—modulators, beam deflectors and optical shutters (chapter 1; [9]). In practical applications of this material, it is essential that the Curie temperature of crystals can be reduced to room temperature by selecting the composition. The necessary property is shown by crystals of the composition $KTa_{0.65}Nb_{0.35}O_3$.

As mentioned above, KTN possesses perovskite-type crystalline structure. Before describing the electro-optic properties of KTN crystals, it would be instructive for further presentation to discuss the electro-optic properties of the ideal perovskite [5].

Near the Curie point, we expand the thermodynamic potential of the crystal in a Taylor series of increasing powers of dielectric displacement D. In the expansion for perovskites, only terms with uneven powers may be present. If the coefficients of the series are defined as derivatives with respect to energy calculated for zero displacement, then the electric field can be represented as

$$E_1 = \beta_1^0 D_1 + \frac{1}{3!} \xi_{11}^0 D_1^3 + \frac{1}{2!} \xi_{12}^0 D_1(D_2^2 + D_3^2) + \dots . \qquad (2.8)$$

The reciprocal β_1^0 of the dielectric permittivity is assumed to obey the Curie–Weiss law

$$\beta_1^0 = k(T - T_C) \qquad (2.9)$$

and the other coefficients are assumed to be constant. The linear dielectric behaviour of the crystal is thus isotropic, but the anisotropy may be due to the higher-order terms only at a temperature close to T_C, where D may take on large values.

The electro-optic behaviour of the perovskite crystal should be described by exactly the same relations, with the exceptions that one of the displacements D must refer to the optical wave and all the others to the low frequencies. The coefficients β, now opacities, are defined by the expressions

$$\beta_1 = \beta_1^0 + \frac{1}{2} \xi_{11}^0 D_1^2 + \frac{1}{2} \xi_{12}^0 (D_2^2 - D_3^2) \qquad (2.10)$$

and

$$\beta_4 = \xi_{44}^0 D_2 D_3. \qquad (2.11)$$

The birefringence B, induced by the dielectric displacement D perpendicular to the light beam, will depend on crystal orientation. For example, if an electric field is applied along $\langle 100 \rangle$, the birefringence will be described by the formula

$$B_3 = \frac{n^3 \varepsilon_0}{4} (\xi_{11}^0 - \xi_{12}^0) D^2 \qquad (2.12)$$

where n is the refractive index and ε_0 is the dielectric permittivity.

If we send the field along $\langle 110 \rangle$, the birefringence will be given by the formula

$$B_3 = \frac{n^3 \varepsilon_0}{2} \xi_{44}^0 D^2. \qquad (2.13)$$

Thus the electro-optic effect in KTN may be anisotropic, in spite of the fact that for zero electric fields these crystals are optically isotropic and do not exhibit birefringence (we mean that the crystals are in the paraelectric phase).

In the course of electro-optic studies, KTN crystals were placed into an electric field that produced a sinusoidal bias in them [5]. The crystal was positioned between nicols, either crossed at an angle of 45° to the optic axis or parallel, and was exposed to collimated monochromatic light (5461 Å). The picture of the crystal was focused onto a photomultiplier. The birefringence B was evaluated from the intensity I of the transmitted light. For crossed nicols, permittance is described by the expression

$$I/I_0 = \sin^2(\pi l B / \lambda) \qquad (2.14)$$

where I_0 is the intensity of the incident beam and l is the optical path length in the crystal. For parallel nicols, the permittance follows the law $\cos^2(\pi l B / \lambda)$. Such periodic functions show maxima and minima at integer values $m = 2lB/\lambda$. Combining this condition with equation (2.12) for the case of a field oriented in the $\langle 100 \rangle$ direction, we obtain

$$m = (l n^3 \varepsilon_0 / 2\lambda)(\xi_{11}^0 - \xi_{12}^0) D^2 + (2l/\lambda) B_0 \qquad (2.15)$$

where B_0 is the natural birefringence of the crystal. The total electro-optic coefficient and the passive birefringence B_0 inherent in the crystal can be determined from the slope and intersection of the line with the ordinate in the plot of m versus D^2. Figure 2.4 represents such a graph for a single crystal of $KTa_{0.65}Nb_{0.35}O_3$ with the field applied along the $\langle 100 \rangle$ and $\langle 110 \rangle$ directions (1 and 2 respectively). The slopes of the straight lines (0.140 for 1 and 0.135 for 2) appeared to be identical within the experimental error, which implies the equality

$$(\xi_{11}^0 - \xi_{12}^0) = 2\xi_{44}^0.$$

Thus the electro-optic effect in the KTN crystal is isotropic, and the crystal orientation is not critical. Curve 1, as distinct from curve 2, does not pass through the origin, which is indicative of the effect of natural birefringence typical of this KTN specimen in the $\langle 100 \rangle$ direction.

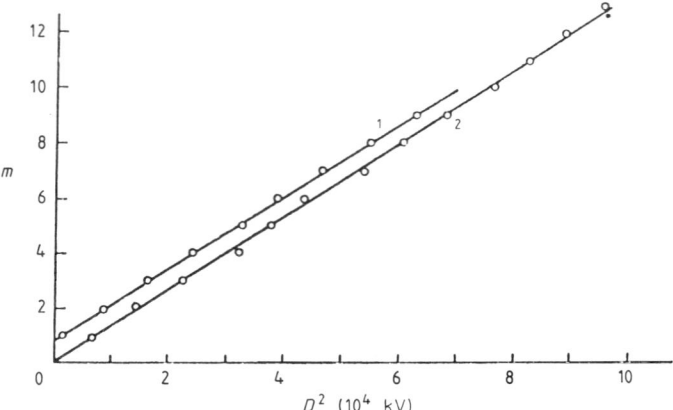

Figure 2.4 Graphic determination of the electro-optic coefficients and of natural birefringence of a $KTa_{0.65}Nb_{0.35}O_3$ single crystal for the $\langle 100 \rangle$ (1) and $\langle 110 \rangle$ (2) directions of the electric field [5].

We have spoken of measurements of the resultant electro-optic effect in KTN crystals. The half-wave voltage and the electro-optic coefficient r_{44} of these crystals were measured in [6]. At room temperature, the KTN ferroelectric belongs to the tetragonal symmetry 4mm. Consequently, this crystal has only three distinct non-zero electro-optic coefficients $r_{13} = r_{23}, r_{33}$ and $r_{42} = r_{51}$. The conventional notation is used and the standard crystal position is such that the z axis is parallel to the polar axis of the crystal [7]. The light propagating along the [100] axis of the crystal and polarized at an angle of 45° to the c axis undergoes an optical delay

$$\phi = 2\pi l(n_a - n_c)/\lambda \tag{2.16}$$

where l is the optical path length and λ is the incident light wavelength. The electric field (E_c) applied along the c axis affects the refractive indices n_c and n_a, which in turn affects the phase difference

$$\Delta\varphi = \pi l n_c^3 r_c E_c/\lambda \tag{2.17}$$

where r_c is specified by the expression

$$r_c = r_{33} - (n_a/n_c)^3 r_{13}. \tag{2.18}$$

The electric field directed along the a axis rotates the ellipsoid of refractive indices by an angle α. In this case, the electro-optic coefficient r_{42} can be determined from the expression

$$r_{42} = [(n_a^2 - n_c^2)/n_a^2 n_c^2](\alpha/E_a). \tag{2.19}$$

Selected for electro-optic studies were perfect specimens of KTN single crystals with sharp peaks in the temperature dependence of the dielectric permittivity at the Curie point ($\varepsilon \geqslant 30\,000$) and a low value of the loss tangent ($\tan \delta \leqslant 0.001$), because wide or repeated maxima on the curve $\varepsilon(T)$ indicate inconsistency of crystal composition over the volume (variability of the ratio Nb/Ta). KTN crystals were polarized at a temperature 20 °C above the Curie point in silicone oil and then slowly cooled in a DC field of 3–5 kV cm^{-1}. The half-wave voltage $V_{\lambda/2}$ was measured by the usual scheme. A He–Ne laser with $\lambda = 6328$ Å served as the monochromatic radiation source. The crystal was placed between a polarizer and analyser crossed at 45° to the c axis. A voltage of frequency 60 Hz was applied to the $\{100\}$ axes. The signal from the phototube was amplified and sent to the cathode oscillograph. The temperature dependence of the half-wave voltage $V_{\lambda/2}$ is depicted in figure 2.5 (the Curie temperature of the crystal was 60 °C). The same figure represents the temperature dependence of r_c determined from the expression

$$V_{\lambda/2} = \lambda \, d / n_c^3 \, l r_c \tag{2.20}$$

where l is the optical path length and d is the distance between the electrodes. To maintain the polar state of the crystal, a constant voltage of several hundred volts was applied to it at a temperature 5 °C below the transition temperature. The electro-optic effect remained linear.

The electro-optic coefficient r_{42} as a function of temperature was established in [6] by employing a somewhat complicated technique (figure 2.6). This coefficient was calculated from equation (2.10) taking into account the refractive indices listed in table 2.1.

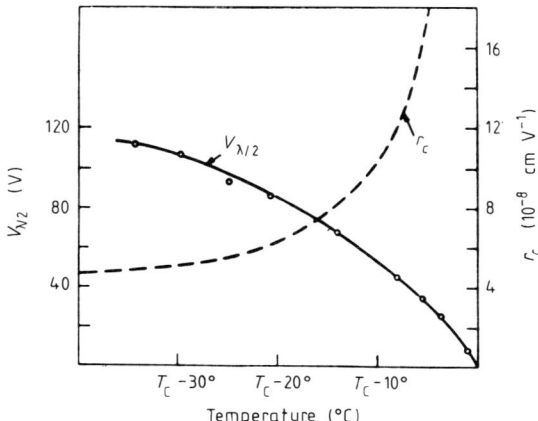

Figure 2.5 Temperature dependences of (1) the half-wave voltage $V_{\lambda/2}$ and (2) the electro-optic coefficient r_c for a KTN crystal with a Curie temperature of 60 °C [6].

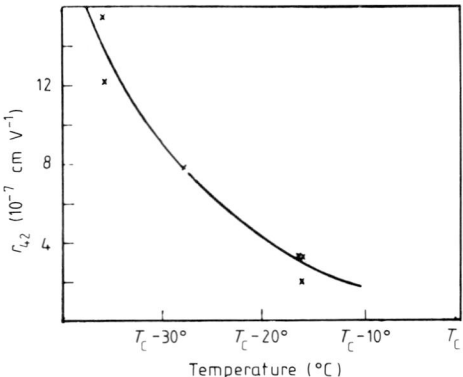

Figure 2.6   Electro-optic coefficient r_{42} of KTN crystals as a function of temperature [6].

Table 2.1   Refractive indices of KTN [6].

$T_C - T$ (°C)	n_a	n_c
16	2.318	2.281
28	2.318	2.277
36	2.318	2.275

2.3   Quadratic electro-optic effect in KTN

The coefficients of the quadratic electro-optic effect g_{ijk} are components of the tensor of rank four defined by the expression [8]

$$\beta_{ij}(P) - \beta_{ij}(0) = \sum g_{ijkl} P_k P_l \tag{2.21}$$

where $\beta_{ij}(P)$ and $\beta_{ij}(0)$ are components of the optical opacity tensor in the presence and absence of polarization P in the crystal.

For an ideal perovskite structure with the symmetry Pm3m, the number of independent quadratic electro-optic coefficients g_{ijk} reduces from 81 to 3 and there remain only the coefficients g_{11}, g_{12} and g_{44}. The values of $(g_{11} - g_{12})$ and g_{44} were measured in [9]. At low frequencies, $(g_{11} - g_{12})$ and g_{44} were determined from the induced phase lag $\Delta\phi$ on a Babinet–Soleil compensator by plotting a graph of the number of lags m of the phase ϕ by π relative to the square of the applied voltage V_{mn}^2. For measuring the quantity $(g_{11} - g_{12})$, the equation

$$V_{mn}^2 = m\left(\frac{\lambda d^2}{n_0^3 (g_{11} - g_{12}) \varepsilon^2 l}\right) \equiv m V_\pi^2 \tag{2.22}$$

was used, which determines the position of the maxima or minima of the transmitted light of intensity $I = I_0 \sin^2(\Delta\phi/2)$. In this equation, d is the distance between the electrodes, l is the crystal length in the propagation direction of light, ε is the static dielectric permittivity, λ is the light wavelength in vacuum, n_0 is the refractive index in the zero external field and V is the voltage applied to the specimen. Equation (2.22) corresponds to the case of a field applied in the [001] direction and light propagation along the [100] direction. For measuring g_{44}, the field was applied along the [110] and the light transmitted along the [001] direction. The values of the static dielectric permittivity and the refractive index needed for obtaining the electro-optic coefficients from the values of $V_{m\pi}^2$ were estimated using conventional techniques. The experimental values of $(g_{11} - g_{12})$ and g_{44} are listed in table 2.2. Along with $KTa_{0.65}Nb_{0.35}O_3$, the table gives quadratic electro-optic coefficients of several other perovskites and represents the absolute values of g_{11} and g_{12} for KTN at room temperature obtained by light beam deviation in a prism.

The values of $(g_{11} - g_{12})$ and g_{44} for paraelectric KTN do not depend on temperature, within experimental error ($\pm 10\%$). For these crystals, the dispersion of $(g_{11} - g_{12})$ measured at room temperature in the range $0.4-2\,\mu m$ is depicted in figure 2.7. The end of the fundamental absorption band in KTN lies near $0.39\,\mu m$ and the strong absorption in the long-wave region of the spectrum near $5\,\mu m$. The dispersion of the refractive index in KTN is noticeable in the interval from 0.4 to $0.7\,\mu m$, and it was taken into account in the calculation of $(g_{11} - g_{12})$. For a DC field and a frequency of 100 MHz, the values of $(g_{11} - g_{12})$ turned out to be the same, and preliminary results show that they remain constant up to a frequency of 10^9 Hz.

The relative independence of quadratic electro-optic coefficients of temperature and of perovskite composition implies their dependence first of all on the oxygen lattice framework which consists of strongly polarized O^{2-} ions. The sharp increase of the electro-optic coefficients in the short-wave band (figure 2.7) suggests that either the electron or the electron–ion contribution dominates in the investigated region of the spectrum (for more details see chapter 9).

The piezoelectric resonance in KTN crystals is small, and therefore the frequency dependence of the electro-optic effect in these crystals is determined by the dispersion of ε caused by the longest wave of optically active lattice vibration. From the infrared spectra of $KTaO_3$, $SrTiO_3$ and $BaTiO_3$ it was found that the dispersion lattice frequencies lie above 200 GHz [10] (in a static field $\varepsilon \approx 20\,000$). A decrease of ε at yet higher frequencies can be expected, since according to modern ferroelectric theories $1/\lambda_l \sim (T - T_C)^{1/2} \sim 1/\varepsilon^{1/2}$, where λ_l is the dispersion wavelength, T_C is the Curie temperature and ε the static dielectric permittivity [9]. The loss tangent in KTN may have a more rigorous frequency limit than λ_l. According to Miller, for these crystals $\tan\delta = 0.25$ at a frequency of 100 GHz and $\varepsilon = 20\,000$. Hence, in the millimetre waveband KTN crystals are apparently suitable for use in electro-optic devices.

Table 2.2 Quadratic electro-optic coefficients g_{ij} ($m^4 C^{-2}$), refractive indices n and Curie temperature T_C for several perovskites ($\lambda = 6328$ Å) [9].

Crystal	$g_{11} - g_{12}$	g_{11}	g_{12}	g_{44}	n_o†	T_C (K)
KTaO$_3$	$+0.16 \pm 0.01$	—	—	$+0.12 \pm 0.01$	2.24 (2–77)	≈ 1
KTa$_{0.65}$Nb$_{0.35}$O$_3$	$+0.174 \pm 0.01$	$+0.136 \pm 0.01$	-0.038 ± 0.01	$+0.147 \pm 0.01$	2.29 (285–310)	≈ 283
SrTiO$_3$	$+0.14 \pm 0.01$	—	—	—	2.38 (4.2–300)	—
BaTiO$_3$	$+0.13 \pm 0.02$	—	—	—	2.4 (408–433)	≈ 401

† The temperature interval (K) in which n has the given value is bracketed.

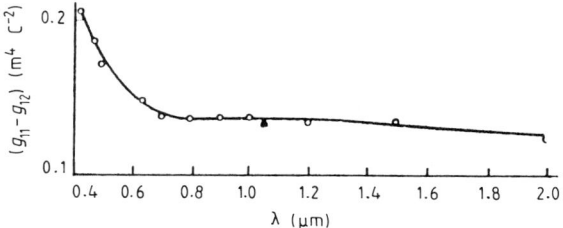

Figure 2.7 Dispersion of the difference of quadratic electro-optic coefficients $(g_{11} - g_{12})$ of a $KTa_{0.65}Nb_{0.35}O_3$ crystal [8].

2.4 Phase diagrams and KTN growth techniques

In the system $K_2CO_3-Ta_2O_5$, four chemical compounds were discovered (chapter 1, [3]), [11]: $K_2O.5Ta_2O_5$, $K_2O.2Ta_2O_5$, $K_2O.Ta_2O_5$ and $3K_2O.Ta_2O_5$, whose melting temperatures were respectively 1645, 1520, 1370 and 1330°C. The first three of them melt incongruently (the decomposition temperatures are indicated). The phase diagram of the system $K_2O-Ta_2O_3$ is similar to that of $K_2O-Nb_2O_5$.

The phase diagram of the system $KNbO_3-KTaO_3$ was investigated in [12] (figure 2.8). The liquidus line of the phase diagram was constructed using thermographic analysis, while for constructing the solidus line the temperature dependence of the conductivity of ceramic samples was examined. The phase diagram obtained for the system $KNbO_3-KTaO_3$ is the graph of a continuous series of solid solutions with the melting points of pure potassium niobate and tantalate, equal to 1039 and 1357°C respectively.

The authors of [4, 13] carried out a detailed study of the system $KNbO_3-KTaO_3$ in the solid state. As a result, they managed to construct the phase equilibrium diagram shown in figure 2.3, and to obtain the

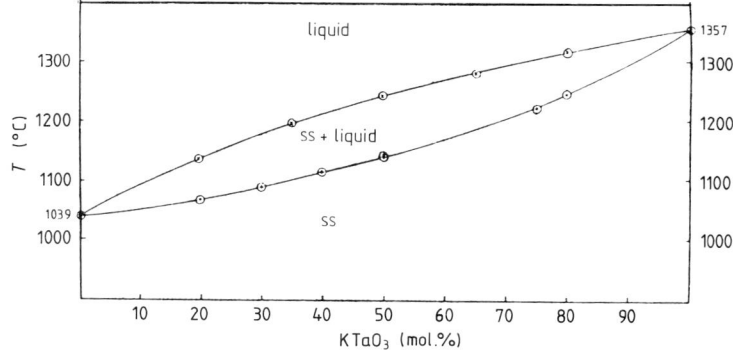

Figure 2.8 Phase diagram of $KTaO_3-KNbO_3$ [4].

dependence of the KTN crystal density on the solid solution composition (figure 2.9).

The starting material for KTN crystallization consists of three components—K_2CO_3, Nb_2O_5 and Ta_2O_5—taken in corresponding percentage. In the melt, $KNbO_3$ and $KTaO_3$ are produced with evolution of carbon dioxide. In the phase diagrams of $K_2CO_3-Ta_2O_5$ and $K_2CO_3-Nb_2O_5$, of interest is the region from 50 to 75 mol.% of K_2CO_3. Taken separately, this region represents the phase diagram of $KTaO_3-3K_2O . Ta_2O_5$, the compounds that do not dissolve in each other. The phase diagram of $Ta_2O_5-K_2CO_3$ is similar to that of $Nb_2O_5-K_2CO_3$ (see figure 1.17). The eutectic point in both cases corresponds to 66 mol.% of K_2O. However, the right-hand side of the diagram shows the formation of a stable compound $3K_2O . Ta_2O_5$ (characteristic maximum), while in the left-hand side there exists the eutectic point of the compounds $2K_2O . 3Ta_2O_5$ and $KTaO_3$, i.e. crystals of both compounds are formed at this point in the crystallization.

The latter fact demands that crystallization be carried out with some excess of K_2O over Ta_2O_5. The magnitude of this excess is determined by a whole number of factors, one of which is the evaporation rate of K_2O. For a closed system, the rate of evaporation from unit melt surface ($g\ cm^{-2}\ h^{-1}$) is specified by the expression (chapter 1, [90])

$$R = 2.6I\exp(-7300/T)$$

and for an open system

$$R = 4.5I\exp(-7300/T)$$

where I is the molar fraction of K_2O in the melt and T is the absolute temperature of the melt surface (K). The evaporation rate of K_2O

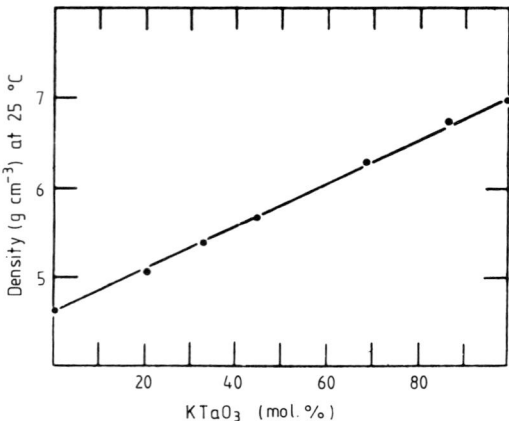

Figure 2.9 Dependence of KTN crystal density on the composition in the system $KTaO_3-KNbO_3$ [13].

measured experimentally for melts with the ratio $Ta/(Ta + Nb)$ within the range 0 to 1 and the K_2O content from 50 to 65 mol.% at a temperature of 965–1365 °C varied from 0.06 to 0.2 g cm^{-1} h^{-1}. The K_2O evaporation rate increases if the crucible surface is blown off with oxygen.

Thus the time of crystal growing from one and the same melt is limited by the amount of excess K_2O. After a complete evaporation of non-stoichiometric K_2O, the melt composition shifts to the eutectic point, and $2K_2O . 3Ta_2O_5$ crystals, which are a non-isomorphic impurity for $KTaO_3$, are formed in the melt. As a consequence, there arises the probability of this impurity being trapped by the growing $KTaO_3$ crystal.

An excess K_2O can also be regarded as a non-isomorphic impurity. Solidification of the excess K_2O proceeds mainly due to intercrystalline confinement of the solution. An insufficiently stable temperature, which leads to a temporary lowering of the melt temperature, may cause solid-phase nucleation at some distance from the crystallization front. The nuclei grow, and part of the non-isomorphic impurity remains in interlayers between crystalline grains. The probability of such a trapping increases if the crystallization front is concave and is formed by small steps.

In the crystallization of a solid solution of KTN, of practical importance is obtaining crystals of the composition $KTa_{0.65}Nb_{0.35}O_3$. From the phase diagram of the system $KTaO_3-KNbO_3$ (see figure 2.8) it follows that in the melt composition $KTa_{0.35}Nb_{0.65}O_3$ (xK_2O) there primarily form $KTa_{0.65}Nb_{0.35}O_3$ crystals.

It is of interest to follow the variations in the composition along the crystal length (such a problem was solved for germanium solution in silicon [12]). The tantalum distribution in a crystallized-out substance is given by the equation

$$m_{cr} = m_0(1 - e^{-A})$$

where m_0 and m_{cr} are the initial masses of the melt and crystal respectively,

$$A = \int_x^{x_0} \frac{dx}{y - x}$$

where x_0, x are the Ta content in the initial and final states of the melt and y is the Ta content in the crystal. The integral A is determined by numerical integration of the phase diagram of $KTaO_3-KNbO_3$ (see figure 2.8). Figure 2.10 presents the dependence of the Ta content in a crystal (C_{Ta}^{cr}) on its concentration in the melt (C_{Ta}^m) [14]. On the basis of this graph, one can judge the Ta distribution in a growing crystal. For example, in a crystallized-out 0.1 part of the melt, the Ta content changes from 68 to 59%. Thus, when obtaining KTN crystals from small amounts of melt one should expect a considerable change in the composition over the crystal length. The latter fact complicates the growth of KTN crystals of a prescribed composition [14].

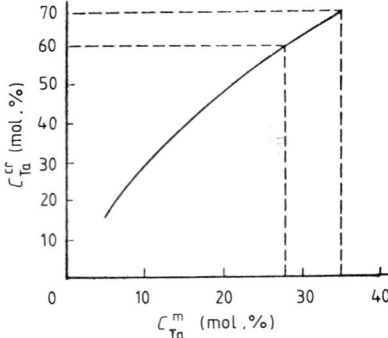

Figure 2.10   Dependence of the content of Ta in the crystal on its content in the melt [14].

Miller (chapter 1, [87]) showed that KTN crystals can be obtained by the spontaneous crystallization method and a modified Kyropoulos method from the melt $(K_2CO_3 + Nb_2O_5 + Ta_2O_5)$ containing 20 mol.% of excess K_2O. Incongruent melting of $KNbO_3$ and $KTaO_3$ complicates significantly the growth of crystals with prescribed optical properties. Later it became clear that the quality of crystals is affected by a whole number of factors.

The first successful attempts to obtain large KTN crystals of high optical quality were made in [2, 15] using the devices designed by the authors. Crystallization was carried out in a platinum crucible which contained 1300 g of melt. The crystallization temperature was 1225 °C. The crucible was mounted on a stand of porous aluminium oxide rotated at a rate of 60 rpm with a reverse every 30 seconds, which promoted a better mixing of the melt. The heating of the working chamber of the furnace was performed by horizontally and vertically positioned heaters made of silicon carbide. The crucible was placed in a muffle in such a way that upper layers of the melt were 50 °C cooler than the bottom of the crucible. The melt cooling rate was $0.1\,°C\,h^{-1}$. The seed crystal was rotated in the upper layer of the melt, and the growth rate was 0.6 cm per 24 h.

Isometric well faceted crystals measuring $5 \times 4 \times 4$ cm³ and possessing high optical quality were grown under strict control over the crystallization conditions. The high purity of the oxides (K_2O, Nb_2O_5, Ta_2O_5) used for preparing initial material is the decisive factor for obtaining high-quality KTN crystals. The presence in the melt of bivalent cations $(Ca^{2+}, Sr^{2+}, Ba^{2+}, Pb^{2+})$ lowers the valency of the Nb^{5+} and Ta^{5+} ions down to $+4$. The appearance of blue coloration connected with the formation of F centres and a significant increase in crystal conductivity are signs of the lowered valency. The same phenomena may also occur with a relatively high crystallization rate or an insufficiently free oxygen circulation above the melt surface.

Free charge carriers may be created in a crystal by substitution of bivalent Ca^{2+} (0.99 Å), Sr^{2+} (1.12 Å), Ba^{2+} (1.34 Å) and Pb^{2+} (1.20 Å) ions for K^+

(1.33 Å) or by formation of oxygen vacancies. Blue coloration of a crystal is due to the light absorption by free carriers in the red region of the spectrum. The concentration of free carriers may be decreased by introduction of tetravalent ions or ions with lower valency instead of the Ta^{5+} or Nb^{5+} ions. A positive effect was obtained with additions of Li^+ (0.68 Å) and Sn^{4+} (0.71 Å) in amounts sufficient to replace Ta^{5+} (0.68 Å) and Nb^{5+} (0.69 A), and also to compensate oxygen vacancies or high-charge potassium-replacing ions. Tin does not colour crystals. But if there is a lack of potassium in the melt then tin goes over to a bivalent state, and the crystals acquire a brown–yellow hue. Besides, an addition of tin to a melt somewhat lowers the niobium content in the crystal as compared to the calculated value. A negative effect upon crystal coloration is produced by the limitation of a free oxygen circulation above the melt surface [2].

The crystal quality is greatly affected by the intensity of melt mixing. When crystals are grown without melt mixing, one can see a tendency towards a decrease of tantalum content in the upper portion of the melt because it is predominantly tantalum that enters the growing crystal. It follows from figure 2.11 that, as the amount of material in the crucible decreases, the Curie temperature increases at first slowly and then rather quickly, this increase being independent of the Curie temperature of the initially formed crystal layers. Melt heating to substantially higher temperatures than the crystallization temperature gives rise to a complete mixing of the melt, and a subsequent crystal grown by the same cycle has practically the same T_C value. An increase of the crystal cross section reduces the influence of melt exhaustion more effectively than mixing [2]. There exists experimental evidence [2, 17] that

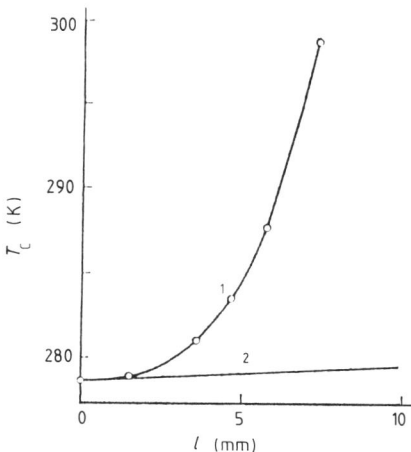

Figure 2.11 Curie temperature variation along the length of a KTN crystal grown (1) without rotation and (2) with crucible rotation [2].

a complete melt mixing may also be reached by rotating a crucible with melt at a rate of 60 rpm with reversal every 30 seconds.

The examination of the phase diagrams of $K_2CO_3-Nb_2O_5$ and $K_2CO_3-Ta_2O_5$ showed that near the melting points of $KNbO_3$ and $KTaO_3$ these compositions decompose. Therefore, crystallization of KTN is possible only from a flux. An excess K_2CO_3 is used as solvent. The presence of a relatively small excess of K_2CO_3 (10 mol.%) and placing a crucible with melt in a furnace with a temperature gradient provide favourable conditions for obtaining KTN crystals [17]. To increase the temperature gradient between seed and melt, one uses a cooled seed holder. Under these conditions, the authors obtained a KTN crystal, measuring 7.5 cm in length, at a growth rate of 125 μm h^{-1}. It was also noticed that in the KTN crystal grown by the Kyropoulos method with a K_2CO_3 concentration in the melt lower than the stoichiometric one, the lower part was colourless but opalesced, which may be due to the presence of a highly dispersed second phase. The domain walls might also give such an effect, but opalescence remained above the Curie point. It was also revealed that with the K_2O content in the melt equal to about 50 mol.% the grown crystal was an insulator, while for 62 mol.% the crystal was a conductor.

Figure 2.12 gives a simplified graph of the relation between the melt composition and the concentration of free charge carriers in a Kyropoulos-grown KTN crystal. Curve 1 characterizes the dependence of the solidus–liquidus line on the composition. The melting temperature varies in the interval $1000 \pm 200\,°C$, depending on the Ta/Nb ratio. Near stoichiometry (50 mol.% of K_2O) the crystal is an insulator. The concentration of free carriers increases, as shown in curve 2, with increasing K_2O concentration in the melt. With approaching eutectics (66 mol.% of K_2O), the tendency to separation of

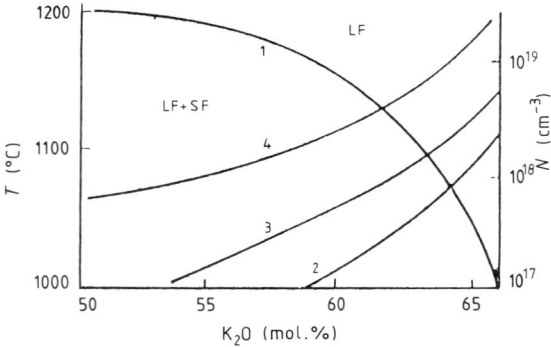

Figure 2.12 Dependence of the solidus–liquidus curve (curve 1) and of the concentration N of free charge carriers in a KTN crystal on the content in the melt of K_2O without lead additions (curve 2) and with lead additions (curves 3 and 4) [17].

$3K_2O$ (Ta, Nb)$_2O_5$ crystals increases, which evidently leads to an increase of oxygen vacancies in the crystal, which can be occupied by potassium ions. For a 50 PPM content of Pb† in the melt, the concentration of free carriers in a grown crystal varies as in curve 3. It is likely that, for 50 mol. % of K_2O in the melt, one Pb^{2+} ion replaces two K^+ ions. This promotes the appearance of additional free carriers which cause structure defects. For a 5000 PPM content of Pb in the melt, the concentration of free carriers in the crystal varies as shown in curve 4. The coefficient of Pb^{2+} distribution in the crystal and melt is greater than unity, whereas for Ca^{2+} it is much less than unity. But the latter increases if together with Ca^{2+} ions Pb^{2+} ions are present in a grown crystal.

---

† 1 PPM corresponds to one fraction of impurity to 10^6 fractions of the basic substance.

3 Lead Zinc Niobate and Lead Magnesium Niobate Ferroelectrics with a Smeared Phase Transition

In the early 1960s, G A Smolensky and co-workers [1–4] discovered a family of perovskite-type ferroelectrics with complex composition. Later on, some of them were obtained as single crystals, which permitted a detailed study of the dielectric, optical and electro-optic properties of these compounds. The ferroelectric crystals of $Pb_3ZnNb_2O_9$ and $Pb_3MgNb_2O_9$ turned out to possess a noticeable quadratic electro-optic effect. The distinctive feature of these compounds is a smeared phase transition which is responsible for the relaxation character of the dielectric permittivity and electro-optic effect. The $Pb_3ZnNb_2O_9$ crystal and its magnesium analogue can have a rather large size and a reasonable optical perfection, which is their undoubted advantage over KTN crystals. The latter fact allows them to serve as electro-optic modulators and light beam deflectors.

A knowledge of the electro-optic properties of such compounds makes it possible to examine the effect of the electronic shell upon the magnitude of electro-optic effects, since the octahedral positions of the crystal lattice of these compounds are occupied by ions with identical radii but distinct electronic shells.

3.1 Ferroelectric crystals of lead zinc niobate

Ferroelectric crystals of $Pb_3ZnNb_2O_9$ (PZN) have rhombohedral symmetry, and at room temperature their lattice constants are $a = 4.061$ Å and $\alpha = 89°55'$. In the paraelectric phase, the crystal symmetry is cubic; the lattice constant $a = 4.066$ Å at $300°C$.

3.1.1 *Dielectric characteristics*

The temperature and frequency characteristics of ε and tan δ in weak AC fields were investigated by a number of authors [chapter 1, 11, 22; 5]. The temperature dependences of ε and tan δ in PZN crystals in the frequency range 10^3–(2×10^9) Hz are depicted in figures 3.1(a), (b). In the range 10^3–10^7 Hz, the maxima of $\varepsilon(T)$ and tan $\delta(T)$ are seen to shift towards higher temperatures with increasing field frequency. In the decimetric waveband (10^8–2×10^9 Hz), the positions of the maxima of ε and tan δ do not depend on the AC field frequency. For PZN, one can distinguish between the low-frequency (relaxational) and the high-frequency dispersion regions of ε, which are distinct in their nature. When the AC field frequency changes by an order of magnitude, the shift of the temperature maximum of ε in the low-frequency dispersion region is 4–5 °C for $Pb_3ZnNb_2O_9$ and 10–12 °C for $Pb_3MgNb_2O_9$ (PMN). It should be pointed out that the ε peaks are sharper in PZN than in PMN crystals [5].

The effect of a biasing DC field upon the dielectric permittivity ε is illustrated in figure 3.2. The measurements were carried out at a frequency of 100 kHz. The peak of the dielectric permittivity was widened and displaced towards higher temperatures with increasing bias.

The temperature dependence of the spontaneous polarization P_s was determined from the dielectric hysteresis loops and the pyroelectric current

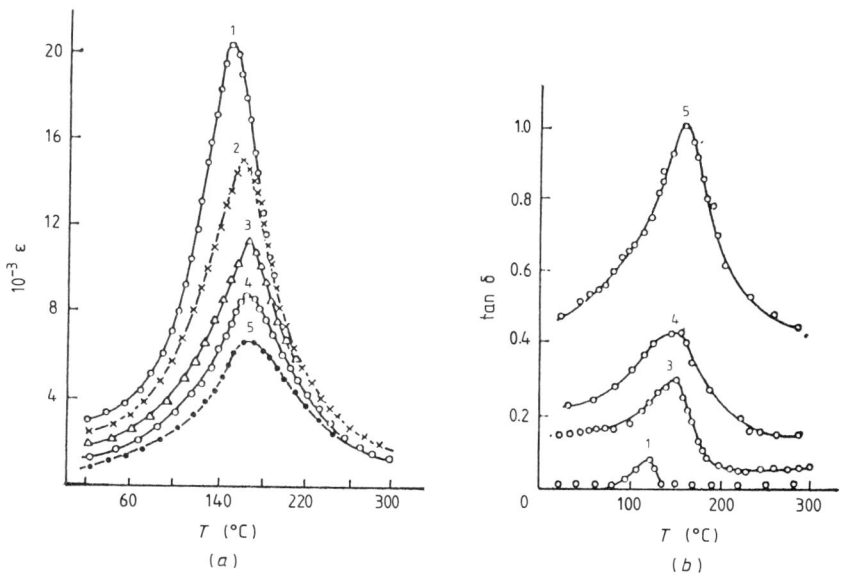

Figure 3.1 Temperature dependence of ε (a) and tan δ (b) of a $Pb_3ZnNb_2O_9$ crystal at different frequencies ([22], chapter 1): 1, 10^3; 2, 5×10^5; 3, 7×10^7; 4, 10^9; 5, 2×10^9 Hz.

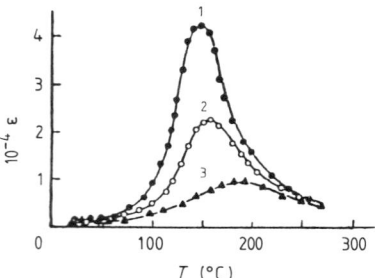

Figure 3.2 The effect of a biasing DC field upon the dielectric permittivity ε of a $Pb_3ZnNb_2O_9$ crystal [5]: 1, 0; 2, 7.5; 3, 15 kV cm^{-1}.

(figure 3.3, curves 1 and 2 respectively). Before the pyroelectric measurements were made, the specimens were exposed to a field of 10 kV cm^{-1} at a temperature of 280°C and were then gradually cooled down to room temperature. An analysis of hysteresis loops has shown that spontaneous polarization holds up to 240°C, whereas according to pyroelectric measurements, the polarization vanishes at 120°C. It was established [22] that the spontaneous polarization determined from the hysteresis loops decreases from 10.2 to 9.5×10^{-6} C cm^{-2} as the temperature rises from -70 to $+80$°C, and with a further temperature increase P_s decreases more rapidly. However, at a temperature of ε_{max} $T_C = 160$°C, at a frequency of 2×10^9 Hz, the spontaneous polarization does not vanish but is approximately equal to 3.3×10^{-6} C cm^{-2}, which is typical of ferroelectrics with a smeared phase transition. At a temperature of 20°C, the coercive field of PZN $E_C = 4.5$ kV cm^{-1}.

Comparison of the results of ε, tan δ and P_s measurements on PZN and PMN crystals shows that the phase transition is less smeared in the former than in the latter. Phase transition smearing in complex composition is explained by composition fluctuations upon the static distribution of different types of ions in identical crystallographic positions. Composition fluctuations

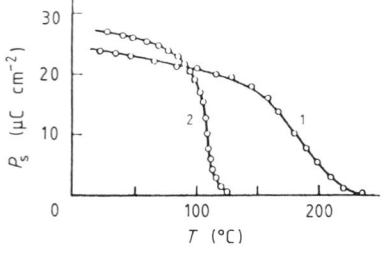

Figure 3.3 Temperature dependence of spontaneous polarization P_s of a $Pb_3ZnNb_2O_9$ crystal obtained from: (1) dielectric hysteresis loops; (2) pyroelectric currents [5].

result in the formation of regions with distinct compositions and, accordingly, distinct local Curie temperatures. The smaller the difference in the local Curie temperatures for regions with distinct compositions, the narrower the width of the transition region (Curie region) in which a paraelectric crystal becomes a ferroelectric. Since the distinction in ion polarizabilities of Zn^{2+} and Nb^{5+} is considerably smaller than that of Mg^{2+} and Nb^{5+}, the Curie region of $Pb_3ZnNb_2O_9$ is narrower than that of $Pb_3MgNb_2O_9$ (chapter 1, [21]).

3.1.2 *Electro-optic properties*
The single-crystal ferroelectric $Pb_3ZnNb_2O_9$ is optically uniaxial and positive, with the optical axis parallel to the $\langle 111 \rangle$ direction. In the paraelectric phase the crystal does not possess optical anisotropy, but acquires this property as an electric field is applied.

The temperature dependence of natural birefringence is plotted in figure 3.4. The birefringence Δn is equal to 0.013 at room temperature. The temperature dependence of Δn is approximately proportional to the variation of the square of the spontaneous polarization P_s. When a biasing DC field is applied, the birefringence is observed above $120\,^\circ C$.

The birefringence of PZN, for a crystal embedded in an electric field, was investigated in [6]. The induced birefringence n is plotted in figure 3.5 (curve 1) as a function of the electric field strength at room temperature. The birefringence in perovskite-type ferroelectrics is known to be proportional to the square of the total polarization [7]. In the initial portion of the dependence $\Delta n = f(E)$, the quadratic law is obeyed, and therefore the field dependence of polarization is linear (curves 2 and 3). For fields ≈ 7 kV cm^{-1}, due to transition of the non-polar regions to the ferroelectric state and due to an increasing degree of dipole moment orientation in the polar regions, the dependence $\Delta n = f(E)$ becomes linear. For a field strength of 20 kV cm^{-1}, the slope of $\Delta n = f(E)$ changes, which corresponds to saturation of polarization in strong fields. The difference in the electro-optic coefficients $(r_{33} - r_{31})$ corresponding to the matrix of electro-optic coefficients for the point symmetry group 4mm was calculated for the fields corresponding to the first

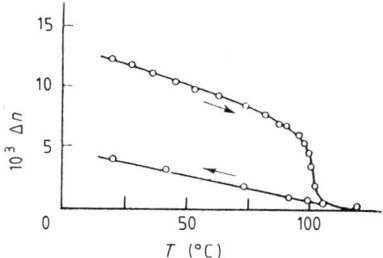

Figure 3.4 Temperature dependence of natural birefringence of a $Pb_3ZnNb_2O_9$ crystal [5].

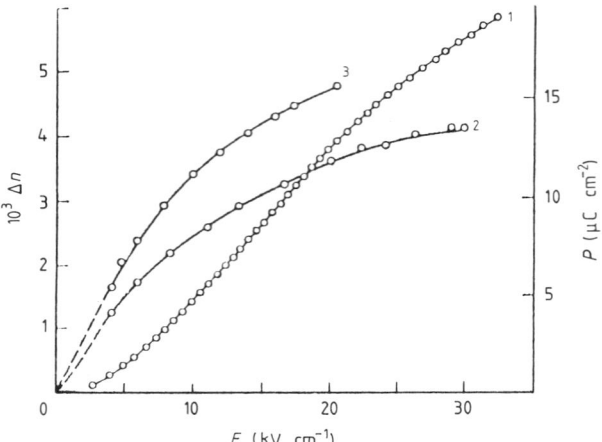

Figure 3.5 Dependences (1) of induced birefringence and (2, 3) of polarization on the electric field strength in a $Pb_3ZnNb_2O_9$ crystal [6]. Curves 1 and 2 were obtained at 20°C; curve 3 at −60°C.

portion of curve 1, figure 3.5. In case the field is applied along the [001] direction, we have for $n_e \approx n_0$

$$r_{33} - r_{13} = 2\Delta n / n_0^3 E. \tag{3.1}$$

For $\lambda = 6328$ Å we have $n_0 = 2.5$, and then $(r_{33} - r_{13}) = 0.85 \times 10^{-5}$ cm (stat V)$^{-1}$, which is 30 times larger than the coefficient r_{36} for KDP crystals.

The plot of the induced birefringence versus the electric field strength in a wide temperature range is depicted in figure 3.6, as calculated in [8].

The linear electro-optic effect in PZN crystals, which occurs in sufficiently strong fields, can be interpreted as a result of linearization of the quadratic effect observed in weak electric fields. For the PZN crystal this was described in [6], and for the PMN crystal in [9].

Figure 3.7 illustrates the temperature dependence of the difference in the induced linear electro-optic coefficients $(r_{33} - r_{13})$ of a PZN crystal. The theoretical values of these coefficients were calculated by the formula [6]

$$r_{ijk} = 2\left(\frac{1}{4\pi}\right)^2 \frac{(n^2 - 1)^2}{n^4} \Delta_{ijkl} P_{0e} \varepsilon_{0e} \tag{3.2}$$

where P_{0e} and ε_{0e} are the polarization and dielectric permittivity values at which the effect is linearized, and the Δ_{ijk} are related to the Miller constants δ_{ijk} by the expression [10]

$$\delta_{ijk} = 2\Delta_{ijkl} P_{0e}. \tag{3.3}$$

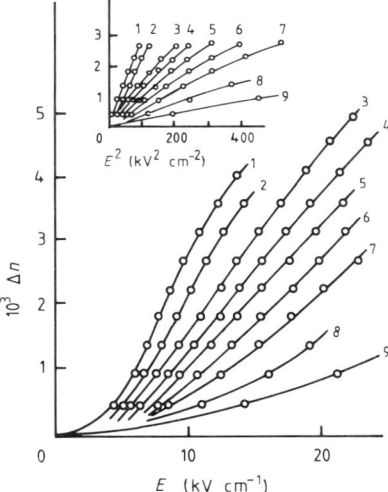

Figure 3.6 Dependence of induced birefringence on the electric field strength in a $Pb_3ZnNb_2O_9$ crystal at different temperatures [8]: (1), -2; (2), 8; (3), 22; (4), 37; (5), 52; (6), 68.8; (7), 85.9; (8), 122; (9), 140°C. The light wave was propagated along the [001] direction; the electric field was applied in the [100] direction.

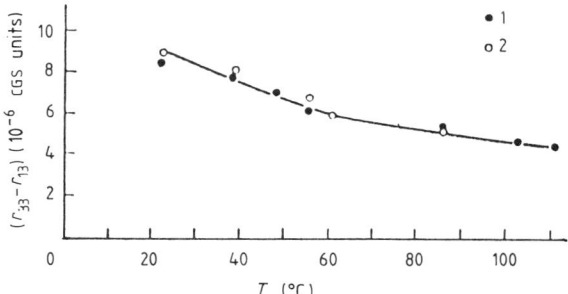

Figure 3.7 Temperature dependence of the difference of induced linear electro-optic coefficients $(r_{33} - r_{13})$ of a $Pb_3ZnNb_2O_9$ crystal [6]: (1) experimental; (2) calculated values.

The Miller constant δ_{ijk} for the PZN crystal is equal to 1.76×10^{-7} cm (stat V)$^{-1}$; the experimental value of P_{0e} is $6 \,\mu C \, cm^{-2}$. The experimentally measured values of $(r_{33} - r_{13})$ agree fairly well with the theoretical calculations.

In ferroelectrics with a smeared phase transition, a noticeable contribution to the electro-optic effect is given, besides the electron and ion polarization, also by the orientational processes, for which reason an important characteristic of a crystal is the frequency dependence of the quadratic electro-optic coefficients (figure 3.8). The discovered dispersion of $(R_{11} - R_{12})$ is due to

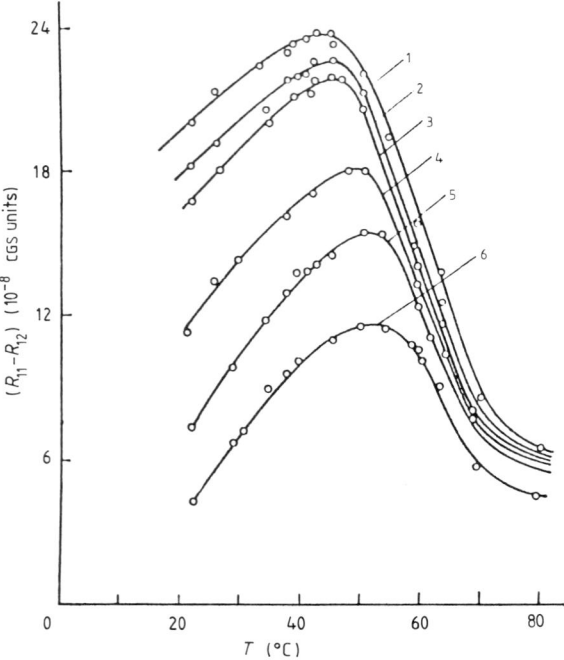

Figure 3.8 Temperature dependence of the difference of quadratic electro-optic coefficients $(R_{11} - R_{12})$ at different frequencies [6]: (1), 80; (2), 8×10^2; (3), 2×10^3; (4), 8×10^3; (5), 8×10^4; (6), 2×10^5 Hz.

the fact that the contribution from orientational processes decreases with increasing electric field frequency and testifies to a considerable contribution from the orientational polarization to the electro-optic effect in ferroelectrics with a smeared phase transition.

The differences in the quadratic electro-optic coefficients $(R_{11} - R_{12})$ for PZN and PMN crystals were calculated by Smolensky *et al* [8]. The matrices for the point group m3m were used in the calculations, which is a certain approximation, as pointed out in [11]. The basic electro-optic characteristics of the investigated crystals are listed in table 3.1. From the table it is seen that PZN possesses higher electro-optic coefficients than PMN. This may be explained by the higher mean Curie temperature of PZN crystals. Besides, a certain role is evidently played by the electron polarizability of ions, which occupy octahedral positions, since the electro-optic effect increases with increasing polarizability. The half-wave voltage $V_{\lambda/2}$ in PZN and PMN at room temperature and at wavelength $\lambda = 6240$ Å is half the value for lithium niobate and an order of magnitude lower than for KDP. Furthermore, the quadratic effect in the crystals in question enables $V_{\lambda/2}$ to be appreciably lowered by application of a DC biasing field. Crystals of PZN and PMN are somewhat

Table 3.1 Room-temperature electro-optic characteristics of PZN and PMN crystals [8].

Crystal	$n_0^3(R_{11} - R_{12})$ (10^{-8} CGS units)	$V_{\lambda/2}$ (kV)
$Pb_3ZnNb_2O_9$	290	1.3
$Pb_3MgNb_2O_9$	218	1.6

inferior in their electro-optic properties to KTN crystals. But the growth technique of optically homogeneous PZN and PMN crystals is much simpler than that of KTN. As is known, the inhomogeneity of KTN crystals is the main obstacle in the way of their practical application.

3.2 Lead magnesium niobate ferroelectrics

3.2.1 *Dielectric properties*
Lead magnesium niobate $Pb_3MgNb_2O_9$ (PMN) is a perovskite-type ferroelectric. Ferroelectric studies of this material were first undertaken by Myl'nikova and Bokov [12], who reported the appearance of a low-frequency peak in the dielectric permittivity ε at approximately 225 °C and the existence of spontaneous polarization up to 355 °C. The PMN properties were later specified in [13]. The temperature dependence of the dielectric permittivity measured there at a frequency of 1 kHz is plotted in figure 3.9 (curve 1), from which ε_{max} is seen to be reached at 265 °C. This value exceeds the one reported in [12], but coincides with the results obtained in [14] on polycrystals. The temperature dependence of the dielectric permittivity above and below the peak ($\varepsilon(+)$ and $\varepsilon(-)$ respectively) is rather well approximated by the Curie–Weiss law

$$\varepsilon(+) = C(T - T_C)^{-1} \tag{3.4}$$

$$\varepsilon(-) = -\tfrac{1}{2}C(T - T_C)^{-1} \tag{3.5}$$

for $C = (3.7 \pm 1.0) \times 10^5$ K and $T_C = 265$ °C.

The dielectric permittivity and polarization of PMN single crystals are plotted in figure 3.10 as functions of an externally applied biasing electric field.

From an analysis of the dielectric constant as a function of an applied biasing DC field, the induced polarization P above the Curie temperature was found to be related to the strength of the electric field E by the expression

$$E = [(T - T_C)(\varepsilon_0 C)^{-1}]P + \xi P^3 + \zeta P^5 + \dots \tag{3.6}$$

where $\xi = +5.6$ V cm^5 μC^{-3}. The ratio $d(\varepsilon^{-1})/dT$ above and below the ε maximum is equal to $-\tfrac{1}{2}$. This fact, as well as the positive sign of ξ, characterize the second-order transition in PMN.

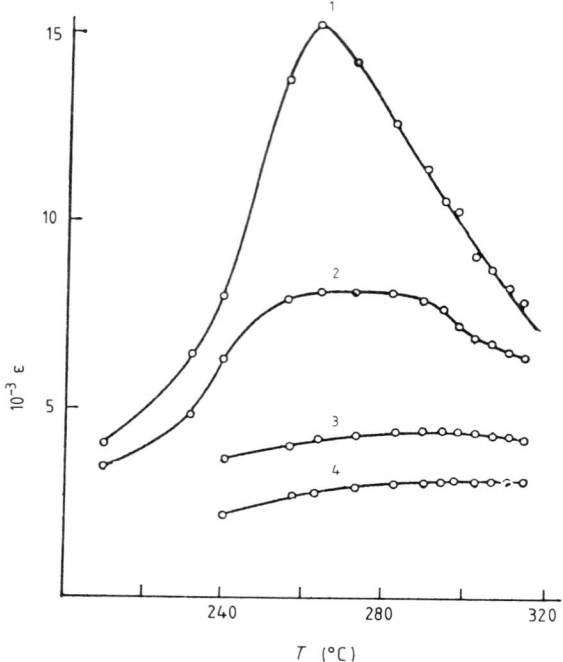

Figure 3.9   Temperature dependence of low-frequency dielectric permittivity of a $Pb_3MgNb_2O_9$ crystal for different electric field strengths [13]: (1), 0; (2), 9.6; (3), 17.8; (4), 26.2 kV cm^{-1}.

The experimental temperature dependence of P_s below 265°C is described by the dependence (in μC^2 cm^{-4})

$$P_s^2 = (3.6 \pm 0.5)[T_C - T]. \tag{3.7}$$

At a temperature $T_C = 265$°C the spontaneous polarization is $\approx 1~\mu C$ cm^{-2}, and only at 275°C does it become close to zero. Such a slow decrease of polarization above the Curie point is explained by the relaxation character of dielectric permittivity.

The quadratic temperature dependence of the inverse dielectric permittivity of lead magnesium niobate

$$1/\varepsilon = A + B(T - T_0)^2 \tag{3.8}$$

where $A = 1/\varepsilon_{max}$ and B are constants and T_0 is the temperature of the maximum dielectric permittivity, was established later in [15], and the dependence $\varepsilon(T)$ was presumed to be typical of all ferroelectrics with a smeared phase transition. The experimental dependence (3.8) can be derived theoretically if one sets the following functional dependence of the distribution

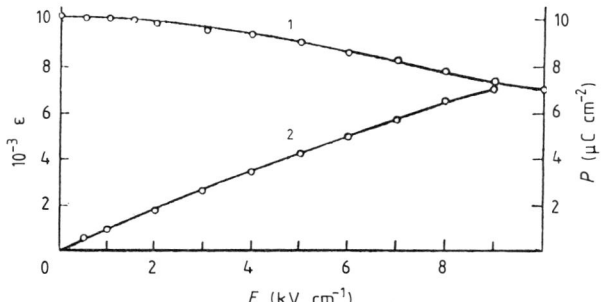

Figure 3.10 Dependence of: (1) dielectric permittivity ε; (2) polarization P of a $Pb_3MgNb_2O_9$ crystal on the biasing electric field [13].

of different crystal regions over the Curie temperature [16, 17]:

$$\varphi(\theta) = \frac{1}{\sqrt{2\pi np(1-p)}} \exp\left(-\frac{n(\theta - \theta_{mean})^2}{2\gamma^2 p(1-p)}\right) \qquad (3.9)$$

or

$$\varphi(\theta) = \frac{1}{\sqrt{2\pi\sigma}} \exp\left(-\frac{(\theta - \theta_{mean})^2}{2\sigma^2}\right) \qquad (3.10)$$

where n is the number of formula units in a crystal region; θ and θ_{mean} are the local and mean Curie temperatures respectively; γ is the value determining θ variation with varying composition; p is the concentration of B ions (e.g. Mg); and σ is the parameter determining the degree of phase transition smearing.

As soon as the specimen temperature decreases, in regions with the highest θ value spontaneous polarization occurs and there appear domainoids, which are polar-phase regions surrounded by the non-polar phase [15]. The electric moments of the domainoids can be aligned along equivalent crystallographic directions of the ferroelectric phase of a crystal, the regions with different directions of the electric moment being separated by potential barriers. The moment reversal is due to thermal fluctuations under which the domainoid vanishes, i.e. the region becomes non-polar, and then appears again with the same or a different direction of the moment. The domainoids with $T = \theta$ alone contribute to the relaxational polarization. The number of such domainoids is determined by equation (3.10). Assuming the temperature corresponding to ε_{max} to differ little from θ_{mean}, and making use of the formula which defines the relaxation part of the dielectric permittivity [18]

$$\varepsilon = \frac{4\pi n\mu^2}{3kT(1 + \omega^2\tau^2)} \qquad (3.11)$$

we derive the expression for $1/\varepsilon$

$$1/\varepsilon = A \exp[(T - T_0)^2/2\sigma^2] \qquad (3.12)$$

where

$$A = \left(\frac{4\pi\mu^2 N}{3kT(1 + \omega^2\tau^2)(2\pi\sigma^2)^{1/2}} \right)^{-1}.$$

Here μ is the value of the electric moment of a relaxing element, ω is the frequency of the applied field, τ is the time constant dependent on the activation energy, temperature and eigenfrequency of the relaxing element in the potential wall and σ is the parameter of phase transition smearing. Expanding equation (3.12) in a power series, we obtain the dependence

$$1/\varepsilon = A + B(T - T_0)^2 + C(T - T_0)^4 + \ldots \qquad (3.13)$$

where $B = A/2\sigma^2$ and $C = A/4\sigma^4$. Equation (3.13) describes the contribution of domainoids to the dielectric polarization; it coincides with expression (3.8), obtained experimentally, up to terms with a second power of $(T - T_0)$.

The inverse dependences of ε for several perovskites, including PMN, were measured at room temperatures for a frequency of 12 MHz [19]. From these dependences, the mean coefficients of nonlinearity $\alpha = (\varepsilon^{-1}\Delta\varepsilon/\Delta E) \times 100\%$ were determined. For PMN, α was estimated to be 1.35% per 1 kV cm^{-1}.

3.2.2 Electro-optic properties

The electro-optic properties of $Pb_3MgNb_2O_9$ were investigated in [11, 13, 14]. These crystals were reported to have a large quadratic electro-optic effect, which makes their application in nonlinear optics promising.

The electro-optic studies were performed at a wavelength $\lambda = 6240$ Å in the temperature range between -30 and $100°C$, which corresponds to the phase transition smearing region for PMN [14]. The experiments were carried out using customary techniques. The planes of crossed nicols made an angle of $\pm 45°$ with the indicatrix axes. The light propagated in the [001] direction; the field was applied either in the [001] or [110] direction. Figure 3.11 presents a plot of the induced birefringence Δn as a function of electric field strength E (E is directed along the [100] axis) at different temperatures. Figure 3.12 is a graph of the ratios $\Delta n/E$ and $\Delta n/E^2$ plotted versus the strength of the field applied along the [100] direction. For comparatively small field strengths, the dependence of Δn on E is quadratic, while for larger field strengths this dependence becomes practically linear, i.e. in PMN, as in PZN, crystals the order of the electro-optic effect decreases. With increasing temperature of the crystal, the region of the fields in which the dependence is quadratic extends, and at $80°C$ in fields up to 18 kV cm^{-1} the effect is exclusively quadratic.

In the absence of the field, a room-temperature PMN crystal belongs to the point symmetry group m3m. The matrices of the quadratic electro-optic

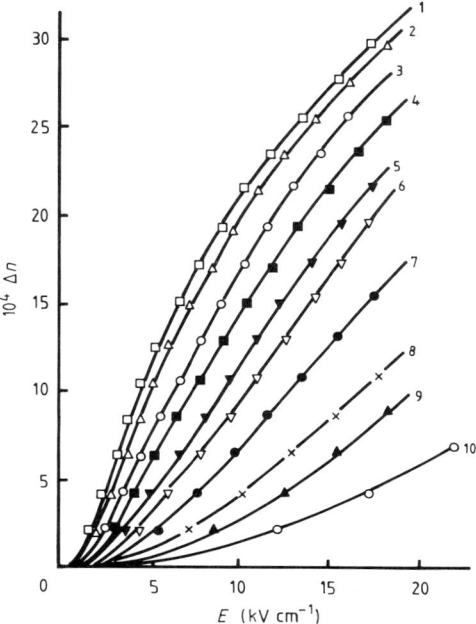

Figure 3.11 Dependence of birefringence Δn on the field strength E in a PMN crystal at different temperatures [14]. The light wave was propagated along the [001] direction, the electric field was applied in the [100] direction, and $\lambda = 6240$ Å. (1), -28; (2), -18; (3), -8; (4), 0; (5), 12; (6), 20; (7), 39; (8), 61; (9), 75.5; (10), 102°C.

coefficients R_{ij} and g_{ij} for this symmetry group each have three non-zero coefficients R_{11}, R_{12}, R_{44} and respectively g_{11}, g_{12} and g_{44}, where R_{ij} describe the electro-optic effect as a function of the square of the applied field and g_{ij} as a function of the square of the polarization.

The phase difference $\Delta\phi$ between ordinary and extraordinary beams for different directions of the applied electric field and for the propagation of light along the [001] direction, as well as the quadratic electro-optic coefficients R_{ij} and g_{ij}, are listed in table 3.2.

Figure 3.13 shows the temperature dependence of the difference $(R_{11} - R_{12})$ for PMN crystals. This quantity is seen to increase sharply with decreasing temperature.

The orientational effect of polarization upon the electro-optic effect has already been mentioned above. In this respect it seems interesting to study the electro-optic effect in strong AC fields. The transverse electro-optic effect in PMN in AC electric fields up to 20 kV cm^{-1} in the frequency region of $20-10^4$ Hz at a wavelength of 6328 Å was investigated in [21] using polarization optical techniques. The temperature varied in the interval from

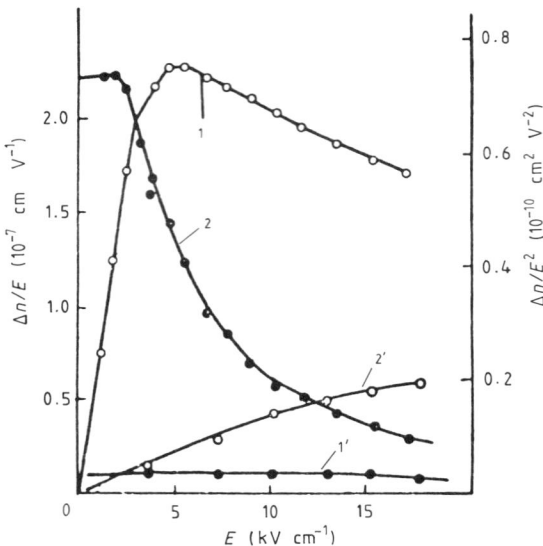

Figure 3.12   Dependence of $(1, 1')$ $\Delta n/E$ and $(2, 2')$ $\Delta n/E^2$ on the field strength at temperatures of -28 (1 and 2) and $61\,^\circ$C ($1'$ and $2'$) [14]. The light wave propagated along [001] and the electric field was applied along [100].

Table 3.2   Formulae for phase difference $\Delta\phi$ between ordinary and extraordinary rays and quadratic electro-optic coefficients R_{ij} and g_{ij} of PMN crystals at room temperature and $\lambda = 6240$ Å $(n_0 = 2.56)$ [14, 20].

Direction of light	Direction of field	Phase difference	R_{ij}, g_{ij} (CGS units)
[001]	[100]	$(\pi l/\lambda)n_0^3(R_{11}-R_{12})E^2$	$R_{11}-R_{12} = 13 \times 10^{-8}$
[001]	[110]	$(2\pi l/\lambda)n_0^3 R_{44}E^2$	$R_{44} = 6.3 \times 10^{-8}$
[001]	[100]	$(\pi l/\lambda)n_0^3(g_{11}-g_{12})P^2$	$g_{11}-g_{12} = 18.3 \times 10^{-14}$
[001]	[110]	$(2\pi l/\lambda)n_0^3 g_{44}P^2$	$g_{44} = 2.8 \times 10^{-14}$

-50 to $+180\,^\circ$C and was maintained with an accuracy of $\pm0.1\,^\circ$C. The quadratic and linear electro-optic coefficients were calculated from the slopes of the corresponding portions of the graphs of the dependence of the birefringence Δn on the field E and polarization P. In AC fields of 5–$7\,$kV cm^{-1}, as well as in DC biasing fields, the quadratic electro-optic effect is linearized. The temperature dependences of the difference in the quadratic coefficients $(R_{11} - R_{12})$ showed relaxational maxima analogous to those presented in figure 3.8. The difference between the linear electro-optic

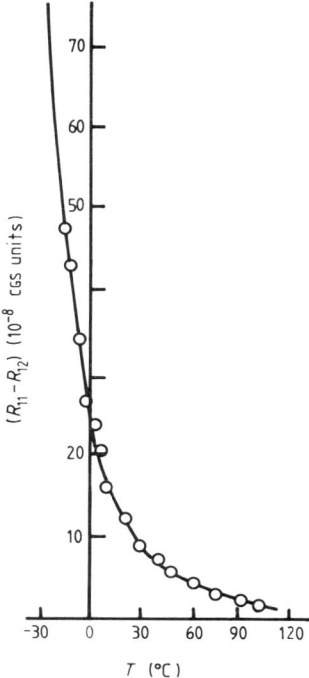

Figure 3.13 The difference of quadratic electro-optic coefficients $(R_{11} - R_{12})$ of a PMN crystal as a function of temperature [14].

coefficients $(r_{33} - r_{13})$ increases with decreasing temperature and does not exhibit maxima in the vicinity of the Curie temperature (figure 3.14).

The absence of temperature maxima of linear coefficients in AC fields can be explained by the fact that in certain fields the thermal motion is not able to reverse the polarization of a considerable part of the domains. Dispersion of the polarization and of the associated electro-optic effect is determined by the mechanism of the in-growth of domains and nuclei of the ferroelectric phase in the electric field.

The behaviour of linear electro-optic coefficients in different regimes of controlling fields generally corresponds to the ferroelectric behaviour of the reversing dielectric permittivity ε_{rev} measured in strong AC fields [19]. Therefore, at lower temperatures the value of $(r_{33} - r_{13})$ measured in strong AC fields can be expected to decrease with increasing coercive force.

Figure 3.15 presents the temperature dependence of the difference in the quadratic electro-optic coefficients $(g_{11} - g_{12})$ with respect to polarization. This value increases with temperature, showing a tendency to saturation. It is known (chapter 1, [73]) that in lead-free oxygen octahedral ferroelectrics the value of g_{ij} is an order of magnitude higher and does not depend on the

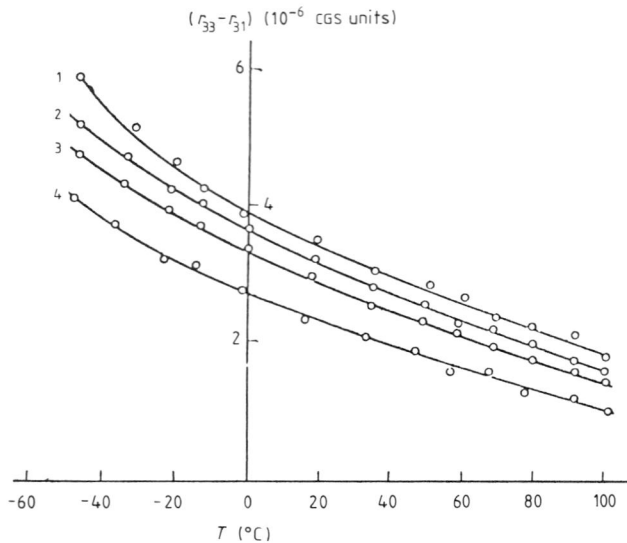

Figure 3.14   The difference of linear electro-optic coefficients $(r_{33} - r_{13})$ as a function of temperature at different electric field frequencies [21]: (1), 20; (2), 10^3; (3), 3×10^3; (4), 10^4 Hz.

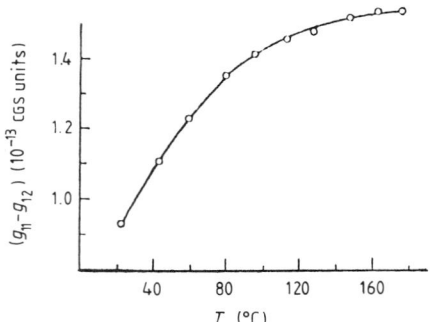

Figure 3.15   Temperature dependence of the difference of quadratic electro-optic coefficients $(g_{11} - g_{12})$ at an AC electric field frequency equal to 10^3 Hz [21].

temperature. In [22] it is assumed that the small value of $(g_{11} - g_{12})$ in lead magnesium niobate is caused by the high lead ion polarizability and by the orientational polarization of polar regions and domains.

The hysteresis studies of the birefringence Δn induced by an electric field in lead magnesium niobate were attempted by Smolensky *et al* [20]. In their experiments, a field was applied along the [100] and light propagated along the [001] direction. Figure 3.16 shows the characteristic electro-optic

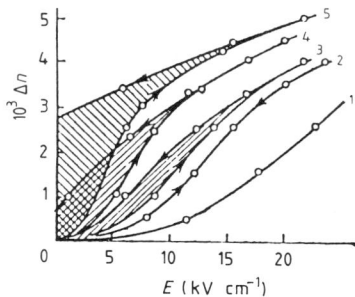

Figure 3.16 Hysteresis of the dependence of induced birefringence of a $Pb_3MgNb_2O_9$ crystal on the field strength at different temperatures [20]: (1), 20; (2), -26; (3), -48; (4), -78; (5), $-110°C$.

hysteresis loops for different temperatures. The hysteresis phenomena of lead magnesium niobate are associated with the smeared phase transition typical of this class of crystals.

3.3 Dispersion of the electro-optic effect

The electro-optic effect, as is well known, is caused by a change in the refractive index due to an electron level shift in a crystal affected by an electric field (figure 3.17). The value of $n_0^3 (R_{11} - R_{12})$ increases with decreasing wavelength when approaching the boundary of the fundamental absorption band of the crystal. But the fairly strong absorption bands in the visible region of the spectrum must also affect the values of the electro-optic coefficients. Kurtz and Robinson [23] showed that the dispersion of electro-optic coefficients in the ultraviolet is determined by the dependence of the ratio $(n^2 - 1)^2/n^4$ on the wavelength, and therefore the anomalous dispersion of the refractive index should lead to anomalous dispersion of the electro-optic coefficients.

Smolensky *et al* [24] examined this question on a sample of $Pb_3NiNb_2O_9$, which has a comparatively large electro-optic effect, absorption bands in the visible region and a smeared phase transition. The absorption spectrum of $Pb_3NiNb_2O_9$ is known to have three wide peaks in the visible and the near infrared spectra. At a temperature of $-190°C$ they lie at 1.25, 0.75 and 0.45 μm and correspond to transitions $^3A_{2g} \rightarrow {}^3T_{2g}$, $^3A_{2g} \rightarrow {}^3T_{1g}^a$, $^3A_{2g} \rightarrow {}^3T_{1g}^b$. In electro-optic studies, the field applied to the crystal made an angle of $45°$ with the polarization planes. In this case, $n_0^3 (R_{11} - R_{12})$ is specified by the expression

$$n_0^3(R_{11} - R_{12}) = \frac{d^2}{l} \frac{\lambda}{V_{\lambda/2}^2} \tag{3.14}$$

where λ is the wavelength of light, d is the distance between the electrodes, l is the optical path length, n_0 is the refractive index in zero electric field and

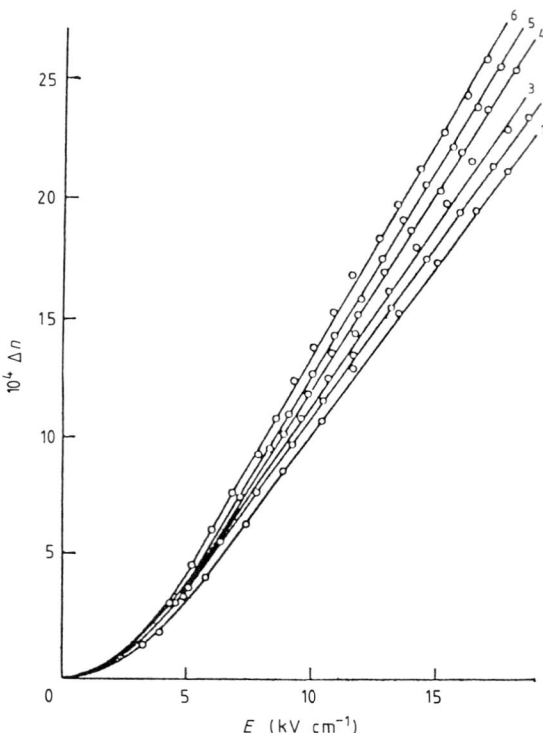

Figure 3.17   Room-temperature dependence of birefringence Δn on the field strength E for different wavelengths λ [14]: (1), 6240; (2), 5600; (3), 5180; (4), 4900; (5), 4580; (6), 4370 Å. The light wave was propagated along [001], the electric field was applied along [100].

$V_{\lambda/2}$ is the half-wave voltage. The dependence of $V_{\lambda/2}$ on the wavelength of light was measured experimentally (figure 3.18). The change of $V_{\lambda/2}$ near the absorption band was found to be anomalous.

The dependence of $(R_{11} - R_{12})$ on the wavelength of light for lead magnesium niobate was experimentally measured in [22]. The electro-optic properties were studied conventionally, and the refractive index of PMN in the visible region of the spectrum (4000–8000 Å) was measured using the minimum deviation method (figure 3.19). Comparison of the obtained dependences of electro-optic coefficients and optical density of lead magnesium niobate [25] shows that an increase of electro-optic coefficients is observed near the boundary of the fundamental absorption band. The plot of the refractive index against the wavelength λ is given in figure 3.20.

The ratio of $n_0^3 (R_{11} - R_{12})$ to $n_0^3 (r_{33} - r_{13})$ as a function of the wavelength λ, within experimental error, is a straight line parallel to the abscissa (curve 3,

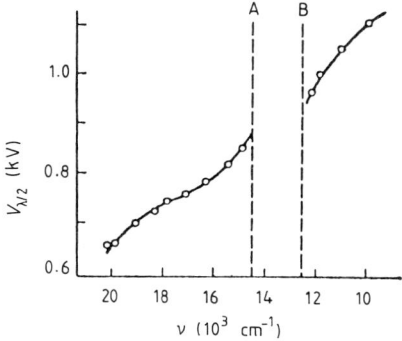

Figure 3.18 Dependence of $V_{\lambda/2}$ on the light wave frequency for $Pb_3NiNb_2O_9$ at $-190°C$ [24].

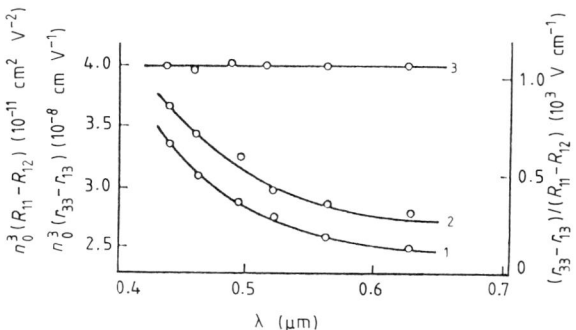

Figure 3.19 Dependence of: (1), $n_0^3 (R_{11} - R_{12})$; (2), $n_0^3 (r_{33} - r_{13})$; and (3), their ratio, on the wavelength of light for a $Pb_3MgNb_2O_9$ (PMN) crystal at room temperature [22].

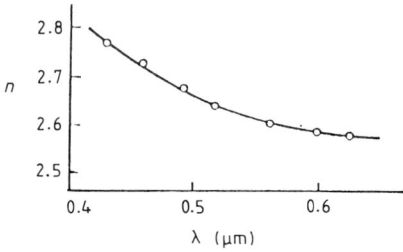

Figure 3.20 Room-temperature dependence of the refractive index n on the wavelength of light λ for a $Pb_3MgNb_2O_9$ crystal [22].

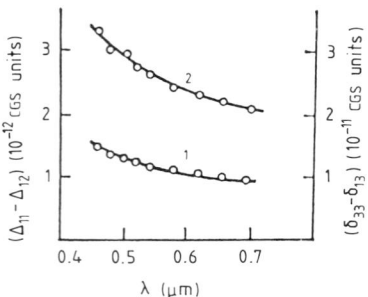

Figure 3.21 Dependences of: (1) the difference of nonlinear Miller coefficients $(\delta_{33} - \delta_{13})$; and (2) the difference of quadratic nonlinear coefficients $(\Delta_{11} - \Delta_{12})$ on the wavelength of light for a $Pb_2MgNb_2O_9$ crystal [22].

figure 3.19). This confirms linearization of the quadratic effect which is proportional [26] to

$$\frac{(n_0^2 - 1)^2}{n_0^4} \Delta_{ijkl}$$

where Δ_{ijkl} are the nonlinear coefficients of the quadratic effect analogous to the Miller coefficients δ [11].

The dispersions of the differences of the nonlinear Miller coefficients $(\delta_{33} - \delta_{13})$ and the quadratic nonlinear coefficients $(\Delta_{11} - \Delta_{12})$ were determined from the dependences of the electro-optic coefficients and the refractive index on the wavelength of light (figure 3.21). It follows from the figure that a significant contribution to the dispersive dependence of the electro-optic coefficients is made by the dispersion of the nonlinear coefficients Δ and δ. One can thus conclude that a simple model of an anharmonic oscillator does not describe all the specific features of dispersion of the electro-optic effect in PMN crystals. The model propsed by Di Domenico and Wemple for perovskite-type ferroelectrics was found, by the authors of [22], to hold also for PMN. These models are treated in detail in chapter 9. The values of the Sellmeier parameters for PMN and some other oxygen octahedral ferroelectrics are listed in table 9.3.

Table 3.3 Difference of piezo-optic coefficients $(\pi_{11} - \pi_{12})$ $(10^{-13} \, cm^2 \, dyn^{-1})$ for PMN and PZN crystals at different temperatures [27].

$T \, (°C)$	PMN	PZN	$T \, (°C)$	PMN	PZN
22	5.2	6.1	54	2.87	3.7
29	4.56	5.45	62	2.87	3.5
34	4.25	4.75	76	2.87	3.2
42	3.4	4.25	90–120	2.87	3.0

The closeness of the dispersion parameters of PMN and other oxygen octahedral ferroelectrics makes it possible to conclude that, in these crystals, a considerable contribution to the dispersion in the visible spectrum is due to transitions between the p band of the oxygen atom and the d band of the transition element atom.

3.4 Elasto-optic effect in PMN and PZN crystals

Elasto-optic studies of PMN and PZN have provided an insight into the nature of smeared phase transitions in these crystals, and have enabled the true electro-optic effect to be identified among the other effects.

For cubic crystals of the symmetry group m3m, which includes PMN and PZN crystals in the paraelectric phase, the number of independent piezo-optic coefficients π_{ij} is 3, the induced birefringence being determined only by the combination of $(\pi_{11} - \pi_{12})$ and π_{44}.

The case of one-dimensional mechanical stress σ along the [100] axis of the crystal was studied in [27] for the light propagation direction [001], when the induced birefringence is specified by the expression

$$\Delta n = \tfrac{1}{2}n_0^3(\pi_{11} - \pi_{12})\sigma. \tag{3.15}$$

The experimental dependences of the optical system transmission intensity I/I_0 on the mechanical load $F = \sigma \cdot s$, where s is the sample area, were determined on electrically insulated samples. The amplitudes of light transmission maxima were found first to decrease with increasing load and then to remain constant as soon as the pressure $F = 6\text{--}7$ kg ($\sigma = 150\text{--}200$ kg cm^{-2}) is attained. This is presumably due to the induced ferroelectric phase in which a-domains are formed. Under a mechanical load $\sigma = 150\text{--}200$ kg cm^{-2} the crystal already consists entirely of a-domains. The dependence of the stress $\sigma_{\lambda/2}$, corresponding to the path-length difference on the wavelength of light in the range 4500–7000 Å, is depicted in figure 3.22, curve 1. Curve 2 gives the difference in the piezo-optic coefficients $(\pi_{11} - \pi_{12})$ plotted against λ in the same range of wavelengths. The increase in $(\pi_{11} - \pi_{12})$ with decreasing λ is connected with approaching the boundary of the fundamental absorption band ($\lambda = 0.2\ \mu$m). The same dependence was observed earlier for electro-optic coefficients.

In PMN and PZN crystals, the dependence of birefringence on the mechanical stress F is linear, but the slope of the straight line depends on the crystal temperature (figure 3.23). At a temperature above 55 °C, the slope of Δn remains practically unchanged up to 95 °C. The change discovered in the slope of the stress-dependent birefringence is connected with a stress-induced orientation of the polar regions and domains, which contributes to the elasto-optic effect. In ferroelectrics with a well defined phase transition, as distinct from the indicated crystals, the corresponding dependence, which is

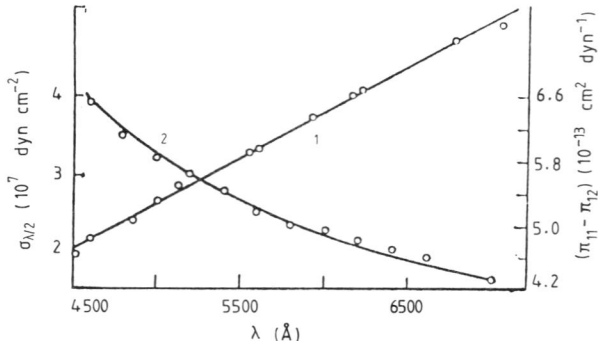

Figure 3.22 Dispersions of the half-wave mechanical voltage $\sigma_{\lambda/2}$ and of the difference of piezo-optic coefficients $(\pi_{11} - \pi_{12})$ for a PMN crystal at room temperature [27].

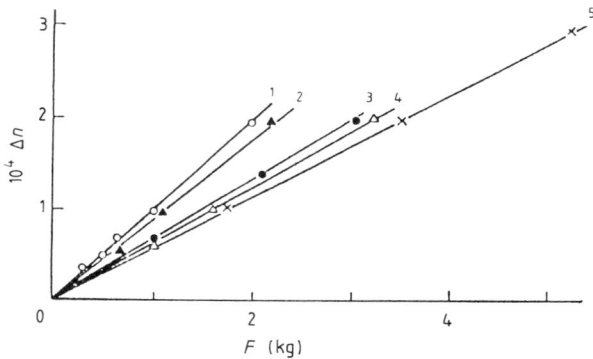

Figure 3.23 Dependence of birefringence Δn on the mechanical load F for a PMN crystal at different temperatures [27]: (1), 24; (2), 28; (3), 43; (4), 47; (5), 54–95°C.

also linear, does not depend on the temperature (except in the vicinity of the Curie point) [28].

The piezo-optic coefficients of PMN and PZN crystals at different temperatures computed from the data of figure 3.23 are listed in table 3.3.

The elasto-optic studies on PMN crystals were developed in [29]. In line with [27], a mechanical load was applied in the [100] and the light propagated in the [001] direction. An electric field was applied to the crystal along the [100] and [010] directions. The values of the stiffness factors necessary for computing the difference in the elasto-optic constants $(p_{11} - p_{12})$ measured by the ultrasonic pulse technique coincided with the values obtained by the dynamic method at a frequency of 1 GHz (table 3.4).

Table 3.4 Elasto-optic characteristics of PMN at room temperature and $\lambda = 6328$ Å [29].

Half-wave mechanical voltage $\sigma_{\lambda/2}$ (kg s cm^{-2})		7.4
Difference of piezo-optic coefficients $(\pi_{11}-\pi_{12})$ $(10^{-13}$ cm^2 dyn$^{-1})$		5.2
Difference of elasto-optic coefficients $(p_{11}-p_{12})$		0.53
Refractive index n		2.53
Absorption coefficient α (cm^{-1})		2
Rigidity constants c_{ij} $(10^{-12}$ dyn cm$^{-2})$	c_{11}	1.67
	c_{12}	0.59
	c_{44}	0.77

Electro-optic properties of crystals exhibiting a smeared phase transition are affected by orientational effects. When a mechanical stress σ is applied perpendicular to the electric field E, the domains are rotated by 90°, some of them being aligned along the electric field ($a_\perp$-domains), and others along the propagation direction of light ($a_\parallel$-domains). The resultant action of mutually perpendicular electric and mechanical stresses has the effect that the effective refractive index decreases in the direction of the electric field and increases in the perpendicular direction. An increase in Δn corresponds to an increase in the electro-optic coefficients. A gradual increase in the load first causes orientation of those polar regions and domains that are fixed weakly; then larger and larger polar regions are subsequently involved in the orientation process. In the pressure range corresponding to a complete orientation of the polar regions and domains, saturation is observed. BaTiO$_3$ crystals exhibited saturation in the pressure range between 200 and 300 kg cm^{-2} [30].

For the case $E \parallel \sigma$, mechanical pressure hampers 90° rotations of domains under the action of an electric field, and one can expect a decrease in the electro-optic coefficients. In practice, such a decrease has not been observed below 80 kg cm^{-2}.

Thus we may state that orientational processes can contribute markedly to the electro-optic effect of PMN and PZN crystals.

3.5 Light scattering in lead magnesium niobate crystals in the region of a smeared phase transition

As has been mentioned above, PMN and PZN crystals show a smeared ferroelectric phase transition. Obviously, the coexistence of phases may cause a substantial light scattering, the light scattering picture changing with varying temperature or with electric field application, which affects the number and size of different phase regions. Thus, the light scattering studies in these crystals may provide information on the properties and parameters of regions of coexisting phases.

A temperature-dependent intensive light scattering was observed in PMN crystals in the temperature range 20–400°C [31]. The coefficient of scattering at an angle of 90° of light with the wavelength $\lambda = 6328$ Å was approximately equal to 4×10^{-4} cm^{-1}. The scattering indicatrices in the plane of polarization of incident light at different temperatures are given in figure 3.24. They indicate a considerable decrease of scattering with increasing temperature. The scattering stops at 360°C, accompanied by an increase of forward transmission of the sample. To establish the origin of scattering, electric fields were applied to crystals along the [100] direction and parallel to the plane of polarization of incident light. As the field was increased, scattering at large angles decreased and the crystal cleared in the direction of incident light. When the sign of the electric field reversed, no hysteresis was observed. An applied electric field possibly leads to an increase of existing polar regions.

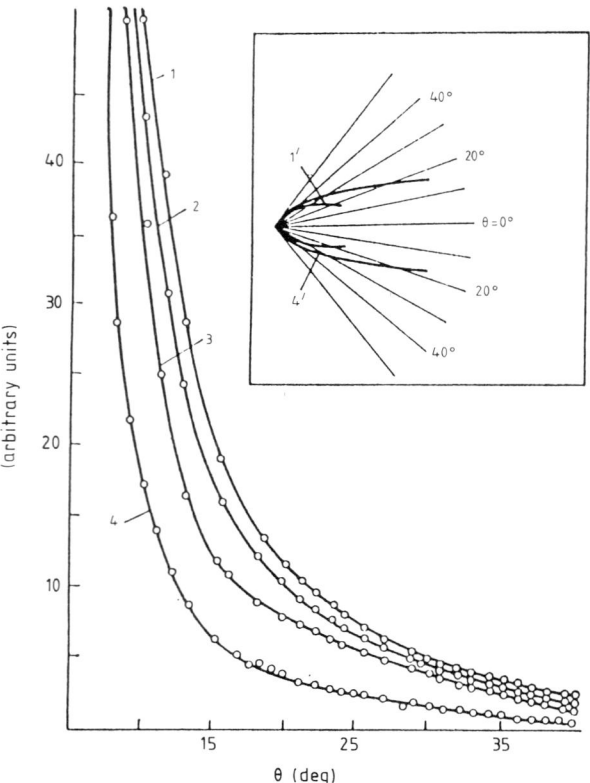

Figure 3.24 Angular distribution of the light intensity *I* in the plane of polarization of incident light at different temperatures [31]: (1), 20; (2), 80; (3), 120; (4), 300°C. Curves 1' and 4' in the top right corner give diagrams for 20 and 300°C respectively.

From the angular distribution of the temperature dependence of light scattering, the sizes of scattering particles were estimated by the Rayleigh–Hans theory on the assumption that they have a spherical shape. In this case, the intensity of light scattered by particles and emerging through a plane face of the sample at an angle θ to the normal of the plane face is described as

$$I = C(T)I_0[(9\pi/2u^3)^{1/2}J_{3/2}(u)]^2 \tag{3.16}$$

where I is the intensity of light emerging at the angle θ, I_0 is the intensity of incident light, $C(T)$ is a certain function of temperature, $u = 2x \sin(\theta/2n)$, $x = ka$, $k = 2\pi n/\lambda$ is the wavevector in a medium, λ is the wavelength of incident light, a is the radius of a scattering particle, and $J_{3/2}$ is the Bessel function of the order of $\frac{3}{2}$. The temperature-dependent part presumably tends to zero when approaching 360 °C. At 120 °C, in samples less than 0.17 mm thick, where the single scattering condition is fulfilled, the mean radius of the scattering particles is ≈ 0.4 μm, according to the above formula. It should be noted that a scattering particle may involve both the polar region and the neighbouring layer of the non-polar phase perturbed by polarization.

It should also be noted that the anomaly in the optical density observed in [25] near the absorption band edge of a PMN crystal could be caused by light scattering, because the contribution of scattering to the optical density may increase on approaching the absorption band.

3.6 PMN crystals in optical radiation control

It has already been stated above that PMN crystals possess a substantial electro-optic effect. The absence of natural birefringence is an important factor in using the crystals as laser beam modulators, since it restricts the modulator aperture and is fairly sensitive to temperature variations [32]. If crystal birefringence is spatially homogeneous, it can be compensated (see e.g. [33]). But spatially inhomogeneous birefringence, which may occur in the crystal growth process, is practically impossible to compensate.

The relation between the quadratic electro-optic coefficients R_{ij} and the dielectric permittivity is [34]

$$R_{ij} = N\Delta_{ij}\varepsilon_\omega^2 \tag{3.17}$$

where

$$N = \left(\frac{1}{4\pi}\right)^3 \left(\frac{n_0^2 - 1}{n_0^2}\right)^2 \tag{3.18}$$

n_0 being the refractive index in zero electric field, Δ_{ij} the Miller coefficients for the quadratic electro-optic effect, ε_ω the dielectric permittivity at the electric field frequency v. Dispersion of the dielectric permittivity and of the refractive index causes dispersion of the quadratic electro-optic coefficients.

Upon a polarization approximately equal to 6.6 μC cm^{-2}, the quadratic electro-optic effect in PMN crystals becomes linearized. The linear electro-optic coefficients r_{ij} can be represented as

$$r_{ij} = 8\pi N \Delta_{ij} P_0 \varepsilon_\omega \qquad (3.19)$$

where ε_ω is the dielectric permittivity for the electric field strength E_0 for which the polarization is P_0.

Phenomenological consideration of the electro-optic effect in crystals with the point symmetry group m3m predicts the existence of quadratic electro-optic effects perpendicular to the [001] and parallel to the [011] directions. The value of the longitudinal effect is determined by the expression $(R_{11} - R_{12} - R_{44})$, which is close to zero for all the known crystals except PMN. The latter exhibit a marked longitudinal electro-optic effect, which makes them promising for application in electro-optic shutters and discrete deflectors. The expressions for the change in the refractive index under longitudinal and transverse quadratic effects are given in table 3.5.

For light propagating along the [010] direction and an electric field applied along the [001] direction, the half-wave voltage is determined by expression (3.14), which can be rewritten in the form

$$V_{\lambda/2} = \left(\frac{\lambda}{n_0^3(R_{11} - R_{12})}\right)^{1/2} \frac{d}{l^{1/2}}$$

where l is the optical path length in the crystal, d is the distance between the electrodes. As a result of the nonlinear dependence of Δn on E, the voltage between the ith maximum and jth minimum is

$$V_{\lambda/2}^{ij} = V_{\lambda/2}^{00} |(2i + 1)^{1/2} - (2j)^{1/2}| \qquad (3.20)$$

where $V_{\lambda/2}^{00}$ is evaluated from (3.14) and $i, j, i + 1$ are positive integers. Equation (3.20) implies that the applied DC bias equal to $\Sigma_{ij} V_{\lambda/2}^{ij}$ can considerably decrease the control voltage. We may note that, under the transverse linear electro-optic effect, $V_{\lambda/2}$ is proportional to the ratio d/l, while under the

Table 3.5 Expressions for induced birefringence Δn in PMN crystals [32].

light wave polarization	field	Δn
	Direction of:	
[001]	[001]	$\frac{1}{2}n_0^3 R_{11} E^2$
[010]	[001]	$\frac{1}{2}n_0^3 R_{12} E^2$
[011]	[011]	$\frac{1}{2}n_0^3(R_{11} + R_{12} + R_{44})E^2$
[010]	[011]	$\frac{1}{2}n_0^3(R_{11} + R_{12} - R_{44})E^2$

quadratic effect $V_{\lambda/2}$ is proportional to $d/l^{1/2}$. In order to reduce the control voltage under the quadratic effect, it is therefore more appropriate to diminish the crystal thickness d rather than to extend its length l.

As a result of linearization of the electro-optic effect, the dependence of $V_{\lambda/2}$ on the crystal size becomes more complicated. It was established earlier that at room temperature the dependence of n on the electric field strength E is linear for $E \geqslant E_0 = 4$ kV cm^{-1}. In order to obtain the expression for $V_{\lambda/2}$ in the case $E > E_0$, one should know the relation between the induced birefringence and the electric field strength for $E > E_0$. When such a field is applied in the [001] direction, PMN crystals become tetragonal with symmetry 4mm, while for a field applied along the [011] direction it becomes orthorhombic. The expressions for induced variation in the birefringence of light propagating along different directions and for a field applied along [001] or [011] directions are presented in table 3.6. From these formulae one can derive the expression for the half-wave voltage $V_{\lambda/2}$ for $E \geqslant E_0$:

$$V_{\lambda/2} = \frac{\lambda}{2n_0^3(r_{33} - r_{13})} \frac{d}{l} + \left(E_0 - \frac{R_{11} - R_{12}}{2(r_{33} - r_{13})}\right)d. \qquad (3.21)$$

The electro-optic characteristics of PMN, listed in table 3.7, were obtained from experimentally measured values of refractive indices. At room temperature, the investigated crystals possess the following properties: $n_0 = 2.56$, at a frequency $v = 5 \times 10^5$ Hz, dielectric permittivity $\varepsilon = 7.400$, $\tan \delta = 0.02$.

For any ratio d/l there exists a thickness d_0 such that $E_{\lambda/2} = E_0$. For the ratio $d/l = 1$ this thickness is equal to 0.16 cm. From figure 3.25, which illustrates the dependence of $V_{\lambda/2}$ on the crystal thickness d under the transverse effect, it is readily seen that with decreasing d the value of $V_{\lambda/2}$ tends to approximately 110 V (curve 1). Under the longitudinal electro-optic effect we have $E_0 \approx 8.3$ kV. For the case of the longitudinal effect, the equation for the half-wave voltage will be, by analogy with the transverse effect,

$$V_{\lambda/2} = \left(\frac{2\lambda l}{n_0^3(R_{11} - R_{12} - R_{44})}\right)^{1/2} \qquad (3.22)$$

Table 3.6　Expressions for induced birefringence Δn in PMN crystals [32] for $E > E_0$.

Direction of light propagation	Field direction	Crystal symmetry	Expression for induced birefringence
[001]	[001]	4mm	$1/2n_0^3[R_{11}E_0^2 + 2r_{33}(E - E_0)]$
[100]	[001]	4mm	$1/2n_0^3[R_{12}E_0^2 + 2r_{13}(E - E_0)]$
[011]	[011]	mm2	$1/4n_0^3[(R_{11} + R_{12} - R_{44})E_0^2 + 2r_{13}(E - E_0)]$
[100]	[011]	mm2	$1/4n_0^3[R_{12}E_0^2 + 2r_{33}(E - E_0)]$

Table 3.7 Electro-optic characteristics of PMN for $\lambda = 0.63$ μm and $l/d = 1$ [32].

R_{ij} (10^{-12} cm^2 V^{-2})	r_{ij} (10^{-10} cm V^{-1})		$V_{\lambda/2}$ (kV)			
			longitudinal effect		transverse effect	
	mm2	4mm	$d < d_0$	$d > d_0$	$d < d_0$	$d > d_0$
$R_{11} = 1.40$	$r_{13} = 69.0$	$r_{33} = 167.0$	$0.52 + 4.9d$	$3.4\sqrt{d}$	$0.11 + 3.4d$	$3.2\sqrt{d}$
$R_{12} = 0.09$	$r_{23} = 3.8$	$r_{13} = 8.2$	—	—	—	—
$R_{44} = 0.67$	—	—	—	—	—	—

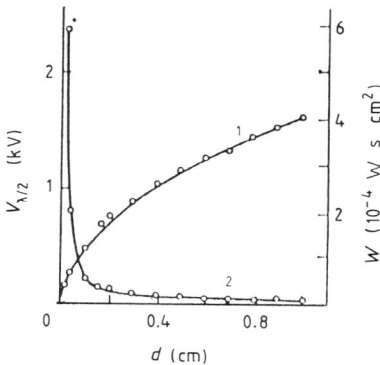

Figure 3.25 Dependences of (1) the half-wave voltage $V_{\lambda/2}$ and (2) the power consumption W on the crystal thickness d in PMN under the transverse effect [32]. Here and in the next figure, the power consumption values are reduced to unit frequency and unit crystal area.

for $d > d_0$ and

$$V_{\lambda/2} = \frac{\lambda}{n_0^3(r_{33} - r_{13})} + \left(E_0 - \frac{R_{11} - R_{12} - R_{44}}{2(r_{13} - r_{23})} \right) d \qquad (3.23)$$

for $d < d_0$.

The plot of $V_{\lambda/2}$ versus crystal thickness d under the longitudinal effect is drawn in figure 3.26. As d decreases, $V_{\lambda/2}$ tends towards a value of 575 V (curve 1).

Figures 3.25 and 3.26 imply that in the case of the quadratic electro-optic effect it is appropriate to employ thin crystals. Since cubic PMN crystals show no natural birefringence, they may be employed as modulators for convergent

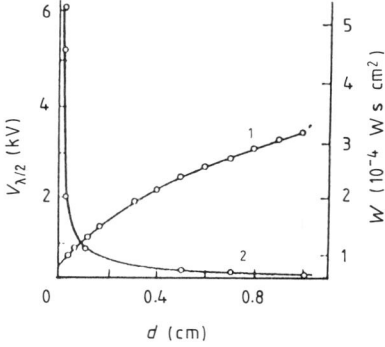

Figure 3.26 Dependences of (1) the half-wave voltage $V_{\lambda/2}$ and (2) the power consumption W on the crystal thickness d in PMN under the longitudinal effect (see figure 3.25) [32].

light beams. This enables the crystal thickness to be substantially reduced without diminishing the angular aperture of the modulator.

The essential deficiency of PMN crystals is their large dielectric permittivity ε_ω. It considerably enlarges the wattage W of the applied field, especially in the case of radio-frequency and microwave fields. But in the low-frequency range, where the dispersion of ε_ω is negligible, the power gain may be substantial. The wattage W as a function of crystal thickness for the transverse ($d/l = 1$) and longitudinal effects is plotted in figures 3.25 and 3.26 (curves 2). The graphs show that for $d < d_0$ the wattage remains unchanged, whereas for $d > d_0$ a nonlinear dependence of wattage on crystal thickness is observed. Estimations of the value of dissipated power give reason to regard PMN crystals as promising for frequencies of about 10 MHz. For these frequencies, the power consumption does not exceed 0.1 W for 70% modulation.

We will now consider the possibility of using PMN crystals for light beam deflection. A change in the propagation direction of light is known to occur on the boundary between regions with distinct refractive indices, and the light beam deflection can be controlled by changing the refractive index of one of the regions.

If the light beam is parallel to the prism base, then according to the refraction laws one can observe the following dependence of the refraction angle on the refractive index

$$\Delta\beta = \frac{2\Delta n \, \sin(\tfrac{1}{2}\alpha)}{\cos[\tfrac{1}{2}(\alpha + \beta)]} \tag{3.24}$$

where Δn is the change in the refractive index, β is the refraction angle, and α is the angle at the prism vertex.

The large electro-optic effect inherent in PMN crystals enables the refractive index to be changed considerably, $\Delta n \approx 5 \times 10^{-3}$. The maximum change in the refractive index is observed when the applied electric field and the electric vector of the light wave both go along the [100] direction. The variation of the refraction angle as a function of the applied electric field for a prism with a vertex angle $\alpha = 31°$ and $d = 2.2$ mm is plotted in figure 3.27. In accordance

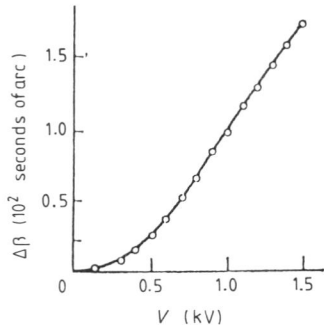

Figure 3.27 Dependence of variation of the refraction angle $\Delta\beta$ on the applied voltage for a PMN crystal [31].

with the wattage-dependent change of the refractive index, the value $\Delta\beta$ exhibits quadratic dependence which becomes linear with a subsequent increase of the field.

Thus, in the electro-optic studies discussed above, PMN crystals showed promise as materials for modulation and laser beam scanning.

3.7 Growth of complex perovskite-type crystals

Complex perovskites were synthesized in the course of a solid state reaction between the initial oxides or corresponding carbonates taken in prescribed proportions. The annealing temperatures were chosen arbitrarily.

The physico-chemical mechanism for the formation of ferroelectrics of complex composition was analysed on an example of several lead-containing perovskites [35]. It was established, using high-temperature X-ray diffraction analysis, that in the case of $Pb_3NiNb_2O_9$, Pb_2FeNbO_6, $Pb_3Fe_2WO_9$, Pb_2MgWO_6 and a number of other compounds an intense binding of PbO starts at $600-700°C$; the first to appear is the pyrochlore structure, and as the annealing temperature is increased the perovskite structure occurs and the pyrochlore disappears.

As has already been mentioned, single-crystal compounds are accessible for more extensive experimental studies. In growing perovskite crystals of complex composition one encounters considerable difficulties because of the lack of phase diagrams, and lack of data on the temperature and the character of melting and on the interaction between compounds and solvents. The first successful growth of single crystals of the complex perovskites Pb_2CoWO_6, $Pb_3CoNb_2O_9$, $Pb_3FeNb_2O_9$, $Pb_3NiNb_2O_9$, $Pb_3MgNb_2O_9$ and some others was first accomplished by Myl'nikova and Bokov [4, 36–38]. They used the method of growing from a flux. Lead oxide was chosen as a solvent. The advantage of this solvent is that it creates a preventive atmosphere that hinders dissociation of the compound under crystallization [39]. For crystallization, the mixtures of oxides with prescribed stoichiometric composition corresponding to the compounds $Pb(B'B'')O_3$ were preliminarily calculated at $800-900°C$. For obtaining single crystals, the mixture thus prepared was mixed with lead oxide in different proportions. This material in capped platinum crucibles was placed in a furnace with carborundum heaters. The furnace was heated up to a temperature of $1200°C$ for 3 h and then the temperature was maintained constant for $1-2$ h. The furnace was cooled to $800°C$ at a constant rate which was varied in different experiments from 60 to 10 K h^{-1}.

The maximum temperature of $1200°C$ was chosen so as not to allow dissociation and, if possible, to maintain the initial lead oxide concentration. To lower the melt viscosity, nearly 3 wt.% of B_2O_3 was introduced.

The composition 50 mol.% Pb–50 mol.% B_2O_3 was recommended [40] as solvent for the crystallization of some lead-containing perovskites ($PbTiO_3$,

Table 3.8 Compositions of typical melts for growing single crystals using the Kyropoulos method (chapter 2, [3])†.

Oxides	Concentration (mol.%)	Concentration (mol.%)	
		in the melt	in the crystal
PbO	55–80	60	60
MgO	5–25	20	20
Nb_2O_5	4–15	6	20
B_2O_3	6–20	14	2

† The last two columns present a particular example of melt compositions and of single crystals grown from this melt.

$PbZrO_3$). Higher-quality crystals were obtained with 75 mol.% of solvent relative to the substance under crystallization.

Kyropoulos-grown $Pb_3MgNb_2O_9$ single crystals of good optical quality and satisfactory size were obtained in [41]. A vertical muffle furnace with horizontal heating elements of silicon carbide was used. The starting material was melted in a 100 cm³ platinum crucible placed in the lower portion of the muffle and heated up to 1150°C. The seed crystal was mounted onto an air-cooled platinum rod which joined with a pulling aggregate and rotated at a rate of 40 rpm. The starting material melted at a temperature one or two degrees higher than the melt crystallization point, and the crystal growth proceeded upon melt cooling by means of increasing the air flux rate or cooling the melt by nearly 0.5°C for an hour. When a crystal attained a size of about 1.0 cm, it was slowly pulled above the melt at a rate of 1.5 mm h⁻¹. When the distance to the melt level was 1.2 cm, the crystal and the furnace were cooled together with the melt for 24 h in order to prevent the crystal from cracking. The crystal boule pulled out of the melt usually appeared to be etched and covered with PbO condensed from the vapours. The inner part of the boule contained no imperfections. Such crystals viewed under a polarization microscope exhibited no thermal stresses.

The compositions of typical melts for growing lead magnesium niobate crystals are listed in table 3.8, as recommended in chapter 2 [3].

PART II

TETRAGONAL
FERROELECTRICS OF
POTASSIUM TUNGSTEN
BRONZE-TYPE STRUCTURE

The numerous oxygen tetragonal ferroelectrics of tungsten bronze type are characterized by distorted tetragonal unit cells, each containing 10 BO_6 octahedra which constitute their crystal lattice. The unit cells have almost identical dimensions of the order of $12.5 \times 12.5 \times 4$ Å^3. Oxygen octahedra stretch along the c axis, which is the ferroelectric polar axis. As the B atom one commonly takes Nb, or sometimes Ta or Ti, as well as a mixture of these.

There are three types of position in the tetragonal bronze framework, defined as triangle (C), tetragon (A1) and pentagon (A2), which all together make up the 10 positions in the unit cell. The chemical formula per unit cell can be written as follows: $[(A1)_2(A2)_4C_4][(B1)_2(B2)_8]O_{30}$. Positions A1, A2 and C can be occupied by alkaline, alkaline earth and rare earth ions whose ratio is determined by electric neutrality of the entire crystal lattice.

Tungsten bronzes exhibit a layer-type structure in which oxygen atoms form slightly strained layers with coordinates $z \approx 0$ and $\frac{1}{2}$. The B atoms are positioned in layers with coordinates $z \approx 0$ and $z = \frac{1}{2}$; the A atoms in layers with $z \approx \frac{1}{4}$. An example of this type of structure may be barium strontium niobate, $Ba_xSr_{1-x}Nb_2O_6$, where $0.25 \leqslant x \leqslant 0.75$. A detailed discussion of this compound is given in chapter 4.

The relating structure belonging to the same family has a distorted lattice of orthorhombic symmetry, the unit cell being approximately equal to $17.5 \times 17.5 \times 8$ Å^3, where $a_{orth} \approx b_{orth} \approx (2a_{tetr})^{1/2}$ and $c_{orth} \approx 2c_{tetr}$. An example of this structure may be barium sodium niobate, $Ba_{4+x}Na_{2-2x}Nb_{10}O_{30}$, to which chapter 5 is devoted.

The number of ions occupying different positions in the structure may serve as the basis for the following classification of tungsten bronzes: 5-0 type (e.g. $Ba_xSr_{1-x}Nb_2O_6$); 6-0 type ($Ba_2NaNb_5O_{15}$); and 5-1 type ($Sr_4KLiNb_{10}O_{30}$). The latter type is treated in chapter 6.

The paraelectric tungsten bronzes comprise mirror planes perpendicular to the c axis. All the atoms are therefore positioned precisely in the $z = 0$ and $z = \frac{1}{2}$ planes. The prototype lattice possesses $4/mmm$ symmetry. For barium strontium niobate ($x = 0.5$), the phase above 408 K presumably exhibits this structure. Atomic displacements in metal relative to the oxygen planes induce spontaneous polarization in tungsten bronzes. For example, in barium strontium niobate, the polar axis is directed along the four-fold axis of the NbO_6 octahedron. The Nb atoms, which fill B1 and B2 positions, are displaced from symmetric positions by 1.06 and 0.048 Å respectively. The Ba and Sr atoms show smaller displacements. Upon polarization reversal the metallic atoms pass through the mirror planes of the prototype lattice, i.e. through the oxygen atomic layers. Thus the indicated materials are displacive-type ferroelectrics.

4 Single Crystals of Barium Strontium Niobate

The high electro-optic coefficients, which greatly exceed those of potassium dihydrophosphate and lithium niobate crystals [1], the high pyroelectric coefficients [2], the very favourable acousto-optic [3] and nonlinear [4] properties of barium strontium niobate (BSN) solid solutions, make these materials promising for applications. The BSN crystals may well be used as a medium for reversible optical memory [5].

4.1 Structural characteristics

Barium strontium niobate solid solutions belong to the class of oxygen octahedral ferroelectrics possessing tetragonal potassium tungsten bronze structure [6]. As in the case of perovskites, the structure is a three-dimensional network of NbO_6 octahedra linked by their corners so as to form alternating five- and four-membered rings (figure 4.1). Spaces inside the rings form three types of voids: tetragonal (A1), pentagonal (A2) and trigonal (C), arranged parallel to one another and to the tetragonal c axis. The structural formula of the compound taking into account the number of different positions (voids) and non-equivalence of the octahedra can be written as $(A1)_2(A2)_4(C)_4(B1)_2(B2)_8O_{30}$. Positions A1 with coordination number 9 and A2 with coordination number 12 can be occupied by Sr and Ba ions (altogether five ions for six A1 + A2 positions). Positions C with coordination number 9 are vacant, which provides electroneutrality. The ten NbO_6 octahedra entering the unit cell composition contain eight Nb ions in B2 and two Nb ions in B1 positions. The Nb ions lie in the horizontal symmetry plane of oxygen octahedra. Voids A1, A2 and C are at a distance $c/2$ from the plane occupied by niobium ions.

A BSN solid solution of composition $Ba_{0.9}Sr_{0.1}Nb_2O_6$ exhibits a cation arrangement (distribution) such that all the pentagon positions A2

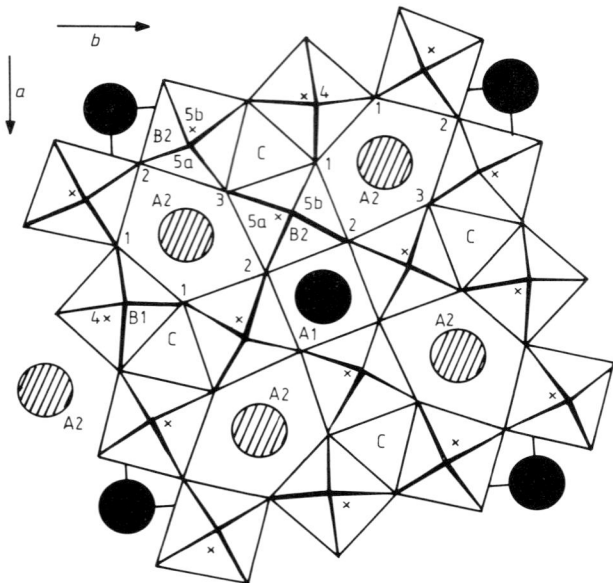

Figure 4.1   Barium strontium niobate structure in the (a,b) plane [6]. Dark circles represent preferable strontium ion positions $(z = \frac{1}{2})$; broken circles represent positions statistically occupied by barium and strontium ions $(z = \frac{1}{2})$. Digits 1–5 and crosses indicate oxygen ion positions in octahedra.

accommodate barium ions and the tetragon A1 positions are filled by an equal number of Ba and Sr ions. Lowering x in the general formula $Ba_xSr_{1-x}Nb_2O_6$ to 0.8 leads to partial replacement of large Ba^{2+} ions $(r = 1.34$ Å) by smaller Sr^{2+} ions $(r = 1.27$ Å) in A1 positions. Above the value $x = 0.55$ there are sufficient Ba^{2+} ions in pentagon positions, and the unit cell volume remains practically unchanged. Variations of x in the range 0.55–0.40 corresponds to a cation distribution such that all the tetragon A1 positions are occupied by Sr ions and the pentagon A2 positions by an equal number of Sr and Ba ions. Such a distribution results in a certain compression of the unit cell due to a high concentration of the smaller Sr^{2+} ions.

If A1 positions are occupied by Sr ions only, then with a change in the Sr/Ba cation ratio the entropy of ion distribution becomes a minimum for $x = 0.67$, as reported in [2].

Interatomic distances in the BSN structure are listed in table 4.1.

The ferroelectric crystals of BSN belong to the polar class 4mm with the space symmetry group $P4bm(C_{4v}^2)$. The tetragonal unit cell contains five formula units. The dependence of cell parameters on the solid solution composition for ceramic samples is depicted in figure 4.2. The results in the

Table 4.1 Interatomic distances (Å) in barium strontium niobate $Ba_{0.27}Sr_{0.75}Nb_2$ $O_{5.78}$ at 298 K [6]†.

Chemical bond	Distance (Å)	Chemical bond	Distance (Å)
Nb(1)–O(1)	1.951 ± 8	Ba/Sr–O(1)	2.707 ± 10
Nb(1)–O(4a)	1.92 ± 2	Ba/Sr–O(1)x	3.035 ± 10
Nb(1)–O(4a)	2.03 ± 3	Ba/Sr–O(3)	2.604 ± 9
		Ba/Sr–O(3)z	2.844 ± 9
Nb(2)–O(1)	1.949 ± 9	Ba/Sr–O(4a)	2.758 ± 9
Nb(2)–O(3)	2.004 ± 8	Ba/Sr–O(5a)	2.87 ± 2
Nb(2)–O(2)	1.958 ± 8		
Nb(2)–O(2)x	2.013 ± 8	Nb(1)–Ba/Sr	3.572 ± 3
Nb(2)–O(5a)	1.88 ± 3	Nb(1)–Ba/Sr	3.643 ± 3
Nb(2)–O(5b)	1.93 ± 3	Nb(2)–Sr	3.387 ± 3
Nb(2)–O(5a)z	2.11 ± 3	Nb(2)–Sr	3.426 ± 3
Nb(2)–O(5b)z	2.00 ± 3		
Sr–O(2)	2.547 ± 8		
Sr–O(2)z	2.919 ± 8		
Sr–O(5a)	2.67 ± 2		
Sr–O(5b)	2.85 ± 2		

† Indices x and z denote translation along x or z axes. Positions of Nb and O atoms are shown in figure 4.1.

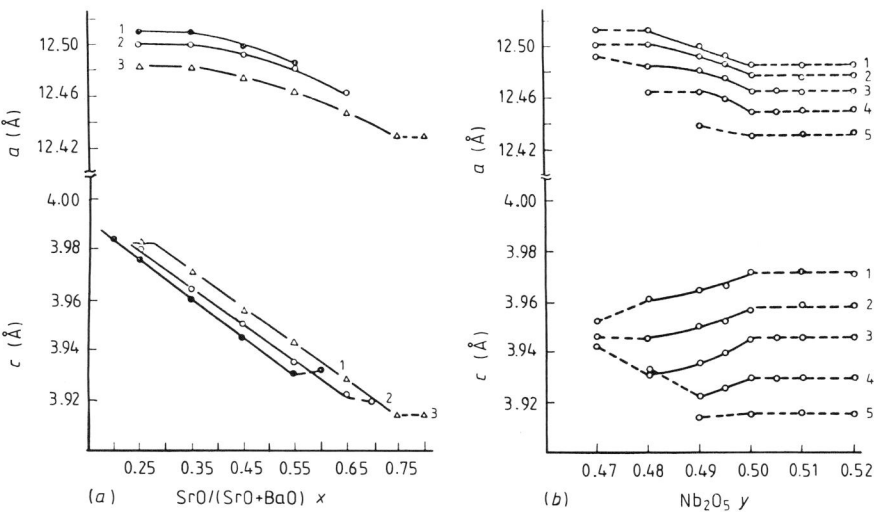

Figure 4.2 Dependence of the parameters of the crystalline structure of the solid solution $(SrO)_x(BaO)_{1-x}(Nb_2O_5)_y$ on the composition [1]. (a) y is equal to (1) 0.48; (2) 0.49; (3) 0.50; (b) x is equal to (1) 0.35; (2) 0.45; (3) 0.55; (4) 0.65; (5) 0.75.

figure imply that both parameters c and a decrease with increasing ratio SrO/BaO. On the other hand, a decreases and c increases as the Nb_2O_5 content increases. The magnitudes of the lattice parameters are strongly affected by the cooling rate of the grown crystal: the higher the crystal boule cooling rate, the smaller the parameter a and the larger the parameter c [7]. The parameters of the cell reported in [7] for $Ba_{0.25}Sr_{0.75}Nb_2O_6$ are as follows: $a = (12.430 \pm 2 \times 10^{-5})$ Å, $c = (3.941\,341 \pm 1 \times 10^{-5})$ Å; unit cell volume $V \approx 604.66$ Å^3. Above the Curie temperature, the space group P4bm is transformed into the non-polar group $P\bar{4}b2(D_{2d}^7)$ with the point symmetry $\bar{4}2m$; these exhibit a monoclinic symmetry [8]. It was assumed [9] that the tetragonal (ferroelectric) phase of BSN is metastable below 850°C. Later [10] the monoclinic phase alone was ascertained to be stable at such temperatures. The calculated and measured values of interplanar distances for both phases are given in table 4.2. The tetragonal–monoclinic transition temperature

Table 4.2 Interplane distances $d(\text{Å})$ and relative intensities I of reflexes for the monoclinic phase $Ba_{0.29}Sr_{0.71}Nb_2O_6$ [8] and the tetragonal phase $Ba_{0.4}Sr_{0.6}Nb_2O_6$ [11].

	Monoclinic		Tetragonal	
hkl	d_{exp}	d_{calc}	d_{exp}	I (relative units)
101	5.030	5.020	3.95	7
210	4.480	4.450	3.45	7
$\bar{1}$11	4.230	4.230	3.45	7
020	3.895	3.880	3.31	1
220	3.160	3.170	3.20	9
$\bar{1}$21	3.080	3.070	3.10	1
301	3.070	3.066	3.00	9
002	2.830	2.830	2.92	7
102	2.730	2.740	2.78	10
401	2.460	2.465	2.59	4
$\bar{3}$02	2.245	2.246	2.43	2
420	2.238	2.240	2.30	1
$\bar{2}$22	2.110	2.114	2.12	2
312	2.075	2.080	1.995	2
501	2.039	2.043	1.860	5
331	1.975	1.976	1.840	1
$\bar{3}$22	1.946	1.943	1.820	3
$\bar{1}$23	1.678	1.679	1.745	3
123	1.678	1.677	1.730	5
313	1.649	1.644	1.695	4
$\bar{2}$23	1.624	1.624	1.682	2
			1.635	4
			1.597	5

depends on the aggregate state of the sample and lies in the vicinity of 1340 °C. The kinetic studies of this transition have shown that at temperatures above 1400 °C the monoclinic phase becomes rapidly and completely tetragonal. In the range 1400–950 °C, the transition rate is substantial, and both phases coexist. Below 950 °C the transition rate slows down markedly.

The monoclinic phase has been registered for BSN compositions with $x = 0.15, 0.20, 0.25$ and 0.29. For the latter composition the lattice parameters were estimated to be $a = 10.997 \pm 0.008$, $b = 7.751 \pm 0.007$, $c = 5.666 \pm 0.005$ Å and $\gamma = 90, 30°$, the space group being $P2_1/a$.

The nature of the chemical bonds in BSN has not been reported in the literature. From comparison of interatomic distances in BSN and $LiNbO_3$ one can only assume an essential covalence of the Nb–O bond in NbO_6 octahedra, which was confirmed using NMR methods in [12]. Employment of a purely ionic model ($Ba^{2+}Sr_{1-x}^{2+}Nb_2^{5+}O_6^{2-}$) in the estimation of the spontaneous polarization P_s has led to values equal to half the experimental ones. This implies a considerable electron contribution to P_s [2].

In conclusion we may note that an introduction of additional atoms to the channels of the crystal lattice may be responsible for the formation of non-stoichiometric solid solutions of BSN. Therefore they may exhibit rather broad-scale variations in the composition and physical properties.

4.2 Ferroelectric properties

4.2.1 *Ferroelectric phase transition*

Crystals of barium strontium niobate belong to ferroelectrics with a smeared phase transition [13, 14]. In the room-temperature ferroelectric phase, all metallic ions in the BSN structure are displaced along the tetragonal polar axis. In the transition to the paraelectric phase, Ba and Sr ions, as well as 20% of Nb ions, are displaced to symmetric positions in the oxygen layers, while the remaining 80% of the Nb ions are positioned with equal probability above and below the oxygen atomic planes. As has already been mentioned, the crystal passes over from the space symmetry group P4bm into P4̄b2. However, as reported in [15], an intermediate state may exist in which all metallic ions lie in the oxygen planes, and the crystal is in the centrosymmetric phase P4/mbm, referred to in [15] as a prototype phase from which the ferroelectric phase can develop merely by ion displacement. The centrosymmetric phase is however missing in the transition from the polar group P4bm into the non-polar group P4̄b2, as reported in [6].

The smearing of the ferroelectric phase transition in BSN crystals is associated with disorder in the Ba and Sr ion arrangement in tungsten bronze-type structures. As a result of this disorder the unit cells become dissimilar, and the parameters specifying the ferroelectric properties of the crystal change from cell to cell or from micro-region to micro-region, which smears the phase transition. Transition smearing depends on the degree of structure disorder, which is in turn specified by the Ba/Sr ion ratio [2]. The ferroelectric phase transition is most pronounced in the composition $Ba_{0.67}Sr_{0.33}Nb_2O_6$, which exhibits the best possible cation ordering.

The position of the phase boundary in BSN solid solutions was determined from the dependence of the Curie temperature on the solid solution composition. The data obtained provide evidence that the Curie temperature of the solid solution $Ba_xSr_{1-x}Nb_2O_6$ increases slowly as the Ba content increases from 50 °C for $x = 0.25$ to about 250 °C for $x = 0.75$. An increase of the oxygen vacancy concentration in crystals, as well as rare earth impurities, cause displacement of the Curie temperature towards lower values [16, 17].

4.2.2   Domain structure

The domain structure of this class of ferroelectrics is very labile. Not only is it affected by external factors, such as temperature, stress and electric fields; it also alters with time without being influenced by any external effects (aging).

The analysis has shown that the domain boundaries of BSN crystals are neither concentric circles nor spirals that reproduce the crystallization front shape, as in the case of polydomain lithium niobate [18] and barium sodium niobate [19] samples. BSN crystals were treated using an etch pit technique. On (010) and (100) planes of BSN cuts, dark and light rectangular etch pits were observed that stretched across one or several striation layers. The dark pits were assumed in [20] to be micro-domains. But in a partially polarized BSN crystal one could not observe such small domains [21]. The polarization state of a ferroelectric has an appreciable effect upon its physical properties: polydomain crystals possess neither pyroelectric nor linear electro-optic properties; their nonlinear properties differ substantially from those of single-domain samples. Real BSN crystals customarily exhibit a slight unipolarity, i.e. domains with one of the possible directions of spontaneous polarization prevail over the others.

It should be emphasized that BSN solid solutions with a low content of barium ($x \leqslant 0.4$) exhibit an unstable domain structure at room temperature. Even very small temperature variations cause partial depolarization [2, 22]. In crystal elements, such a depolarization raises the half-wave voltage in low-frequency fields and affects phase relations at different points of the crystal. An introduction into BSN crystals with $x < 0.4$ of rare earth elements, in particular lanthanum, removes depolarization. The crystal remains single domain for a long time, but impurities raise the Curie temperature [17].

4.2.3   Dielectric properties

The dielectric characteristics of BSN solid solutions were investigated in the fundamental work by Glass [2].

The dielectric permittivity of BSN crystals is quite sensitive to the solid solution composition (figure 4.3). Crystals with a low content of barium, $0.16 < x < 0.25$, exhibit the highest ε_c values. The temperature-dependent behaviour of the dielectric constants along the c and a axes at a frequency of 10 kHz and the AC field strength of 3 V cm^{-1} indicate that the dielectric anomaly is weaker along the a than along the c axis, and that ε_a is much lower than ε_c. The higher the Curie point of a compound, the larger its dielectric permittivity ε_c, which decreases correspondingly at room temperature [24, 25]. The maximum values of the dielectric permittivity, ε_{max}, correspond to the phase transition point. The temperature dependences of the dielectric permittivity in weak fields, at which the initial polarizability is weakly connected with domain reorientation, play the most important part in the discovery and analysis of ferroelectric phase transitions. Broad permittivity peaks of BSN crystals with a low barium content imply a smeared phase transition. The permittivity peaks become sharper as the barium content increases (figure 4.4).

The peak permittivities of BSN crystals measured in the frequency range between 50 Hz and 20 MHz exhibit a shift towards higher temperatures and a decrease in magnitude with increasing frequency, the shifts being accompanied by peak smearing (figure 4.4) [2, 26].

The permittivity peak is smeared and shifted towards higher temperatures as soon as a DC electric field is applied [2]. We may note that the dielectric permittivity values measured experimentally depend to some extent on electrode material, sample thickness and degree of unipolarity, heating and

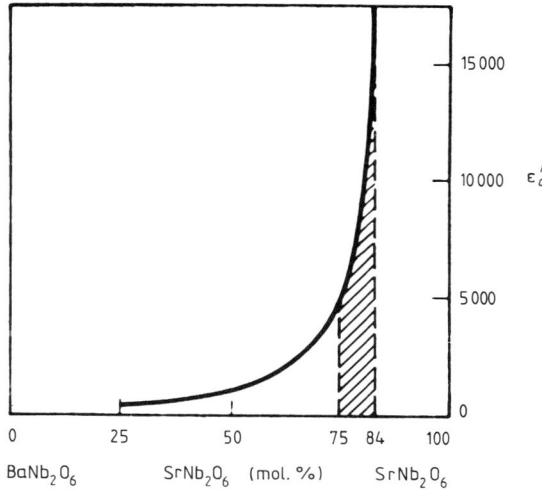

Figure 4.3 Dielectric permittivity ε along the polar axis c as a function of the $Ba_xSr_{1-x}Nb_2O_6$ crystal composition at room temperature [23].

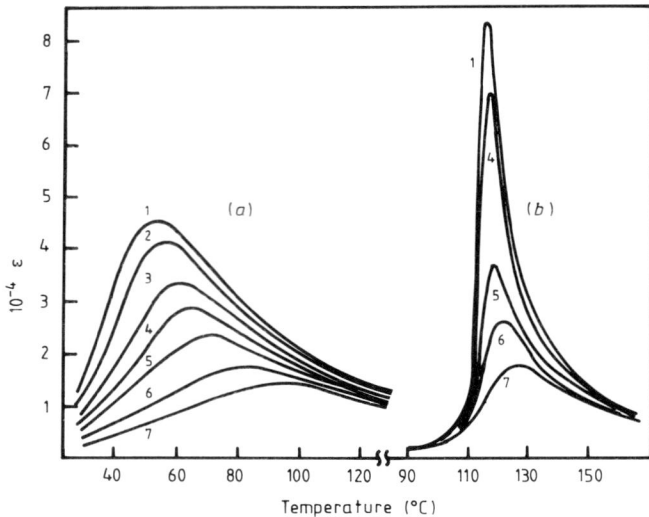

Figure 4.4 Temperature dependence of dielectric permittivity of $Ba_{0.25}Sr_{0.75}Nb_2O_6$ (a) and $Ba_{0.5}Sr_{0.5}Nb_2O_6$ (b) crystals for different frequencies: (1) 50; (2) 5×10^2; (3) 5×10^3; (4) 5×10^4; (5) 5×10^5; (6) 5×10^6; (7) 2×10^7 Hz [2].

cooling rates, as well as on the preheating conditions [2, 27]. But for all types of electrode materials, for any sample thickness and for various Ba/Sr ratios, the ε_{max} values below T_C are higher for unpolarized than for polarized crystals. The ε_{max} value changes monotonically with the degree of crystal unipolarity. The increase of ε_{max} in unpolarized BSN crystals is explained by the existence of domain walls.

The dielectric permittivity ε and the loss tangent $\tan \delta$ were measured in [28–30] as functions of temperature in crystal elements cut out perpendicularly to the c axis. The calculations were performed using the well known formulae

$$\varepsilon = 4\pi Cd \times 0.9/S \tag{4.1}$$

where C is the sample capacitance in pF, d is the sample thickness in cm and S is the electrode area in cm^2, and

$$\tan \delta = \frac{C_{meas} \tan \delta_{meas}}{C} - \frac{C_{ad} \tan \delta_{ad}}{C_{meas}} \tag{4.2}$$

where C_{meas} and C_{ad} are the measured and additional capacitances (pF), $\tan \delta_{meas}$ and $\tan \delta_{ad}$ are the measured and additional losses.

The data reported in [30] have shown (see table 4.3) that an increase of barium content in a solid solution from $x = 0.25$ to $x = 0.67$ accounts for the Curie temperature increase from about 50 up to 155 °C.

Table 4.3 Dielectric permittivity ε, loss tangent $\tan \delta$ (at room temperature) and Curie temperature (under cooling) of $Ba_xSr_{1-x}Nb_2O_6$ [30].

x	Impurity (wt.%)	ε	$\tan \delta$	T_C (°C)
0.25	—	5000	0.017	48
0.50	—	402	0.026	118
0.54	—	500	0.05	124
0.67	—	530	0.122	155
0.54	La_2O_3-1.0	1200	0.10	84
0.54	La_2O_3-0.5	1260	0.095	94
0.54	Tm_2O_3-0.5	670	0.06	118
0.54	Y_2O_3-0.5	800	0.086	121

A short-time annealing of single-domain BSN samples at a temperature of 200 °C does not have any effect upon the dielectric properties. But a 5 h annealing at 800 °C completely removes polarization. Annealing in oxygen with the addition of water vapour does not influence the dielectric properties, which agrees with the low content of OH groups in BSN crystals reported in [29].

Dielectric studies of single-domain BSN crystals revealed a decrease of ε and $\tan \delta$ with time at room temperature (the aging effect). The temperature dependence of ε and $\tan \delta$ is depicted in figure 4.5, which shows that the stabilization takes at least 24 h.

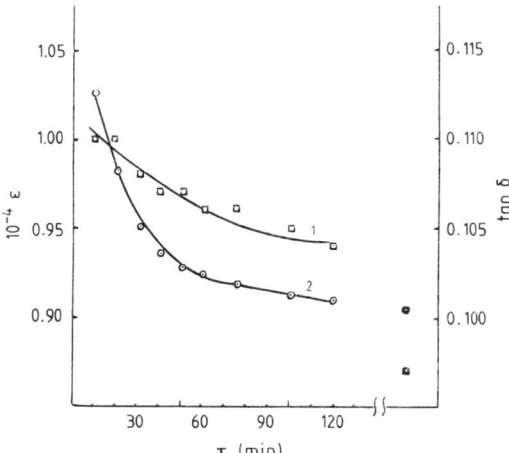

Figure 4.5 Time dependences of (1) ε; (2) $\tan \delta$ of single-domain BSN samples.

4.2.4   *Spontaneous polarization*

In AC electric fields, ferroelectrics exhibit dielectric hysteresis.

Dielectric hysteresis loops in stoichiometric BSN crystals with $x = 0.25$ in the temperature range between 20 and 90 °C were examined in [28]. The data obtained assisted in the determination of coercive fields E_c and spontaneous polarization P_s (figure 4.6). At room temperature, $E_c = 2.3$ kV cm^{-1} and $P_s = 17$ C cm^{-2}. The E_c and P_s values increase upon cooling through the Curie point. Dielectric hysteresis loops were also investigated at different strengths of the applied electric field (figure 4.7) [31]. The loop was

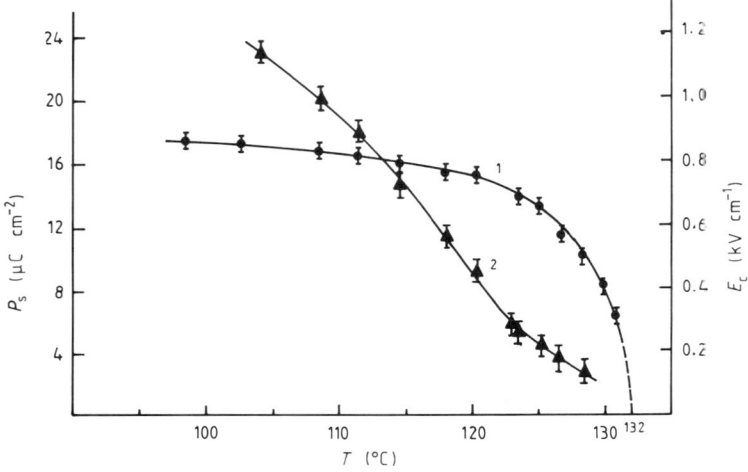

Figure 4.6   Temperature dependences of (1) spontaneous polarization P_s and (2) coercive field E_c of BSN crystals ($x = 0.25$) [33].

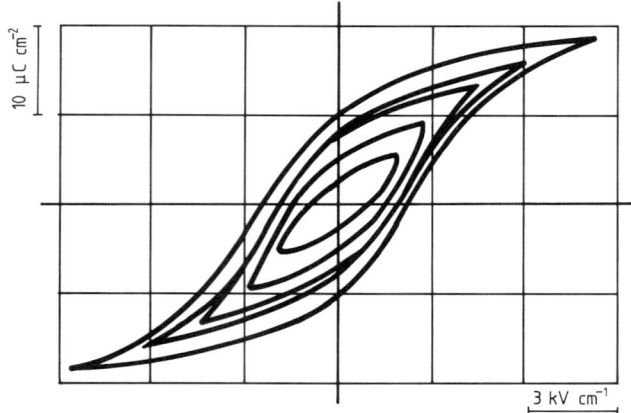

Figure 4.7   Hysteresis loops of a BSN crystal for different electric field strengths [31].

established to be completely formed only at a strength of 9 kV cm^{-1}. The temperature dependence of the hysteresis loop shape (figure 4.8) shows that the hysteresis vanishes not at a temperature of about 50°C, at which the permittivity of this crystal is maximum, but at about 90°C [32]. At this temperature, evidently, the crystal completely passes over to the paraelectric phase.

For the composition $Ba_{0.54}Sr_{0.46}Nb_2O_6$, the dielectric hysteresis loops were recorded in [33] at temperatures of 104, 120 and 131°C; from these loops, the spontaneous polarization P_s and the coercive field E_c were estimated.

Spontaneous polarization can be measured in single-domain crystals only. But direct static measurements of spontaneous polarization from the bound charge value are difficult to perform in a single-domain crystal because of the relatively high electric conductivity and the presence of free charges in the surrounding atmosphere. Spontaneous polarization is therefore determined from the temperature dependence of the pyroelectric current. Two types of pyroelectric measurements are commonly made: either the crystal is subjected to heating by modulated light pulse, and a pyroelectric alternating current $i = (dP/dT)(dT/dt)$ is registered in the external circuit (Chynoweth's dynamic method), or the charge occurring at crystal faces with increasing temperature is continuously integrated and the change in crystal polarization $[\Delta P]_{T_1^2}$ is subsequently measured from the feedback capacitor charge. The values of P_s measured at 300 K by charge integration techniques for five different BSN compositions are listed in table 4.4. The experimental error reached 3% [2]. The values of spontaneous polarization P_s' at 300 K, obtained by the dynamic

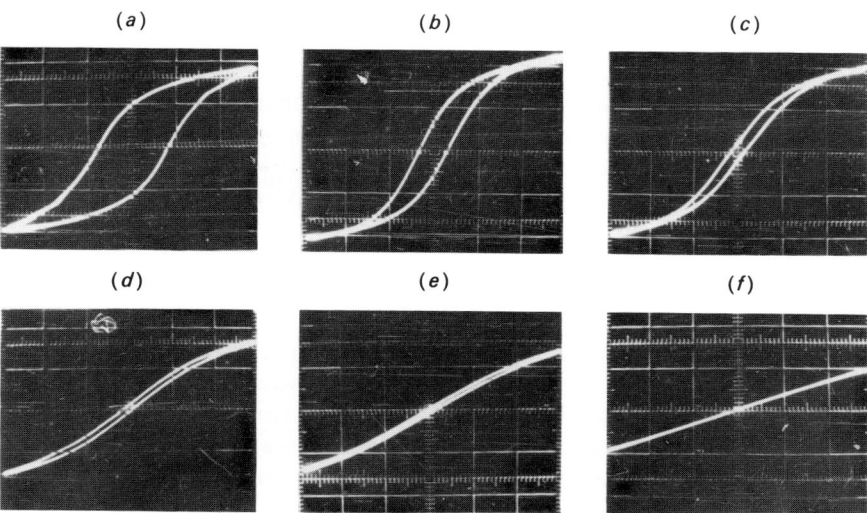

Figure 4.8 Dielectric hysteresis loops of BSN crystals for 20 (*a*), 39 (*b*), 48 (*c*), 62 (*d*), 70 (*e*) and 90°C (*f*) [32].

Table 4.4 Spontaneous polarization P_s ($\mu C\ cm^{-2}$) of $Ba_xSr_{1-x}Nb_2O_6$ crystals measured by different methods [2]†.

x	P_s (300 K)	P_s (10 K)	P_s' (300 K)	P_s'' (10 K)
0.27	19.5 ± 0.6	31.3 ± 1.0	17.0 ± 1.0	33.1 ± 2.0
0.33	23.3 ± 0.7	31.9 ± 1.0	—	33.6 ± 2.0
0.40	27.4 ± 0.8	33.7 ± 1.0	—	34.4 ± 2.1
0.52	29.2 ± 0.9	35.0 ± 1.1	25.0 ± 1.0	36.0 ± 2.2
0.72	30.5 ± 3.1	34.3 ± 3.4	—	39.5 ± 2.4

† The P_s are obtained by the charge integration method, the P_s' by the dynamic method [2], and the P_s'' from the Abrahams relations [34].

method, and those calculated from Abrahams *et al*'s relations [34], are also tabulated. The data in table 4.4 show that the P_s values are higher than the P_s' values by approximately 15%, but agree fairly well with P_s''.

Observations of the hysteresis loops at 300 K for BSN with $x = 0.25$ yielded $P_s > 30\ C\ cm^{-2}$. This value is appreciably higher than the one obtained by other methods, which is due to high room-temperature dielectric losses. At 300 K, spontaneous polarization P_s of crystals with $x = 0.27$, calculated within the purely ionic model ($Ba_x^{2+}Sr_{1-x}^{2+}Nb_2^{5+}O_6^{2-}$), is equal to $8.8 \pm 3.5\ C\ cm^{-2}$. This value is about half the value obtained experimentally, which implies a substantial electronic contribution to P_s [2].

The magnitude of the coercive field for BSN with $x = 0.25$ at $33\,^\circ C$ was determined in [35]. It turned out to be equal to $1.3\ kV\ cm^{-1}$.

4.2.5 Electrical conductivity

A knowledge of the electrical conductivity of crystals provides an insight into the nature of defects and their activation energy. The electrical conductivity of ferroelectrics is specifically due to the presence of domain structure and phase transitions. Electrical measurements were made on stoichiometric samples with $x = 0.25$, whose face surfaces were covered with ohmic palladium contacts [31]. Silver electrodes are not recommended since silver has a strong tendency to diffuse inside the crystal. The measurements were made in the temperature range $25-800\,^\circ C$, the heating rate being $150\,^\circ C\ h^{-1}$. The temperature near the crystal was controlled by a platinum to platinum–rhodium thermocouple. The electrical conductivity was measured both on direct current and alternating current with a frequency of 100 Hz (figure 4.9). In the vicinity of $400\,^\circ C$ the curve $lg\ \sigma = f(1/T)$ shows a characteristic break which separates two rectilinear regions: the high-temperature region with chiefly intrinsic conduction and the low-temperature region with impurity conduction. This type of ferroelectric exhibited no anomalies in the phase transition region ($T_C = 50\,^\circ C$). Note that above $400\,^\circ C$ the behaviour of electrical conductivity on direct and alternating currents coincides. The ferroelectric nature of the crystal manifests itself below this temperature on

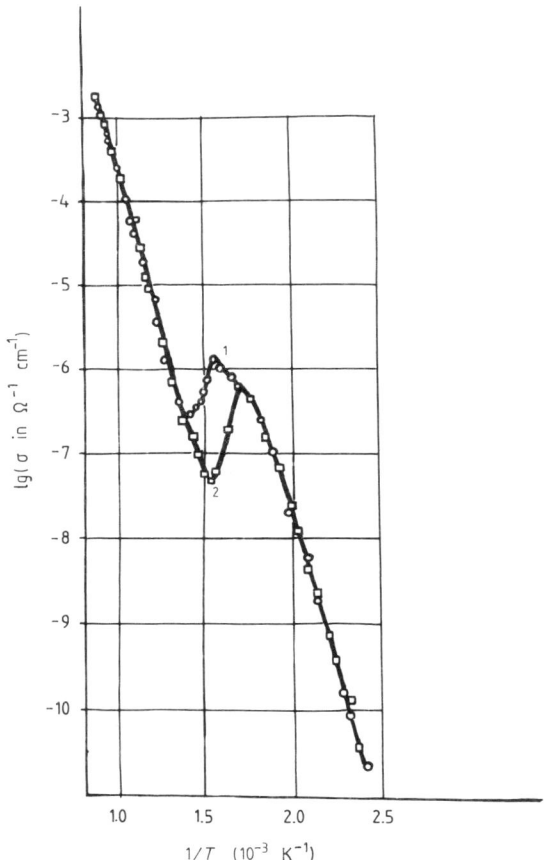

Figure 4.9 Temperature dependence of electric conductivity measured along the *c* axis of BSN single crystals, of sizes (1) 0.9 and (2) 4.71 mm along the *c* axis [31].

alternating current. The activation energy of charge carriers calculated from the slope of the rectilinear portion of the curve $\lg \sigma = f(1/T)$ is equal to 1.5 eV over the entire temperature range.

The electrical conductivity σ of stoichiometric single crystals of BSN at room temperature is about 10^{-10}–10^{-12} Ω^{-1} cm^{-1}, as reported by Glass [2]. Ceramic BSN crystals have $\sigma \approx 10^{-13}$–10^{-14} Ω^{-1} cm^{-1} [36].

For restored ceramic samples of $Ba_{0.6}Sr_{0.4}Nb_2O_6$ with the deviation from stoichiometry in oxygen equal to 0.2 mol.%, the electrical conductivity σ was 5×10^{-2} Ω^{-1} cm^{-1}, the direct velocity of charge carriers being equal to about 10 cm^2 V^{-1} s^{-1}. At temperatures below 300°C, the electrical conductivity of BSN crystals, as established in [36], is of electronic character;

the activation energy depends on the degree of restoration and lies within the range 0.26–0.63 eV. The decrease of activation energy with increasing concentration of oxygen vacancies is explained by the interaction among structure defects [37].

4.3 Optical and electro-optic properties

Ferroelectric BSN solid solutions are optically negative uniaxial crystals with a broad transmission region ranging from 0.3 to 6 μm (figure 4.10). The room-temperature dispersion of the refractive indices of ordinary (n_o) and extraordinary (n_e) waves between 0.4 and 1.6 μm is presented in figure 4.11 [39]. The refractive index of the ordinary wave n_o is practically independent of the Ba/Sr ratio in the solid solution and of the temperature, whereas the refractive index of the extraordinary wave n_e is extremely sensitive both to compositional and to temperature changes (figures 4.11 and 4.12). The temperature coefficient dn_e/dT found experimentally is equal to 3×10^{-4} $°C^{-1}$ in the range 30–50°C [35, 39]. The natural room-temperature birefringence ($n_e - n_o$) for solid solutions of BSN with $x = 0.25$, 0.5 and 0.75 for $\lambda = 0.6328$ μm makes up respectively -0.013, -0.039 and -0.055 (± 0.002)(table 4.5), which does not enable this material to be used for second harmonic generation since phase matching cannot be realized even at temperatures much below 0°C.

The refractive indices of BSN crystals measured for Nd:YAG laser wavelengths $\lambda_1 = 10\,642$ Å and $\lambda_2 = 5321$ Å are tabulated in table 4.6. The processing of all the experimental data on n_o and n_e using the least-squares techniques gave the parameters E_0 and E_d of the single-oscillation model [40]. Making use of these parameters, one can estimate n_o and n_e of BSN

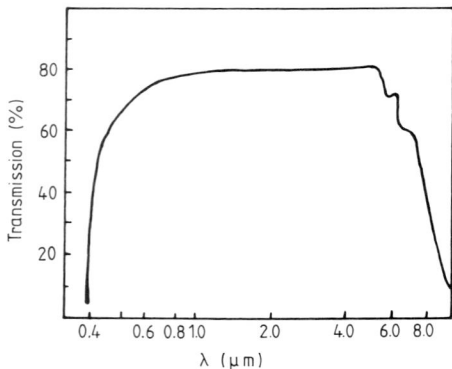

Figure 4.10 Optical transmission of a colourless $Ba_{0.55}Sr_{0.45}Nb_2O_6$ crystal [38].

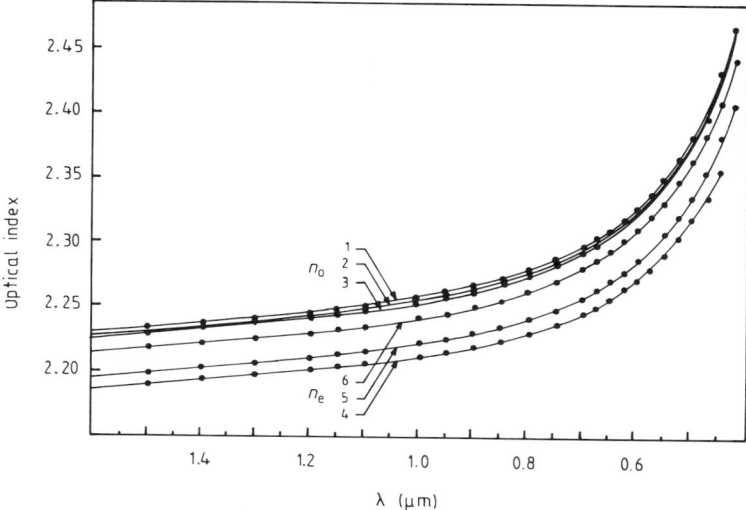

Figure 4.11 Room-temperature dispersion of the refractive indices n_o (an ordinary beam) (1, 2, 3) and n_e (an extraordinary beam) (4, 5, 6) for $Ba_xSr_{1-x}Nb_2O_6$ crystals with $x = 0.25$ (1, 4); 0.50 (2, 5) and 0.75 (3, 6) [39].

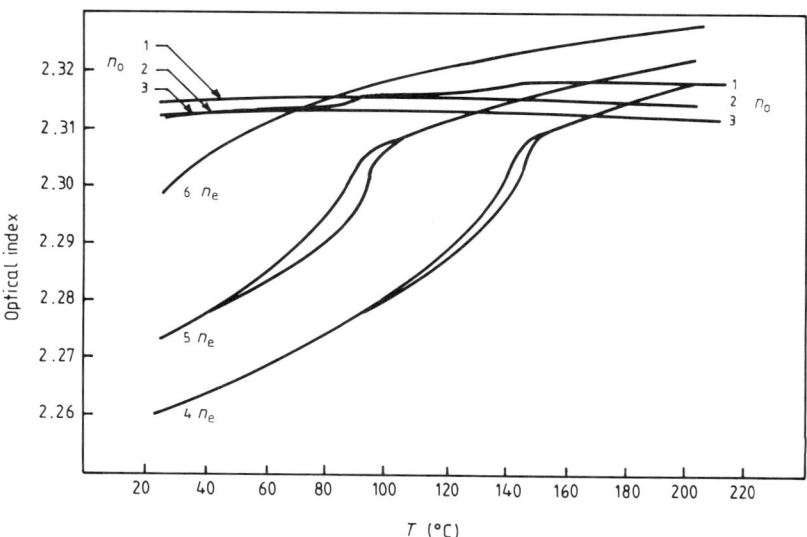

Figure 4.12 Temperature-dependent variation of the refractive indices of ordinary and extraordinary beams (n_o and n_e) for a $Ba_xSr_{1-x}Nb_2O_6$ crystal with $x = 0.25$, 0.50 and 0.75 [39] (curve numbering corresponds to that of figure 4.11).

Table 4.5   Refractive indices of $Ba_xSr_{1-x}Nb_2O_6$ crystals for $\lambda = 6328$ Å [1].

x	n_o	n_e
0.25	2.3117	2.2987
0.50	2.3123	2.2734
0.75	2.3144	2.2596

Table 4.6   Refractive indices n_o and n_e of $Ba_xSr_{1-x}Nb_2O_6$ crystals at wavelengths of Nd:YAG laser radiation (λ_1) and its second harmonic (λ_2) [41].

	λ_1		λ_2	
x	n_o	n_e	n_o	n_e
0.25	2.249	2.236	2.356	2.343
0.39	2.250	2.222	2.354	2.324
0.54	2.251	2.214	2.355	2.311
0.67	2.252	2.210	2.357	2.303

Table 4.7   Parameters E_0 and E_d (eV) describing dispersion of refractive indices n_o and n_e of $Ba_xSr_{1-x}Nb_2O_6$ crystals in the visible and near IR ranges [41].

	n_o		n_e	
x	E_0	E_d	E_0	E_d
0.25	6.23	24.40	6.20	23.94
0.39	6.32	24.82	6.32	24.03
0.54	6.33	24.88	6.45	24.36
0.67	6.31	24.80	6.55	24.65

crystals for any wavelength $\lambda = hc(eE)^{-1}$ in the visible and near IR regions of the spectrum by the formula $n^2(E) - 1 = E_0 E_d(E_0^2 - E^2)^{-1}$. The calculated values are presented in table 4.7 [41].

Refractive indices of BSN crystals correlate well with the values of n_o and n_e found for other compositions of the same crystals [39], and the values of E_0 and E_d are rather close to those for other niobates [42]. Table 4.6 illustrates that the birefringence of BSN crystals with $x = 0.67$ is three times larger than that for the composition $x = 0.25$. But this value is insufficient for phase matching. As the content of Ba atoms increases, the ordinary refractive index increases weakly, whereas the extraordinary refractive index decreases.

For BSN crystals possessing the point group symmetry 4mm, the matrix of electro-optic coefficients has three independent non-zero terms: $r_{13} = r_{23}$, $r_{42} = r_{51}$ and r_{33}. A considerable electro-optic effect occurs when the electric field is applied along the polar c axis and the light propagates normally to this direction. In a polarized crystal of $Ba_{0.25}Sr_{0.75}Nb_2O_6$ the room-temperature transverse linear electro-optic effect is two orders of magnitude larger than that in $LiNbO_3$ [1]. The electro-optic coefficients are $|r_{33}| = 13.4 \times 10^{-8}$, $|r_{13}| = 6.6 \times 10^{-9}$ and $|r_{51}| = 4.2 \times 10^{-9}$ cm V^{-1}. For $\lambda = 0.6328$ μm and the incident light polarized at 45° to the two principal crystallographic axes, the half-wave voltage $V_{\lambda/2}$ makes up 37 V for a direct current, 40 V for 1 MHz and 24 V for 15 MHz (table 4.8) [1]. As the barium content in solid solution increases (see table 4.8), these values increase but remain much lower than those for $LiNbO_3$ and $LiTaO_3$ [18]. The control voltage can be made still lower by using electro-optic elements with optimal geometric relations. The effective electro-optic coefficients of amplitude modulation $[(n_e/n_o)^3 r_{33} - r_{13}]$ at a radiation wavelength $\lambda = 6328$ Å for various BSN compositions are presented in table 4.9.

The longitudinal linear electro-optic effect in crystals with the point group symmetry 4mm reaches its maximum in the normal direction to the crystal plate if this plate makes an angle of 55° with the optical axis [13]. For the linear electro-optic effect in a thin crystal 45° BSN plate ($x = 0.3$), the half-wave voltage is approximately equal to 1 kV at room temperature. As the temperature increases, this value decreases rather rapidly and reaches

Table 4.8 Half-wave displacement voltage $V_{\lambda/2}$ of $Ba_xSr_{1-x}Nb_2O_6$ for different frequencies v of the electric field and $\lambda = 6328$ Å [1].

v (MHz)	$V_{\lambda/2}$ (V) for x		
	0.25	0.50	0.75
0	37	250	—
1	80	676	1340
15	48	580	1236
100	—	580	—

Table 4.9 Effective electro-optic coefficients of the amplitude modulation $[(n_e/n_o)^3 r_{33} - r_{13}]$ of $Ba_xSr_{1-x}Nb_2O_6$ crystals for $\lambda = 6328$ Å [1].

Electric field frequency (MHz)	$[(n_e/n_o)^3 r_{33} - r_{13}]$ (10^{-8} cm V^{-1}) for x		
	0.25	0.5	0.75
0	13.8	2.05	—
1	6.4	0.76	0.38
15	10.6	0.9	0.41
100	—	0.9	—

its minimum (100 V) at 55°C, i.e. in the phase transition region (figure 4.13) [15]. The high half-wave voltage of the room-temperature longitudinal linear electro-optic effect is determined to a great extent by strong nonlinear polarization characteristics of the material.

A rather strong quadratic electro-optic effect [13, 15, 35] takes place in BSN crystals along with the longitudinal linear electro-optic effect. The linear effect is observed only at temperatures below the ferroelectric phase transition, whereas the quadratic effect, which occurs at temperatures somewhat below the transition temperatures, exists high above the Curie temperature when the material is in the non-polar phase (figure 4.13). This transformation of the linear electro-optic effect into the quadratic one in the phase transition region might be associated with the transition of the crystal to the centrosymmetric phase where the linear effects are equal to zero, but as reported in [6], the BSN crystals have no centrosymmetric phase. It should be noted that the linear electro-optic effect depends on the P_s direction, and a polydomain crystal can exhibit only the quadratic effect. The presence of the quadratic effect alone therefore suggests either that the remanent polarization is due to unequal domain sizes or that the number of domains is small [15].

According to Wample and Di Domenico [42], for tungsten bronzes (including BSN) the values of the quadratic electro-optic coefficient g_{33} are determined from the ratio $g_{33} = (g_{11})_p / \eta$, where η is the relative packing

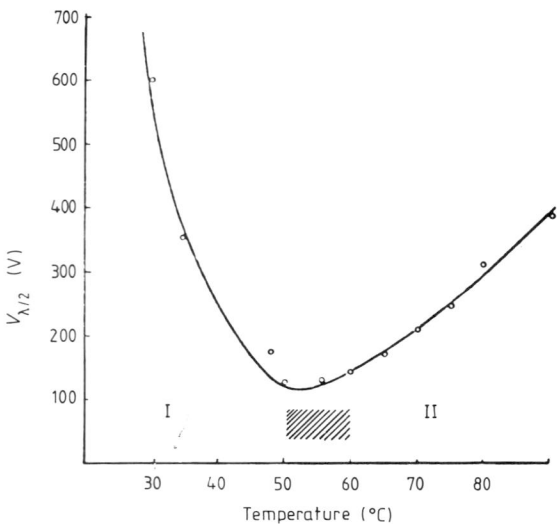

Figure 4.13 Temperature dependence of the half-wave voltage for a $Ba_{0.3}Sr_{0.7}Nb_2O_6$ crystal plate [15] cut out at an angle of 45° to the c axis. I is the region of the linear and II of the quadratic electro-optic effect.

Table 4.10 Half-wave voltage $V_{\lambda/2}$ of $Ba_xSr_{1-x}Nb_2O_6$ crystals of different compositions for $\lambda = 0.6328$ μm at room temperature [30].

Crystal composition	Impurity (wt.%)	$V_{\lambda/2}$ (V)
$Ba_{0.25}Sr_{0.75}Nb_2O_6$	—	100–120
$Ba_{0.54}Sr_{0.46}Nb_2O_6$	—	420
$Ba_{0.67}Sr_{0.33}Nb_2O_6$	—	500
$Ba_{0.25}Sr_{0.75}Nb_{2.05}O_{6.12}$	—	85–100
$Ba_{0.25}Sr_{0.76}Nb_{2.05}O_{6.13}$	—	75–120
$Ba_{0.25}Sr_{0.72}Nb_2O_{5.97}$	—	60–70
$Ba_{0.54}Sr_{0.46}Nb_2O_6$	La_2O_3-1.0	320
$Ba_{0.54}Sr_{0.46}Nb_2O_6$	Tm_2O_3-0.5	390
$Ba_{0.54}Sr_{0.46}Nb_2O_6$	Y_2O_3-0.5	420
$Ba_{0.54}Sr_{0.46}Nb_2O_6$	Y_2O_3-1.0	370

density defined as the ratio of the number of B ions in unit BSN crystal volume to the number of B ions in unit volume of an ideal perovskite-type crystal, and $(g_{11})_p$ is the coefficient of the quadratic electro-optic effect in a perovskite-type crystal. For $Ba_{0.25}Sr_{0.75}Nb_2O_6$ the value of g_{33} was estimated to be 0.14 m^4 C^{-2} [29], which is in close agreement with the experimental results of other papers. The half-wave voltage estimated from the coefficients obtained turned out however to be anomalously low, which is presumably due to the disordered BSN structure, where P_s and ε_c in the crystal volume change from cell to cell.

The half-wave voltage $V_{\lambda/2}$ was measured on stoichiometric BSN single crystals with a small deviation from stoichiometry, as well as on La-, Tm- and Y-doped crystals [28, 30] (table 4.10). The measurements were made using the dynamical method [18]. The analysis of the experimental results shows that the lowest $V_{\lambda/2}$ values are exhibited by BSN crystals grown in the melt with a shortage of strontium (1.5 mol.%) with respect to stoichiometry. As is seen from table 4.10, the addition of 1 wt.% of La_2O_3 lowers the half-wave voltage by 100 V. Thulium and yttrium produce a slight effect upon the electro-optic properties of BSN crystals. The positive effect of lanthanum is explained by the specific structure of its electron shells.

4.4 Device use

4.4.1 *Laser beam modulators*
A laser beam modulator prepared from a BSN single crystal with $x = 0.5$ is described in [43], the composition $x = 0.25$ being inapplicable for this purpose because of the high value of the dielectric permittivity ε. A BSN $0.3 \times 0.4 \times 10$ mm^3 single crystal with the larger side directed along the x

axis was used for the modulator. The crystal capacitance was 20 pF, and the crystal was switched into a 50 Ω wide-band synchronizing circuit. The reflected power was less than 1 % up to 300 MHz, and voltage pulses had a rise front up to 3 ns and a rapid fall. The control voltage for the light wavelength $\lambda = 0.633$ μm was 22 V. The intensity ratio of the transmitted to the absorbed light was 15 dB. The diameter of the laser beam on the crystal surface was 0.1 mm.

Nomura *et al* [44] pointed out great difficulties in controlling a BSN modulator at frequencies higher than several hundreds of kHz due to an extremely low level of modulator impedance caused by a high dielectric permittivity. To achieve the necessary level of impedance and to raise the frequency of piezoelectric resonance above video frequencies, the authors prepared modulators with a volume of 0.45 mm². Gold electrodes were deposited on the crystal.

The modulation was measured by means of a potentiometer and He–Ne laser radiation with a wavelength $\lambda = 6328$ Å and an output power of 20 mW. An electric field was applied along the *c* axis (the optic axis); the light beam propagated perpendicularly to this direction. As follows from the curves of figure 4.14, the modulation remains constant while the frequency of the applied field is less than the frequency of the piezoelectric modulator resonance, the latter increasing with decreasing modulator size. Thus, the modulator size determines the maximum frequency for which the modulation remains constant. It should be noted that the modulation depends on the strength of the applied field. To compensate in BSN crystals for the refractive index variation due to temperature fluctuations, two BSN crystals of identical size and orientation were placed one by one in the way of the light beam. The beam was modulated by a video signal by means of the first BSN crystal, passed through the second compensating crystal, shone onto a photodiode

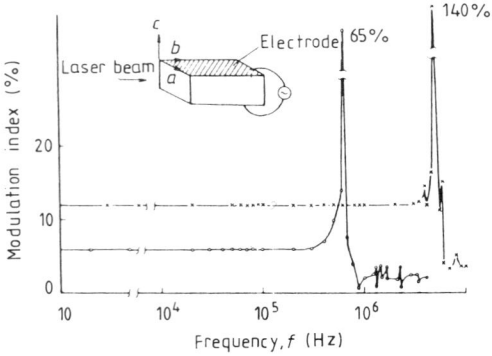

Figure 4.14 Dependence of modulation index on the applied field frequency. Crystal sizes were $3.2 \times 2.4 \times 2.0$ mm³ (1) and 0.45 mm³ (2); the modulating voltage was equal to 30 V [44].

and modulated a TV picture. The transmission of the whole system made up about 99% for $\lambda = 6328$ Å if an anti-reflection coating was used.

A modulator made of 20 elements (0.25 mm^3) arranged in series had optical losses of about 20% and a control voltage of about 10 V. The modulation of such a system remained constant up to 10 MHz.

An effective laser beam modulation by BSN crystals with $x = 0.25$ was reported in [45]. The measurements were carried out by the optical polarization method at a wavelength $\lambda = 6328$ Å of a He–Ne laser. Spontaneous birefringence of crystals was compensated by a quartz wedge. In contrast to previous papers, relatively massive samples were used ($3 \times 3 \times (3-4)$ and $7 \times 7 \times (3-4) \text{ mm}^3$). The crystals exhibited no growth ridges (striae), and the gradients of the refractive indices n_o and n_e did not exceed 1×10^{-5} and $5 \times 10^{-5} \text{ cm}^{-1}$ respectively. The residual light beam in nicols crossed at an angle of $45°$ to the optical crystal axis was 2–4% for a beam aperture not exceeding $4 \times 4 \text{ mm}^2$. The facets of the crystal plates were oriented perpendicularly to the [100], [010] and [001] directions within $\pm 2°$ accuracy. Light was propagated along the [010] or [100] direction; an electric field was applied along the c axis. Modulation was carried out up to a frequency of 2×10^4 Hz. The half-wave voltage increased up to 70 V because at a temperature of 35–45 °C the domain structure of the indicated crystals becomes unstable. But an applied biasing voltage of $\approx 500 \text{ V cm}^{-1}$ decreased the half-wave voltage to 50 V. For a beam aperture equal to $4 \times 4 \text{ mm}^2$ the modulation depth made up 98%. The results obtained show that for crystals of the above mentioned size the effective modulation frequency did not exceed 2×10^4 Hz.

As has already been mentioned, along with the transverse electro-optic effect, BSN crystals also possess the longitudinal electro-optic effect, which can also be used to obtain laser radiation modulation [15, 46].

To calculate a longitudinal half-wave voltage for an arbitrary beam propagation direction in a crystal, not only the values but also the signs of the electro-optic coefficients should be known. The absolute values of these coefficients are usually known from the literature whereas the signs have to be found. The only exceptions are lithium niobate [47] and lithium tantalate [48], for which both the values and the signs are established.

For BSN crystals, the coefficient r_{33} exceeds substantially the other coefficients, and therefore the calculated range of variation of $V_{\lambda/2}$ connected with the uncertainty in the signs of the other coefficients is relatively small. It should be noted that the values of the minimum half-wave voltage for some BSN crystals are much less than those of KDP crystals ($V_{\lambda/2} = 4$ kV), which are widely used for modulation of laser radiation on the basis of the longitudinal electro-optic effect.

The direction of light propagation in a crystal corresponding to the minimum half-wave voltage has a finite birefringence whose value depends on the crystal temperature and the light wavelength. If an exceeding range of variation $\Delta\Gamma$ is undesirable, certain restrictions are imposed upon admissible temperature fluctuations and the angular aperture. Table 4.11 gives a list of the angular apertures and admissible temperature variations $(\Delta T)_{\text{adm}}$ for modulators with a thickness of 0.2 mm. The estimates are presented for a maximum deviation of the path length difference of $\Delta\Gamma \approx \lambda/10$. Table 4.11 also shows the modulator capacitance C per unit area (mm^2).

For BSN crystals with a high room-temperature dielectric permittivity ε, of primary importance is the upper boundary of the effective modulation, because at a high frequency the electrodes are heated, which leads to an increase in the crystal temperature and, therefore, to a change in the birefringence value.

Hulme [46] examined this question on an example of a square-shaped modulator, with side a and thickness d, in which light propagated normally to the plate. The angular aperture $\Delta\psi$ of a modulator is known to be characterized by the half-angle at the beam cone vertex inside the crystal. Then the path length difference $\Delta\Gamma'$ of beams traversing the crystal plate within a cone $2\Delta\psi$ can be written in the form

$$\Delta\Gamma' = 2\pi(d/\lambda)(\mathrm{d}B/\mathrm{d}\psi)\Delta\psi \qquad (4.3)$$

where $\mathrm{d}B/\mathrm{d}\psi$ is the variation of the value of the birefringence with the angle for directions close to the normal of the crystal plate.

For a longitudinal modulator with parallel transparent electrodes, the capacitance can be determined by the flat capacitor formula $C = \varepsilon\varepsilon_0 a^2/d$. One then assumes that the electrode resistance per unit area R (Ω cm^{-2}) and the sinusoidal voltage with an amplitude $V_{\lambda/2}$ are applied to the crystal. The heat generated in the crystal disperses with heat transfer coefficient Q (watts per unit volume and per degree of temperature increase), the heat due to dielectric losses being disregarded. The mean modulator temperature then increases as

$$\Delta T = (\omega^2 V_{\lambda/2}^2 C^2 R)/(96Qa^2). \qquad (4.4)$$

This increase of crystal temperature results in an additional path length difference

$$\Delta\Gamma'' = 2\pi(d/\lambda)(\mathrm{d}B/\mathrm{d}T)\Delta T. \qquad (4.5)$$

The total path length difference for an oblique beam is then given by the sum

$$\Delta\Gamma = \Delta\Gamma' + \Delta\Gamma''. \qquad (4.6)$$

Equations (4.3)–(4.6) imply that the maximum effective frequency of the modulator is determined by

$$\omega_{\max} \approx \left(\frac{6Q}{R\Delta\psi}\right)^{1/2} \frac{\lambda\Delta\Gamma}{\pi\varepsilon_0 a} \frac{1}{\varepsilon V_{\lambda/2}[(\mathrm{d}B/\mathrm{d}T)(\mathrm{d}B/\mathrm{d}\psi)]^{1/2}}. \qquad (4.7)$$

Table 4.11 Characteristics of room-temperature longitudinal modulators of oblique cuts† for $\lambda = 6328$ Å [46].

Crystal	$(V_{\lambda/2})_{min}$ (kV)		α (grad)	Figure of merit‡ $(V^{-1} K^{1/2} rad^{1/2})$	Aperture		C (Pf mm^{-2})
	calculation	experiment			(grad)	$(\Delta T)_{ad}$ (K)	
LiNbO$_3$	2.3	2.5	55	5.7	± 0.5	± 12	3
LiTaO$_3$	11	19	55	2.5	± 9	± 10	4
Ba$_2$NaNb$_5$O$_{15}$	0.65–0.96	—	40–50	3.3	± 0.4	± 6	7
Ba$_{0.25}$Sr$_{0.75}$Nb$_2$O$_6$	0.075–0.10	—	55	2.8	± 8	± 2	130

† The normal to the crystal plate makes an angle α with the optical axis (c axis) of the crystal.
‡ The figure of merit is given by the expression $10^3 [\varepsilon V_{\lambda/2}(dB/dT \cdot dB/d\Psi)^{1/2}]^{-1}$.

The last term in the product depends exclusively on the nature of the crystal and can be used as the figure of merit in the comparison of different crystals for the purpose of longitudinal modulation. Table 4.11 presents numerical values of this parameter for certain crystals.

The estimation of the maximum effective frequency of a modulator for the particular case of a digital detector of light, in which $Q = 10^3$ W m^{-3} K^{-1}, $\lambda = 6 \times 10^{-7}$ m, $\Delta\Gamma = \lambda/5$, $R = 50\ \Omega$ cm^{-2}, $\Delta\psi = 2 \times 10^{-2}$ rad, and $a = 5 \times 10^{-3}$ m, provides the value $\omega_{\mathrm{max}} = 5.2 \times 10^5$ s^{-1}.

To increase the angular aperture, the optical range of wavelength and the effective frequency of the modulator, and to decrease the thermal sensitivity, one can employ the compensation pattern depicted in figure 4.15. Two identical flat modulators 1 and 1′ oriented in a similar manner are separated by a half-wave plate 2 with oscillations directed at 45° to the directions of the modulator oscillations. The dependence of the total path length difference $\Delta\Gamma$ on the modulator temperature is thus reduced, since it is now determined by the thermal sensitivity of the half-wave plate. The effective frequency increases substantially since the plates of electro-optic crystals are in good thermal contact with the half-wave plate and will depend only on correspondence between the electrode resistance and the rest of the scheme. Restrictions on the angular aperture and on the optical frequency band depend on the angular and chromatic properties of the half-wave plate.

The experimental results obtained on a 45° BSN plate with $x = 0.30$ have shown that the room-temperature half-wave voltage is much higher than that predicted by the theory if one takes the values of electro-optic coefficients reported in [15]. Such a difference is caused by nonlinear variations of the birefringence due to the electric field. These variations are associated with the dependence of the dielectric permittivity on the field (figure 4.16). Using the dependence shown in figure 4.16, one can calculate the correction graph and introduce an additional factor to the expression which determines the true path length difference when a laser beam traverses a crystal.

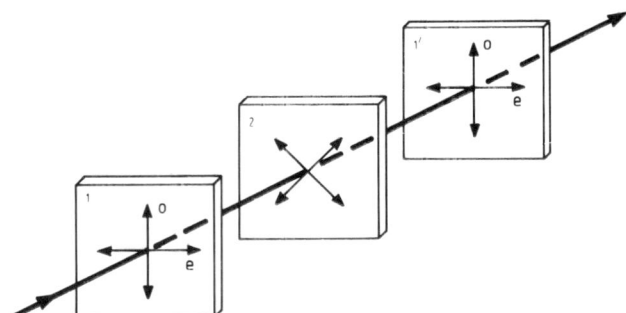

Figure 4.15   Three-element compensation modulator [46]. 1 and 1′ are electro-optic crystal plates; 2 represents a half-wave plate. Arrows indicate oscillation directions in modulator elements.

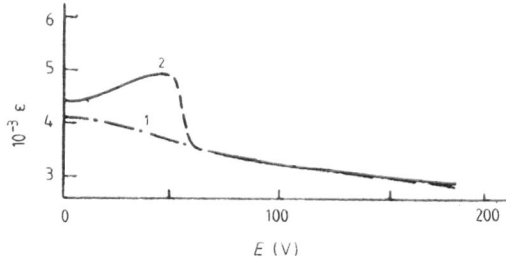

Figure 4.16 Dependence of the dielectric permittivity of a *z*-cut plate of a BSN crystal ($x = 0.25$) on the biasing field ($T = 25\,°C$) [15]: in curve 1 E is parallel and in curve 2 E is perpendicular to the P_s direction of the crystal.

The transition from linear to quadratic electro-optic effect, as has already been mentioned, is observed either near the BSN phase transition temperature or as soon as a polydomain state is established in the crystal, i.e. when the crystal contains approximately an equal number of domains with oppositely oriented spontaneous polarization vectors. This is rather probable, since the domain size in a BSN crystal is rather small [49]. Being dependent on the direction of P_s, the linear electro-optic effect is attenuated and the quadratic effect alone is observed. The BSN crystal depolarization can be judged by the fact that, as soon as a field is applied to the crystal below the transition temperature, the linear effect occurs first and then slowly falls to zero. Near the transition temperature the fall is rapid and high above the transition temperature the linear effect is not observed at all.

The half-wave voltage for the quadratic electro-optic effect is determined by the equation [15]

$$V_{\lambda/2} = \left(\frac{4\lambda d\eta^3}{n^2 \varepsilon_0^2 \varepsilon (g_{11} - g_{12})\varepsilon_c - g_{44}\varepsilon_a} \right)^{1/2} \tag{4.8}$$

where g_{ik} are quadratic electro-optic coefficients, $g_{11} = 0.17$, $g_{12} = 0.04$, $g_{44} = 0.12 \text{ m}^4 \text{C}^{-2}$ and η^3 is the relative packing density, the rest of the notation being the same as that used before.

The value of the quadratic effect rises with increasing temperature, and the half-wave voltage drops, reaching 400 V at 90 °C. The experimental values of $V_{\lambda/2}$ are in satisfactory agreement with the calculated ones if we take the values of the electro-optic coefficients reported by Lenzo *et al* [1].

Thus the use of the linear longitudinal modulator on the basis of BSN crystals with $x = 0.30$ is restricted by the nonlinearity of the dielectric permittivity and by depolarizing effects that lead to a considerable increase of the half-wave voltage. In low-frequency light modulators it is therefore more reasonable to use the quadratic longitudinal effect which is unaffected by the above mentioned factors.

The main defect of linear and quadratic longitudinal modulators is the restriction upon the upper limit of the effective frequency due to a high electric capacity of the crystal. For a typical modulator with a thickness of 0.5 mm and an area of 10 mm^2, the capacitance is 10^4 pF. For electrodes with a resistance of 20 Ω and a dissipated power of 50 mW, the upper frequency boundary reaches 28 kHz for a 100% modulation depth. As has already been mentioned, this limit can be widened by the use of a compensating pair consisting of two identical BSN crystal plates (figure 4.15).

4.4.2 Discrete deflectors

A discrete deflector was created on the basis of a BSN crystal with $x = 0.25$ [21]. A 100 μm thick plate was cut out of a single-domain barium strontium niobate crystal $Ba_{0.25}Sr_{0.75}Nb_2O_6$ perpendicular to the optic axis. The diffraction grating was made on the crystal surface by means of chrome electrode deposition, one of the electrodes covering completely the whole crystal side and the other having a saw-shaped lattice with spacing Λ, prism height $h = 800$ μm and junction plane width $b = 100$ μm (figure 4.17). The experimental set-up was constructed from a radiation source (He–Ne laser with $\lambda = 6328$ Å), a collimator, a long-focus lens, an electromechanical scanning device for scanning the diffraction spectrum, a photomultiplier with an input slit of 10 μm and an oscillograph. A voltage was applied to the crystal in the form of rectangular pulses with a maximum duration of approximately 10^{-3} s and a relative pulse duration not less than 5 $\times$ 10^{-3} s in order to exclude modulation of the refractive index along the crystal thickness due to a possible accumulation of space charges. The duration of the leading edge of optical response to an applied field pulse was limited to the time of plate polarization reversal and was approximately equal to 10^{-5} s. The duration of the trailing edge was determined by the electrode resistance and was approximately equal

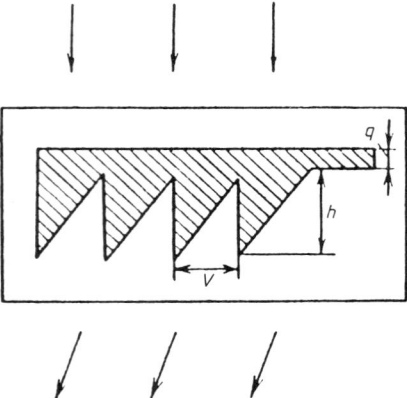

Figure 4.17 Diffraction deflector [21].

to 10^{-6} s. Within the interpulse period (≈ 0.01 s) the sample was partially depolarized.

For the lattice spacing $\Lambda = 660$ μm the diffraction angle for first-order reflex was $\theta_1 = \lambda/\Lambda = 0.96 \times 10^{-3}$ rad. The reflex intensity maximum I_1 equal to 90% of the energy falling on the deflector was reached for a pulse amplitude of $U_1 = 22$ V, which is 1.5 times as large as the expected value for an ideal saw-shaped phase grating induced in a single-domain crystal.

The intensity maximum I_2 ($\theta_2 = 2\lambda/\Lambda$) of the second-order reflex corresponded to $U = 35$ V; for the third order ($\theta_3 = 3\lambda/\Lambda$) the maximum I_3 was reached for $U = 46$ V.

It should be noted that an extraordinary beam is diffracted on the grating. An ordinary beam undergoes practically no diffraction, since the corresponding electro-optic coefficient r_{13} is 20 times smaller than r_{33} for an extraordinary beam. The diffraction angle can be increased by a decrease of the diffraction grating spacing. An experimental value $\theta_1 = 2.17 \times 10^{-3}$ rad for $\Lambda = 300$ μm and $I_1 = 0.71$ was reported by Elinson *et al* [21].

A discrete deflector was made also as a prism of a single-domain BSN crystal with $x = 0.25$ [35]. A prism with vertex angle $\varphi = 29.5°$ and basis length 3 mm was cut out in such a manner than the highest electro-optic coefficient r_{33} was used. The larger parallel sides of the prism were coated with gold electrodes, which were afterwards covered with silver paint. The distance between the electrodes was 5 mm (figure 4.18). A He–Ne laser with $\lambda = 6328$ Å working in a single-mode regime TEM$_{00}$ served as the light source. The prism was set up in such a way that the laser beam inside it could be parallel to the base. As soon as a field was applied, the refractive index n_e of the extraordinary beam was changed by a quantity Δn_e. The light beam deviation angle θ caused by the change in the refractive index was calculated by the formula

$$\theta \approx -\frac{2 \sin(\varphi/2)}{\sqrt{n_1^2 - n_e^2 \sin^2(\varphi/2)}} \Delta n_e = -0.628 \Delta n_e \tag{4.9}$$

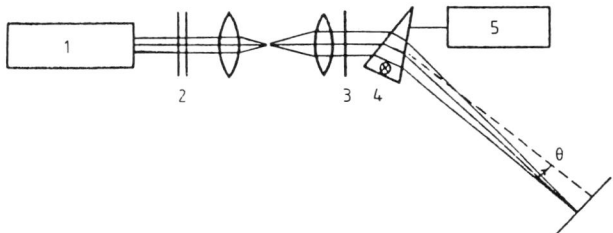

Figure 4.18 Experimental set-up for beam deviation [35]: 1 is a He–Ne laser; 2 is the attenuator; 3 is the polarizer; 4 is a prism; 5 is the prism drive feed. The directions of the c axis of the prism, of the polarization of light and of the applied electric field are perpendicular to the plane of the figure.

where n_1 is the refractive index of the alien medium (for air $n_1 = 1$), $n_e = 2.30$ ($\lambda = 6328$ Å). The deviation angle θ as a function of the applied field is presented in figure 4.19. In these experiments the field was gradually changed by the cyclic law. The so-called butterfly hysteresis loop obtained in this case is similar to the one observed for the electro-optic effect in Rochelle salt and for the inverse piezoelectric effect in BaTiO$_3$ (see [8], chapter 2). Such loops are known to occur under a change of ferroelectric polarization followed by hysteresis of other properties. The authors determined the coercive force of the crystal, $E_c = 1.3$ kV cm^{-1}, and the remanent polarization $P_s = 6$ μC cm^{-2} at a temperature of 33 °C. The laser beam induced spot-like distortions of the prism surface. The authors associate these distortions with the formation of polydomain regions in the crystal, which can evidently be removed by an applied biasing DC field.

In conclusion we may note that the applicability of deflectors on BSN crystals is limited. This is due to two facts: firstly these crystals possess

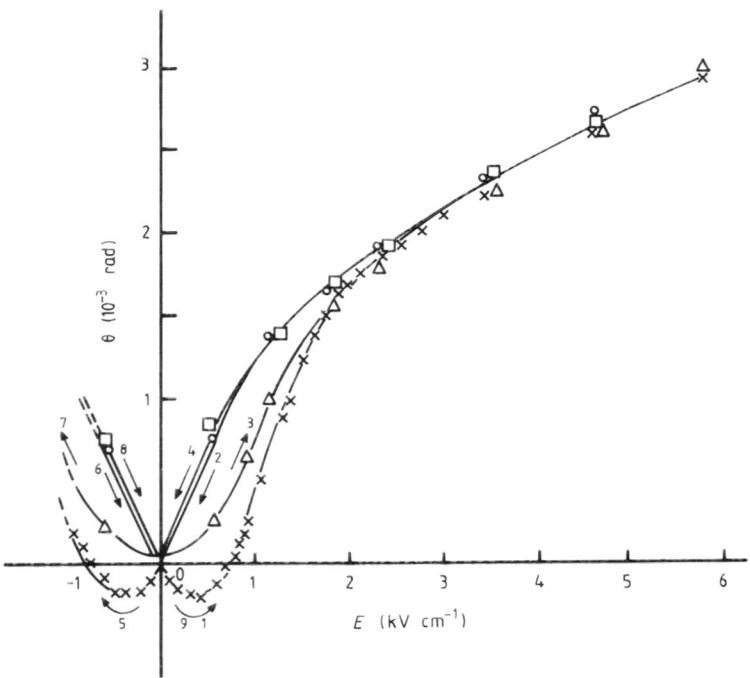

Figure 4.19 Hysteresis curves of the dependence of the beam deflection angle θ on the field applied to the BSN crystal (at 33 °C) [35]. Digits and arrows indicate succession of cycles of E variation. The curves in the negative E region are almost symmetrical with the curves in the positive region.

photosensitivity in the UV and in the visible region of the spectrum; and secondly the high value imposes restrictions upon the switching rate of the deflector.

4.5 Phase diagrams

Phase equilibrium in the ternary system $BaO-SrO-Nb_2O_5$ was examined using differential thermal and X-ray diffraction analyses [11]. Four phases with potassium tungsten bronze structures were identified (I, II, III and IV), phase I being a tetragonal solid solution $Ba_xSr_{1-x}Nb_2O_6$ (BSN I) and phase II its orthorhombic modification (BSN II). Phase III was found to correspond to the solid solution $3BaO.5Nb_2O_5$ and phase IV to the isomorphous solid solutions $2SrO.5Nb_2O_5-BaO.3Nb_2O_5$ which melt peritectically (figure 4.20).

A detailed analysis of the pseudo-binary phase diagram of $BaNb_2O_6-SrNb_2O_6$ (figure 4.21) made it possible to establish the existence region of $Ba_xSr_{1-x}Nb_2O_6$ for x ranging within the limits $0.2 < x < 0.8$ [11]. Later, Kopylov *et al* [23] indicated the homogeneity region $0.16 < x < 0.75$. The solidus and liquidus curves on the phase diagram in the existence regions of solid solutions BSN I are close to each other, which shows the possibility of obtaining these crystals by melt crystallization.

Subject to Nb_2O_5 concentration (figure 4.22), the high-temperature portion (25% of BaO) of the isopleth shows that the region of compositions from 46 to 52% of Nb_2O_5 corresponds to phase I. An increase of Nb_2O_5

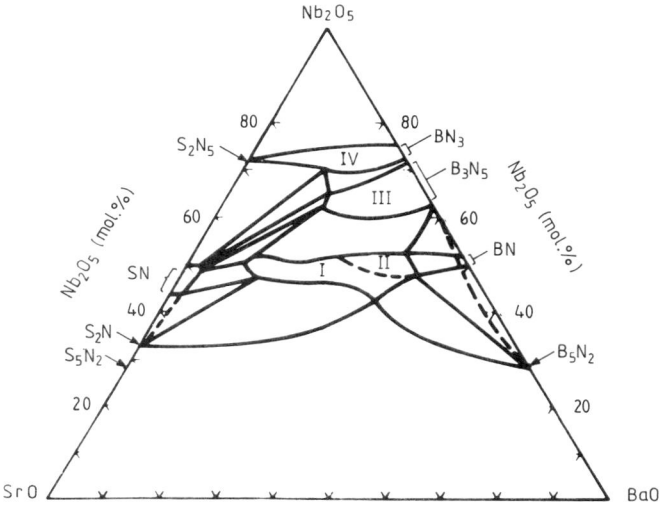

Figure 4.20 Triple phase diagram of the system $BaO-SrO-Nb_2O_5$ [11].

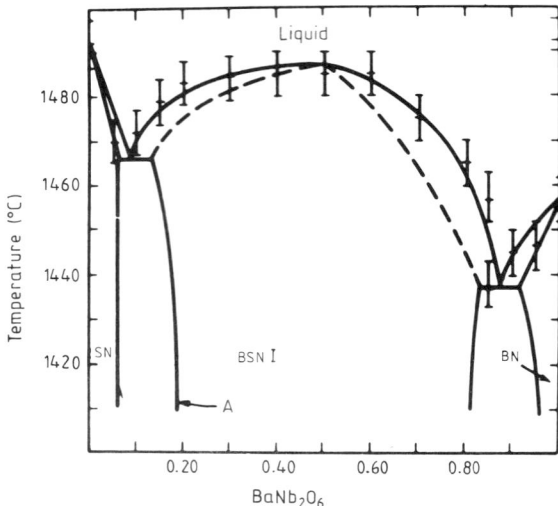

Figure 4.21   Equilibrium pseudo-binary diagram of the state of the system $BaNb_2O_6$–$SrNb_2O_6$ [11]. BSN I stands for $Ba_xSr_{1-x}Nb_2O_6$. The rest of the notation is the same as in figure 4.20.

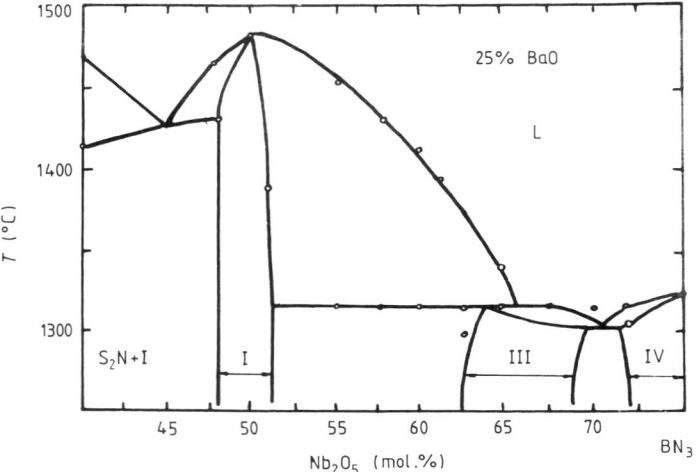

Figure 4.22   The high-temperature part (25% of BaO) of the isopleth (the notation is the same as in figure 4.20) [11].

concentration from 52 to 60% leads to the two-phase region (BSN I + BSN III). A rather steep run of the solidus curve of the isopleth shows that BSN I crystal growth from the melt in the two-phase region can readily cause a violation of known types of crystal composition upon insignificant melt temperature fluctuations.

In the crystals of BSN solid solutions first obtained, the variability of their composition over the slab was noticed. A chemical analysis of crystals and melts [38] of solid solutions $Ba_xSr_{1-x}Nb_2O_6$ showed that with increasing x ($x > 0.4$) the composition of the grown crystals goes farther and farther from the initial melt with decreasing barium content (figure 4.23). Detailed X-ray fluorescence studies in the range $0.25 < x < 0.75$ showed that the grown crystals exhibit a slight excess of BaO and SrO over the stoichiometric formula. It was established that, for the melt composition $Ba_{0.25}Sr_{0.75}Nb_2O_6$, the content of Ba^{2+} ions in crystals varies from 0.256 to 0.272, that of Sr^{2+} ions varies from 0.761 to 0.788 and the ratio $(BaO + SrO)/Nb_2O_5$ lies within the limits 1.050–1.098. A somewhat different picture was observed in the analysis of melt-grown $Ba_{0.5}Sr_{0.5}Nb_2O_5$ crystals. One could also observe here an excess of Sr^{2+} in the crystals (from 0.532 to 0.537) as compared with the initial melt, but simultaneously the crystals are depleted in Ba^{2+} (0.472–0.488). Therefore, although in this range of melt compositions there exists a certain excess of $(BaO + SrO)$ with respect to Nb_2O_5 (from 1.016 to 1.044), it is much lower than for Sr-rich compositions. Typically, for the entire indicated range of melts the content of niobium in crystals is lower (1.908–1.982) than in the initial melts [50]. Solubility up to 4% of excess $(BaO + SrO)$ and 1% of excess Nb_2O_5 was found in $Ba_xSr_{1-x}Nb_2O_6$ crystals. This circumstance results in the variability of crystal composition.

The composition of BSN crystals grown by the Czochralski method from a stoichiometric melt of $2.5BaO–7.5SrO–10Nb_2O_5$ and the remanent melt was analysed [52] using the X-ray fluorescence method [53]. The results of the analysis reproduced in table 4.12 suggest that BSN crystals grown by the Czochralski technique from a melt of the indicated composition exhibit a

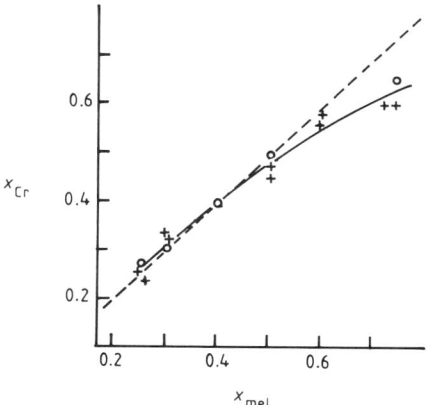

Figure 4.23 Dependence of the composition of grown $Ba_xSr_{1-x}Nb_2O_6$ crystals on the composition of the initial melt: $\bigcirc$, [38]; $+$, [50].

Table 4.12   Composition of BSN crystals grown from the melt $Ba_{0.25}Sr_{0.75}Nb_2O_6$ [52].

Sr	Ba	Nb	O
0.700 ± 0.026	0.259 ± 0.012	1.908 ± 0.046	5.729
0.733 ± 0.019	0.2775 ± 0.010	1.936 ± 0.034	5.851
0.726 ± 0.020	0.2775 ± 0.012	1.974 ± 0.054	5.938
0.701 ± 0.019	0.272 ± 0.011	1.994 ± 0.040	5.958
0.724 ± 0.022	0.285 ± 0.014	1.938 ± 0.048	5.854
0.719 ± 0.018	0.287 ± 0.011	1.942 ± 0.038	5.861

small increase in barium content and a decrease in the amount of strontium as compared with the content of these elements in the melt, i.e. an increase in barium content is compensated by a decrease in the amount of strontium. The content of niobium in these crystals is somewhat lower than in the melt. In all the crystals investigated, the overall number of Ba^{2+} and Sr^{2+} ions in a unit cell within experimental error was estimated to be equal to 5. This means that Ba^{2+} and Sr^{2+} ions are statistically distributed in the structure with an occupancy factor equal to $\frac{5}{6}$. The examinations suggested that, for concrete crystal growing conditions for BSN with $x = 0.25$, the initial compound must have excess niobium pentoxide, namely $Ba_{0.25}Sr_{0.75}Nb_{2.05}$ $O_{6.125}$.

4.6   Non-uniformity of the chemical composition

A change of composition of BSN crystals along the length or diameter is extremely undesirable because the most important BSN characteristics, such as Curie temperature, refractive indices and phase transition temperature, depend strongly on the composition. Therefore, even a slight non-uniformity of this kind may noticeably degrade the physical characteristics of the crystal samples.

The uniformity of the chemical composition of BSN single crystals was analysed [54] using the micro-probe X-ray spectral method. In crystals grown from melts of stoichiometric compositions $x = 0.25$, $x = 0.5$ and $x = 0.25$ with 1.2 mol.% of Nb excess over the stoichiometry, variations in the Ba/Nb and Sr/Nb ratios along the length and diameter of the slab were determined (figure 4.24).

The micro-probe X-ray spectral analysis of the indicated compositions of crystals made it possible to establish that the central sub-seed regions of the grown crystals show anomalies in the element distribution, which are different for different compositions. So for crystals with a large barium content ($x = 0.5$) the barium concentration increases noticeably in the sub-seed region, whereas

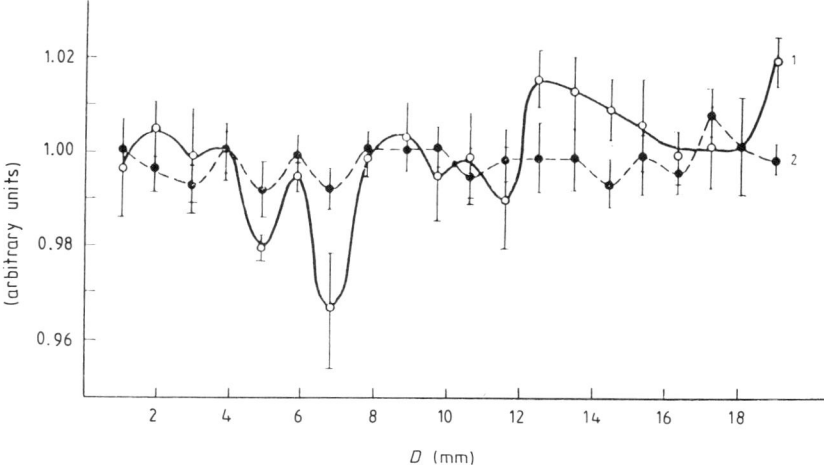

Figure 4.24 Variation of the ratios Ba/Nb (1) and Sr/Nb (2) along the diameter of a BSN crystal slab of composition $Ba_{0.25}Sr_{0.75}Nb_2O_6$ [54].

for samples grown from melts of the composition $x = 0.25$ the central region exhibits a slight decrease of barium content. The most uniform composition in the direction of growing is exhibited by those parts of crystal boules where the diameter of growing crystals is stable. An introduction to the initial melt composition of a certain niobium excess does not practically change the behaviour of the Ba/Nb and Sr/Nb ratios along the length or cross section of the crystal boule within experimental error, which for this method is 2–3%.

The variation of the composition over the length of a crystal slab was estimated [51] from the value of the refractive index. It was shown that in the course of crystal growing the composition $Ba_{0.25}Sr_{0.75}Nb_2O_6$ becomes rich in barium and accordingly, when pulled, the melt becomes rich in strontium. The measured gradient of the refractive index reaches the value $\Delta n \approx 8 \times 10^{-4}$ per 1 cm of crystal length, which corresponds to a cation composition variation of approximately 1% and a melt temperature variation in the sub-crystal region of 0.3–0.4 °C. It is an experimentally established fact that the value of Δn increases if, when being pulled from the melt, the crystal slows down its rotation. Note that the cation composition variations reported in [51] are somewhat smaller than those reported in [50].

It is well known that high-quality crystals with constant values of the refractive indices can be obtained from a congruently melting composition, because it is only in this case that the crystal composition is practically independent of temperature variations at the crystallization front and of the decrease of the melt mass during crystallization. An analysis of the Curie temperature as a function of the composition of ceramic samples and crystals grown from melts of the compositions $0.25 < x < 0.75$ made it possible, by analogy with the binary systems $LiNbO_3$ [55] and $LiTaO_3$ [56], to obtain a congruently melting BSN composition $Ba_{0.54}Sr_{0.46}Nb_2O_6$ [50]. This analysis has also shown that with increasing $BaNb_2O_6$ content the solid solution Curie temperature rises gradually from $56\,°C$ for $x = 0.25$ to $250\,°C$ for $x = 0.75$. An analysis of the Curie temperature as a function of Nb_2O_5 concentration along the 25% BaO isopleth on ceramic samples between 46 and 60 mol.% of Nb_2O_5 has revealed that the Curie temperature rises from 50 to $200\,°C$ with increasing concentration of Nb_2O_5 [11]. Noticeable changes of T_C within the range $48–52$ mol.% of Nb_2O_5 in ceramic samples were also reported by Megumi *et al* [7], but according to this paper the Curie temperature of single crystals remained almost unchanged within a fairly broad range of Nb_2O_5 concentrations (figure 4.25). The broken curve in figure 4.25 joins compositions for which ceramics and crystals have equal T_C.

Attempts to grow crystals have revealed that single crystals grown from the above mentioned congruently melting composition exhibited striations, which stimulated Megumi and his colleagues to seek for a 'true' congruent composition of BSN solid solution. For this purpose they used X-ray fluorescence and differential thermal analyses. They also determined the lattice parameters and Curie temperatures as functions of the crystal composition. The authors investigated the compositions of initial melts, of

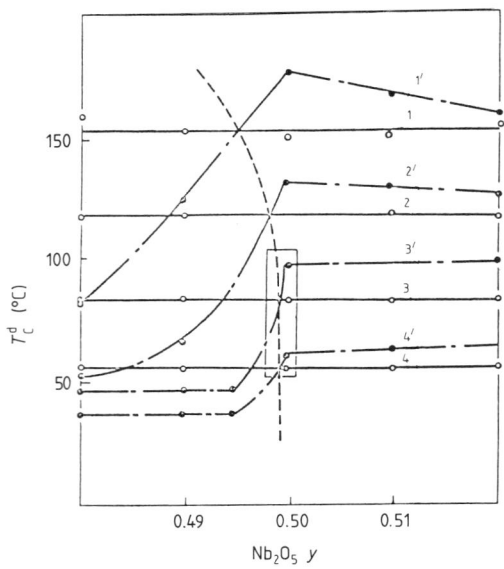

Figure 4.25 Dependence of T_C of single-crystal (1–4) and ceramic (1′–4′) samples $(BaO, SrO)_x(Nb_2O_5)_y$ on the content of Nb_2O_5 [7]; x is equal to 0.4 (1, 1′), 0.5 (2, 2′), 0.6 (3, 3′) and 0.7 (4, 4′).

crystals grown from these melts and of remanent melts, because when solid solution crystals were grown from a congruently melting composition the composition of the growing crystal gradually changed, inducing changes in the melt composition. It should be noted that, at early stages of crystal growth, the relative changes in the melt composition exceed substantially the relative changes in the crystal composition. The melts $(Sr_xBa_{1-x}O)_{1-y}(Nb_2O_5)_y$ were examined within the ranges $0.40 < x < 0.70$ and $0.48 < y < 0.52$ (figure 4.26). The analysis of these curves suggests that in crystal pulling from melts with $x_0 > 0.61$ the melt is depleted in strontium, and when $x_0 < 0.61$ the content of strontium increases. The results are shown by the analysis of the compositions of the grown crystals. So the lower portion of the crystal grown from a melt of the composition $x_0 = 0.60$, $y_0 = 0.4995$ contained more barium (less strontium) than the initial melt, and the sample prepared from a melt with $x_0 = 0.62$, $y_0 = 0.4995$ contained less barium (more strontium) than the initial one.

The data from X-ray fluorescence analysis, X-ray diffraction and dielectric measurements have led to the conclusion [7] that the congruent composition of BSN differs from the stoichiometric one and that it corresponds to $y = 0.4993$ and $x = 0.61$ in the formula $(Sr_xBa_{1-x}O)_{1-y}(Nb_2O_5)_y$. The deviation of the congruent composition from stoichiometry towards a deficit of Nb_2O_5 was only 0.07 mol.%. It is of interest that the congruent compositions $LiNbO_3$ and $LiTaO_3$ are displaced from stoichiometry towards excess Nb_2O_5 and Ta_2O_5 by 1.4 and 1.25 mol.% respectively. Five 'congruent' compositions for the system $BaO-Na_2O-Nb_2O_5$ can be found in the literature, the distinctions in them being beyond experimental error [57].

A detailed thermographic study of the system $Sr_xBa_{1-x}Nb_2O_6$ [7] gave the melting temperature maximum near $x = 0.61$ and not $x = 0.46$ as reported

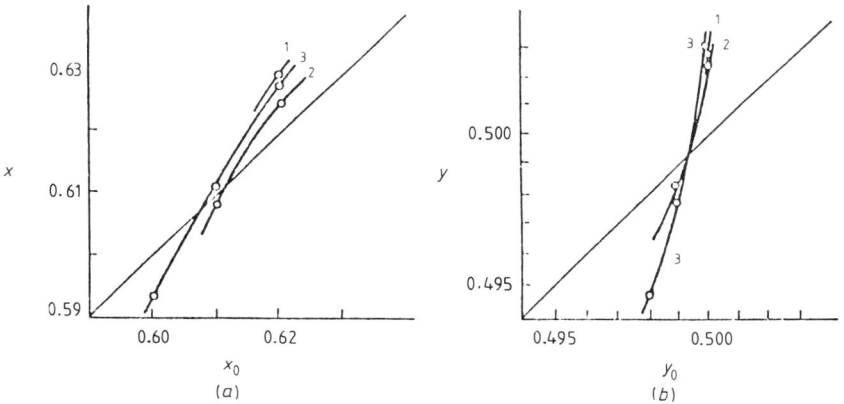

Figure 4.26 Interdependence of the compositions of initial (x_0, y_0) and remanent (x, y) melts of $(Sr_xBa_{1-x}O)_{1-y}(Nb_2O_5)_y$ [7]. (a) y_0 is equal to 0.494 (1), 0.499 (2), 0.500 (3); (b) x_0 is equal to 0.60 (1), 0.62 (2) and 0.620 (3).

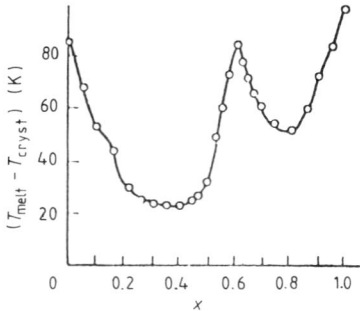

Figure 4.27   Difference between melting and crystallization temperatures in a series of $Sr_x Ba_{1-x} Nb_2 O_6$ solid solutions [7].

earlier in [11]. Besides, the melting and crystallization temperatures of solid solutions along the binary region of $BaNb_2O_6-SrNb_2O_6$ (figure 4.27) did not coincide. Overcooling is an ordinary effect in the course of crystal growing, but the dependence of the degree of overcooling on the solid solution composition, which appears to be maximum for extreme members of the series, i.e. for $BaNb_2O_6$ and $SrNb_2O_6$, and for a composition close to congruent, are of interest here. The reason for such an anomalous behaviour remains unclear. It should also be pointed out that for compositions with high strontium content the degree of overcooling is somewhat higher (on the average by approximately 30 °C) than for compositions with low strontium content (figure 4.27).

4.7   The effect of isomorphous replacement and alloying additions upon the structure of BSN solid solutions

The growth of optically homogeneous BSN crystals encounters substantial technological difficulties due to stoichiometry violation and the occurrence of a high-temperature phase transition between monoclinic and tetragonal symmetries of the crystal lattice. Stoichiometry violation and a possible formation of a second phase may be responsible for optical inhomogeneities in the crystal. The optical inhomogeneity may also depend on the domain structure if the crystal is not entirely single domain. The latter fact is a consequence of phase transition smearing and a relatively low Curie temperature (for the composition $Ba_{0.25}Sr_{0.75}Nb_2O_6$ one has $T_C = 50$ °C). Attempts were made to avoid the factors mentioned above by way of introducing small amounts of rare earth and lead oxides. An introduction of 2 wt.% of rare earth oxides affects the crystal morphology, stabilizes the domain structure, lowers the Curie temperature and improves the crystal growing conditions [16]. An introduction of 1 wt.% of PbO reduces the dielectric losses [58].

The influence of isomorphous replacement in niobium and barium sublattices and of alloying additions upon stabilization of tetragonal single phase was investigated in [30, 59]. The elements Ti, Sn, Zr, Ta, V, Mo, W, La, Y and Mg were used as isomorphic additions. The initial components were oxides and metal carbonates. An exception was barium nitrate since its relatively low melting temperature made it possible to lower the synthesis temperature down to 900 °C. The replacement of Nb and (Ba, Sr) in barium strontium niobate with $x = 0.25$ by the elements listed above was carried out, allowing for 5, 10 and 15 mol. % without stoichiometry violation. The samples were prepared using conventional ceramic technology.

Examinations have shown that an addition to a BSN composition of niobium-replacing elements, such as Ti, Sn, Zr or W, leads to a reduced content of monoclinic phase, and 15 mol. % of titanium completely stabilizes the tetragonal phase. On the other hand, when Nb is replaced by V and Mo, the content of the monoclinic phase increases (figure 4.28(*a*)).

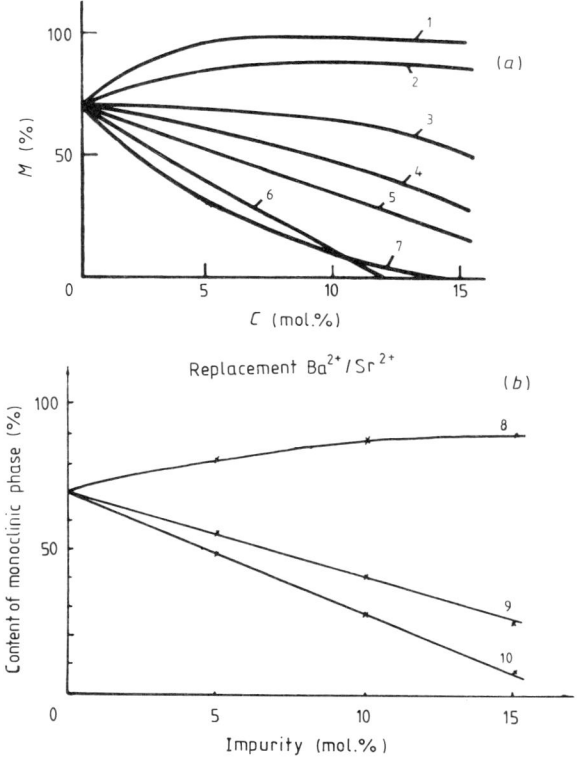

Figure 4.28 The influence of isomorphic replacement of Nb (*a*) and (Ba, Sr) (*b*) by different elements upon the content of the monoclinic phase in $Ba_{0.25}Sr_{0.75}Nb_2O_6$ ceramic samples [59]: (1) V^{5+}; (2) Mo^{6+}; (3) Ta^{5+}; (4) Zr^{4+}; (5) W^{6+}; (6) Ti^{4+}; (7) Sn^{4+}; (8) Mg^{2+}; (9) Y^{3+}; (10) La^{3+}.

Of all the elements isomorphically replacing barium and strontium (figure 4.28(*b*)), lanthanum is the best for promoting a decrease in the content of monoclinic phase in barium strontium niobate; its content increases when Ba and Sr are replaced by Mg.

To see the effect of rare earth elements upon the quality of BSN crystals, an attempt was made to grow BSN crystals of congruent composition ($x = 0.54$) with additions of La, Y and Tu [30]. Crystallization was carried out under conditions close to optimal for obtaining 'pure' BSN crystals of sufficiently high optical quality. It was established that such crystals have a tendency for being tetrahedral, and are prone to cracking and growth striations. The latter testifies to the fact that stoichiometry violation in BSN crystals is not compensated by the indicated elements. The dislocation density found by etching techniques turned out to be much higher in La-doped than in pure BSN crystals, the dislocation density configuration showing parallel lines of etching pits. Thus an introduction of 1 wt.% of rare earth elements into BSN crystals is not able to improve crystal quality, but facilitates to some extent their growth conditions.

4.8   Point and linear defects in BSN crystals

The most essential effect upon the optical characteristics of BSN crystals is produced by inhomogeneities of the spatial distribution of defects connected with the non-uniform distribution of impurities and the violation of stoichiometric composition. Deviation from stoichiometry leads to the appearance in a crystal of point defects and has a direct effect upon their overall concentration.

Using the Raman spectroscopy method, Scott and Burns [60] revealed that after several days of annealing the congruent composition of $LiNbO_3$ crystals becomes unstable, and below 800°C the second phase of the composition $LiNb_3O_8$ appears, so that at room temperature the phase of stoichiometric composition appears to be stable. We should note that at a high cooling rate (≈ 100°C h^{-1}) the appearance of the second phase in lithium niobate crystals is not observed. This suggests that, under conventional growing and annealing techniques, microscopic volumes of the second phase are observed in BSN crystals, which may have a negative effect on the optical quality of crystals and may require certain changes in the technology.

It was noticed [61] that oxygen octahedral ferroelectrics practically always exhibit non-stoichiometry, and most often this is a deficit in oxygen. Growth in a reducing atmosphere yields non-transparent dark-blue BSN crystals whose colour is associated with an oxygen deficit and partial reduction of niobium to a tetravalent state.

At the present time no reliable data are available on the character of the oxygen defect distribution in the BSN crystal structure. So it is not known whether they exist in the form of vacancies or whether they form disordered planes. But the analysis of oxygen annealing of crystals suggests that the predominant defects are oxygen vacancies.

The microstructure of BSN crystals was examined using etching and decoration methods [54]. The decoration method, involving electron or optical microscopy, allows one to reveal such active elements of crystal surface as point defects and their clusters, as well as elements of geometrical micro-relief on the surface with different electrical properties [62].

In BSN crystals, decoration was applied to z cuts (perpendicular to the c axis) and to fresh chips which were made by cleaving a crystal boule perpendicularly to the c axis immediately prior to decoration (figure 4.29). It should be noted that the surface of those chips was very imperfect, due to the absence of cleavage planes in these crystals.

Non-uniformity of decorating particle precipitation on the chip surface makes it possible to reveal a spatial electric heterogeneity in the crystal, which may be either due to composition inhomogeneity (microscopic volumes of the second phase) or due to inhomogeneous polarization of the crystal, i.e. its polydomain nature. Comparison of the revealed structure of a BSN chip surface with the domain structure of ferroelectric triglycine sulphate (TGS), which was investigated in detail using decoration [62], showed that lens- and

Figure 4.29 A typical picture of decoration of a fresh BSN crystal chip by anthraquinone vapours. Enlargement × 225 [54].

cigar-like electrically charged regions of BSN crystals are similar in shape to domains in TGS. The dimensions of these regions are comparable with the dimensions of TGS domains and are of the order of several tens of microns. The arrangement of charged regions is regular: they stretch in the direction of the ferroelectric axis. But a typical anthraquinone texture inherent in domains of opposite signs is not observed on BSN chips. Charged regions of chips are edged by neutral regions on which the decorating substance is not crystallized.

In ferroelectric crystals the domain configuration can be established by etch pit techniques, since the etching rates of oppositely charged ends of domains are different. Selective etching of BSN crystals [20] did not reveal the domain structure typical of $LiNbO_3$ and $Ba_2NaNb_5O_{15}$. BSN crystals showed only large dark octagonal pits identified as micro-domains disseminated in the crystal base with a reverse polarization. Etching of polarized samples showed, however, that the density of dark pits did not change. Ito and Furuhata [63] interpreted pits of such a shape as edge dislocations.

The domain structure of single crystals of BSN solid solutions was disclosed in the analysis of the mechanism of polarization reversal by an applied electric field (transparent electrodes were deposited onto z cuts of a crystal). The domain structure of BSN is reproduced in figure 4.30, where one can see domains of one sign disseminated into a region of domains with an opposite sign.

The selective etching technique enables the crystal dislocation structure to be characterized by etch pit configurations. The stress fields due to dislocations induce local changes in the refractive index and density, which may cause

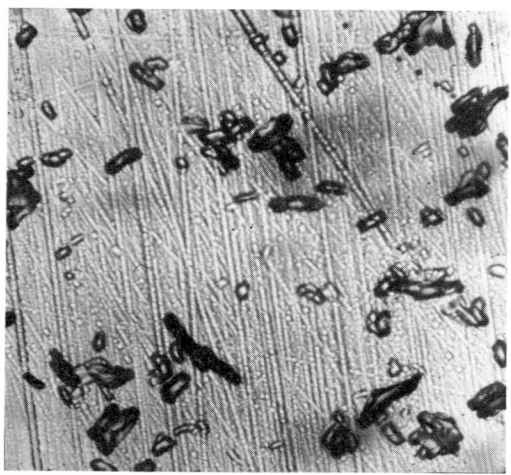

Figure 4.30   Microphotograph of the domain structure of a $Ba_{0.25}Sr_{0.75}Nb_2O_6$ crystal. Enlargement $\times 148$.

an additional light scattering. A marked effect upon the optical properties may also be produced by rearrangement of impurity atoms in the stress fields of dislocations. The etch pit technique is therefore one of the principal quality control methods for BSN crystals.

The dislocation structure of BSN crystals revealed by the etch pit technique exhibits certain regularities (figure 4.31). The highest dislocation density is observed in the central region below the seed (figure 4.31(a)). Dislocation clusters in this region are presumably [64] due to the imperfection of seeding and can be reduced considerably by using a dislocation-free seed crystal, but cannot be completely removed since dislocations around the seed crystal contour remain and are inherited by the growing crystal. These dislocations evidently occur at the moment of seeding as a result of a thermal shock. The use of thin (1.5–2 mm) seeds substantially narrows the defective region in the centre of the crystal, but even the crystals grown under optimal conditions have a 'core'.

A rather high dislocation density is observed at the surface of a crystal slab, which is obviously due to crystal lattice distortion caused by precipitation of platinum specks coming from the crucible surface during crystal growth [64]. Dislocation density also increases in places of fusion or of a sharp uncontrollable change in the diameter. Dislocations occurring in this region intergrow along the entire boule length. This suggests that if some surface defects occur in the course of crystal expansion, it is necessary to melt the entire expansion cone up to the seed crystal. Grown BSN crystals sometimes

(*a*)   (*b*)

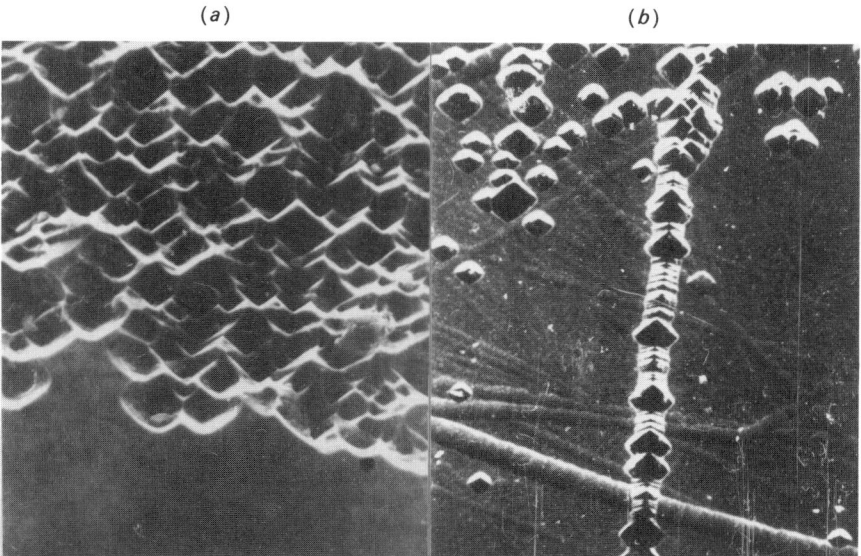

Figure 4.31 Dislocation structure of the sub-seed region of a BSN crystal (*a*) and the dislocation series coming from it (*b*). Enlargement: (*a*), ×600; (*b*), ×300.

exhibit dislocation rows going from the crystal centre to the periphery (figure 4.31(*b*)). For the BSN crystallization conditions described below the number of this type of dislocation rows is not large. Clusters of dislocations are also encountered at block boundaries.

As far as the shape of etch pits is concerned, they are most often truncated square pyramids and more seldom octagonal ones. In polarized light near these pyramids one can observe rosettes of photoelastic stresses.

For crystals of BSN solid solution pulled under optimal conditions, large regions are absolutely free of dislocations.

Using X-ray diffraction analysis (the method of epigrams and the method of photography of reciprocal lattices), a new type of structure defect possessing dispersibility different from that of the average lattice was revealed in BSN crystals with $x = 0.25$ [65].

The extension of these defects along the c axis was found to be (5×10^{-5})–(1×10^{-4}) cm, which makes 600–1200 unit cell spacings of the lattice. The dimensions in the perpendicular direction proved to be much smaller. On the basis of the results obtained it was concluded that such quasi-one-dimensional defects, whose concentration is high, are not due to the crystal growing conditions but are rather inherent in the material. The authors point out, however, that without an exact picture of the distribution of barium and strontium atoms and their vacancies over the sites of a three-dimensional lattice one cannot establish how different types of atoms are distributed in the defects discovered.

It was reported in the same paper that, in the BSN crystal lattice, of all the examined compositions ($x = 0.25$, 0.50, 0.54 and 0.67) there exists a three-dimensional ordered alternation of atoms (superlattice).

Many types of structure imperfections change the crystal lattice spacing in microscopic regions. These variations can be studied by X-ray methods because the volume in which X-ray radiation is diffracted is very small [66]. For barium strontium niobate solid solutions, the dependence of lattice parameters on the composition was established in several papers (e.g. [50, 67]). The most detailed study of crystal and ceramic samples was performed in [7]. The parameters a and c of a unit cell in a system of BSN solid solutions were found to depend not only on the solid solution composition but also on thermal treatment regimes. So the parameters a and c increase with increasing barium content and with decreasing strontium content respectively, whereas with increasing niobium content and increasing cooling rate it is only the parameter c that increases while the parameter a decreases. An analysis of the obtained dependence of lattice parameters on the content of Nb_2O_5 in a solid solution shows that the parameters a and c undergo noticeable variations in the region of a niobium deficit and are practically independent of niobium content above stoichiometry ($y > 0.5$).

Precision measurements of the relative variation of the parameters a and c in the BSN crystal lattice along the length and cross section of a crystal slab made it possible to establish [54] that, for samples of stoichiometric composition with $x = 0.25$, the relative variation of the parameter a over the cross section of a crystal slab exceeds those of c (figure 4.32). Sharp jump-like variations of the value exceeding the experimental error by several times ($\pm 1.6 \times 10^{-5}$) are typical of the parameter a, whereas the variations of the parameter c are within the experimental error ($\pm 1.3 \times 10^{-5}$). For crystals grown from a material with 1 mol.% of excess niobium, a gradual increase of the parameters a and c from the centre of the crystal to the periphery is observed, the c variations exceeding those of a. In crystal regions directly adjoining the core, both parameters show an anomalous increase. X-ray topography of such a sample revealed the diffusion region whose dimensions coincided with those of observed anomalies of the parameters.

Measurements of the relative a variations over the length of the crystal slab revealed anomalies in the area where the crystal boule surface showed

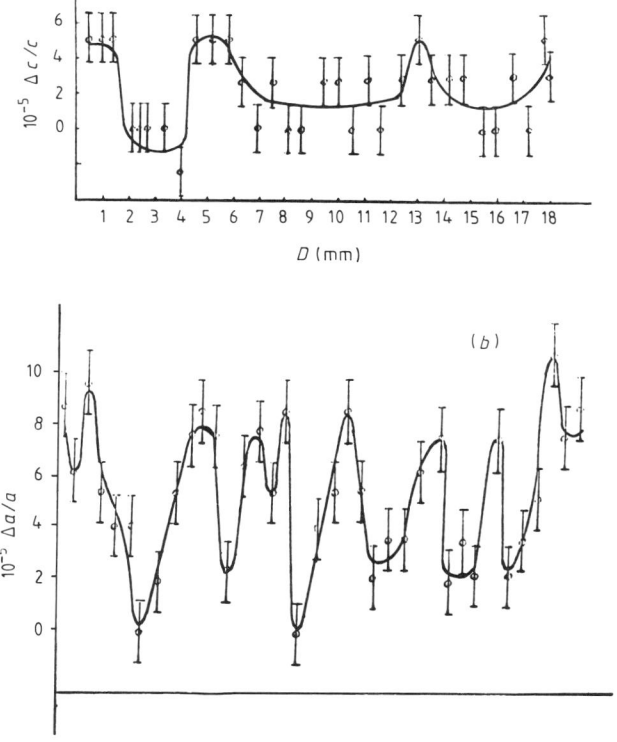

Figure 4.32 Relative variations of the parameters c (a) and a (b) of the lattice of a $Ba_{0.25}Sr_{0.75}Nb_2O_6$ crystal along the boule diameter [54].

sharp variations in the diameter accompanied by changes in the faceting. Outside these areas, the relative *a* variations appeared to be within experimental error (figure 4.33).

If the lattice parameter variations are considered to be associated with stoichiometry violations, the approximate estimate gives a value of the order of 0.1 mol.% towards an increase of barium content in defect areas of the specimen.

The sharp anomalies of the lattice parameters observed in stoichiometric BSN crystals lead apparently to substantial strains in the volume of a crystal slab which, in turn, are responsible for cracking. In this sense an excess of niobium should be thought of as a favourable factor since in this case the parameter variations are smooth, and therefore the strain distribution in the crystal boule is more uniform, which lowers the probability of cracking. Indeed, it is experimentally evidenced that stoichiometric BSN crystals are more prone to cracking than crystals grown from melts with excess niobium. Cracks in stoichiometric BSN crystals are as a rule parallel to the growth *c* axis.

The above mentioned lattice parameter variations in the crystal core are obviously due to a sharp change in the diameter of a crystal, growing on a seed, at the initial moment of its growth. If the diameter of a grown crystal is made somewhat smaller than the seed diameter, then no anomalous increase of the parameters is observed in the areas directly below the seed. A simultaneous increase of *a* and *c* implies that such regions are barium-rich. Constancy of the lattice parameters over the length (within 30 mm) of a crystal with a constant diameter indicates that BSN crystals of solid solutions grown under optimal conditions can be free of striations.

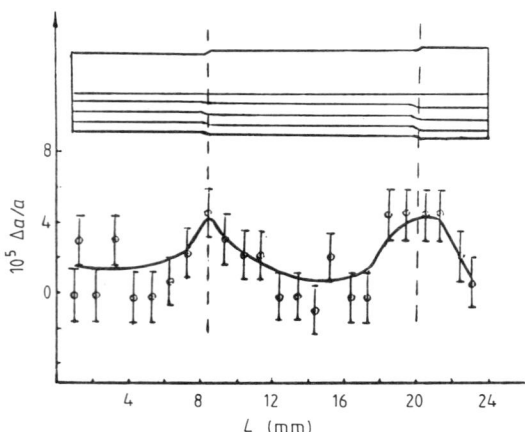

Figure 4.33   Relative variations of the crystal lattice parameter *a* along the length of a crystal slab of the composition $Ba_{0.25}Sr_{0.75}Nb_2O_6$ [54].

4.9 Growth of BSN single crystals

Barium strontium niobate single crystals are grown by the Czochralski method. Heating is achieved either in resistance furnaces or by high-frequency currents. From iridium crucibles one obtains dark amber crystals which can be then partially decoloured by means of a long-term oxygen annealing. Platinum crucibles allow us to obtain colourless specimens. Initial reagents are customarily barium and strontium carbonates and niobium pentoxide. Nitrates are used more rarely. All materials should be highly pure.

A preliminary solid-phase synthesis of the starting material is performed at temperatures of the order of 1100–1300°C for 2–5 h [8, 68]. X-ray phase analysis is employed to make sure that the reaction is completed. To obtain a tetragonal single phase, it is recommended that a high-temperature synthesis at temperatures above 1400°C is used. Even minute traces of a monoclinic phase in the starting material lead to crystallization defects [68].

A component of the pseudo-binary barium niobate–strontium niobate system is $BaNb_2O_6$, and therefore the kinetics and the mechanism of the formation of this joint are of undoubted interest. Thermography, X-ray phase and chemical analyses were used [69] to study the interaction between solid-phase $BaCO_3$ and Nb_2O_5 with the mole ratio 1:1. It was shown that prior to the formation of $BaNb_2O_6$ there occurs a synthesis of the compound $BaNb_{0.8}O_3$ ($5BaO.2Nb_2O_5$). This is explained by the fact that $BaCO_3$ in the presence of Nb_2O_5 is already rapidly decomposed at 700°C, whereas the binding of the produced barium oxide with niobium pentoxide goes at a sufficient rate only at much higher temperatures. After a ten minute soaking at 1300°C, the yield of $BaNb_2O_6$ reaches 95%. When synthesis proceeds in a platinum crucible, one should take into consideration that the reactivity of the BaO unreacted with platinum rises especially around 1300°C. Differential thermal analysis shows that barium oxide losses due to such an interaction may reach from 5 to 10 mol.% and lead to stoichiometry violation in the grown crystal [70]. The presence in the starting material of unreacted barium oxide should be completely excluded. Thus, a long-term high-temperature synthesis is the most reliable way of obtaining high-quality starting BSN material.

An analysis of $Ba_xSr_{1-x}Nb_2O_6$ as the starting material within the range $0.25 < x < 0.67$ showed a strong dependence of monoclinic phase content in the starting material on the Ba/Sr ratio. So an increase of barium content leads to a decrease in the amount of monoclinic phase in the starting material (from 70% for $x = 0.25$ to a perfectly tetragonal single phase for $x \geqslant 0.50$). Stabilization of the tetragonal single phase with increasing $BaNb_2O_6$ content in a solid solution is evidently due to an increase of structure ordering with increasing x in $Ba_xSr_{1-x}Nb_2O_6$ (the structure becomes completely ordered for $x = 0.67$). It should be noted that crystals grown from a starting material with different contents of monoclinic phase were single phase and had a tetragonal structure.

In this connection it was supposed [68] that at melting temperature the monoclinic phase transforms into tetragonal, but the presence of monoclinic phase in the starting materal was found to have a negative effect upon the optical quality of crystals. To obtain a tetragonal single phase, it was recommended that the synthesis time should be prolonged to several tens of hours, which is practically very difficult.

The X-ray phase analysis of melts, rapidly and slowly cooled through the melting temperature, showed that in the former case the monoclinic phase content is approximately equal to 90%, while in the latter case a tetragonal single phase is formed (figure 4.34). On the basis of these results it was assumed that the monoclinic phase percentage in a cooled melt is determined by the crystallization rate. At a crystal pulling rate not higher than 6 mm h^{-1} only a tetragonal single phase is formed. There exists a certain probability of monoclinic phase micro-inclusions, which can be due to instability of growth device operation, that induces instantaneous melt temperature oscillations and, therefore, oscillations of crystallization rate. Besides, if the melt contains a near order corresponding to the monoclinic phase, then monoclinic phase micro-inclusions are possible, even in the case of a smooth crystallization front, by way of a direct in-growth of a second phase into the crystal.

To achieve a complete substance homogenization, the melt is soaked as a rule for 1–1.5 h prior to the growth procedure. BSN crystals are pulled at a rate of 5–12 mm h^{-1}, the rotation rate being 20–80 rpm [71]. Crucible rotation is used very rarely. The seed crystal is aligned along the tetragonal axis; this axis must not decline from the growth axis by more than 1°.

According to Brice *et al* [38] the instability of the diameter *d* of a growing BSN crystal is related to the melt temperature fluctuations (ΔT) by

$$\Delta d = \Delta T K_{\mathrm{L}} d^{3/2} / (4\theta_{\mathrm{m}} K_{\mathrm{S}}^{1/2} \delta_T \varepsilon^{1/2})$$

where K_{L} and K_{S} are the heat conductivities of liquid and solid phases respectively, θ_{m} is the difference between the melting and environmental temperatures, δ_T is the thickness of the adjoining liquid layer, and ε is the rate of heat removal through unit surface of solid phase, as the environmental temperature increases by 1°C. This formula implies that the stability of the crystal diameter can be increased either by decreasing the environmental temperature (increasing the difference θ_{m}) or by speeding up the heating removal rate ε. This had led to attempts to stabilize the diameter of a growing crystal by oxygen blasting in transverse and axial directions [38]. The transverse blast did facilitate diameter stabilization but lowered the optical quality of the specimen: there occurred noticeable striations and cracking. The results were somewhat better with an axial oxygen blast preliminarily heated to 400–500°C. Such modifications of the Czochralski method

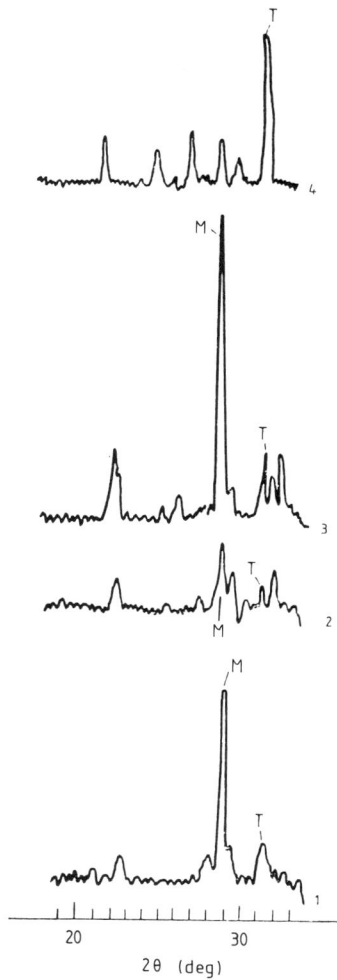

Figure 4.34 Diffractograms of BSN solid solutions of the composition $x = 0.25$: (1) starting material; (2) gradually and (3) sharply cooled melts; (4) crystal. Reflections corresponding to the monoclinic and tetragonal phases are denoted M and T respectively.

complicated the technological process and did not result in an appreciable improvement in the optical quality of the crystals.

Measurements of the heat emitted by a growing crystal allowed Megumi *et al* [71] to derive an empirical expression relating the crucible temperature T to the crystal length L for a constant diameter of a growing crystal: $T = T_0 + A \exp(-L/L_0) + BL$, where T_0, A, L_0 and B are constant quantities depending on the shape and size of a crucible as well as on its thermal packing. The variability of the temperature during growth makes it possible, according to this expression, to obtain BSN crystals of controlled diameter and of maximum dimensions $d = 32$ mm, $L = 20$ mm.

The first crystals obtained from BSN solid solutions were reported in [72, 73]. The facets of crystals grown along the c axis were as a rule well pronounced, which is generally rather rare in the Czochralski method (figure 4.35). Their form has 24 facets of four prisms: $\{110\}$, $\{120\}$, $\{100\}$ and $\{130\}$. The facets of the two latter simple forms are developed better on crystals with a high strontium content. So, for the composition $Ba_{0.75}Sr_{0.25}Nb_2O_6$, four facets of the prism $\{100\}$ degenerate into narrow strips, and the facets $\{110\}$ noticeably prevail. For the composition $Ba_{0.25}Sr_{0.75}Nb_2O_6$, the facets of the four prisms develop approximately identically. The crystal facets are perfect and specular. With an enlargement of $\times 200$ they do not show a micro-relief. Perfection of the edges increases with increasing barium content. For a slow separation of a crystal from the melt, the lower part of the slab shows pronounced facets of pyramids $\{120\}$ and $\{111\}$ and pinacoid $\{100\}$. An increase of barium content in the solid strengthens the tendency to stable facet formation [74].

Growth of single crystals of barium strontium niobate solid solutions by the Czochralski method is realized as a rule by using induction heating on a standard crystallization set-up (figure 4.36). The rotation rate of the upper water-cooled rod may change smoothly from 1 to 100 rpm, the pulling rate being varied between 0.1 and 30 mm h^{-1} with an accuracy of $\pm 1\%$. The maximum run of the rod is 250 mm. It may be moved either by hand or automatically. The crucible is moved inside the inductor manually by means of the lower rod. A high-frequency generator makes it possible to realize induction heating of the platinum crucible measuring from 50 to 100 mm in diameter. The maximum generator power on the inductor makes up 25 kW.

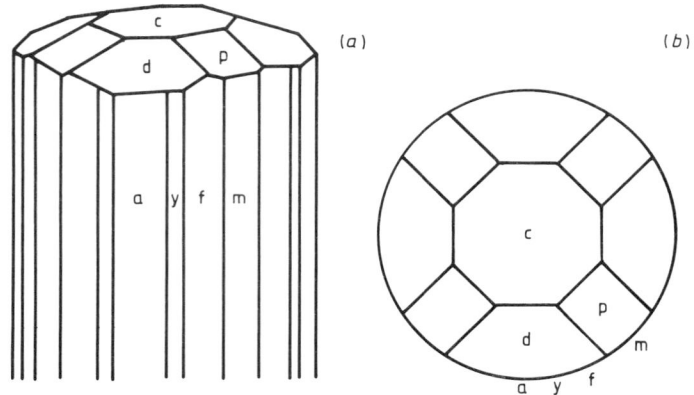

Figure 4.35   The general form (a) and the lower part (b) of a BSN crystal in the case of slow separation from the melt [75].

Figure 4.36 The general form of a 'Donets-1' crystallization set-up.

The generator power was stabilized and controlled in the course of crystal growth by a precision temperature controller which enabled the temperature to be maintained at a level of 1500 °C with an accuracy of ± 0.3 °C. The servo system operation was proportional integral–differential. The programmer on the control desk made it possible to change smoothly the generator power within a given time interval.

To grow colourless single crystals of BSN, platinum crucibles only are used. Crystallization of BSN melts, accompanied by an increase in the substance volume, leads to a considerable deformation of crucibles after only several growth processes, which suggests that crucible walls should be made thicker. Figure 4.37 illustrates one of the versions of a crucible with thermal packing (*a*) and axial distribution of temperature above the melt (*b*). With such packings, the axial temperature gradient above the melt can be varied from 40 to 110 °C cm^{-1}. The radial temperature gradients in the immediate vicinity of the melt surface in the packings are small (< 0.3 °C), and the temperature fluctuations in the melt and in the gas volume above the melt are no higher than ± 0.6 and ± 0.25 °C respectively.

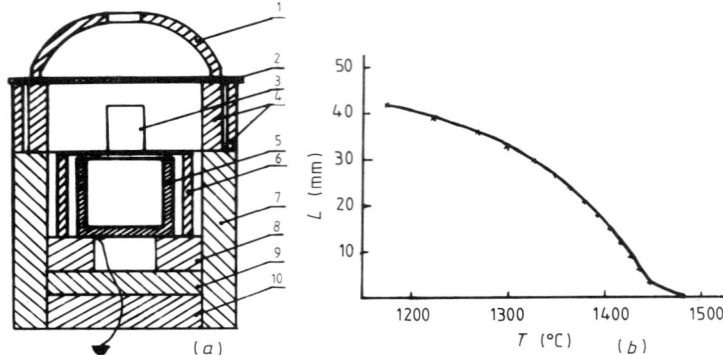

(a)   (b)

Figure 4.37 Crucible for growing BSN single crystals (a) and axial distribution of temperature (b): 1, ceramic lid; 2, platinum disc; 3, sight hole; 4, ceramic rings; 5, platinum crucible; 6, 7, ceramic tubes; 8, 9, ceramic discs; 10, refractory stand.

It should be noted that the indicated experimental values of fluctuations do not agree with the results reported in [64]. However, close values (up to 0.2 °C) were obtained for these quantities in the calculations using the curves presented in [64] for the dependences of the corresponding temperature fluctuations on the diameter of the opening in the upper lid of the thermal packing (figure 4.38).

In the method described, the seed crystal is fixed by a platinum wire to a massive platinum rod. Such a crystal holder is necessary for a good heat removal through the seed crystal in the process of growth, since the controlled

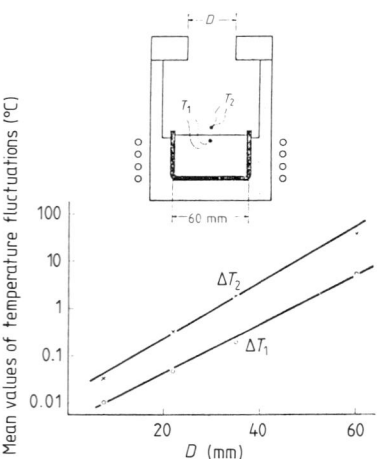

Figure 4.38 Dependences of melt temperature fluctuations ΔT_1 and of gas temperature fluctuations ΔT_2 above the melt on the hole diameter D in the heat-insulating crucible lid [64].

growth is hindered by the low heat conductivity of BSN. Seed crystals are cut in the form of $2 \times 2 \times 25$ mm^3 parallelepipeds. The declination of the growth axis from the tetragonal axis ($< 1°$) is under X-ray control.

The morphological properties of BSN solid solutions do not allow a smooth narrowing of a seed crystal or, therefore, the removal of the inner and surface defects of the seed. To minimize the defects inherited by the growing crystal from the seed, the latter is cut out of the most perfect portion of the crystal boule. The surface defects occurring in the crystal cutting are removed by etching of the seed in a polishing etchant, for which concentrated sulphuric acid is used at a boiling temperature of 300 °C. To remove a damaged 15 μm layer of the seed, it suffices to hold the seed crystal in the acid for 30 min.

Before the growth process, the melt is maintained at a level of 3–5 mm from the crucible edge. When rather large BSN crystals are grown in small-sized crucibles, the melt level is significantly lowered, which leads to a change in the temperature conditions of crystallization and therefore to a change in the crystal composition along the diameter. To eliminate the indicated negative effects, larger crucibles should be used.

Besides, an increase of the melt mass (over 1.5 kg) raises the thermal persistence of the melt and stabilizes the temperature.

It should be noted that a repeated use of one and the same melt with additions of new portions of starting material is undesirable, since this may cause uncontrollable variations in the composition of crystals grown from incongruent melts.

The specific features of the growth of BSN single crystals include persistence in retaining their shape and diameter within a certain temperature interval. For example, a slight lowering of the melt temperature does not result for some time in a change in the diameter of the growing crystal, but changes sharply with a significant change of the temperature regime (up to several degrees). The diameter depends on the axial temperature gradient above the melt, on the gradient in the melt and on the degree of overcooling. Crystal expansion is a controllable process. The crystal diameter may be slowly extended up to a desirable size through a sufficiently slow melt temperature lowering. A typical pattern of BSN crystal expansion is presented in figure 4.39.

Obtaining long BSN single crystals of a given diameter requires a continuous melt temperature lowering. Otherwise the growing crystal will separate from the melt surface. For each crystal diameter there exists an optimal rate of melt temperature lowering which is chosen empirically in each particular case with allowance made for temperature gradients, pulling and rotation rates, and the diameter of a pulled-out crystal boule.

The effects described obviously originate in the melt overcooling, which for BSN solid solutions reaches several tens of degrees, and also in the crystal tendency to faceting in the process of growth. The results were reproducible for crystals measuring 35 mm in diameter and 50 mm in length.

(*a*)

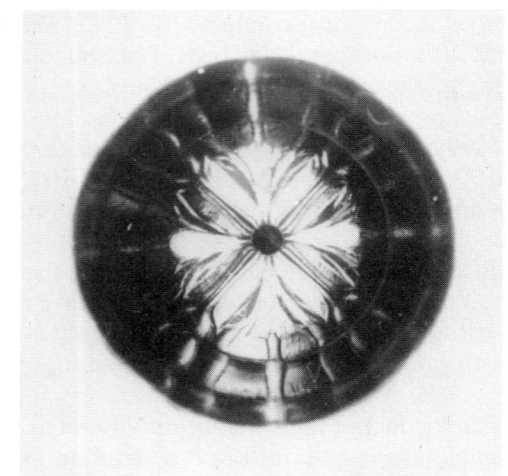

(*b*)

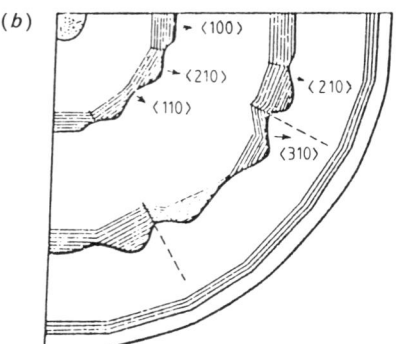

Figure 4.39   Photograph of the upper portion of a crystalline BSN boule
with morphological peculiarities appearing under boule expansion (*a*)
and schematic picture of the expansion zone of the crystal (*b*) [64].
The crystallographic indices indicate face growth directions.

The morphology of the side surface of the crystals depends on the direction
of growth. Crystals grown along the [001] direction have 24 facets of four
prisms, $\{100\}$, $\{110\}$, $\{120\}$ and $\{130\}$; for the composition $x = 0.25$ all the
faces develop approximately equally (figure 4.40). Pulling BSN crystals in the
[110] direction leads to a pear-like shape with clearly pronounced faceting
(figure 4.40). The growth of crystals with such a shape is unstable due to a
high anisotropy of the growth rates of different faces. They are very sensitive to
the slightest violations of the temperature conditions. Even very small changes
in the melt temperature result in an uncontrollable growth of the crystal boule.

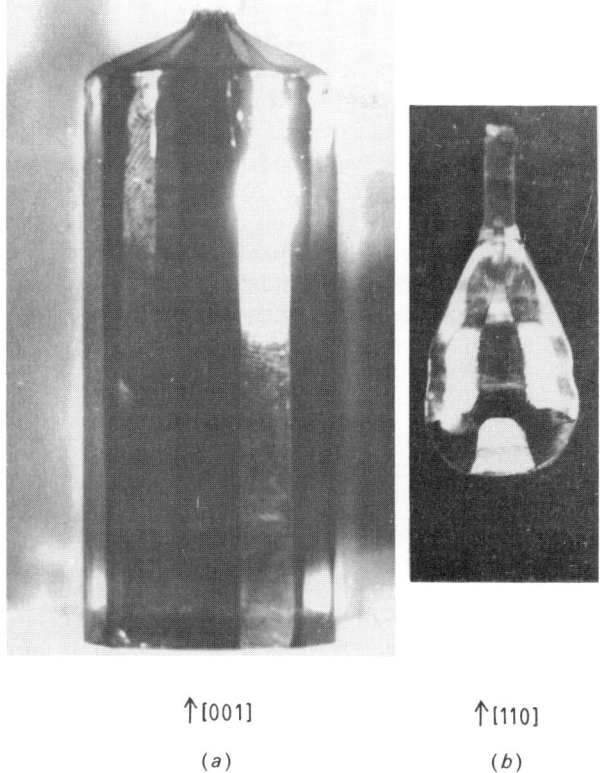

↑[001]   ↑[110]

(a)   (b)

Figure 4.40   BSN crystals grown along the [001] and [110] directions.

Variation of the growth regimes, in particular the temperature gradients above the melt, as well as the introduction of some doping impurities, have an effect upon the morphology of BSN crystals. So, a lowering of axial temperature gradients leads to a broadening of {110} facets on the side of the crystals pulled out in the [001] direction, and for sufficiently low values of the temperature gradient the crystal may acquire a square cross section. The crystal shape may be similarly changed by an introduction of 0.5 and 1 mol.% of lanthanum oxide, and the higher the La concentration the more developed the {110} facets. The same amount of yttrium, thulium and cerium introduced to a melt does not affect the crystal morphology. Crystals containing Y and Tu are colourless. Nd-doped crystals acquire a violet colour typical of Nd. Cerium-containing crystals are coloured dark cherry.

4.9.1   *Growth of BSN crystals by the Stepanov method*
This method is known in the literature as the EFG method. Single crystals of BSN : Ce solid solution have recently been reported as suitable for reversible

holographic information recording (see chapter 8). Plate-shaped crystals are most convenient for such devices. Preparation of plates by cutting Czochralski-grown boules leads to considerable losses of expensive material. On the other hand, single crystals of a given cross section can be grown directly from a melt using the Stepanov method, as reported in [75]. The shapers for growing single crystal plates of BSN:Ce were made of 0.5 and 1 mm thick platinum sheets. The lengths of the shapers were 20 and 35 mm; the heights 20 and 25 mm; and the gap widths 0.5 and 1 mm. The shapers were either placed on the bottom of a platinum crucible or fixed on a ceramic ring independently of the crucible. The axial temperature gradient was measured in the absence of the crystal in the usual manner. The growth conditions for BSN:Ce single crystals for the Czochralski and Stepanov methods are given in table 4.13.

In growing large BSN:Ce crystals by the Czochralski method, the melt level falls considerably, which affects greatly the thermal conditions at the crystallization front. For example, as the melt level in the crucible falls by 20 mm, the axial temperature gradient above the melt decreases by $25-30°C$ cm^{-1} for a crucible measuring 80 mm in diameter. The radial temperature gradient remains practically unchanged. With an increase in the crystal length, the heat removal from the crystallization front decreases due to the small thermal conductivity of BSN. For all these reasons the shape of the crystallization front changes from convex for an expanding crystal to concave as it separates from the melt. As the melt level in a crucible falls to a certain critical value h, the crystal growth becomes unstable. A small variation in the power applied to the heater leads either to a crystal separation from the

Table 4.13 Conditions for growth of BSN:Ce single crystals by the Czochralski and Stepanov methods.

Method of growth	Axial temperature gradient above the melt (grad cm^{-1})	Direction of growth	Pulling rate (mm h^{-1})	Revolution rate (rpm)	Crystal size (mm)
Czochralski method	85	[001]	6–10	10–25	Length up to 60; diameter up to 35
Stepanov method	85–120	[001]	15–30	—	Length up to 110; width up to 35; thickness up to 1.5–2.5

melt or to a sharp crystal expansion. In our experiments, the value h was 25–30 mm (for a crucible diameter of 80 mm and a crystal diameter of 35 mm).

In the growth of BSN : Ce single-crystal plates by the Stepanov method, the crystallization front is at the level determined by the position of the upper edge of the shaper. The axial temperature gradient does not practically change in the process of growth; it is unaffected by the melt level variation when the shaper position is fixed. Therefore, one can choose such thermal conditions that the growth of a BSN : Ce single-crystal plate will proceed with a constant power applied to the heater, except in the period of seeding and expansion of the crystal. Figure 4.41 shows the graphs of variation of the power applied to the heater in BSN : Ce crystal growth by the Czochralski and Stepanov methods. A stationary growth is seen to proceed with a constant power applied to the heater, whereas to maintain a constant diameter of a single-crystal boule the power should be lowered considerably. It should be noted that in the Stepanov method the crystal pulling rate is two–three times higher than that in the Czochralski method. It is known that the 'rotational' growth striation is absent in Stepanov-grown crystals. Besides, the use of a shaper fixed relative to the crucible significantly reduces the melt temperature variations at the crystallization front.

However, Stepanov-grown BSN crystal plates sometimes exhibited striations due to faceting in the process of growth, which is typical of BSN crystals.

Microscopical examination of the edges revealed a step-like structure. Large steps consisted of a series of small ones. The height of the large steps was about 6 μm, the width $\sim$280–310 μm, and the height of the small steps ranged between 0.5 and 2.5 μm, their width being approximately equal to the height. The mechanism of the appearance of steps on a plate face was proposed in [75]. As a result of anisotropy of the growth rates of different faces, the liquid–solid interface declines from horizontal. In this region the melt rises. On reaching the critical height, the melt is rapidly crystallized.

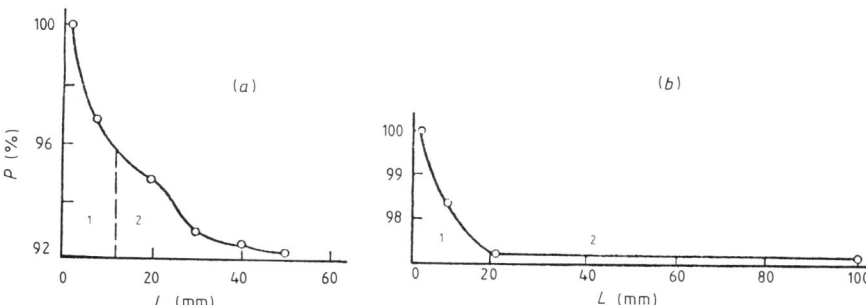

Figure 4.41 Variation of power applied to a heater of crystallization set-up in growing single crystals by the Czochralski (*a*) and Stepanov (*b*) methods: 1, crystal expansion region; 2, crystal stationary growth region [75].

(a)
(b)

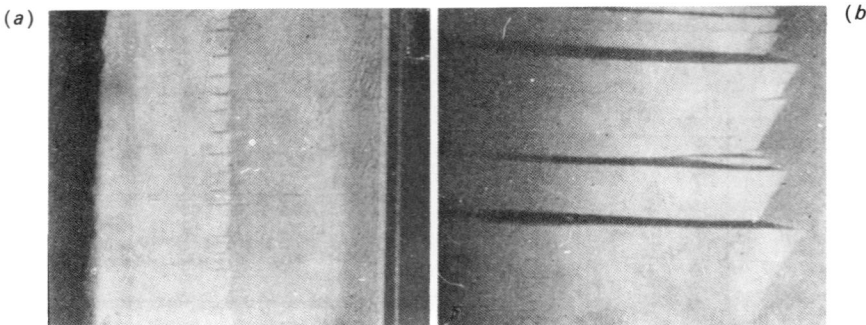

Figure 4.42   Development of a step-like face on the surface of a plate grown by the Stepanov method: (*a*) top of the plate; (*b*) middle part of the plate (× 24) [75].

This results in step formation on the surface of such a face. Figure 4.42 shows the kinetics of the evolution of a step-like face. After an optical treatment of a plate, the region of a step-like face exhibits striae with a period equal to the width of large steps, while in the region of a smooth face no striae are observed. To eliminate such imperfections, single-crystal plates of BSN : Ce should be grown such that their side surfaces coincide with the crystallographic planes with simple indices $\{h, k, 0\}$. Polarization–optical studies of Stepanov- and Czochralski-grown BSN : Ce crystals revealed a higher optical homogeneity of Stepanov-grown samples.

The homogeneity of the chemical composition of BSN : Ce crystals along the growth axis was verified by an X-ray micro-analyser. The Ba/Nb and Sr/Nb variations remained the same (within experimental error, $\pm 0.4\%$) in both cases. Thus the Stepanov method shows prospects in growing BSN : Ce single crystals.

4.10   Growth defects

One of the characteristic types of optical inhomogeneity in BSN crystals is striations associated with the variations of crystal composition upon fluctuations of instantaneous growth rate due to thermal variations at the crystallization front.

The shape of growth striae in a crystal is directly associated with the shape of the crystallization front. If the front is only slightly different from flat, the regular striae in the crystal are practically parallel to one another over the entire crystal cross section and are parallel to the front plane. If the front is strongly convex or concave, the growth striae may form a system of embedded cones with vertices in the centre of the crystal and the opening angle up to

170°. When a BSN crystal is pulled along the [001] axis and the crystallization front is convex towards the melt, one observes formation of (001) facets in the centre of the interface. When the front is concave, {001} facets are formed near the side surfaces of the growing crystal. In the investigated BSN crystals with $x = 0.25$ and 0.50, the distance between regular striae ranged between 8 and 80 μm. According to the results of optical studies [19], in broad growth striae (≈ 80 μm) the variation of the refractive index Δn was approximately equal to 5×10^{-6}, which corresponds to a variation of rare earth cation content within a stria of not less than 0.01 mol.%, whereas the composition variation within narrow striae reached 0.07 mol.%.

The variations of thermal conditions responsible for striations in the Czochralski method can be grouped as follows: (i) 'regulation' variations, which are due to imperfections in the regulation (control) and in the thermal insulation of the growth system; (ii) 'convective' variations, which are due to melt convection and partially to radiation and thermal conductivity of the system; (iii) temperature variations at the crystallization front due to crystal rotation in an asymmetric thermal field, pulling rate variations, vibration etc.

Sharp temperature variations, for example in the course of crystal expansion, may result in substantial variations of the basic composition on 'regulation' growth striae. It was shown [76] that in the region of a clearly pronounced 'regulation' striation the Ba/Sr ratio in a BSN crystal changes by several per cent, whereas the niobium content in the striae region remains unchanged to within 0.1%, i.e. within experimental error. Thus in this area of the diagram of the ternary system $BaO-SrO-Nb_2O_5$, the variable components are BaO and SrO.

'Regulation' oscillations of temperature are eliminated by improving the thermal insulation of a growth system and increasing the temperature maintenance accuracy on the crucible wall. The temperature feedback control is preferable to power feedback control because it takes into account the resultant change of external conditions. To obtain optically perfect BSN crystals, the 'regulation' variations should not exceed 0.1–0.2°C, as shown by experiment.

There exist two ways of eliminating convective fluctuations: either the temperature gradient in the melt [77] or the melt height in the crucible [19] should be reduced. The former way leads to a reduction of admissible growth rate and crystallization instability. According to the literature, BSN crystal growth can be realized within the range of temperature gradients from 10 to 100°C cm^{-1}, the most typical gradient being equal to 40°C cm^{-1}. To lower the gradients, various thermal screens are employed [78, 79]. Although the lowering of the melt height reduces convective flows, it amplifies segregation phenomena in a crystal (the striation contrast increases). Convective fluctuations in a melt can also be suppressed using a platinum screen mounted in it at a certain optimal distance h, where $0.9L < h < 0.5L$ (L is the melt depth in the crucible). In the latter case, the temperature oscillations of the melt do not exceed 1 °C, and BSN crystals exhibit no regular striations. But such a screen is very inconvenient in practice. Another way to suppress convective fluctuations was proposed [76] which enabled temperature fluctuations in the melt to be lowered to 0.04–0.05 °C under optimal conditions. Using appropriate techniques, one can go from turbulent convection to periodic oscillations and then to a completely suppressed convection in the melt.

Simultaneous measurements of temperature fluctuations at different points of the melt during crystal growth [64] demonstrated that the amplitude of temperature fluctuations in a melt is small compared with that in a gas adjoining the melt and the growing crystal. The authors concluded that the temperature fluctuation intensity in the melt depends strongly on the temperature fluctuations of the surrounding medium, and that the melt temperature can be levelled by diminishing temperature fluctuations of the surrounding gas. If these fluctuations do not exceed ± 0.25 °C then, even growing from a non-congruent melt, the crystal does not exhibit striations. To weaken air convective flows above the melt, the vision slit in the heater is closed by quartz glass and the heater itself is covered by platinum and alundum lids.

In Czochralski-grown crystals, rotational striations may occur due to thermal field asymmetry in the melt or if the crystal rotation axis does not coincide with the symmetry axis of the thermal field. If the crucible is motionless, the distance between striae is equal to the pulling-to-rotation rate ratio v/ω and remains unchanged for constant v and ω. The shape of striations is associated with the shape of the crystallization front. Striae disappear for $\omega = 0$ [51, 68, 80]. But in this case the growth becomes extremely unstable; a bend appears in the crystal which affects the diameter size. In such regions strong mechanical stresses appear, worsening the optical quality and inducing cracking. Furthermore, rotationless crystal pulling requires a very high purity of starting materials, and elimination of any temperature fluctuations and of radial asymmetry in the thermal gradient in the melt-filled crucible. If all these requirements are met, one obtains striationless BSN crystals up to 10 mm in diameter and 50 mm in length [76]. Another way to eliminate rotational striations is to increase the crystal rotation rate [81]. An optimal choice of growth conditions gives a flat crystallization front at rotation rates between 50 and 60 rpm, and for crystals not exceeding 10–12 mm in diameter the rotation rate equals 120 rpm. The diameter of the pulled crystal remains unchanged over rather wide temperature limits (up to 10 °C), which is due to rigid faceting.

An attempt was made [80] to explain striations in BSN crystals by the presence of K and Ca impurity ions whose content in the melt makes up hundredths and thousandths of a per cent, since it was established using electronic probes that the distribution of the basic elements Ba, Sr, Nb and O over the crystal is uniform within experimental error, while the distribution of alkaline and alkaline earth impurities has a periodic nature. However, the crystals grown from a starting material with K and Ca ion impurities not exceeding respectively 0.0004 and 0.0001% were not completely free of striations. Besides, such a concentration of K and Ca ions could not account for a rather large difference in the value of the refractive index observed within a stria.

The change of striation with crystal expansion is illustrated in figure 4.43. At an early stage of growth, the direction of the stria is parallel to the surface of the growing crystal, but as it expands the striae become curved and along them there occur large strains. These boundaries are places of the appearance of dislocations which have the shape of radial straight lines. It is supposed that there exist two mechanisms of the appearance of radial dislocations on expansion striations. One of them is due to local strains caused by lattice displacement which results from a large temperature gradient on the interface. The other is due to thermal strains occurring upon cooling in a thermal field with a radial temperature gradient. It remains unclear, however, which of the two mechanisms prevails in the generation of dislocations.

In single crystals grown under stable conditions, dislocations are as a rule rectilinear and are inherited by the growing crystal. This leads to the fact that dislocations inclined to the growth axis of the crystal gradually come onto its side surface. Dislocations parallel to the growth direction are traced over the entire length of the single crystal.

The generation of dislocations can be controlled and their density in the grown crystals can be reduced to $10-10^{-2}$ cm^{-2} by using perfect seed crystals, lowering the temperature gradients down to $10°C$ cm^{-1} and avoiding the heating of the platinum crucible walls.

To decrease striations in already grown crystals, a long-term high-temperature annealing is recommended [82, 83]. But upon a long-term post-growth annealing at temperatures below $1340°C$, BSN crystals may exhibit optical inhomogeneities due to transition from a tetragonal to a monoclinic phase, for which reason such a possibility should always be taken into account.

Figure 4.43 Schematic picture of growth striae in the BSN crystal cross-section parallel to the growth axis [64].

A multi-step post-growth annealing of single crystals of BSN solid solutions is recommended. At the first stage, the diffusive annealing is carried out directly in the growth chamber immediately after the crystal separation from the melt surface without lowering the temperature of the melt-filled crucible. Annealing lasts for 15 h. It enables the stria contrast to be lowered and the optical homogeneity of the crystal to be thus increased. Then the temperature in the annealing region is sharply reduced to 1000 °C, i.e. the crystal is quenched. It prevents the appearance of the second phase in the crystal. A subsequent gradual annealing down to 150 °C for 24 h is realized by a slow decrease of the generator power, after which the set-up is switched off and the crystal is cooled down to room temperature in the process of inertial cooling of the whole crystallization system. After such a multi-step annealing, large BSN crystals are usually dark blue because niobium has gone over to a tetravalent state due to a shortage of oxygen in the growth chamber atmosphere. For the sake of decolouring, the crystals are additionally oxygen annealed at a temperature of 800–900 °C for 32 h (including heating and cooling). The annealed crystals are clear and decoloured.

Whereas growth striation is responsible for optical inhomogeneity along the growth axis, the formation of a 'cord' (a channel with a distinct chemical composition along the growth axis) induces optical inhomogeneity in the perpendicular direction. In this case, birefringence inhomogeneity occurs in the central part of the boule even if the crystal is grown from a congruent melt and the crystallization front is convex towards the melt.

The region with a sharp change of the refractive index in the central area of the crystal, the so-called growth column, is a very frequent defect in BSN crystals which is hard to remove. The appearance of a 'growth column' in BSN crystals is first of all due to seed defects and thermal stresses along the seed boundary during seeding. The boundary between the 'growth column' and the bulk crystal is a very high dislocation density area (figure 4.44). Mechanical stresses in the 'growth column' region and in crystals with such a defect are appreciable [23].

Figure 4.45 (curve 1) presents the stress distribution in the (001) plane over the diameter of the plate depicted in figure 4.44(a). The extremes of the stresses correspond to the 'growth column' boundaries. The character of stress distribution outside the 'growth column' corresponds to that in the inclusion region. The distribution of stresses in the region of a single dislocation or a successive series of boundary dislocations in the direction perpendicular to the plane of sliding is characterized by contraction on one side of the plane of sliding and expansion on the other side. When dislocations form a ring, which is the case in the 'growth column' region (figure 4.44(a)), the distribution of stresses characterizes the 'growth column' as an inclusion. Alternation of expansions and contractions at the 'growth column' boundary often induces crystal cracking along the growth axis. The distribution and magnitude of stresses in BSN crystals free of 'growth column'

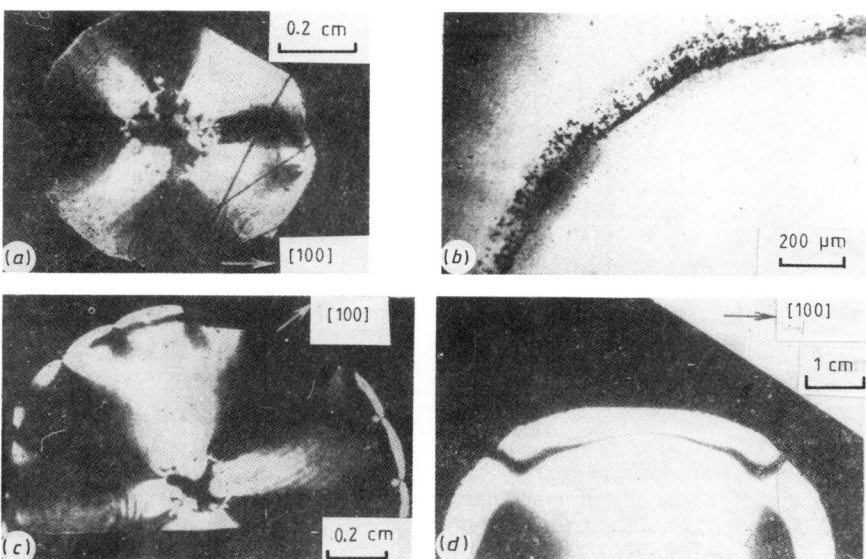

Figure 4.44 Photographs of transverse cuts of BSN crystals with the 'growth column' in polarized light (*a*); dislocation clusters at the boundary of the 'growth column' (*b*) [23].

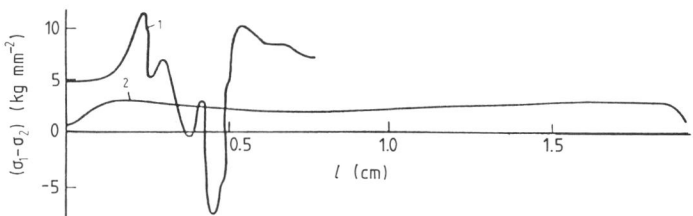

Figure 4.45 Stresses in the (001) plane of a BSN crystal. Distribution of stresses over the diameter of the crystal plate (1) with the 'growth column' and (2) without the 'growth column' [23].

are represented in figure 4.45 (curve 2). The character of the distribution of tensile stresses points to their quenching origin. The magnitude and nature of the distribution of stresses are different for different crystal cross sections perpendicular to the growth axis. Figures 4.46(*a*), (*b*) present the distribution of stresses for different cross sections of a crystal. The highest absolute values of the stresses are observed in the places of largest crystal expansion. The character of their distribution indicates that, in quenching and expansion, the plastic strain area is immediately next to the crystallization front. The concentration of stresses in the expansion cone causes additional dislocations,

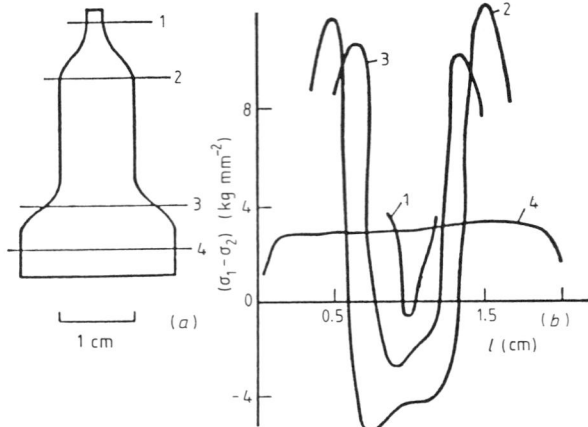

Figure 4.46 Distribution of stresses in various cross sections of a BSN crystal: (*a*) the appearance of the crystal, digits number plates cut out in this cross section; (*b*) distribution of stresses over the diameter of corresponding plates [23].

and the presence of expanding and contracting forces may be responsible for crystal cracking along the growth axis even in the absence of a 'growth column'. The magnitude of stresses in the expansion cone is a function of the cone opening angle, and as this angle decreases, so does the stress concentration.

Pyramid facets {201} and {111} may appear at low temperature gradients on the periphery of the crystallization front, which is either flat or slightly convex towards the melt. The places of separation of these facets are regions of stress concentration whose magnitude and distribution depend on the extent to which the prism facets {100} or {110} are pronounced. The character of stresses associated with faceting of BSN crystals is reproduced in figure 4.44(*c*), (*d*). Stresses at side facets lead to dislocation density increase on the crystal periphery and to cracks whose plane is parallel to the side surface of the crystal (figure 4.44(*c*)).

In BSN crystals having no 'growth column', the stresses in the (001) plane remain constant practically throughout the entire cross section.

4.11   Obtaining a stable single-domain state

BSN crystals are made single domain (polarized) by application of a DC electric field along the ferroelectric axis at temperatures close to the Curie point. Of importance here is the allowance made for remanent unipolarity: if voltage is applied in the direction of unipolarity, polarization proceeds faster and more effectively, without worsening the optical quality of the crystal. The electrode material may affect the magnitude of the half-wave voltage of a sample [7, 45]. Electrodes are customarily made of aquadag, silver-containing paste, vacuum-deposited silver and gold. For this purpose, BSN samples with $x = 0.25$ were used [5, 45] at temperatures ranging from 65 to 150°C in a field equal to 10^4 V cm^{-1} with subsequent cooling below the field. For crystals with a Ba/Sr ratio set at 1, the polarization proceeded in a field of 15 kV cm^{-1} at 100°C [17]. Furuhata [27] achieved polarization of 1 mm thick BSN crystals by applying an electric field of about 20 kV cm^{-1} at room temperature for one minute. The degree of polarization was controlled by the minimum value of the half-wave voltage at room temperature.

To obtain a stable single-domain state, one should note that, for the composition $Ba_{0.25}Sr_{0.75}Nb_2O_6$, room temperature corresponds to a phase transition region. For this reason, the domain structure of a crystal is unstable at this temperature. This explains the unsuccessful attempts to obtain a stable single-domain state by a mere crystal cooling below the field down to room temperature, and by application of a field stronger than coercive at room temperature.

The choice of polarization conditions depends on the crystal use. For example, crystals intended for pyroelectric detectors should be polarized with high voltages, large hold times and high temperature, lest the single-domain state should be broken by short-range overheating through the phase transition temperature. Crystals intended for electro-optic elements should be polarized under softer conditions lest optical inhomogeneities should appear. But the half-wave voltages will not have a minimum value.

The polarization process was investigated [31] on BSN cubic crystals with an edge of 6 mm. Silver electrodes were deposited onto facets perpendicular to the c axis of a crystal. The quality of the electrodes was checked by measuring ε and tan δ of the sample; in the case of a good electrical contact between the sample surface and the electrodes for the composition $x = 0.25$, $\varepsilon = (5-10) \times 10^3$. The electric circuit of the polarizing set-up is given in figure 4.47. The current running through the crystal in the course of

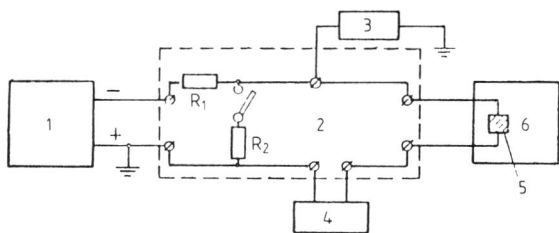

Figure 4.47 Schematic drawing of the set-up for BSN crystal polarization: (1) high-voltage source; (2) voltage divider; (3) kilovoltmeter (3); (4) microammeter; (5) crystal; (6) thermal bath.

polarization was controlled by a photo-compensative microammeter measuring a minimum current of 6×10^{-9} A. In the course of polarization, a sample was placed in a silicon liquid in order to avoid an electric disruption on the surface; the liquid was heated up to approximately 150°C. Completeness of polarization was controlled by half-wave voltage measurements at a frequency of 50 Hz.

The current density variations in the course of polarization of a BSN crystal with $x = 0.25$ in a silicon liquid at 150°C and $E = 5$ kV cm^{-1} are plotted in figure 4.48(a). The current running through the crystal increases exponentially

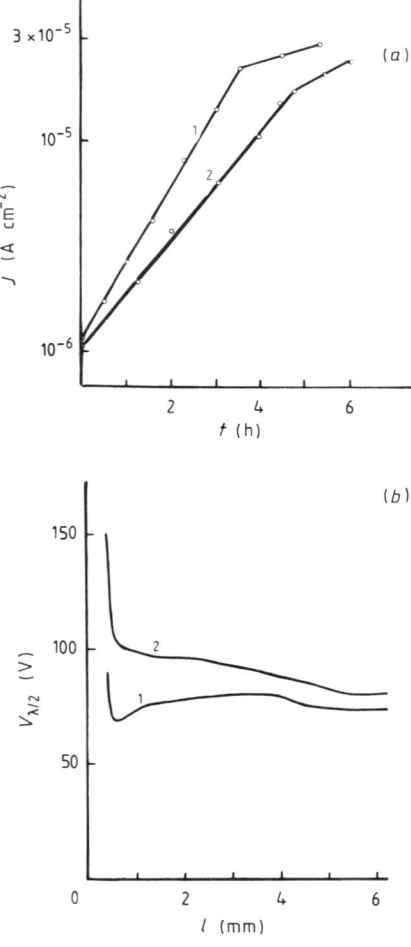

Figure 4.48 Time dependence of current through a BSN crystal of composition $x = 0.25$ under polarization in a silicon liquid (a) and distribution of the half-wave voltage $V_{\lambda/2}$ along the c axis of this crystal (b).

with time (time constant $\tau = 1.0 \times 10^4$ s). The rate of current increase slows down at the final stage of the process. During six hours of polarization, the current density through the sample increases by more than an order of magnitude. A similar dependence for a crystal of the same composition in polarization in the air ($T = 200\,^\circ$C, $E = 5$ kV cm^{-1}) is plotted in figure 4.49, where one can see a slight increase of the current.

An increase of the current through a sample surrounded by a silicon liquid is explained by the fact that it has a restoring effect, i.e. it induces oxygen loss by a crystal. As is known, oxygen vacancies in oxygen octahedral ferroelectrics are donors, and an increase of their concentration and the gradient of the distribution over the crystal leads to an increased stability of the single-domain state. Therefore, BSN polarization in a silicon liquid provides higher crystal unipolarity as compared with polarization in the air at the same temperature. The plots of figures 4.48(b) and 4.49(b) suggest

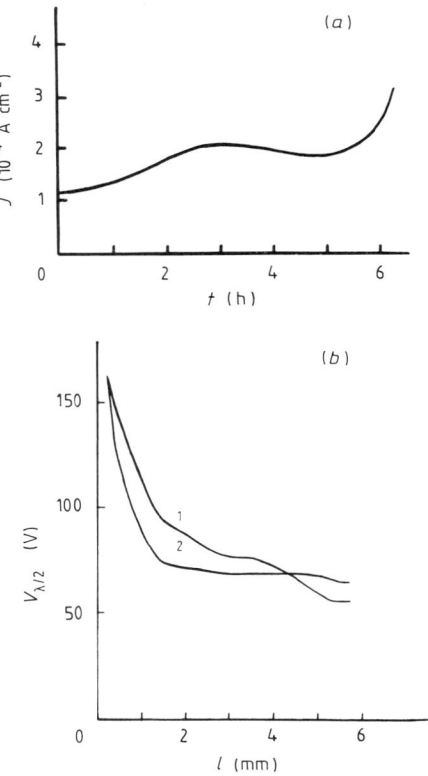

Figure 4.49 Time dependence of current through a BSN crystal of composition $x = 0.25$ under polarization in air (a) and distribution of the half-wave voltage $V_{\lambda/2}$ along the c axis in this crystal (b) (1, immediately after polarization; 2, after aging).

that the samples polarized in a silicon liquid exhibit lower half-wave voltage than those polarized in the air.

Crystals polarized at high temperatures ($\approx 200\,^\circ$C) have in some places a low half-wave voltage. But an analysis of the optical quality of such samples revealed a substantial variation of the refractive index along the c axis and a variation of $V_{\lambda/2}$ in different crystal areas.

Stability of the single-domain state of BSN crystals stored in the dark at room temperature was examined [84] by a short-time heating through the Curie temperature and with exposure to light. Under natural aging conditions, the value of $V_{\lambda/2}$ changes insignificantly over the crystal, except near the positive electrode, where it increases (figure 4.50). The mean value of $V_{\lambda/2}$ makes up 85 V immediately after crystal polarization and 98 V after 480 h. The most significant $V_{\lambda/2}$ changes occur within the first 100 hours, subsequent variations being small.

Stability of short-time overheating was examined with pyroelectric detectors on the basis of BSN crystals with $x = 0.25$. Pulses of a powerful CO_2 laser with $\lambda = 10.6$ μm were detected. In several experiments, the temperature on the surface of pyro-active material reached 330 K.

Investigations [85, 86] revealed that crystals preserve high pyroelectric qualities up to beam destruction.

The action of light was tested on crystals which had undergone natural aging for 60 days after polarization (the light source was a mercury lamp).

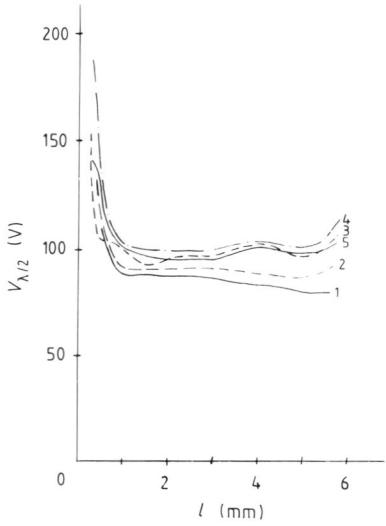

Figure 4.50 Distribution of the half-wave voltage $V_{\lambda/2}$ along the c axis of a BSN crystal with $x = 0.25$: 1, immediately after polarization; 2, after 22 h; 3, after 66 h; 4, after 136 h; 5, after 480 h under conditions of natural aging [84].

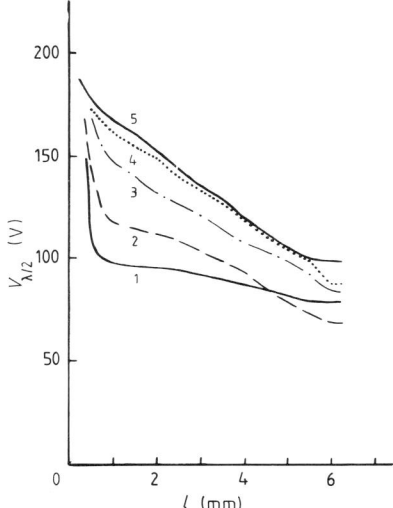

Figure 4.51 Light-induced distribution of $V_{\lambda/2}$ along the c axis of a BSN crystal: 1, corresponds to sample measurements immediately after polarization; 2, within 1, 3–4, 4–11 and 5–18 h of exposure [84].

The half-wave voltage along the c axis of the crystal versus light doses is plotted in figure 4.51. The graph implies that exposure to light leads to an increase of $V_{\lambda/2}$ and increases its inhomogeneity along the c axis of the sample. Such an inhomogeneous aging is evidently due to the non-uniform impurity distribution arising under polarization. Exposure to light leads to the appearance of non-equilibrium carriers which may cause destabilization of the single-domain state [85].

Thus, an appropriately chosen thermoelectric treatment may give a single-domain state of BSN crystals which is stable to natural aging and short-time overheating above the Curie temperature.

The basic properties of barium strontium niobate crystals are listed in table 4.14.

Table 4.14 Basic properties of BSN crystals.

Property	Experimental data	References
Chemical formula	$Ba_xSr_{1-x}Nb_2O_6$	
	$0.16 \leqslant x \leqslant 0.75$	
Melting temperature ($^\circ$C)	1470–1490	[11]
Molecular weight ($x = 0.25$)	381, 858	—
Density of single crystals (g cm^{-3})	5.4	[72]
	5.1 ± 1	[6]

cont.

Table 4.14   *Continued.*

Property	Experimental data		References
Hardness (Mohs scale)	5.5		[72]
Class of symmetry	4mm		[72]
Space group	P4bm		[6]
Parameters of the cell ($T = 298$ K)			
a (Å)	$12.430\,24 \pm 0.000\,02$		[6]
c (Å)	$3.913\,41 \pm 0.000\,01$		[6]
Number of formula units in a cell	5		[6]
Curie temperature (°C)	$x = 0.25$	50	[72]
	$x = 0.50$	130	
	$x = 0.75$	200	
Refractive indices ($\lambda = 6328$ Å)	$x = 0.25$	2.3117	[39]
n_o	$x = 0.50$	2.3123	
	$x = 0.75$	2.3144	
n_e	$x = 0.25$	2.2987	
	$x = 0.50$	2.2734	
	$x = 0.75$	2.2596	
Birefringence Δn ($\lambda = 6328$ Å)	$x = 0.25$	-0.013	[39]
	$x = 0.50$	-0.039	
	$x = 0.75$	-0.055	
Domain of transparency (μm)	0.3–6		[38, 72]
Electro-optic coefficients			
$[(n_e^3/n_o^3)r_{33}-r_{13}]$ cm (stat V)$^{-1}$			[1]
$v = 1$ Hz	$x = 0.25$	4.15×10^{-5}	
	$x = 0.50$	6.15×10^{-6}	
	$x = 0.75$		
$v = 1$ MHz	$x = 0.25$	1.92×10^{-5}	
	$x = 0.50$	2.28×10^{-6}	
	$x = 0.75$	1.14×10^{-6}	
$v = 15$ MHz	$x = 0.25$	3.2×10^{-5}	
	$x = 0.50$	2.7×10^{-6}	
	$x = 0.75$	1.24×10^{-6}	
Half-wave voltage $V_{\lambda/2}$ (V)	$x = 0.25$	48	[1]
	$x = 0.50$	580	
	$x = 0.75$	1236	
Loss tangent tan δ			
electric field $\parallel c$ axis	0.23		[38]
electric field $\parallel a$ axis	0.04		
Dielectric permittivity ε ($T = T_{room}$)	$x = 0.25$	3400	[1]
	$x = 0.50$	450	
	$x = 0.75$	118	
Electric conductivity, σ (Ω^{-1} cm^{-1})	10^{-13}–10^{-14}		[36]
Spontaneous polarization P_s (μC cm^{-2})	$x = 0.25$	17	[33]
($T = T_{room}$)	$x = 0.50$	25	
Coercive field E_c (kV cm^{-1}) ($T = T_{room}$)	$x = 0.25$	2.3	[33]

Table 4.14 *Continued.*

Property	Experimental data		References
Pyroelectric coefficients γ (μC cm^{-2} K^{-1})	$x = 0.27$	0.28	[2]
	$x = 0.52$	0.065	
	$x = 0.75$	0.030	
Latent heat of fusion Q_m (kcal g^{-1})	120		[87]
Specific heat C_p (kcal g^{-1} K^{-1})	0.12		[87]
Heat conductivity K (W cm^{-1} K^{-1})	0.008		[87]
	(1370–1470°C)		
	0.006 (T_{room})		
Heat transfer coefficient (W cm^{-1} K^{-1})	0.10		[87]
Thermal expansion coefficient α			
(10^{-6} grad^{-1})			
$\alpha \parallel a$ axis	10		[88]
$\alpha \parallel c$ axis	9 (200–900°C)		[87]
Coefficients of electromechanical coupling			
K_{31}	0.137		[63]
K_{33}	0.48		
K_p	0.24		
Coefficients of elastic compliance			
(10^{-12} m^2 H^{-1}) at constant field			
$S_{11}^{E,T}$	5.1		[63]
$S_{12}^{E,T}$	-1.75		
$S_{13}^{E,T}$	-2.25		
$S_{33}^{E,T}$	10.8		
at constant charge			
$S_{33}^{D,T}$	8.41		[63]
$S_{11}^{D,T}$	5.0		
$S_{12}^{D,T}$	-1.88		
$S_{13}^{D,T}$	-1.76		
Coefficients of elastic rigidity (10^{11} H m^{-2})			
at a constant charge $c_{33}^{D,T}$	1.56		[63]
at a constant field $c_{11}^{E,T}$	2.79		
$c_{12}^{E,T}$	1.34		
$c_{13}^{E,T}$	0.86		
$c_{33}^{E,T}$	1.29		
Piezoelectric coefficients (10^{12} C N^{-1})			
d_{31}^T	-18.0		[63]
d_{33}^T	95.0		
Piezoelectric coefficients (C m^{-2})			
e_{31}^T	0.736		
e_{33}^T	9.16		

5 Barium Sodium Niobate Single Crystals

First reports on the new ferroelectric material barium sodium niobate appeared in the literature in 1967 [1, 2]. These crystals possess considerable advantage over the known nonlinear crystals. They are resistant to UV radiation [3] and exhibit no optically induced inhomogeneities of the refractive index that take place in $LiNbO_3$ and $LiTaO_3$. The stability under UV radiation is assumed to be due to the barium sodium niobate structure [4].

Stability under intense laser radiation and high nonlinear coefficients made it possible to double the frequency of powerful continuous IR laser radiation more effectively on these crystals than on $LiNbO_3$ [5–7].

Thus the combination of extremely high ferroelectric, piezoelectric, optical, electro-optic, nonlinear and elastic properties of barium sodium niobate enables these crystals to be applied both in laser communication systems and in other fields of science and technology [8–16].

5.1 Structure and phase transitions

The ferroelectric $Ba_2NaNb_5O_{15}$ (BNN) has a tetragonal potassium tungsten bronze (TPTB)-type structure similar to the BSN structure (see chapter 4). In this structure, slightly distorted NbO_6 octahedra are linked with their vertices and form quasi-linear chains parallel to the [001] direction (figure 5.1). In the perpendicular direction, there also exist octahedra chains, not linear but saw-like ones (octahedra rings). The spaces between both types of chains form trigonal (C), tetragonal (A1) and pentagonal (A2) voids. The number of occupied A and C positions is determined by the chemical nature of cations and meets the requirement of electroneutrality.

Barium sodium niobate has an occupied TPTB-type structure. The term 'occupied' implies that all the positions A1 and A2 are occupied with cations, while the positions C are vacant. In a 'completely occupied' TPTB structure,

C positions are occupied. An example of a substance with a completely occupied structure may be a relative compound $K_3Li_2Nb_5O_{15}$ (chapter 6). The general formula of the compounds with a TPTB-type crystal structure has the form $(A1)_2(A2)_4C_4(B1)_2(B2)_8O_{30}$.

In a tetragonal unit cell (figure 5.2) of $Ba_2NaNb_5O_{15}$, four A2 positions are occupied with Ba^{2+} cations (ionic radius 1.35 Å), two A1 positions with Na^+ ions (ionic radius 0.95 Å), and B1 and B2 positions are occupied with Nb^{5+} cations [9]. Thus an ideal tetragonal unit cell of barium sodium niobate must have two formula units, which corresponds to $Ba_4Na_2Nb_{10}O_{30}$. But the chemical and diffraction studies of a crystal grown from a stoichiometric melt [18] suggest that the best approximation is the formula $Ba_{4.13}Na_{1.74}$ $\square_{0.13}Nb_{10}O_{30}$, where $\square$ are sodium ion vacancies. A gravimetric analysis carried out on samples of three different crystals has shown that the ratios Ba/Nb and Na/Nb lie within the limits (4.2–4.4)/10 and (1.5–1.7)/10 respectively.

Above the Curie temperature, the centrosymmetric tetragonal phase (point group 4/mmm) is stable [19, 20]. On passing through the Curie point (585°C), a crystal undergoes a ferroelectric phase transition to a non-centrosymmetric tetragonal phase with the symmetry 4mm (space group P4bm). During further cooling to 260°C, a non-ferroelectric phase transition occurs in the course of which the tetragonal structure undergoes slight orthorhombic distortion, and the crystal symmetry becomes not 4mm but mm2. In this transition, the diagonal of a tetragonal cell becomes an edge a (or b) of the orthorhombic cell which contains four formula units [9]. Such a transition is usually called a TO or an OT transition and is denoted

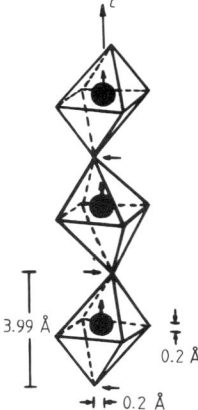

3.99 Å

0.2 Å

0.2 Å

Figure 5.1 Chains of NbO_6 octahedra in the structure $Ba_2NaNb_5O_{15}$ [17]. Vertical arrows indicate niobium ion displacements in the ferroelectric phase; horizontal arrows show displacements of NbO_6 octahedra (distortions are enlarged).

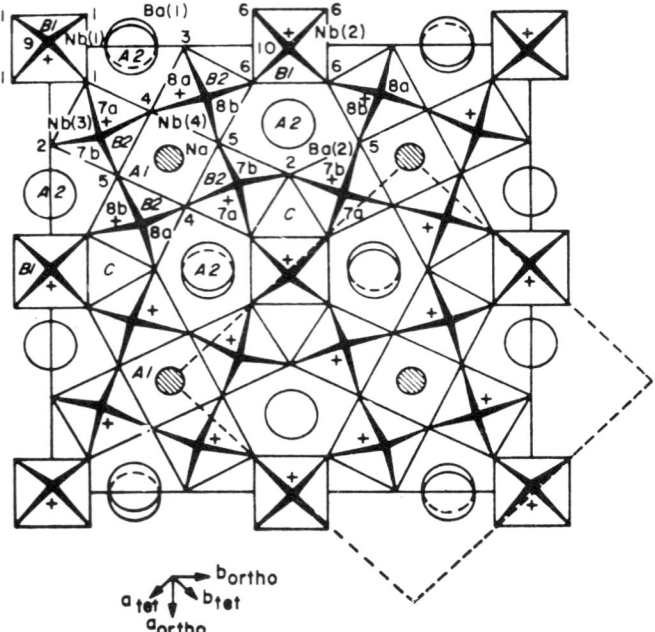

Figure 5.2 Projection of the $Ba_2NaNb_5O_{15}$ structure onto the (001) plane [18]. The tetragonal unit cell is distinguished by broken lines. Positions of oxygen atoms in opposite octahedron vertices are shown by crosses.

(110/110/001). The polar axis in BNN crystals, both in the tetragonal ferroelectric and in the orthorhombic phase, is the c axis, which is in agreement with 4mm and mm2 symmetries. The diagram of BNN structural variations is presented in figure 5.3.

The non-ferroelectric transition at 260 °C is accompanied, due to structure distortion, by micro-twinning of crystals. The transition temperature changes as a function of the composition. So, a crystal of the composition $Ba_{2.085}Na_{0.711}Nb_5O_{15}$ undergoes an OT transition (the same as the ferroelectric transition) at temperatures approximately 20 °C lower than the stoichiometric crystal $Ba_2NaNb_5O_{15}$ [21]. The temperature at which a BNN

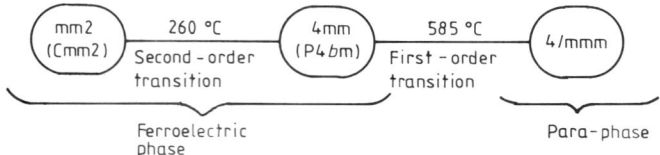

Figure 5.3 Diagram of structural transformations of BNN crystals.

unit cell becomes orthorhombic depends mainly on the Nb/Ba ratio. If this ratio exceeds 2.8, the tetragonal phase may exist even at room temperature. The investigations carried out later on [22] have established that the crystals remain uniaxial tetragonal up to liquid nitrogen temperature when the Nb_2O_5 content increases up to 53 mol.%. A large excess of Nb_2O_5 leads, however, to Nb^{5+} ion implantation into A positions (figure 5.2), which increases light scattering by the crystal.

The dilatometric studies of BNN single crystals [21, 23, 24] revealed the character of temperature variation of the shape of a unit cell of a crystal up to 900 °C (figure 5.4). In the orthorhombic phase, the lattice expansion rate along the *a* axis was found to be higher than that along the *b* axis, and since the spacing along the *b* axis exceeds the one along the *a* axis [25], the orthorhombic lattice distortion decreases with approaching OT transition. Obviously, this accounts for small spatial variations under an OT transition. With a subsequent temperature rise, the cell expands along the *a* and *b* axes and contracts along the *c* axis. Above T_C, the crystal expansion along the *a* and *b* axes proceeds by one and the same law, which reflects the two-dimensional isotropic behaviour of the tetragonal phase, whereas along the *c* axis the temperature-dependent variations are small.

The composition variations inside the BNN homogeneity region affect substantially the lattice parameters [26, 27] and have an effect upon the atom displacement under an OT transition, which is shown by the change in the slope of the expansion curves along the *a* and *b* axes at the phase transition temperature.

The lattice parameter also varies along the *c* axis under an OT transition [21]. The responsibility for these anomalies along the *c* axis under OT

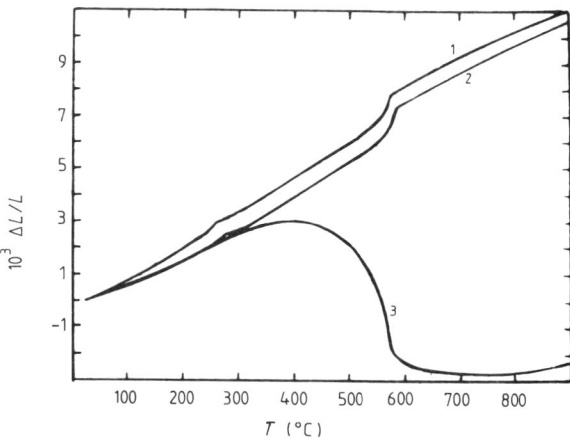

Figure 5.4 Curves of axial expansion of BNN crystals with temperature: (1), *a* axis; (2), *b* axis; (3) *c* axis [24].

transition may lie with the reconstruction of NbO_6 octahedra chains inducing doubling of the parameter c.

Optical analyses have suggested that the transition from the tetragonal to the orthorhombic phase proceeds within a broad temperature interval (200–300 °C), and a BNN crystal changes from uniaxial to biaxial.

Transition smearing and the small variation of the expansion curve slope in the phase transition region allowed Vere *et al* [23] to assume that the non-ferroelectric OT transition in BNN is a second-order transition, which was experimentally confirmed by Scott *et al* [26] on the basis of the results of differential thermal analysis.

The OT transition in BNN is accompanied by elastic anomalies, but does not induce anomalies of the dielectric properties. This phase transition may therefore be identified as a ferroelastic (ferroelectric) transition [28], and barium sodium niobate may be referred to ferroelastics.

Thus, BNN crystals have a room-temperature orthorhombic symmetry (space group Cmm2) [1, 9]. The parameters of an orthorhombic cell at 25 °C are $b = 17.625\,60 \pm 0.000\,05$, $a = 17.591\,82 \pm 0.000\,01$ and $c = 3.994\,915 \pm 0.000\,004$ Å. The room-temperature unit cell volume is 1238.688 Å^3; the crystal density $\rho = 5.407\,60$ g cm^{-3} [18].

The results obtained by Jamieson *et al* (chapter 4, [6]) have shown that the BNN parameter c is twice as large as the value mentioned above. The doubling of c is indicated by the saw-like arrangement of NbO_6 octahedra (figure 5.1). The X-ray studies [19] have confirmed the c doubling. Thus the true value of the parameter c is equal to 7.9898 Å, and a unit orthorhombic cell contains eight formula units.

5.2   Ferroelectric properties

5.2.1   *Dielectric properties*

The temperature dependences of the dielectric permittivity ε and the loss tangent tan δ were conducted on BNN crystals grown from melts of stoichiometric ($Ba_2NaNb_5O_{15}$) and non-stoichiometric ($Ba_{2.085}Na_{0.711}Nb_5O_{15}$) compositions [29].

Measurements were made at a frequency of 1 kHz from room temperature to 60–80 K above the Curie point. The mean heating and cooling rate was ≈ 150 K h^{-1} and fell to 30 K h^{-1} in the vicinity of the ferroelectric transition. Such a heating regime prevented the samples from cracking. The quantities ε and tan δ were calculated by formulae (4.2).

Under heating from room temperature, the ε and tan δ values increased gradually, and no anomalies were observed in the transition from orthorhombic to tetragonal phase. For the samples examined, tan δ reached its maximum near 440 °C and then fell steeply to the minimum at the Curie point (figure 5.5). This dependence has the shape of a curve typical of the 'impurity

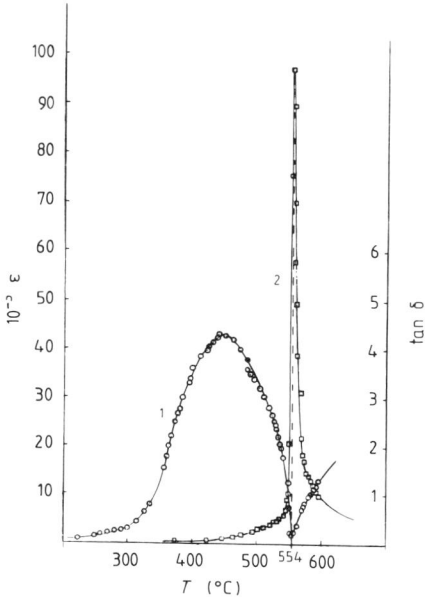

Figure 5.5 Temperature dependences of (1) tan δ, (2) ε for BNN crystals [29].

ion–vacancy' and 'cation vacancy–anion vacancy' losses [30]. The dielectric permittivity ε of specimens grown from stoichiometric melts increased during heating from 40 at room temperature up to $\approx 10^5$ at the Curie point. For specimens grown from non-stoichiometric melts, the ε values were as a rule still higher at the ferroelectric transition point.

The Curie temperature of BNN crystals as a function of the melt composition is shown in figure 5.6 [31, 32]. The Curie temperature of crystals grown from non-stoichiometric melts proved to range from 554 to 568 °C; for stoichiometric melts T_C reached 580 °C.

The inverse dielectric permittivity $1/\varepsilon_z$ of BNN crystals is a linear function of temperature in the interval ranging between 100 K above and below the Curie point. For them the Curie–Weiss law $\varepsilon_z = C_0/(T - T_C)$ is obeyed above the Curie point [29, 33, 34]. The slope ratio C_0/C_1, where C_0 is the Curie–Weiss constant and C_1 the inverse value of the slope $(d\varepsilon_z/dT)^{-1}$ for $T < T_C$, is in the range 6–7, implying that this is a first-order transition [35]. The temperature dependences $1/\varepsilon$ for crystals obtained from stoichiometric and non-stoichiometric melts are plotted in figure 5.7.

It should be noted, that, under a slow crystal cooling from high temperatures, the Curie point falls by 4–5 K, while the dielectric permittivity slightly increases. Temperature hysteresis was also observed for a cooling rate of about 0.1 K min^{-1}, when the crystal was close to thermal equilibrium [34].

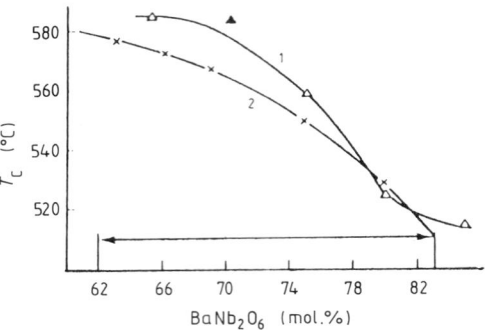

Figure 5.6 Dependence of the Curie temperature of BNN crystals on the melt composition according to (1) [31] and (2) [32]. Arrows indicate the existence region of BNN solid solution.

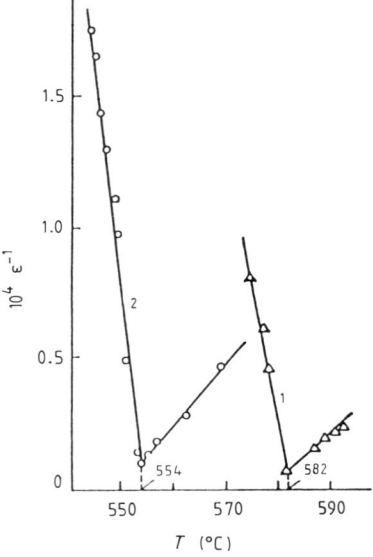

Figure 5.7 Temperature dependences of $1/\varepsilon$ for BNN crystals grown from (1) $Ba_2NaNb_5O_{15}$ and (2) $Ba_{2.085}Na_{0.711}Nb_5O_{15}$ melts [29].

The low-frequency (1.6 kHz) room-temperature dielectric permittivities of BNN crystals measured along the X, Y and Z axes were estimated to be $\varepsilon_x = 238 \pm 5$, $\varepsilon_y = 228 \pm 5$, $\varepsilon_z = 43 \pm 2$. The dielectric permittivity of BNN was measured [13] along the c axis for different frequencies. The ε_z values obtained were as follows: 41.7 ± 0.2 (for 1 kHz) and 29.9 ± 0.3 (for 20–100 MHz). These results have proved the invalidity of the assertion made in [1] concerning the frequency independence of the dielectric permittivity of barium sodium niobate.

The degree of crystal polarization has an effect upon the frequency dependence of the dielectric permittivity ε. The dependence $\varepsilon_c(v)$ is plotted in figure 5.8, which shows that the dispersion of the dielectric permittivity for a polydomain crystal is of relaxation character and for a single-domain crystal is of a resonance character due to the piezoelectric effect. For low frequencies ($\approx 10^3 - 10^4$ Hz) in a single-domain crystal ε_c increases only slightly. The high-frequency dielectric permittivity is more critical to the degree of polarization and may be used for its estimation [36]. The dependence of the high frequency on polarization is explained by the size and nature of the domains in unpolarized specimens. It has also been established [13] that the high-frequency dielectric permittivities of polarized and unpolarized crystal regions are substantially distinct.

Room-temperature electric conductivity of BNN crystals was measured before and after polarization. The results, listed in table 5.1, show that after high-temperature polarization the electric conductivity increases by almost two orders of magnitude. The oxygen with water vapour transmitted through the thermostat does not affect significantly the electric conductivity of the crystals. The low-temperature polarization obtained in silicon oil also increases electric conductivity by two orders of magnitude. It should be noted that the difference in the values of electric conductivity is somewhat larger in the polarized than in the initial specimens.

Typical temperature dependences of the logarithm of electric conductivity are shown in figure 5.9. The curves have two regions described by the exponential functions that correspond to the low-temperature (extrinsic) and high-temperature (intrinsic) conductivity regions. The anomalous behaviour

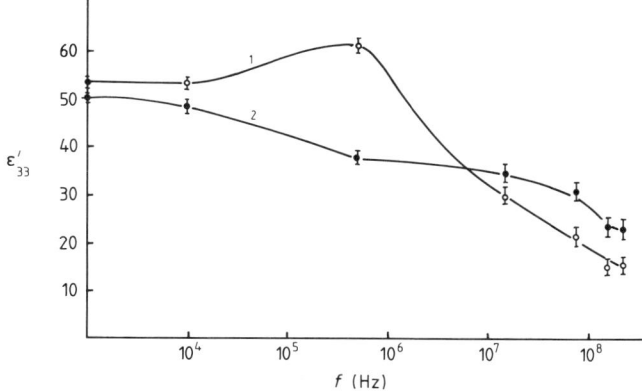

Figure 5.8 Frequency dependence of ε_c for (1) a single-domain and (2) a multi-domain BNN crystal [36].

Table 5.1   Electric conductivity σ ($10^{-14}\,\Omega^{-1}\,cm^{-1}$) of polydomain and single-domain† BNN crystals (300 K).

Polydomain	Single domain	T (°C)	E (kV cm^{-1})	Medium
0.54‡	6.7	625	—	Air
0.46	31	630	—	$O_2 + H_2O$
0.63	38	630	—	$O_2 + H_2O$
0.60	17	160	3.4	Oil
0.55	41	160	3.4	Oil
0.84	38	150	3.4	Oil

† The last three columns show the polarization conditions.
‡ The crystals were grown from a melt of congruent composition ($Na_{0.72}Ba_{2.09}Nb_{5.02}O_{15}$). The other crystals were grown from a stoichiometric melt.

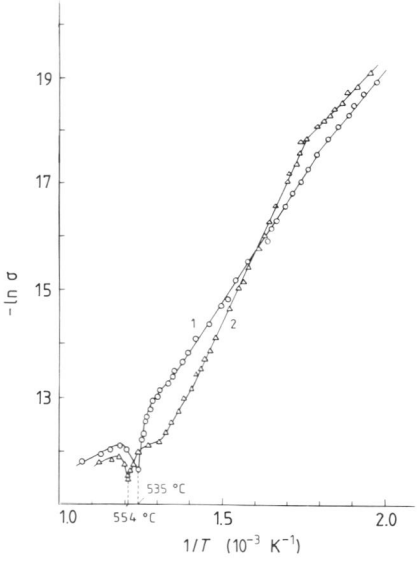

Figure 5.9   Temperature dependences of electric conductivity of BNN crystals grown from (1) $Ba_2NaNb_5O_{15}$ and (2) $Ba_{2.085}Na_{0.711}Nb_5O_{15}$ melts [29].

of the curves $\ln\,\sigma = f(1/T)$ at high temperatures is due to an increase of dielectric losses in the ferroelectric transition region. In the intrinsic conductivity region the activation energy in crystals of both compositions was approximately equal to 0.8–1.0 eV. The activation energy of conductivity in the extrinsic region of crystals grown from stoichiometric melts was 0.35–0.45 eV, whereas for crystals grown from non-stoichiometric melts it was 0.2–0.3 eV. This implies that in crystals of non-stoichiometric composition there are more defects (obviously cation vacancies) than in crystals grown from a stoichiometric melt. The latter is confirmed by the fact that to non-stoichiometric crystals there corresponds the higher break temperature on the curve $\ln \sigma = f(1/T)$. There exists a certain correspondence between the break temperature on the conductivity curve and the non-ferroelectric phase transition under which twins disappear. One may suppose that a certain contribution to the extrinsic conductivity is made by the twin boundary conductivity.

The electric conductivity of barium sodium niobate was also measured in [34]. Apart from the temperature region near the dielectric peak, the electric conductivity σ_z was approximately described by the formula $\sigma_z = M \exp(-E/kT)$, where E (the activation energy) ranged within the limit of 0.96 ± 0.2 eV for $T < T_C$ and 1.1 ± 0.1 eV for $T > T_C$. The coefficient M varied for different crystals from 1 to $20\,\Omega^{-1}\,cm^{-1}$. The conductivity of specimens had a local maximum at a temperature 2 K lower than the dielectric peak. The activation energy values obtained in this paper from the conductivity curves are in close agreement with the estimates, $E = 1.00$ eV for $480-770°C$ and $E = 1.33$ eV for $770-900°C$, reported by Van Uitert *et al* [37].

As compared with the results obtained with direct current, the frequency dielectric losses lead to considerably higher M values and local extremum heights near T_C [34]. The authors explain the variability of M by defects in crystals, poor electric contacts and a possible surface conductivity.

The effect of annealing in oxygen with water vapours upon the dielectric properties of BNN crystals was experimentally established [38]. Annealing in oxygen with water vapours below the field was found to lead to a Curie temperature reduction of $17-19$ K and to a dielectric permittivity reduction at the Curie point. This effect was observed for crystals grown from both stoichiometric and non-stoichiometric compositions. The phenomenon observed can be explained by occupancy of the vacancies in the structure. At the same time, the anisotropy of elastic properties rises and the dielectric permittivity along the polar axis falls.

Measurements of electric conductivity of the indicated crystals in $O_2 + H_2O$ revealed the growth of sample resistivity at the Curie points. For non-stoichiometric samples it increased by 0.2 of the initial value, whereas for stoichiometric samples it was doubled. The growth of sample resistivity upon heating in $O_2 + H_2O$ is explained by a reduced number of cation and perhaps anion vacancies in the BNN structure. The vacancy mechanism of high-temperature conductivity in these crystals plays a dominating role.

5.2.2 Domain structure and striations

The character of the division of a ferroelectric crystal into domains depends on the presence of defects and strains in the crystal, on its electric conductivity, and on the conditions under which a transition to a ferroelectric state occurs, namely the cooling rate and temperature uniformity throughout the crystal (domain nucleation will start in crystal regions with lower Curie temperature).

The type of domain structure is determined by the number of spontaneous polarization orientations admitted by the point group symmetry of the crystal.

In barium sodium niobate, the direction of spontaneous polarization which agrees with the point group symmetry mm2 and 4mm is possible along the *c* axis. The direction of spontaneous polarization in neighbouring domains should therefore differ by 180° (180° domains) [19].

The domain structure of BNN can be discovered using chemical etching of polished surfaces perpendicular to the growth axis. Hydrofluoric and orthophosphoric acids can be used as etching agents for these crystals [33, 39]. The structure is revealed because the etching rates differ for domains with opposite polarization directions [40].

In Czochralski-grown BNN crystals, the distribution of ferroelectric domains was first established by Van Uitert *et al* [9]. The domain structure is a set of embedded concave discs (a file of 'plates') whose axis coincides with the growth axis. When this configuration is cut by a plane perpendicular to the growth axis, the domain structure acquires the form of concentric rings (figure 5.10) [41]. The neighbouring plates are oppositely orientated domains with thickness varying from 2 to 10 μm [13, 22]. The domains naturally repeat the interface surface and growth layers.

Like lithium niobate, BNN is a uniaxial ferroelectric with a non-piezoelectric high-temperature phase. Hence, only optically indistinguishable antiparallel

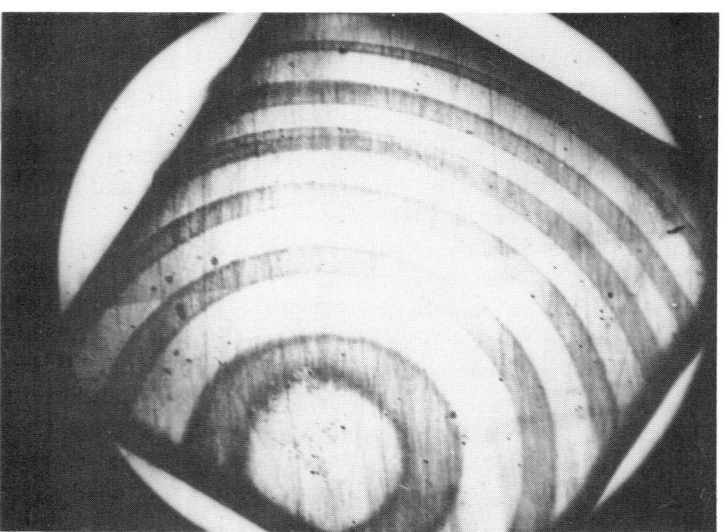

Figure 5.10 Domain structure of BNN in the plane perpendicular to the ferroelectric axis [41]. Enlargement × 50.

domains exist in BNN. At room temperature the domain structure in all the above mentioned crystals is 'frozen', i.e. it does not respond to the external electric field. Thus, while investigating electro-optic effects, one can treat the domain structure as fixed.

Domain walls can be observed under a polarization microscope when an electric field is applied along the $[010]_p$ direction of the orthorhombic cell. In the neighbouring domains the optical indicatrices turn by a small angle, and the domain walls readily show up. It is seen (figure 5.11) that the domains have a needle-like shape and are localized relative to striation.

The domain structure of BNN, as well as that of some other ferroelectric niobates, can be considered as the sum of irregular and lamellar components, namely needle-like micro-domains elongated in the direction of the polar axis Z and unipolar layers reproducing the shape of striations. In crystals with moderately expressed striations, micro-domains are the background for unipolar bands, henceforth called domain lamellae (figure 5.11(a)). When striations are pronounced, micro-domains can be identified at the boundaries of domain lamellae or cannot be seen at all (figure 5.11(b)).

The geometry of domain lamellae depends on the interconnection of domain structure and striations. As was independently suggested in [42], this interconnection is due to the polarizing action (direct or indirect) of the concentration gradient grad c. Here c denotes the self (intrinsic) or impurity (extrinsic) component of the solid solution of which the striations are built; figure 5.12 illustrates this statement.

The polarizing effect of grad c is proportional to its projection on the polar axis. For this reason, in crystals grown precisely at right angles to the z axis, neither domain lamellae nor field-driven diffraction will appear.

(a) (b)

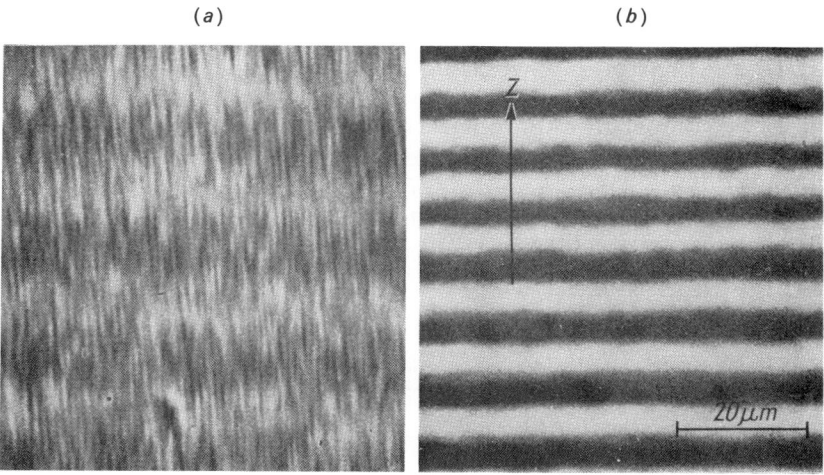

Figure 5.11 Microphotographs of (a) domain structure and (b) growth striae in a BNN crystal [42]. The arrow indicates the direction of spontaneous polarization.

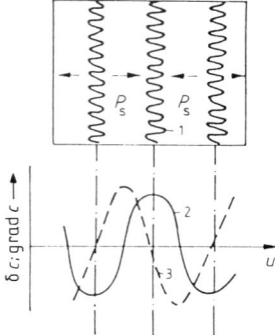

Figure 5.12   Scheme of the relative position of domain walls (1), growth striations (2) and grad c (3) for crystals grown along the polar axis Z [42].

Unipolarity of the domain structure in striation-containing crystals is practically equal to zero, irrespective of the law of component concentration variation within the limits of each stria. Domains are fixed best of all in large growth layers (lamellae) corresponding to a significant composition variation [39]. Striation-free samples were reported to be a random arrangement of domains. Thus all BNN crystals grown from the melt are polydomain, and the character of the domain structure is directly associated with growth conditions.

5.3   Optical properties

BNN crystals at room temperature are biaxial and optically negative, i.e. $n_e - n_o < 0$. Optical axes lie in the (ac) plane, and the angle $2V$ between the optical axes is $13°$ [19, 39]. Using the experimental values of the refractive indices, the estimation of the angle $2V$ gives $14°14''$. The refractive indices at $30°C$ and wavelengths used in electro-optics and nonlinear optics are tabulated in table 5.2, as reported in [1]. Owing to the small magnitude of barium sodium niobate structure distortion at room temperature, the difference between the two refractive indices of ordinary beams $n_a^o(n_x)$ and $n_b^o(n_y)$ is about 0.002 [34].

Table 5.2   Refractive indices of BNN crystals for several wavelengths [1].

λ (μm)	n_x	n_y	n_z
0.532	2.373	2.370	2.256
0.6328	2.326	2.324	2.221
1.064	2.263	2.261	2.175

The dispersion curves of BNN refractive indices reported in [34] are reproduced in figure 5.13. The authors also investigated the temperature dependence of the refractive index from room temperature to 650°C (figure 5.14). As temperature increases in the orthorhombic phase, the refractive indices n_x^o and n_y^o decrease and n_z^e increases. In the tetragonal phase the refractive index n_y^o is almost temperature independent, whereas n_z^e keeps increasing until it becomes equal to n_y^o at the Curie temperature.

Figure 5.15 illustrates the temperature dependences of the birefringence $\Delta n = n_z^e - n_y^o$ for $\lambda = 1.064$ and $0.532\ \mu$m. The slope of the curve for $\lambda = 1.064\ \mu$m varies in the region of 450°C from $-8 \times 10^{-5}(°C)^{-1}$ to $-1 \times 10^{-4}(°C)^{-1}$. Above this temperature birefringence falls very rapidly to zero at the Curie point.

BNN crystals are clear from the self-absorption band edge near 0.370 μm to $\sim 5.0\ \mu$m [34, 39] (figure 5.16). Of particular interest is the appearance of a pale pink colour in BNN crystals during growth in air. The absorption spectrum of such crystals is strained on the side of the self-absorption band edge to about 0.6 μm and has no maxima typical of F centres [38].

The transmission spectra are influenced by the crystal polarization state (figure 5.17). On the differential spectrum of polarized and unpolarized crystals, one can distinguish four broad absorption bands with maxima at 395, 440, 490 and 560 nm. The spectrum of crystals polarized at high temperature exhibits a small absorption peak at 2.230 μm, which is perhaps due to H$^+$ ion diffusion into the crystal [44]. Besides, all the samples showed a narrow absorption band with a half-width of 50 cm^{-1} for $V \approx 3500$ cm^{-1}. Using a polarizer, it is resolved into three bands located at 3425 cm^{-1} (σ polarization), 3486 cm^{-1} (π polarization) and 3492 cm^{-1} (σ polarization) with the intensity ratio $4:2:1$ [34, 35].

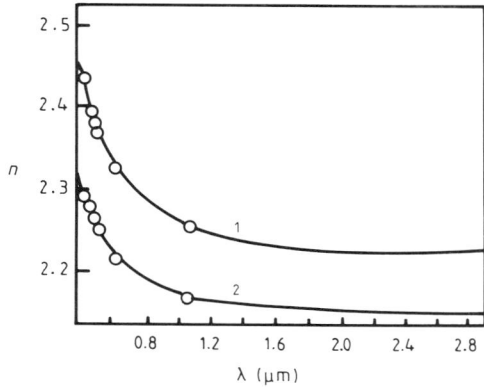

Figure 5.13 Dispersion of refractive indices (1) n_x and (2) n_z of a BNN crystal at room temperature [34].

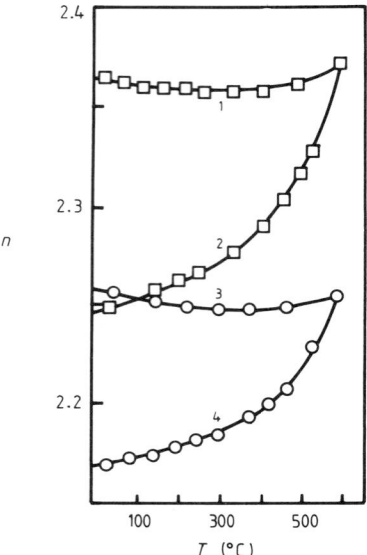

Figure 5.14 Temperature dependences of the refractive indices of an ordinary beam n_y^o (1, 3) and an extraordinary beam n_z^e (2, 4) for a BNN crystal. Curves 1 and 2 correspond to $\lambda = 1.064\ \mu$m; curves 3 and 4 correspond to $\lambda = 0.532\ \mu$m [34].

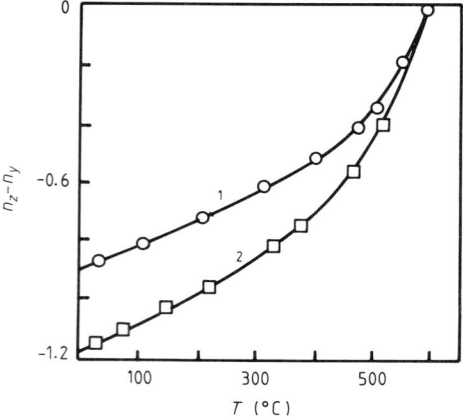

Figure 5.15 Temperature dependences of birefringence ($n_z^e - n_y^o$) for BSN crystals: (1) $\lambda = 1.06$ and (2) $\lambda = 0.53\ \mu$m [34].

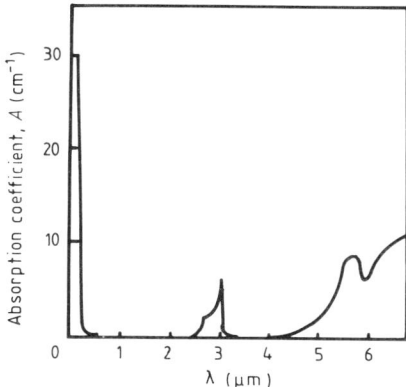

Figure 5.16 Absorption spectrum of BSN crystals [43].

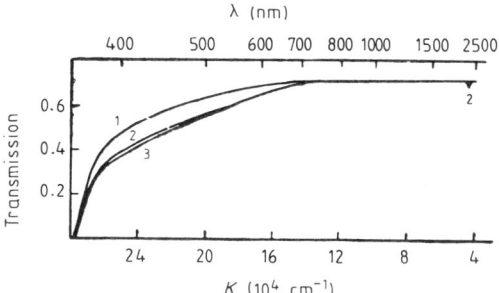

Figure 5.17 Transmission spectrum of a BNN crystal: (1) polydomain; (2) polarized above the Curie point; and (3) polarized at 160°C.

These striations lie in the region of OH oscillations, which suggests that they are due to the presence of hydroxyl groups in the crystal. The appearance of OH absorption bands results from proton embedding into ionized vacancies in the BNN structure, as in the case of alkali halide crystals and $LiNbO_3$. The places of hydroxyl group embedding were discussed in [45]. The authors give somewhat distinct wavelength values for OH absorption bands—3500 and 3450 cm^{-1}—and refer these bands to different optical centres. They assume that OH groups may occupy vacant interstitial positions of 3500 cm^{-1} and replace oxygen ions (3450 cm^{-1}).

A possible reason for the building-in of OH groups into the lattice is crystal growing in a moist air or hydrolysis of initial components. Polarization in the presence of water vapour leads to an increase of proton content in the crystal, and the intensity of the indicated absorption bands in single-domain crystals increases. The calculations [38] show that the hydrogen ion concentration in a crystal increases in the course of polarization from 10^{17} cm^{-3} to 10^{18} cm^{-3}. But the amount of hydrogen impurity is small compared with the total number of vacancies in the structure.

A considerable hydrogen content in BNN crystals may contribute to the optical stability of the material. It was experimentally established [46] that materials with a high hydrogen ion concentration possess a better resistivity to optical damage.

All the crystals containing NbO$_6$ as the basic structural element were reported [47] to have a comparatively low resistivity to the action of powerful light. For example, in lithium niobate crystals the local variations of the refractive index at wavelengths $\lambda = 0.690$ and $\lambda = 0.488$ μm occur at just a power density of the order of several W cm^{-2}. But for BNN crystals this level increases up to kW cm^{-2}. It is now conventional to think that this difference in resistance is due to different location depths of electron traps, that is, structural voids A1 and A2 between NbO$_6$ octahedra. Occupation of these traps by electrons is due to photo-ionization by powerful light beams. However, at $-80\,^\circ$C, BNN crystals exhibit optical inhomogeneities [9]. This happens when smaller traps, i.e. C voids, are occupied.

5.4   Nonlinear optical properties

For the point group mm2, to which BNN belongs, there exist three independent nonlinear coefficients d_{31}, d_{32}, d_{33} [1]. These coefficients were estimated from the results of measurements of second harmonic output on a wavelength of 0.532 μm using focused Nd:YAG laser radiation ($\lambda = 1.064$ μm) with continuous pumping and periodic Q modulation [1, 34]. The coefficients d_{ij} of barium sodium niobate were estimated relative to the coefficient d_{11} of quartz (table 5.3). Some difference in the nonlinear coefficients reported by different authors is obviously due to the different quality of the crystals employed.

The effective nonlinear coefficients of barium sodium niobate are twice as large as those of LiNbO$_3$ and 20 times higher than those of KHPO$_4$ [5]. Second harmonic generation in BNN crystals can be realized both at room temperature and at higher temperatures. In the latter case, the phase matching angle θ_{phm} is 90° and SHG proceeds without birefringence. For $\theta_{\text{phm}} \neq 90^\circ$ a

Table 5.3   Relative values of nonlinear coefficients d_{ij} of BNN for $\lambda = 0.532$ μm [34].

d_{ij}	d_{ij}/d_{11} (SiO$_2$)	d_{ij}	d_{ij}/d_{11} (SiO$_2$)
d_{31}	40 ± 2	d_{15}	40 ± 2
d_{32}	40 ± 4	d_{24}	38 ± 2
d_{33}	55 ± 4		

shortage of phase matching is caused by the fact that the Poynting vector of the reference wave declines by a certain angle from the second harmonic wavevector due to birefringence, which leads eventually to restriction of the effective length of the crystal. Therefore, for an optical utilization of nonlinear properties of the medium, the angle θ_{phm} should be $90°$. The phase matching temperatures T_{phm} at which a $90°$ matching is reached were estimated in [34] as $T_{phm}^{31} = 100°C$ and $T_{phm}^{32} = 88.7°C$ for the coefficients d_{31} and d_{32} respectively. Let us compare these temperatures with those observed experimentally: $T_{phm}^{31} = 101°C$ and $T_{phm}^{32} = 89°C$. For the majority of the examined samples, the T_{phm}^{31} and T_{phm}^{32} values range from 90 to 110 and from 80 to $100°C$ respectively. The authors of [34] have also pointed out that the phase matching temperature of each successive boule pulled from a stoichiometric melt was lower than that of the preceding boule. This is an indication of composition variation which is possibly due to Na concentration lowering in the melt.

The degree of BNN polarization also has a substantial effect upon the temperature dependence of SHG. Figure 5.18 represents the temperature dependences of the second harmonic intensity for BNN crystals of different degrees of polarization [36]. The radiation source was a He–Ne laser with the fundamental wavelength $\lambda = 1.152 \ \mu m$. For a single-domain crystal, three series of peaks are observed that correspond to second harmonics of the

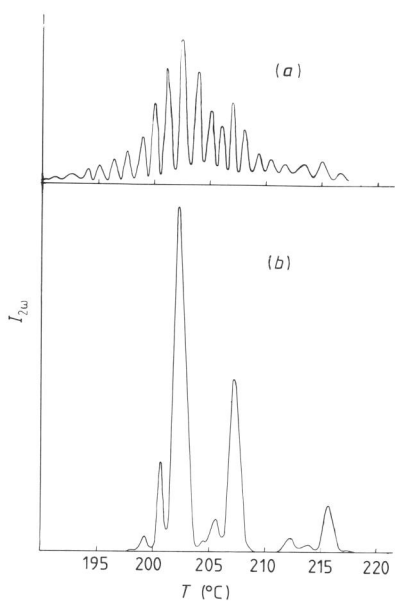

Figure 5.18 Temperature dependences of SH intensity of partially (*a*) and completely (*b*) polarized BNN crystals [36].

waves close to the fundamental laser-radiated wavelength. Such SHG curves were calculated [10] as a function of $d \, \Delta n$, where d is the diameter of the laser beam and Δn is the birefringence gradient. It was established that for $\Delta n < 10^{-4}$ the half-width of the peak of SHG temperature dependence is independent of the laser beam diameter.

The domain structure of the investigated BNN crystal was found to be of irregular character, i.e. there is no pronounced boundary between polydomain and single-domain parts of the crystal, and the temperature dependence of SHG of a polydomain crystal has the same shape as that for a crystal with a large birefringence gradient [10].

The crystal quality can be characterized by its effective length l_{eff}:

$$l_{\text{eff}} = k / 1.125 \, \Delta T$$

where ΔT are half widths of SH peak, $k = 0.51 \pm 0.2$ for $\lambda = 1.15 \ \mu\text{m}$ [13]. For a single-domain crystal (figure 5.18(b)), l_{eff} coincides with its geometrical sizes in the beam propagation direction, which confirms the high optical quality of the crystal.

From the series of SH maxima corresponding to different wavelengths generated by a He–Ne laser, one can estimate the coefficient $\Delta \lambda / \Delta T$ corresponding to the displacement of the second harmonic wavelength $\Delta \lambda$ when the phase matching temperature changes by ΔT. The He–Ne laser is known to generate, near 1 μm, the wavelengths of 1.118, 1.153, 1.199 and 1.207 μm [48]. The maxima of second harmonic intensities lie at temperatures of 202.5, 207.5 and 216 °C (figure 5.18) and are separated by distances of 5 and 8.5°, respectively. If we assume that the first SH maximum corresponds to the wavelength $\lambda = 1.118 \ \mu$m and the SH maxima of the fundamental waves 1.199 and 1.207 μm are not resolved, then the estimated value $\Delta \lambda / \Delta T = 6.6 \times 10^{-3} \ \mu\text{m K}^{-1}$, i.e. when temperature increases by 1 K, the wavelength satisfying the phase matching increases by $6.6 \times 10^{-3} \ \mu$m.

5.4.1   SH *frequency doubling on ferroelectric domains in* $Ba_2NaNb_5O_{15}$ *crystals*

Barium sodium niobate is a uniaxial ferroelectric with a centrosymmetric paraelectric phase, and therefore it possesses only 180° domains which are optically indistinguishable. The latter is valid only for linear properties, while quadratic nonlinear susceptibility changes the sign at domain boundaries. Second harmonic generation for the two principal types of domain structure of BNN crystals—lamellar and micro-domain type structures—has been studied in [49]. The experimental data are explained on the basis of the theory of nonlinear light scattering on domains [50]. Figure 5.19(a) shows the angular distribution of the second harmonic for a crystal with a disordered domain structure containing needle-like micro-domains. A laser beam passed through the sample perpendicularly to the polar Z axis in the direction of the X axis. The oscillation direction of the light wave electric vector coincided with the Z axis. The only nonlinear coefficient acting under such a polarization of an incident beam is d_{33}. In accordance with this, the radiation at a wavelength of 0.53 μm had a polarization coinciding with that of the incident beam. In this case, one observed a smeared spot of second harmonic elongated in the XY plane (the scattering angle φ as seen from figure 5.19(*a*) reached 15°) and contracted in the Z axis direction (the scattering angle in this direction did not exceed 2°).

Nonlinear light scattering obeys the law of conservation of momentum:

$$q = -2k_1 + k_2 \tag{5.1}$$

where q is the vector representing the action of the domain structure upon the beam, and k_1, k_2 are the wavevectors of the fundamental wave and second harmonic. In a rough approximation [50], the possible directions of the vector q are perpendicular to the domain walls, and its modulus is inversely proportional to the domain size L in a given direction:

$$|q| \simeq \pi/L. \tag{5.2}$$

The scattering angles of the second harmonic are determined, according to the condition (5.1), by the construction of a vector triangle with the use of the surfaces of the wavevectors $2k_1, k_2$ (figure 5.19(*b*)). Using the estimate (5.2), the experimental data can be explained by second harmonic scattering on micro-domains whose transverse dimension ranges from 1 to 3 μm, and the dimension in the direction of the polar Z axis is of the order of 10 μm. These values are in close agreement with the data obtained by the Merz method (figure 5.11(*a*)).

Quite a different character of SHG is observed in BNN crystals with lamellar domain structure. The domain lamellae in BNN imitate the growth striation

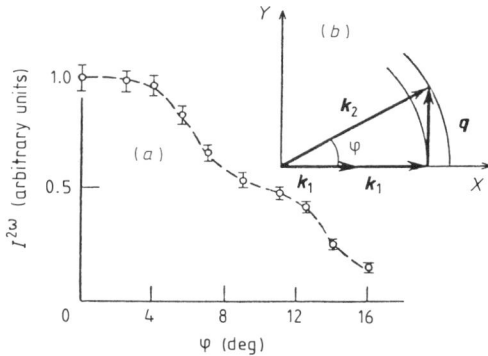

Figure 5.19 SHG in a BNN crystal with micro-domains. Dependence of SH intensity on the scattering angle φ in the (XY) plane (*a*). Scheme of scattering: full curves show regions of surfaces of the wavevectors $2k_1$ and k_2 (*b*) [42].

configuration. Under certain growing conditions these striations form a periodic structure. As distinguished from scattering on micro-domains, in crystals with a periodic arrangement of domain lamellae, SHG proceeds at strictly definite angles since only $|q|$ values which are multiples of $2\pi/\Lambda$ are allowed (Λ is the domain structure period). The direction of the vector q is perpendicular to the layer surfaces.

Second harmonic scattering on a lamellar domain structure ('nonlinear diffraction') was observed in the propagation of the fundamental wave in the XZ plane of a crystal. The oscillation direction for the fundamental wave coincided with the Y axis, and therefore the nonlinear coefficient d_{32} was responsible for the SHG process. Under similar conditions, single-domain BNN crystals exhibit phase matching for oo–e type processes. The angle between the direction of a matched SHG and the Z axis is $\theta_c = 75°$ (all the angles are given for light waves in a crystal at room temperature).

The experimental dependence of the second harmonic intensity $I^{2\omega}$ on the angle θ between the incident beam and the Z axis is depicted in figure 5.20(a). The specimen was an X-cut plate; the domain structure period $\Lambda = 12$ μm. To investigate the dependence $I^{2\omega}(\theta)$, the crystal was rotated around the Y axis. The second harmonic intensity was negligibly small in the entire range of angles θ (from 70 to 90°) as compared with the intensity near $\theta = 81°$. The nonlinear interaction at $\theta = 81°$ was non-collinear. The direction of the fundamental wave made an angle of $\beta = 1.2°$ with the direction of the second harmonic. The vector triangle (5.1) calculated for a first-order scattering (when $|q| = 2\pi/\Lambda$) gave the values $\theta_1 = 82.1°$ and $\beta_1 = 1.12°$. Thus the data

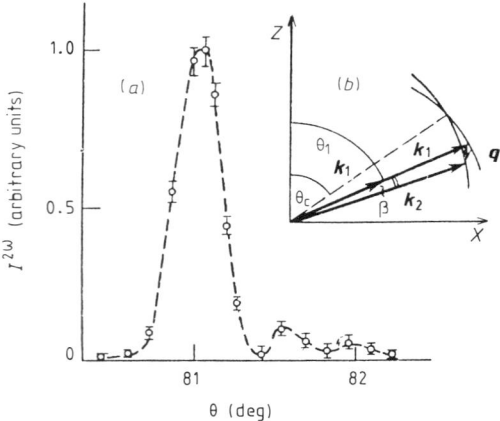

Figure 5.20 SHG in a crystal with a lamellar domain structure: the dependence $I^{2\omega}(\theta)$ in the neighbourhood of $\theta = 81°$ (a); scheme of a quasi-synchronous process for $\theta = \theta_1$ (b). Broken lines indicate the direction of synchronous SHG for a single-domain crystal [42].

obtained are explained fairly well by the theory of nonlinear light scattering by domains.

For a disordered domain structure, the interaction is not synchronous; $I^{2\omega}$ increases (in the approximation of a given field) in proportion to the crystal length, whereas in a crystal with a period domain structure a peak at $\theta = 81°$ corresponds to the effective 'quasi-synchronous' interaction when $I^{2\omega}$ increases in proportion to the length squared. For a perfectly periodic structure with infinitely thin flat domain boundaries, $I^{2\omega}(\theta_1) = (4/\pi^2)I^{2\omega}(\theta_c)$, where $I^{2\omega}(\theta_c)$ is the second harmonic intensity for a single-domain crystal under phase matching conditions.

At least two factors had to lower the $I^{2\omega}(\theta_c)$ value. First, BNN samples always contain a certain number of micro-domains. Second, the measured angular peak width of a quasi-synchronous SHG made up 22′, whereas the calculated value was only 12′. Peak broadening is explained by the difference in domain thicknesses at a level of 5% of the mean value.

5.5 Electro-optic properties

The matrix of BNN linear electro-optic coefficients in the orthorhombic phase has five independent non-zero coefficients $r_{13}, r_{23}, r_{33}, r_{42}, r_{51}$. Accordingly, the indicatrix equation has the form [34]

$$(1/n_x^2 + r_{13}E_z)x^2 + (1/n_y^2 + r_{23}E_z)y^2 + (1/n_z^2 + r_{33}E_z)z^2$$
$$+ 2r_{42}E_y yz + 2r_{51}E_x xz = 1.$$

In the tetragonal phase, only three of the five coefficients are independent, since the symmetry of the point group 4mm is responsible for the equalities $r_{13} = r_{33}$ and $r_{42} = r_{51}$.

The linear electro-optic coefficients of BNN are listed in table 5.4. The signs of the coefficients r_{ij} were not determined, but r_{13}, r_{23} and r_{33} were found to be of the same sign [1].

The high-frequency linear electro-optic effect in BNN, as in LiNbO$_3$, depends slightly on the melt composition. However, the differences in the values of the coefficients are associated not with the change of crystal composition but rather with different degrees of their polarization. A more detailed consideration is given below.

Table 5.4 Linear electro-optic coefficients of BNN (10^{-12} m V^{-1}) for $\lambda = 0.633$ μm [13, 22].

ν (MHz)	r_{13}	r_{23}	r_{33}	r_{42}	r_{51}
0	15 ± 1	13 ± 1	48 ± 2	92 ± 4	90 ± 4
90	6.1 ± 0.3	6.1 ± 0.3	24.3 ± 1.2	—	—
100	7	8	29	79	95

The quadratic electro-optic coefficients of oxygen octahedral ferroelectrics are related to the linear coefficients as follows (chapter 4, [42]):

$$g_{12} = \frac{r_{i3}}{2\varepsilon_0(\varepsilon_z - 1)P_s} \qquad g_{11} = \frac{r_{33}}{2\varepsilon_0(\varepsilon_z - 1)P_s} \qquad g_{44} = \frac{r_{ij}}{2\varepsilon_0(\varepsilon_j - 1)P_s}$$

where ε_j and ε_0 are the dielectric permittivities of the crystal and vacuum, and P_s is the spontaneous polarization. Using these formulae and the experimental data $\varepsilon_x = 238$, $\varepsilon_y = 228$, $\varepsilon_z = 42$ and $P_s = (0.40 \pm 0.01)$ C m^{-2}, one can evaluate the quadratic electro-optic coefficients g_{ij} (m^4 C^{-2}):

$$g_{12} = 0.052 \pm 0.006 \qquad i = 1 \qquad g_{11} = 0.17 \pm 0.01$$

$$g_{44} = \begin{cases} 0.11 \pm 0.01 & i, j = 4.2 \\ 0.11 \pm 0.01 & i, j = 5.1. \end{cases}$$

The obtained g_{ij} values are comparable with the quantities measured for other oxygen octahedral ferroelectrics. It should be noted that $g_{11} - g_{44} \approx g_{12}$.

When light propagates normally to the c axis, the half-wave voltage $V_{\lambda/2}$ for a crystal with a point symmetry group mm2 is expressed by the relation [13]

$$V_{0j} = \lambda/(n_3^3 r_{33} - n_j^3 r_{j3}) \qquad j = 1, 2.$$

When light propagates along the X and Y axes, the difference in the half-wave voltage due to anisotropy of the refractive indices is about 9%.

The experimental values of the half-wave voltage are measured from 1250 to 1950 V. Such a broad range of $V_{\lambda/2}$ values would naturally be ascribed to the variation of crystal composition. But it was demonstrated [13] that, for two crystals whose phase-matching temperatures differ by 30 K, the half-wave voltages differ by not more than 2%. The difference in the $V_{\lambda/2}$ values can be explained only by taking account of the fact that the resistance and dielectric permittivities of the crystal are associated with the completeness of its polarization, and may affect the values of electro-optic coefficients at low and high frequencies. Variations of the resistance and dielectric permittivity over the crystal volume lead to an uncontrollable distribution of electric fields inside the crystal. In this situation, the electro-optic coefficient determined even in a completely polarized part of the crystal may differ from that in another part of the crystal.

Noting that the resistance for a direct current of a polarized sample is much higher than that of an unpolarized sample, Nash *et al* [13] calculated the half-wave voltage in different crystal regions in a partially polarized sample. So, for a light beam propagating in a polarized area (region 1, figure 5.21) normal to the c axis the apparent half-wave voltage, according to [13], is expressed by the formula

$$V_0(1) = V_0 \frac{x_1 + (\rho_2/\rho_1)x_2}{x_1 + x_2}$$

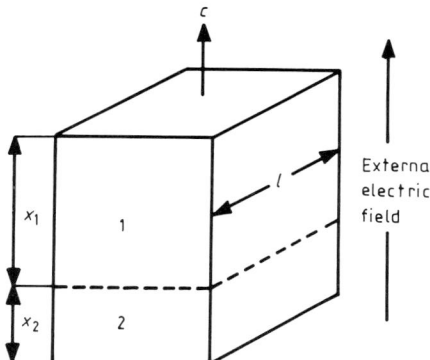

Figure 5.21   Model of a partially polarized BNN crystal [13]. Region 1 is polarized completely; region 2 partially.

where V_0 is the half-wave voltage of a single-domain sample, x_1 and ρ_1 are the length and resistivity of the polarized region of the sample and x_2 and ρ_2 are those for the unpolarized region. Since $(\rho_2/\rho_1) < 1$, the apparent half-wave voltage is smaller than the ideal one. For a beam propagating in a partially polarized region (region 2) normal to the c axis, we have

$$V_0(2) = V_0 \frac{x_2 + (\rho_1/\rho_2)x_1}{x_1 + x_2}$$

where V_0 is the true value of the half-wave voltage. Since usually $\rho_1/\rho_2 > 1$, then $V_0(2) > V_0$. Thus, the observed half-wave voltage can be either larger or smaller than the V_0 which characterizes completely polarized crystals. According to [13], the V_0 value for BNN crystals makes up 1830 ± 40 V and is not dependent on the crystal composition.

By the magnitude of the half-wave voltage, one can judge the degree of crystal polarization. It has been established [36] that to obtain a completely single-domain state of a crystal it should be annealed, prior to application of an electric field, for an hour at a temperature 100 K higher than the Curie point. From figure 5.22 one can see that polarization is carried out near the Curie temperature (figure 5.22(a)); the crystal remains polydomain since its half-wave voltage increases on moving away from the electrodes (figure 5.22(b)). Crystal annealing at a temperature of 700°C for an hour and a subsequent application of an electric field (figure 5.23(a)) provides complete polarization (figure 5.23(b)).

5.5.1   *Isotropic diffraction*
To describe quantitatively the linear electro-optic effect in a ferroelectric crystal containing antiparallel domains, one can introduce according to [50] a space modulation function M equal to $+1$ for domains with definite

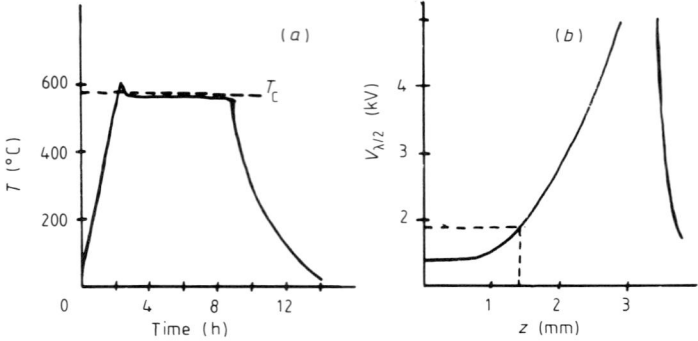

Figure 5.22   Polarization regime for a BNN crystal (*a*) and distribution of $V_{\lambda/2}$ along the *c* axis of this crystal (*b*) [36].

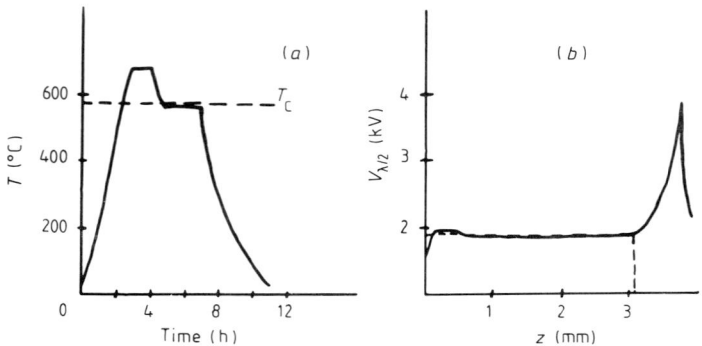

Figure 5.23   The same as in figure 5.22, but for another polarization regime [36].

direction of spontaneous polarization P_s and -1 for oppositely oriented domains. The linear electro-optic coefficients r_{pk} are then replaced by coefficients $r_{pk}M$ changing sign at the domain wall.

The function M can be resolved into three components: a constant one, describing the mean unipolarity of a crystal; an irregular one, corresponding to micro-domains in barium sodium niobate; and a regular one, M_1, which represents the lamellar domain structure. As striations are located in the XY plane, the M_1 component depends on Z only. This component is responsible for electro-optic diffraction, and the associated effective electro-optic coefficients are $r_{pk}M_1(z)$.

When the light beam propagates in BNN approximately in the striation plane, one can observe, even in the absence of applied field, isotropic light diffraction. This diffraction is due to the variations of the refractive indices $n(z)$, caused in turn by crystal composition variations $c(z)$ [42], i.e. by growth striations.

The external electric field E_3 applied to a BNN crystal gives rise to another volume phase grating $\Delta n_E(z)$. This is electro-optic by nature:

$$n_{pE} = \frac{-n_p^3 r_p M_1(z) E_3}{2} \qquad (5.3)$$

where $p = 1, 2, 3$ and n_p is the principal refractive index. The diffraction pattern is controlled by the sum of two gratings, $\Delta n(z) = \delta n(z) + \Delta n_E(z)$.

The particular feature of light diffraction by striations and domains is substantially an anharmonic profile of a phase grating. If the boundaries between domain lamellae are flat, then $\Delta n_E(z)$ is of meandering rather than of sine type. For light modulation, the case of Bragg diffraction is of most interest. Owing to the angular characteristics of Bragg diffraction, the diffraction pattern will be formed by the first spatial harmonic of the sum grating $\Delta n_{(1)}(z)$ when the angle of incidence is a Bragg angle φ ($\sin \varphi = \lambda/2n\Lambda$, where λ is the light wavelength in vacuum).

The amplitude of the first harmonic is

$$\Delta n_{(1)} = [\delta n_{(1)}^2 - n^3 r E M_{1(1)} \delta n_{(1)} \cos \beta_{(1)} + \tfrac{1}{4} n^6 r^2 E^2 M_{1(1)}^2]^{1/2}. \qquad (5.4)$$

Here the angle $\beta_{(1)}$ corresponds to the space shift of the domain structure with respect to striations, and $M_{1(1)}$ and $\delta n_{(1)}$ are the amplitudes of the first harmonics of $M_{(1)}(z)$ and $\delta n(z)$.

The Bragg diffraction takes place when the parameter Q is large: $Q = 2\pi \lambda L/n\Lambda^2 > 2\pi$. In the case of an anharmonic grating, this condition is still correct for the first harmonic of a grating. If the angle of incidence is exactly the Bragg angle, the intensity of diffraction light I_1 is given by

$$I_1 = I\left(1 - \frac{q^2}{2\eta^2}\right) \sin^2(qL/2) \qquad (5.5)$$

where I is the intensity of incident light, $\eta = 2\pi\lambda/n\Lambda^2$ and $q = 2\pi\Delta n_{(1)}/\lambda$.

The Bragg diffraction by domains in BNN has a distinct feature: the dependence $I_1(E)$ is nearly symmetric relative to $E = 0$ (curve 1 in figure 5.24(a)), and the intensity of diffracted light in the absence of applied field is rather high. All curves in figure 5.24 correspond to the sample with lamellar domain structure displayed in figure 5.11(b). The last two formulae, (5.4), (5.5), the experimental dependence $I_1(E)$, the known values of the refractive index $n_3 = 2.22$ and the electro-optic coefficient $r_{33} = 46 \times 10^{-12}$ m V^{-1} were used to calculate some parameters of the sample: $\delta n_{(1)} = 1.2 \times 10^{-4}$, $M_{1(1)} = 1.12$, $\beta_{(1)} = -1.04\pi/2$. The value of $\delta n_{(1)}$ is sufficient for marked diffraction at zero field (the so-called light diffraction by growth striae). The symmetric shape of the $I_1(E)$ graph is due to the quarter-period relative shift of the growth and electro-optic gratings (for the same reason $\beta_{(1)} \approx -\pi/2$). The magnitude of $M_{1(1)}$ is near its maximum value $4/\pi$, the latter corresponding to the domain lamellae with flat boundaries, when the dependence $M(z)$ is a meander.

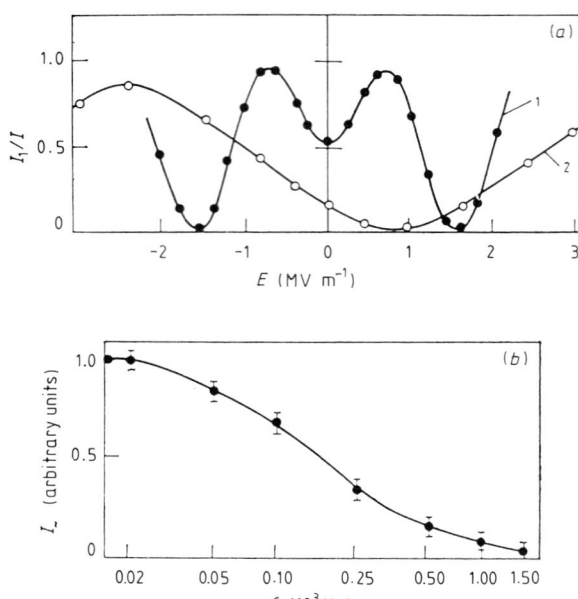

Figure 5.24 (*a*) Intensity of the first-order diffraction maxima versus the amplitude of applied AC field. Isotropic diffraction, Bragg incidence of the light wave polarized in the XZ plane. Parameters of the sample: $\Lambda = 11 \ \mu$m, $L = 1.4$ mm, $Q = 20.7$. (1) Frequency 800 Hz, sample unpoled; (2) frequency 20 Hz, poled sample. (*b*) The modulation signal $I_\sim$ versus frequency after the poling procedure. Amplitude of applied AC field 2 kV cm^{-1} [49].

Curve 2 in figure 5.24(*a*) was taken after polarization of the sample. As a result of this procedure, the conductivity of the sample increased by a few orders of magnitude, so that the DC relaxation time dropped to the value 4.5×10^{-3} s and the dielectric loss tangent at 1 kHz increased from 0.003 to 0.21.

The dependence $I_1(E)$ is quite different for single- and multi-domain states of a crystal. After polarization, the relative spatial shift of two gratings is about half a period ($\beta_{(1)} \approx \pi$). At sound frequencies a drastic dispersion of the electro-optic effect is observed (figure 5.24(*b*)), while in the polydomain crystal it is practically independent of frequency up to hundreds of MHz. The significant reduction of I_1 measured at zero field after the poling procedure needs attention as well. This fact is associated with the reduction of $\delta n_{(1)}$ by approximately a factor of two. After depolarization of the sample by heating it up to the Curie point and cooling to room temperature, the parameters $\delta n_{(1)}$, $M_{1(1)}$, $\beta_{(1)}$ were restored with an accuracy of 2 to 10% as compared to their initial values.

The modulation depth of diffracted light was measured at 6.4 MHz and a bias field of 2.5×10^5 V m^{-1} in the polydomain state of the sample. The depth was $(92 \pm 5)\%$ of its value at 800 Hz and the same amplitude of modulation field (10^5 V m^{-1}).

Some measurements were taken at 800 MHz. The BNN specimen was 4 mm long and had a cross section of 2×2 mm^2. The interaction length L was considerably less than the specimen length, $L = 1.4$ mm. The period of the domain structure was 10 μm. To exploit the linear part of the $I_1(E)$ dependence, a bias field of 400 to 450 V was applied to the crystal. The modulation depth amounted to 30% at a driving power of 4 W (figure 5.25), the bandwidth being 40 MHz.

The temperature behaviour of the diffraction was studied using a BNN sample with $\Lambda = 14$ μm, $Q = 18$. The results are plotted in figure 5.26. The maximum value of $r_{33}M_{1(1)}$ corresponds to a temperature of 1 K below the Curie point.

The study of light diffraction in strained BNN crystals sheds light on their domain structure and physical properties. Of great importance is the spatial shift of domains relative to the striae. As the measured value of $\beta_{(1)}$ is nearly $-\pi/2$, the domain walls are located at minima and maxima of refractive index variations. As the shift of grad c with respect to composition variations is a quarter of a period, the location of domain walls and striations supports the hypothesis about the polarizing action of the concentration gradient.

The drastic change in the isotropic diffraction characteristic after the polarizing procedure is a consequence of the different origin of the electro-optic grating in polarized samples as compared to unpolarized ones. In the latter case, the grating is formed by conductivity variations $\sigma(z)$. The

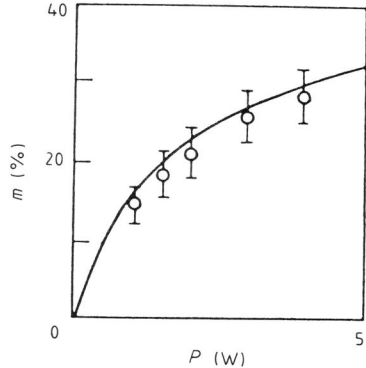

Figure 5.25 The dependence of modulation depth (light polarized in the XY plane) on the power of the SHF signal. The full line is calculated using the value of the electro-optic coefficient r_{23}^s [49].

redistribution of applied field results in a non-uniform stationary field $E \propto \sigma(z)^{-1}$, which forms the grating. As the redistribution of applied field is complete at 20 Hz, curve 2 in figure 5.24(a) offers an opportunity to calculate the conductivity variations. The amplitude is 25% of the mean σ value.

The unpoled crystal with a perfect lamellar domain structure is expected to demonstrate electro-optic diffraction without any strong dispersion up to the frequency of mechanical resonance for a domain lamella. This frequency is 350 MHz [42] if $\Lambda = 14$ μm. Data on isotropic diffraction at 6.4 MHz support this statement. As the resonant frequency for a single-domain sample of the same dimension is 1.3 MHz, the clamping effect should result in a 40% reduction of the signal. The reduction was not observed in polydomain BNN samples. At 800 MHz the crystal is clamped, and therefore the corresponding value of the electro-optic coefficient was used in the calculation of the theoretical curve in figure 5.25. The possibility of SHG light modulation demonstrates the proposed nature of field-driven diffraction and gives an outlook for utilizing unpoled BNN crystals in optical modulation techniques.

The analysis of temperature dependences displayed in figure 5.26 leads to some important conclusions as well. The similarity of curve 1 to the $r_{33}^{-1}(T)$ curve in [34] for a single-domain sample shows that $M_{1(1)}$, the main parameter of domain lamellae, is practically independent of temperature. Thus, the lamellar domain structure is formed in the vicinity of the Curie point and does not rearrange during the cooling process.

The temperature plot of I_1 (curve 2 in figure 5.26), as well as the noticeable reduction of the so-called 'growth variations of refractive indices' δn after poling, shows that diffraction in the absence of applied field has a complex

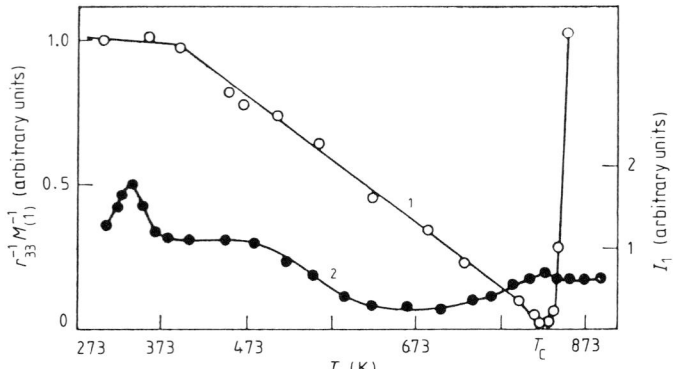

Figure 5.26   Temperature dependence of $r_{33}^{-1} M_{1(1)}^{-1}$ (1) and the intensity of the first-order diffraction maximum in the absence of applied field (2) for isotropic diffraction and light polarized in the XZ plane [49].

nature, the δn value being related to the ferroelectric properties of a material. Four contributions to the observed δn magnitude were resolved [42].

5.6 Phase diagram of the system $BaO-Na_2O-Nb_2O_5$

The formation of the tetragonal potassium tungsten bronze (TPTB)-type structure $Ba_2NaNb_5O_{15}$ was studied on the basis of the phase diagrams of the ternary system $BaO-NaO-Nb_2O_5$ and the pseudo-binary system $BaNb_2O_5-NaNbO_3$. The state diagram of the latter (figure 5.27) was examined in [26] using differential thermal analysis and the quenching method.

In the region of the phase diagram close to $NaNbO_3$, the measurements of the lattice parameters in quenched specimens and the analysis of the phase composition showed that solid solutions on the basis of $NaNbO_3$ exist with up to 2 mol.% of $BaNbO_6$. The field of the liquid phase of these solid solutions stretches from 1416°C to eutectics at 1356°C. The composition at an invariant point was fixed at 23 ± 2 mol.% of $BaNb_2O_6$.

By means of differential thermal analysis, $Ba_2NaNb_5O_{15}$ was established to crystallize in the wide field of the liquid phase that stretches from 23 to 66.7 mol.% of $BaNb_2O_6$. The diffractograms taken on powder specimens showed, however, that TPTB solid solutions are actually stable up to 62 mol.% of $BaNb_2O_6$ at a solidus temperature of 1356°C. In compositions containing

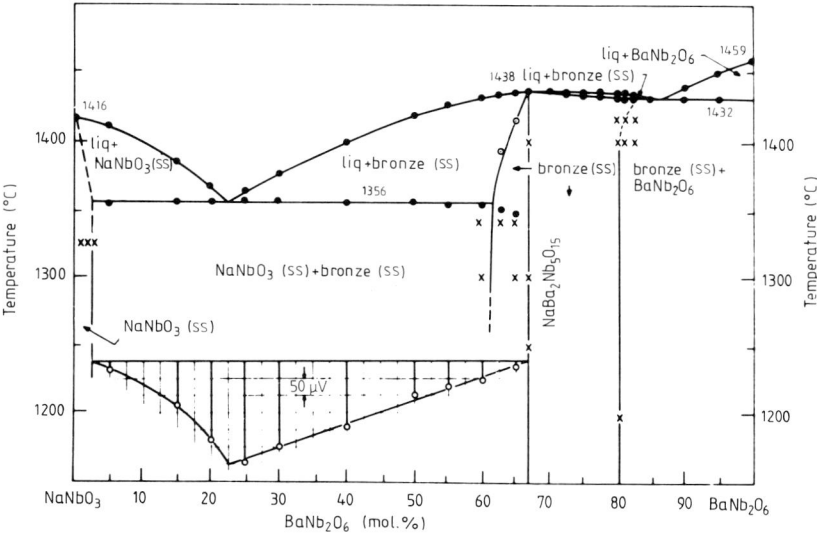

Figure 5.27 Diagram of the state of the pseudo-binary system $NaNbO_3-BaNb_2O_6$ [26].

63 and more mol. % of $BaNbO_6$, X-ray phase analysis revealed the presence of $NaNbO_3$. This indicates that a true equilibrium was not reached in the cycles of differential thermal analysis.

A study in the region of the pseudo-binary diagram from 66.7 to 85.5 mol. % of $BaNb_2O_6$ showed the existence in these limits of a TPTB phase, for which the liquidus and solidus lines differ in temperature by a value comparable with the experimental error (± 2 K). By means of differential thermal analysis it was established that the boundary of the solid solution lies at the level of 1432 °C of the solidus line, the $BaNb_2O_6$ content being equal to about 83 mol. %, whereas in quenched specimens the region of TPTB phase decreases down to the boundary which lies below 1400 °C and corresponds to 80 mol. % of $BaNb_2O_6$.

The closeness of the liquidus and solidus curves, especially in the region of TPTB solid solutions, made it impossible to establish exactly the composition at 1432 °C. The composition at the invariant point corresponds approximately to 85 ± 2 mol. % of $BaNb_2O_6$.

The pseudo-binary phase diagram shows that TPTB phases on the basis of barium sodium niobates have a wide field of compositions embracing the region from 23 to 85 mol. % of $BaNb_2O_6$, in which solid solutions containing from 62 to 83 mol. % of alkaline earth component can, in principle, be grown in the form of single crystals.

The phase equilibrium in the ternary system $BaO-Na_2O-Nb_2O_5$ was investigated in [27, 51, 52]. The most detailed analysis was given to that part of the phase diagram near the stoichiometric compound $Ba_2NaNb_5O_{15}$ ($10Na_2O-40BaO-50Nb_2O_5$). The phase boundaries of the BNN homogeneity region were examined and the fields of bronze-type phases differing only in size and shape were presented. Triple eutectics were not determined, and the field of solid solutions, according to [52], had only approximate boundaries due to insufficient accuracy of the methods applied. The TPTB phase in the solid state may contain a relatively large excess Nb_2O_5, as established in the chemical analysis.

Barraclough *et al* [51] reported that the BNN liquidus and solidus barium sodium niobate phase surfaces are almost flat (the liquidus and solidus temperatures of single-phase specimens lie within the limits 1430–1455 °C). A considerable solubility of Nb_2O_5 in BNN means that the excess Nb atoms are readily located in pentagonal voids A2 (see figure 5.2) of the crystal lattice. In this portion of the phase diagram the slope of the solidus line is small, and therefore the crystal composition changes radically with temperature for the melts from this region. This results in the appearance of large composition gradients, which may explain, according to [52], the broadening of dielectric permittivity peaks. Carruthers and Grasso [27] pointed out that on the side of excess Nb_2O_5 the liquidus and solidus curves are relatively close, and the solidus line falls sharply on the cation-rich side of the solid solution. This steep slope of the solidus line corresponds to the small region of BNN compositions with deficient Nb_2O_5, which was established by all the authors who examined this ternary system. The limited existence region of the compositions depleted in Nb_2O_5 is explained by relatively high ion binding energies in NbO_6 octahedra and, therefore, by a relatively high vacancy-formation energy in niobium.

The X-ray studies [51] showed that Nb_2O_5-rich samples contain $BaNb_2O_5$ as a second phase. Giess *et al* [52] reported that, although the solubility of Nb_2O_5 in BNN is appreciably high, effective high-temperature ferroelectrics are solid solutions with compositions between $Na_{1.2}Ba_{1.9}Nb_5O_{15}$ ($12Na_2O-38BaO-50Nb_2O_5$) and $Na_{0.46}Ba_{2.27}Nb_5O_{15}$ ($4.6Na_2O-45.4BaO-50Nb_2O_5$) which lie on (or almost on) the line $NaNbO_3-BaNb_2O_6$.

The solubility of the excess Na in BNN is insignificant [27]. So, crystals grown from a Na_2O-rich melt ($20Na_2O-30BaO-50Nb_2O_5$) had the composition $11.9Na_2O-36.9BaO-51.2Nb_2O_5$ and opalesced, evidently due to the formation of $NaNbO_3$ in the course of annealing since the distribution factor of $NaNbO_3$ is small, according to [27], even for the stoichiometric composition because the solidus line falls steeply in this region. As a result, growing crystals undergo constitutional overcooling. Such an assertion agrees with observations [53], where the grown crystals were reported to exhibit broad Na-rich striations parallel to the growth axis.

Finding a congruently melting composition was one of the principal goals of the paper devoted to the study of the ternary system $BaO-Na_2O-Nb_2O_5$, because crystals grown from a congruent composition exhibit no growth striations. To determine a congruent composition, one needs precise information on the homogeneity region as well as on the liquidus and solidus surface shapes in this region. As is known, for such a composition the liquidus and solidus lines coincide. It has already been mentioned that BNN liquidus and solidus surfaces are flat inside the homogeneity region, which poses additional obstacles for an exact determination of a congruent composition.

It was found [27] that the temperature maximum on the liquidus line makes up approximately $1440°C$ and the congruent composition lies near $8Na_2O-43BaO-49Nb_2O_5$, which corresponds to the formula $Ba_{2.15}Na_{0.8}Nb_{4.9}O_{14.8}$. In a more recent paper [54], the congruence of such a composition was called into question, and the conclusion was drawn that the congruently melting composition has an excess Nb_2O_5 and lies on the boundary of the homogeneity region. But in crystallized melts of such a composition a considerable amount of second phase is observed, which does not confirm congruence of this composition.

Crystals of congruently melting compositions were reported [27] to be obtained from stoichiometric melts. But on the other hand it was stated [51] that all stoichiometric melts give Na_2O-depleted and Nb_2O_5-rich crystals. The reasons for these controversies are not quite clear, although they may be due to different methods and different accuracy used in determining the sample composition. The crystals grown from a melt which underwent a preliminary double recrystallization, and primarily had a stoichiometric composition, are free of growth striations. An analysis of these crystals has given the composition $7.2Na_2O-42.2BaO-50.6Nb_2O_5$, which corresponds to the formula $Ba_{2.04}Na_{0.92}Nb_5O_{15}$ [55].

The congruence of BNN melting was also considered in [32], where another congruent BNN composition was proposed which differs little from the stoichiometric composition and corresponds to the formula $Ba_{2.11}Na_{0.72}Nb_{5.06}O_{15.12}$.

Giess *et al* [52] reported obtaining striation-free crystals of the composition $Ba_{2.07}Na_{0.86}Nb_5O_{15}$ $(8.6Na_2O-41.4BaO-50Nb_2O_5)$ from a Na_2O-rich melt $(15Na_2O-37.5BaO-47.5Nb_2O_5)$. Crystals were obtained by a slow cooling of a melt-filled crucible and showed up a less optical scattering than Czochralski-grown crystals. But the crystals had a strong tendency towards cracking and possessed cellular structure, the cell walls being obviously rich in the sodium which was in excess in the melt [53].

Thus, according to the available data, striation-free crystals can be grown from a number of BNN melts, three of which—$Ba_{2.15}aN_{0.8}Nb_{4.9}O_{14.8}$ [37], $Ba_{2.11}Na_{0.72}Nb_{5.06}O_{15.12}$ [51] and $Ba_{2.04}Na_{0.92}Nb_5O_{15}$ [32]—are being proposed as congruently melting compositions.

The limited homogeneity region and high energy of niobium vacancy formation in BNN testify to a complicated growth of crystals depleted of niobium. This suggests that the first of the three indicated compositions has a smaller probability for congruent melting. But according to [54], the second and third of the above mentioned melts do not guarantee perfectly striation-free crystals either. A truly congruent composition should apparently be rather close to a stoichiometric one and possess a small excess of barium, which can be compensated by a deficit of sodium.

The search for a congruent composition still remains an urgent task in the technology of obtaining optically high-quality BNN crystals. One should remember that the determination of a congruent composition depends both on the purity of starting agents and the growth conditions, including the composition of the gas atmosphere in the growth chamber (chapter 4, [63]).

5.7   Growth of barium sodium niobate crystals

Single crystals of BNN can be obtained using Kyropoulos [56] and Bridgman [57] methods as well as spontaneous polarization [52, 58]. An attempt was made [59] to grow BNN crystals using the floating-zone method. Each of these methods has its own advantages and disadvantages. However, the basic method for obtaining BNN single crystals, which yields large high-quality crystals for device use, is the Czochralski method [41–43].

The growth of single crystals of BNN encounters a number of difficulties, such as the choice of initial melt composition for obtaining optically homogeneous crystals. Besides, grown crystals are polydomain ones, undergo twinning in transition from a tetragonal into an orthorhombic phase and are prone to cracking due to strong lattice stresses along the c axis that occur near 400–600 °C (figure 5.3) in the course of cooling of a crystallized slab.

Single crystals of barium sodium niobate were obtained for the first time by Van Uitert and collaborators [1, 2, 60]. Now such crystals are grown by leading firms in the USA, England and Japan. In the USSR this problem has also been successfully solved [41–43; chapter 4, 57].

The melting temperature of BNN is 1438 °C [26], and therefore only platinum crucibles can be used as melt containers. Iridium and platinum–rhodium crucibles are not employed, since the crucible material penetrates into a growing crystal which acquires a pale brown colour [36, 61]. BNN crystals are grown in an oxidizing (air, oxygen) or neutral (argon, nitrogen) atmosphere. In the latter case, the grown crystals should be annealed in oxygen.

Platinum crucibles employed for this purpose become deformed because the growth temperatures are close to the platinum melting temperature (1770 °C). So, Van Uitert [62] noticed that after 50 h of work a crucible with 1.5 mm walls almost doubles its volume due to melt expansion during hardening. To avoid such a marked deformation, Van Uitert proposed to use iridium crucibles coated with platinum on all sides. Another outcome is the use of thick-walled platinum crucibles [41, 63]. Sodium and barium carbonates and nitrates and niobium pentoxide are used as starting materials, all the agents having a high degree of purity. Prior to single-crystal growth, carbonates or nitrates are decomposed and barium sodium niobate is synthesized in the solid phase. To this end, the mixture of agents is fired at a temperature of 1000–1200 °C for 4–6 h. The powder thus prepared melts without a significant change in its composition.

BNN crystals are very sensitive to the temperature regime of growing. Van Uitert *et al* (chapter 4, [78]) investigated the influence of various temperature gradients during growth upon the optical properties of the crystals. The authors do not give the values of the temperature gradients during growth; they only describe the growth conditions. In all cases the upper part of the crucible was at the level of the upper inductor turn. Several temperature gradients were tested: (i) 'large-sized'—the crystal was pulled into a non-heated space; (ii) 'medium-sized'—a ceramic 7.5 cm screen was mounted above the crucible, and the whole crystallizer was placed inside a Pyrex dome; (iii) 'small-sized'—the conditions are the same as in the first case but inside the growth chamber a temperature of approximately 250 °C was maintained, the temperature gradient above the melt being approximately 200 K cm^{-1} with conditions excluding phase transition during growth; (iv) 'moderate-sized + small-sized'—the conditions are the same as in the third version, but moderate vertical and radial temperature gradients were created using a reflecting platinum ring fixed above the crucible.

Crystals grown with a 'large' temperature gradient (version (i)) were prone to cracking and exhibited noticeable and uncontrollable striations. Crystals containing regions free of striations and scattering centres were obtained at small temperature gradients (conditions (ii) and (iv)).

The study of growth striations with a vertical temperature gradient showed that striations in BNN crystals decrease if the heat losses from the melt surface are reduced. In view of this fact, and proceeding from the assumption that striations result from composition variations due to temperature fluctuations on the crystal–melt interface which are caused by the instability of thermal flows in the melt, Zupp *et al* [61] concluded that BNN crystals should necessarily be grown under conditions of a low temperature gradient. Indeed, the lowering of the temperature gradient in the melt made it possible to decrease convection, and crystals grown under such conditions had a less pronounced striation. Since melts heated by high-frequency currents exhibit significant convection due to a non-uniform crucible heating, it was proposed [61] to use a resistance furnace as a heating source. In this case, convection fluxes in the melt are suppressed by creating in the crucible a region of almost unchanged temperature from bottom to top.

For a stronger suppression of convection, the crucible was half-filled with melt and then capped with a platinum lid with a hole. The seed crystal was fixed to a sapphire rod and pulled at a rate of 6.6 mm h^{-1}, the rotation rate being 38 rpm. Immediately after separation from the melt, the crystal was taken out of the furnace at a rate of 28.5 mm min^{-1}. Using such technology, non-cracking crystals with 6 mm diameter were obtained [61], while larger crystals were prone to cracking. A limited crystal diameter is one disadvantage of this method, because crystals of such a diameter are of limited use. Besides, employment of a resistance furnace complicates observations of the crystal growth; temperature control in the course of growth is also complicated due to a large thermal inertia of the system. A low temperature gradient on the crystal–melt interface hampers the removal of crystallization heat during growth. This is especially important for BNN because crystals with tungsten bronze-type structure have a low heat conductivity. To provide an optimal gradient in a resistance furnace is rather difficult, whereas by using high-frequency heating one can vary the temperature gradients between wide limits.

To decrease the temperature gradient in the melt, Brice *et al* [64] proposed to use a platinum screen hung in the melt on a platinum wire. Such a separating screen suppresses thermal convection near the crystallization front, and crystals grown under such conditions exhibit minimum striations. It should be noted, however, that in this case it is difficult to maintain a constant diameter of the crystal, and this complicates the growth process.

When using the Czochralski method for growing BNN crystals, one should constantly control the power applied to the crucible since the crystal tends to grow very unsteadily, which leads to a spontaneous sharp increase or decrease of its diameter or even a separation from the melt. This is particularly the case with small temperature gradients on the interface. This is caused by a release of latent crystallization heat, which does not have time to be removed from the crystallization front due to the low heat conductivity of the crystals.

Stabilization of growth conditions, which is especially important in growing a crystal to a required diameter, is promoted by the gas cooling of the crucible bottom; this procedure changes the temperature distribution in the melt [54]. The main effect occurring on the interface as a result of crucible bottom cooling consists of a reshaping of the crystallization front with the formation of an interface with a convex side facing the melt. Besides, when the crucible bottom is cooled, the melt motion is stabilized and convection becomes more regular. But the effect of crucible bottom cooling is insignificant in magnitude and should be applied only in specific cases: i.e. under conditions of small temperature gradient and almost flat interfaces.

A serious problem for BNN technology is the tendency of these crystals towards cracking. As has already been mentioned, the main reason for cracking is an anomalously large variation of the lattice parameter along the c axis in the region of the ferroelectric transition between 400 and 600°C (figure 5.4). Clearly, the presence of a temperature gradient along the c axis provokes cracking in passing through this region.

It was demonstrated [32] that, due to a different content of $BaNb_2O_6$, the central and peripheral regions of the crystal have different thermal expansion coefficients. Therefore in the course of cooling there appear large additional stresses that intensify longitudinal (in the case of growth along the c axis) crystal cracking.

The content of $BaNb_2O_6$ in the centre and on the periphery of the crystal makes up respectively 66 and 69 mol.%, according to [32]. In our opinion, such a difference in the compositions can only be due to different thermal conditions of crystallization in the centre and on the periphery of the crystal, because in the course of cooling the diffusion in the solid phase can be ignored. The main reason for this phenomenon may be a concave shape of crystallization front due to inappropriately chosen temperature gradients. Growth of large-sized crystals may be accompanied by melt overheating in the centre due to a low heat conductivity of the crystal and a poor removal of crystallization heat from the interface. One can avoid a composition variation along the crystal diameter and the associated additional tendency towards cracking by maintaining a flat crystallization front in the process of growing.

The cracking of BNN crystals perpendicular to the growth axis (for c axis crystals) may be induced, according to [32], by striations that occur upon large growth rate fluctuations.

Ballman *et al* [24] point to another reason for BNN crystal cracking. The high-temperature strains that appear during crystal growth may cause small-angle boundaries and mechanical twins that are also responsible for cracking.

The tendency of BNN crystals for cracking is different for different compositions. According to Bonner *et al* [32], BNN crystals of stoichiometric composition have a strong tendency towards cracking, the minimum cracking being exhibited by crystals with 69 mol.% of $BaNb_2O_6$, which corresponds to the formula $Ba_{2.04}Na_{0.92}Nb_5O_{15}$. In their opinion it is only in this case that we deal with a congruent melting of the starting material, which leads to equality of the coefficients of thermal expansion in the centre and on the periphery of the grown crystal.

The use of seeds of different orientations shows that *a* axis crystals are much more prone to cracking than *c* axis crystals. This is connected with the fact that radial temperature gradients due to cooling will induce non-symmetric strains in *a* axis crystals (figure 5.4).

To reduce cracking in grown BNN crystals, Ballman *et al* [24, 65] employed an additional resistance furnace placed above the crucible. Such a heater made it possible, first to maintain crystal temperature during growth above 600 °C, and second to soften the crystal after-growth cooling regime by placing the crystal in the central part of the furnace whose temperature was decreased at a controlled rate from 600 to 400 °C. In a crystallizer of this construction, the authors of [65] studied the influence of the cooling rate upon crystal cracking. They found the empirical dependence of the optimal cooling rate on the crystal diameter. Ballman and Carruthers divided the whole temperature range from room temperature to the melting point into three temperature intervals and indicated the cooling laws for them. In the first interval, from the crystallization temperature to 600 °C, the cooling rate must obey the law $200/R^2$ K h^{-1}, where R is the crystal radius in cm, and in the second interval, from 600 to 400 °C, it must obey the law $150/R^2$ K h^{-1}. In the third interval, from 400 °C to room temperature, the cooling is no longer critical, and the crystal can be taken out of the furnace as soon as this temperature is reached. The indicated dependences of the cooling rate were established for *c* axis crystals. It was emphasized that the most critical is the cooling rate in the second temperature interval, especially for crystals grown along the *a* (or *b*) axis. For such crystals, the cooling rate in the second interval should be lowered to $5/R^2$ K h^{-1}. If the growth direction differs from that of crystallographic axes, the optimal cooling rate in this interval depends on the disorientation angle and changes linearly from $150/R^2$ to $5/R^2$ K h^{-1}. But even such low rates caused cracking, which led the authors to the final conclusion that for a complete avoidance of cracking the cooling rate in the second interval should not exceed ~ 30 K h^{-1} for *c* axis crystals and 5 K h^{-1} for *a* axis ones.

Zharikov *et al* [41] and Valyashko *et al* [66] used a platinum screen heated by a high-frequency field as an additional heater. This was a two-section inductor, allowing the combination of the growth process with primary annealing.

The above mentioned technological studies have shown that BNN crystals are very sensitive to thermal conditions of growth, and therefore a thorough experimental choice of thermal gradients both in and above the melt is needed. It is desirable that the temperature gradients be small since crystals obtained under these conditions exhibit less optical defects. The various technological improvements proposed by different authors (bars, reflecting screens, additional heaters etc) do not prevent completely the appearance of the defects in grown crystals. An interesting innovation is air-cooling of the crucible bottom, which permits the creation of more stable thermal conditions of growth. The tendency of crystals towards cracking can be reduced by an appropriate choice of the melt composition, by creation of a flat crystallization front and by using a slow cooling of a grown crystal, especially in the ferroelectric phase transition region ($600-400\,^\circ$C).

Below we present the technology of growing BNN single crystals which was used in [41]. Crystallization was carried out on the installation 'Donets-1'; crystals were pulled in air from platinum crucibles $\varnothing = 70$ mm and $h = 70$ mm in size and with a wall thickness of 5 mm. Such thick-walled crucibles are less subject to deformation. The crystallizer construction was analogous to that depicted in figure 4.37. BNN crystals $20-30$ mm in diameter and $40-60$ mm in length were grown (figure 5.28). The optimal growth parameters for crystals of the indicated size were: a pooling rate of 5 mm h^{-1} and a growth rate of 20 rpm. Large crystals cracked, as a rule, in the course of annealing or in treatment, which can evidently be explained by the appearance of significant stresses due to large temperature gradients along the diameter and length of the crystal. It was established that the size of the

(*a*)  (*b*)

Figure 5.28 BNN crystal grown from the composition $Ba_{2.09}Na_{0.72}Nb_{5.02}O_{15}$ under isothermal conditions (*a*) and at a vertical gradient ≈ 100 K cm^{-1} (*b*).

crystal diameter is more critical for cracking. Zharikov *et al* [41] grew BNN crystals from the stoichiometric composition $Ba_2NaNb_5O_{15}$ and two congruent compositions $Ba_{2.11}Na_{0.72}Nb_{5.06}O_{15.12}$ and $Ba_{2.04}N_{0.92}Nb_5O_{15}$ (according to [51] and [32] respectively). Growing was carried out under comparable conditions. Crystals obtained from the melt composition $Ba_{2.04}Na_{0.92}Nb_5O_{15}$ exhibited properties very close to those of crystals grown from stoichiometric melts. Experiments showed that crystals grown from the melt composition $Ba_{2.11}Na_{0.72}Nb_{5.06}O_{15.12}$ are more prone to cracking than crystals grown from a stoichiometric melt. This is possibly explained by the fact that crystals grown from a non-stoichiometric melt possess a larger amount of vacancies and embedded atoms, whose concentration leads to dislocations and stress fields near them. These stresses just cause cracking of crystals of non-stoichiometric composition. The grown crystals were annealed directly in the growth device. After separation from the melt the crystals were annealed without temperature lowering during elevation for 2–3 h, and then the temperature was decreased at a rate of $60–100$ K h^{-1}.

5.8   Twinning

Let us consider twinning in barium sodium niobate in more detail because crystals exhibiting twinning can find very limited practical application. As has already been mentioned (§ 5.1), micro-twinning proceeds upon cooling during a phase transition from a tetragonal (4mm) to an orthorhombic (mm2) modification at a temperature of approximately 260 °C. The twinning process acts as a mechanism which weakens the stress occurring in the crystal due to the phase transition. The size of the unit cell along the *c* axis is almost constant under the phase transition, but the parameters *a* and *b* of the unit cell are equal in the tetragonal and distinct in the orthorhombic phase. As a result, a transition with a simple orientation reversal of crystallographic axes would induce great stresses. These stresses are removed by crystal twinning.

Twinning occurs along $(100)_T$ and $(010)_T$ planes. It divides the crystal into a number of rectangular prismatic blocks (figure 5.29) which are parallel to the $[001]_T$ axis of tetragonal symmetry. As reported in [26], the $[100]_O$ direction of an orthorhombic unit cell lies at an angle of 45° to the $[100]_T$ direction of a tetragonal cell. Twin blocks are so aligned that the $[100]_O$ direction of one block is perpendicular to the $[010]_O$ direction of neighbouring blocks (figure 5.30).

Since twinning is induced by stresses, the distribution of twins in a material free of defects must be uniform. Structure violations create additional local stresses and are thus responsible for a non-uniform density of twin blocks. These violations are striations, cellular structure, concentration of point defects and defects spreading inside the crystal from the seed [23, 53].

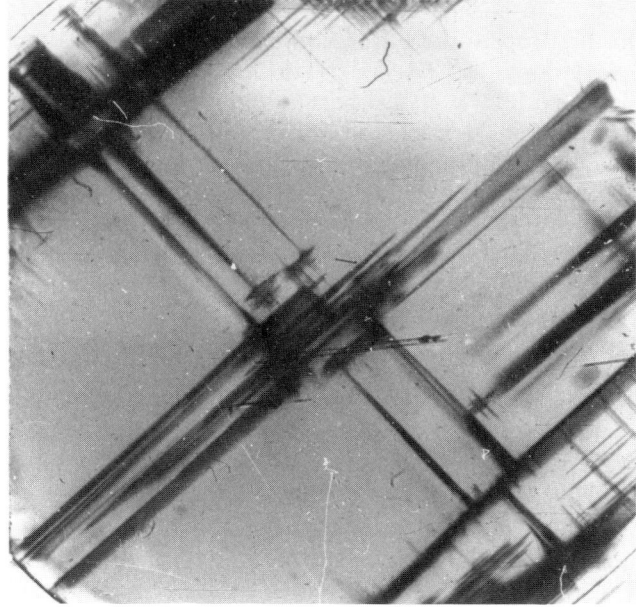

Figure 5.29   Twins in a BNN single crystal [41]. Enlargement × 10.

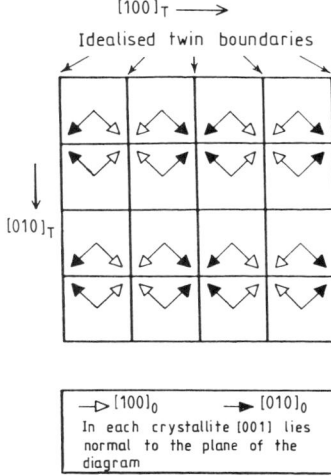

Figure 5.30   Schematic drawing of twin blocks in BNN crystals [23].

Micro-twinning in barium sodium niobate can be suppressed by crystal cooling from the temperature of the non-ferroelectric phase transition with pressure applied along the $[100]_o$ axis (the a axis) or $[010]_o$ (the b axis) of the orthorhombic cell. The crystal becomes free of twins and has a smaller parameter a of the orthorhombic cell, which is parallel to the load axis. Twin blocks can be removed by applying an external pressure of about 70 kg cm^{-2} at 250°C with subsequent cooling down to room temperature under this load. According to [33], detwinning may be performed by applying a sufficiently high pressure also below phase transition temperatures: from 2100 to 3500 kg cm^{-2} at 25°C, 700 kg cm^{-2} at 175°C and less than 7 kg cm^{-2} at 300°C. However, such loads should be applied with care because such high pressures (especially at room temperature) may lead to crystal destruction.

Mechanically, detwinned BNN crystals are stable, and when cut and polished they do not exhibit twins. But in a crystal repeatedly heated up to 300°C twin blocks appear again.

A detailed study of the detwinning process [23] made it possible to establish that an application of the indicated load does not always result in a desirable detwinning. For each particular sample there exists a particular optimal pressure under which micro-twins completely disappear. It depends on the density and distribution of the twin blocks, which are determined by the presence of defects and impurities in the crystal. Smaller loads affect weakly, or not at all, the twinned structure, while loads above the optimal values increase the density of the twin blocks.

The magnitude of optimal loads for detwinning increases with increasing density of twin blocks. Figure 5.31 presents a histogram of optimal loads obtained in [23] for 20 crystals with different densities of twin blocks. It was concluded that it is practically very difficult to reach complete detwinning of samples with a high density of twin blocks, since the applied voltages do not affect the crystal defects which are the main cause of twinning.

In a number of papers it was established that the temperature and rate of cooling produce no direct effect upon the efficiency of detwinning. This emphasizes the decisive role of crystal defects in the twinning process. When a crystal is cooled from the phase transition temperature, twins appear first of all in the areas containing defects. When a crystal is heated through the phase transition temperature, the twins fixed by defects disappear later than the other ones. The attempt to remove defects by a long-term annealing at high (700–1300°C) temperatures was unsuccessful. Thus a large number of defects makes the state of a detwinned crystal unstable, and when heated up to a temperature even substantially lower than the phase transition temperature, the crystal again exhibits twins in the vicinity of defects. The stability of detwinning, and therefore a high quality of optical elements, can be reached only in regions with a low density of twins and defects responsible for twinning.

Along with mechanical detwinning, twinning in barium sodium niobate can be eliminated by creating conditions under which the tetragonal phase

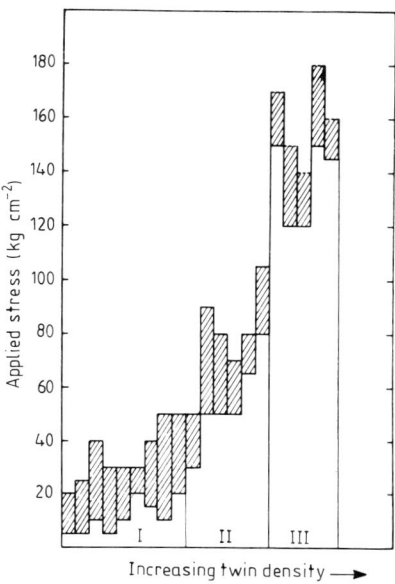

Figure 5.31 Histogram of optimal mechanical loads for BNN crystal detwinning as a function of twin block density [23]. Region I corresponds to crystals with a low density of twin blocks; region II to crystals where some regions have an anomalously high twin density; and region III to crystals with an anomalously high twin density over the entire volume.

remains stable at room temperature. This may be reached through a change of the crystal composition and replacement of part of Ba^{2+} or Na^+ cations by other ions. As mentioned above, the temperature of the transition from tetragonal to orthorhombic phase depends on the ratio Nb/Ba, and the room-temperature tetragonal phase may exist at Nb/Ba $\geqslant 2.8$ (chapter 4, [78]). As reported in [22], BNN crystals free of twins may be obtained with a 10 wt.% excess of Nb_2O_5. But crystals of such compositions are strongly strained and crack easily both during growth and during cooling. Besides, they scatter light and their T_{phm} for SHG is below room temperature, which is rather undesirable from the viewpoint of practical application [21].

Crystals are usually made detwinned after the polarization process, since heating a crystal up to the temperature of the non-ferroelectric transition was found to have no effect upon its single-domain state. A sample intended for detwinning is usually given a parallelepiped shape with edges directed along the a, b and c axes (figure 5.32). The facets to which mechanical stress is applied are made plane parallel to avoid the development of micro-cracks on surface bendings when a compressive stress is applied. Sample surfaces perpendicular to the c axis of the crystal should be polished in order that

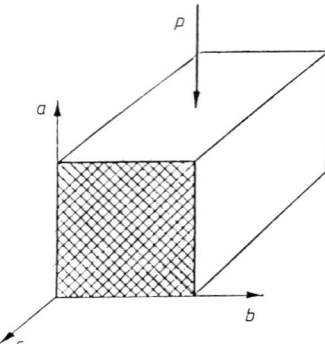

Figure 5.32   Drawing of a BNN crystal intended for detwinning. Twinning planes are indicated by fine lines.

the appearance and disappearance of twins under phase transition could be observed, from which one can determine the optimal heating temperature and choose the magnitude of loading necessary for detwinning. The heating rate is about $60-80$ K h^{-1}. After the visually observed disappearance of twins, the temperature increase is stopped, and mechanical load is applied to the crystal. The magnitude of the initial load is $60-70$ kg cm^{-2}. The crystal is then cooled to a temperature $30-40\,°C$ lower than the phase transition point. If twins appear again, the heating cycle is repeated and the load is increased by $20-30$ kg cm^{-2}.

The load necessary for detwinning is chosen individually for each sample, and at first the applied load should not be large because an exceeded optimal pressure leads to an increased twin density. The range of loads allowing micro-twinning to be suppressed lies usually in the limits $70-170$ kg cm^{-2}. After the required suppression value is chosen, the crystal is cooled down to room temperature under the same load. The detwinning thus performed is stable, which enables the crystal to be treated mechanically without fear of a repeated twinning. One should remember, however, that crystal heating through the phase transition temperature again induces twinning.

5.9 Polarization

For use as electro-optic and nonlinear optic elements in laser systems, BNN crystals should be single domain. To obtain single-domain crystals, the poling procedure is used, i.e. heating and cooling to room temperature in a DC electric field applied along the ferroelectric polar axis. This process induces not only domain orientation reversal but also variations in the defect structure of the material.

Several regimes of BNN crystal poling have been proposed. For example, it was recommended [2] that polarization proceed at a temperature several degrees (5–10 K) below the Curie point in a DC electric field of about 250 V cm^{-1} along the c axis during a time determined by the sample size, i.e. approximately one hour per cm of sample length. In a later paper [35] it was recommended that the time of the process be increased up to three hours per cm of sample length. Rice *et al* [33] report that it is preferable to pole samples without any heating at all. The values of electric field strength presented in the literature also differ significantly (e.g. 100 V cm^{-1} according to [33] and 250 V cm^{-1} according to [2]).

There exist two ways of BNN crystal poling: the high-temperature method in which a sample is heated through the Curie point and a relatively weak electric field $E = 100–100$ V cm^{-1} is applied; and the low-temperature method in which a sample is heated up to 200°C in fields $E = 10^3–10^5$ V cm^{-1}.

Low-temperature BNN poling is described in [67] and [68]. During this process, the current running through the crystal increases. The authors explained this fact in terms of dielectric aging, and associated it with the formation of oxygen vacancies, which in their opinion facilitates the domain orientation reversal. The time dependences of the current density for several samples obtained in [44] are presented in figure 5.33. For the majority of samples, three stages of the process are observed (curve 1): at the first stage

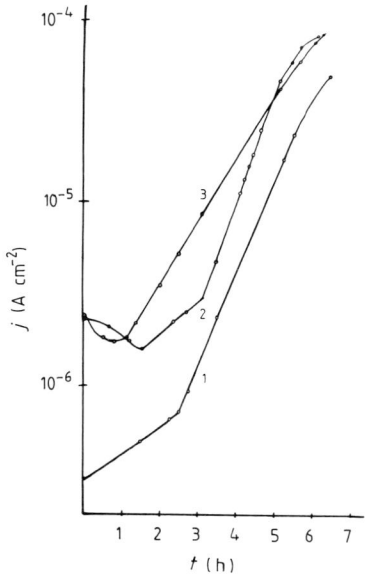

Figure 5.33 Time dependences of the densities of currents flowing through BNN samples in the course of polarization at 160°C with a field of 3.4 kV cm^{-1} [44].

the current decreases; at the second and third it increases by an exponential law with the time constant $\tau_1 = 10^4$ s and $\tau_2 = 3 \times 10^3$ s respectively. But in some cases, either the first or the second stage is not observed, as shown by curves 2 and 3. The completeness of polarization was controlled by the magnitude of the half-wave voltage and the output of the second harmonic in different regions of the samples. This control has shown that the process stops only after the onset of the third stage. It is known that in some crystals (e.g. in $BaTiO_3$ and TiO_2 [69]) a high-temperature increase of electric conductivity with time in strong electric fields is due to oxygen losses. In BNN crystals, oxygen losses may also take place when a sample subjected to poling is placed in a reduction medium (silicone liquid) and is exposed to an electric current. This may be confirmed by the polarization-induced variations in the transmission spectrum (see figure 5.17).

During high-temperature polarization [36], a sample is heated by 50–60 K above the Curie point ($\approx 630°C$). A constant current density of the order of 5×10^{-3} A cm^{-2} throughout the sample is maintained during the whole process, the voltage increasing almost by an order of magnitude with increasing resistance of the sample.

Also during high-temperature polarizaton [35, 36], hydrogen diffusion into the crystal is observed, which is accompanied by a displacement of the visible boundary from the positive to the negative electrode. The boundary becomes visible due to a change of the refractive index in the crystal region wherein the hydrogen diffusion has taken place. According to [35], the penetration of hydrogen ions into a crystal promotes domain orientation reversal and onset of the single-domain state. After the visible boundary has passed through the whole crystal, the crystal is cooled down to room temperature under the field. It should be noted that, after the visible boundary has reached the negative electrode, the voltage on the crystal falls somewhat and then exhibits periodic oscillations. At first these oscillations have a small amplitude and a complicated shape (figure 5.34(a), (b)). In about 25 min the oscillations become sinusoidal with an amplitude of approximately 20 V cm^{-1}, which makes up 25% of the voltage applied to the crystal (figure 5.34(c)). For samples with dimensions of $5 \times 2.5 \times 5$ mm^3 along the a, b and c axes respectively, the oscillation period is equal to 5 s, whereas for samples with dimensions of $10 \times 5.4 \times 8.4$ mm^3 along the same directions, the period increases up to 20 s. If the voltage on the crystal is kept constant, then the current oscillations will not tend to dampen (figure 5.34(c)). So their amplitude and period remained unchanged for three hours. Temperature lowering leads to a decrease of the oscillation amplitude and an increase of the period. They vanish at a temperature of 350°C. Since the oscillations are regular, their origin cannot be associated with domain orientation. In all probability, they are due to the negative differential resistance of the sample in the near-electrode region.

In the course of poling, if it proceeds at a high temperature or in a silicone

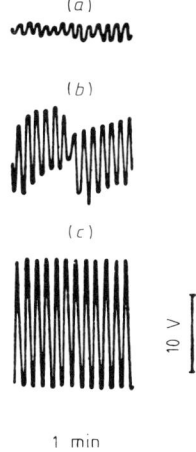

Figure 5.34 Time dependence of the voltage of a BNN sample one minute (*a*), 15 minutes (*b*) and 40 minutes (*c*) after high-temperature polarization [44]. The sample temperature is 535°C; the current density is 1.6 mA cm^{-2}.

liquid, the crystals become brown. The transmission spectra of such crystals for wavelengths from 0.35 to 2.0 μm are given in figure 5.17. The crystals poled under different conditions are seen to have distinctions in the visible region of the spectrum.

Arsenev *et al* [45] associated the absorption of BNN crystals in the visible region with Nb^{5+} reduction up to Nb^{4+}. This process may proceed under high-temperature or low-temperature polarization carried out in a reduction medium, although niobium ions may be reduced for different reasons. As stated in [35], under high-temperature polarization the H$^+$ protons diffuse into the crystal, and therefore to meet the condition of electroneutrality the cations in this crystal should be reduced. A Nb^{5+} ion changes valency for Nb^{4+} rather easily, which affects the absorption spectra in the visible. As mentioned above, under low-temperature polarization in oil, crystals lose oxygen, which induces niobium ion reduction.

The analysis of photoconductivity of BNN crystals has shown that in poled samples the light-induced high electric conductivity does not disappear immediately after the light is off but remains for a long time. So, after the action of light with $\lambda = 0.440$ μm and intensity of 1 mV cm^{-2}, the current relaxes to the initial value for more than 24 h. The voltage cut-off does not affect the relaxation time (figure 5.35).

The long-term relaxation of photoconductivity in polarized BNN crystals is not a unique phenomenon but is observed in some semiconductors (ZnO, PbO, Cu$_2$O [70]). According to the majority of models, it is connected with the existence of microscopic potential barriers caused by various

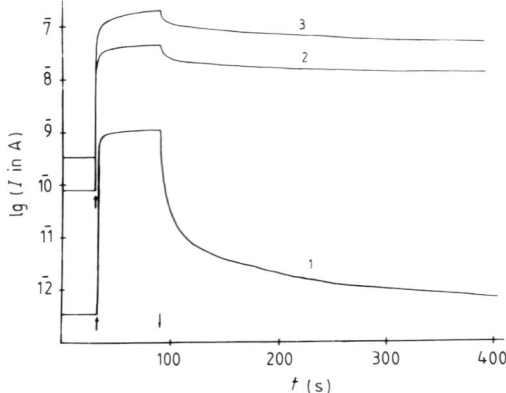

Figure 5.35   Kinetics of photoconductivity of BNN crystals: (1) polydomain; (2) polarized above the Curie point; (3) polarized at $160\,^{\circ}$C [44]. Arrows indicate the beginning and the end of crystal exposure.

inhomogeneities, which may be due to a fixing of oxygen vacancies and impurities at block boundaries in the electrolysis accompanying polarization.

As mentioned above, the domain structure of BNN crystals is established by etching in the mixture $2HF + 3HNO_3$. On sample surfaces perpendicular to the growth axis, alternating dark and light concentric circles are observed. A typical picture of the crystal domain structure is represented in figure 5.10. But the etching technique determines the domain structure only near the etching surface, whereas the degree of polarization of the whole crystal remains undetermined. The latter problem can be solved by measuring the half-wave voltage $V_{\lambda/2}$ for the transverse electro-optic effect. It should be noted that the difference in the composition of BNN crystals does not affect significantly the magnitude of the half-wave voltage which is 1830 ± 40 V [13]. All other $V_{\lambda/2}$ values that one comes across in the literature [1, 33, 34, 39] are a consequence of an incomplete polarization of the crystals.

The degree of BNN polarization can also be controlled using the second harmonic, whose intensity changes when the crystal is rotated around the c axis. For a single-domain crystal, the signal of the second harmonic produces an oscillogram with zero intensities in the minima, whereas in a polydomain specimen the minimum values of the second harmonic output are not equal to zero [34]. The completeness of crystal polarization can also be judged by the temperature dependence of the SH intensity [36, 71, 72].

5.10   Optical imperfections

Some defects of BNN crystals are associated with the growth conditions, and can therefore be eliminated by an appropriate choice of these conditions.

The defects hamper the device use of BNN in electro-optics and nonlinear optics, since they induce local changes of refractive indices in the crystals [39] and thus lower their optical quality. So, a BNN crystal used as a nonlinear medium for SHG should have very small optical losses. Losses equal to 2.4% were observed to decrease the radiating power of the second harmonic by half. If a crystal does not absorb radiation with a wavelength of 1.06 μm, then the scattering and diffraction of light on the refractive index gradients, which may be due to defects in the crystal, become a dominating mechanism of the losses [73].

Let us consider in more detail the defects that lower the optical quality of BNN crystals and limit their application in quantum electronics and nonlinear optics.

5.10.1   Growth striations

The most widespread type of defect typical of many oxide crystals grown from the melt by the Czochralski method is growth striation, which induces periodic variations of the refractive index along the growth axis. In SHG experiments, the polarization of light scattered by growth striae was found to be different [22]. This leads to a scattering of the second harmonic, and therefore to a decrease of its intensity. The striae are the main optical defects in BNN crystals which are difficult to eliminate during growth. In these crystals, two types of striae occur: thin striae, 3–10 μm, and broad striae, 25–70 μm [53]. According to [42], striation does not depend on the ferroelectric nature of the material; instead the presence of striae in a crystal, as has already been mentioned, stabilizes a certain domain structure (figure 5.11). The ability of striae to fix domains depends on the stria thickness. In this respect, the most effective are broad striae. Brehm *et al* (chapter 4, [80]) showed that striae also exhibit inhomogeneity in their mechanical properties.

Striae contain, as a rule, different amounts of impurities. A mass spectrum analysis [39] has shown, however, that it is the composition rather than the impurity concentration that mainly varies in growth layers of BNN crystals. Many of the researchers engaged in growing BNN crystals support this opinion [22, 34, 41, 51, 62].

Striation is customarily associated with temperature fluctuations at the crystal–melt interface; these are due to an unstable thermal convection in the melt [74, 75]. The appearance of inhomogeneities in the composition of BNN crystals is explained by the fact that barium sodium niobate has a wide range of existence, and the crystal composition is affected by temperature fluctuations at the crystallization front.

Linares and Sigsway [53] believe that striations result not only from temperature fluctuations but also from crystal faceting, the former being responsible for thin and the latter for broad striae. In their opinion, striation due to faceting can be eliminated by growing crystals in a direction 5° from the [100] direction. Temperature fluctuations can be decreased by weakening the temperature gradient in the melt as well as by lowering the melt level in the crucible. It should be noted that the complete elimination of temperature fluctuations in the melt under induction heating is impossible, because of the strong melt convection caused by a non-uniform crucible heating. Temperature fluctuations, and therefore crystal striation, can be decreased through crucible heating in a resistance furnace. But in this case it is very difficult to control temperature during growth because the inertia of such heaters is great.

A detailed analysis shows that striations can only be completely eliminated by using a severe control over the temperature conditions of growth, temperature gradient in and above the melt, growth rate of the crystals and their composition. But the technological solution of all these problems encounters serious difficulties, which can be avoided by growing BNN crystals in the direction of the *a* axis [33]. In elements cut out of such a crystal, light will propagate perpendicularly to the striae. This prevents light diffraction on the striae, and aberration becomes much weaker. But a comparison of the second harmonic powers for crystals grown along the *c* and *a* axes [76] has shown that in the first case the efficiency of transformation into the second harmonic is higher. This can be explained by the fact that in an *a* axis crystal the phase matching temperature varies along the specimen length due to a composition gradient along the growth axis (*a* axis). Thus, for an effective transformation of laser radiation into the second harmonic, the growth striation in crystals should necessarily be reduced.

5.10.2   Cellular structure

A typical defect in BNN crystals is cellular structure accompanied by filament-like formations: 'whiskers' parallel to the growth axis of the crystal. These 'whiskers' are formed in the corners of cellular blocks and spread inside the crystal, inducing a variation of the refractive index along the growth direction. Cellular structure, and therefore 'whiskers', are formed in crystals at high pulling rates especially from non-stoichiometric or impurity-containing melts [53]. The walls of cellular blocks are at the same time ferroelectric domain walls, because they are not observed in polarized crystals [77]. Electron beam probing has shown that the walls of the blocks are Na-rich and Ba-depleted. Iridium and platinum (crucible materials), which get into the crystal due to their partial oxidation, may segregate among blocks [53].

5.10.3   Dislocations and other optical imperfections

Dislocations in crystals are revealed by etching of polished crystalline plates. A mixture of the acids HF and H_3PO_4 in the ratio $3:1$ at a temperature of about 65°C (with an etching time of 20–40 min), as well as a mixture of HF and HNO_3 in the ratio of $2:3$ at a boiling temperature of 120°C (etching time of 2–7 min), proved to be the best etching reagent for BNN crystals.

The etching pits, which indicate the places where dislocations come out, have the shape of a frustum of a pyramid with sides parallel to the $(100)_O$ and $(010)_O$ planes of an orthorhombic cell. The mean dislocation density for stoichiometric and non-stoichiometric specimens was approximately the same and was equal to 10^3 cm^{-2} [41].

The indicated etching reagents made it possible to reveal some other defects too. Apart from dark pyramidal pits, there exist small light flat-bottomed pits which form rectangular 30 to 80 μm spirals (figure 5.36). For some crystals such formations fill a considerable part of the crystalline plate surface. Similar etch patterns were also observed in [23], but they had a different shape and larger size, which may be due to distinctions in the growth conditions and in the purity of starting agents. Affected by polarization, spiral configurations disappear, although pyramidal parts remain.

In BNN crystals there may also occur gradual variations of the refractive index along the growth axis induced by a change of the composition in this direction. This is explained by the fact that during growth a crystal forces back extraneous impurities with the distribution coefficient $K < 1$. These impurities are accumulated in the melt, and the crystal when pulled becomes gradually enriched with them. Besides, when a crystal is pulled out of a non-congruent melt, the latter changes its composition and causes lengthwise variations of the crystal composition, since barium sodium niobate is a compound of variable composition.

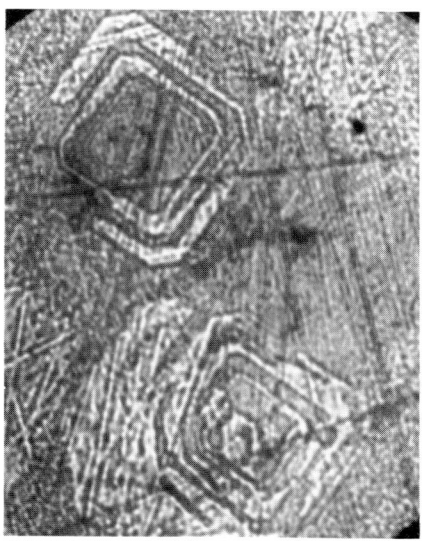

Figure 5.36 Spiral-shaped etch patterns on the (001) plane of a BNN crystal [41]. Enlargement × 570.

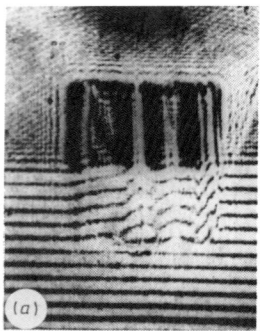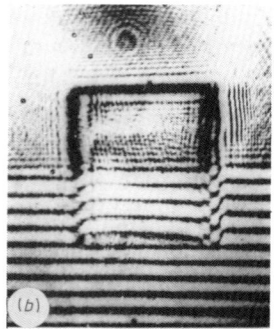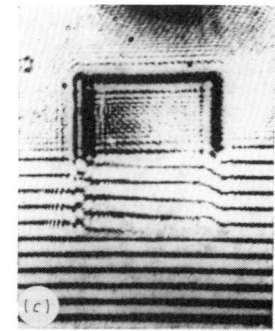

Figure 5.37   Dark and interference patterns (upper and lower halves, respectively) of BNN crystal elements before detwinning (*a*), after detwinning (*b*) and after an additional polishing (*c*) (chapter 4, [57]).

Voronov *et al* (chapter 4, [57]) investigated the optical homogeneity of BNN crystal elements at all stages of polarization and detwinning. It was noticed that after detwinning the side surfaces of the element which are parallel to the applied pressure are deformed, thus inducing inhomogeneity of the refractive index over the crystal volume. A thorough polishing of strained surfaces removed this optical defect (figure 5.37).

Thus, an exacting choice of the chemical composition and technological conditions of growing allows one to obtain optically perfect BNN crystals with a 100% transformation of YAG : Nd laser radiation into the second harmonic.

Some physical and physico-chemical properties of single crystals of barium sodium niobate are listed in table 5.5.

Table 5.5   Basic properties of BNN crystals.

Property	Values	References
Stoichiometric formula	$Ba_2NaNb_5O_{15}$	—
Molecular weight	1002, 1908	—
Density (g cm^{-3})	5.4076	[18]
Hardness (Mohs scale)	5.5	[34]
Melting point (°C)	1430	[12, 15, 34]
Point symmetry group at a temperature T (°C)		
$T < -160$	4mm	[79]
$-163 < T < 300$	mm2	[34, 73]
$300 < T < 560$	4mm	[34, 73]
$T > 560$	4/mmm	[34, 73, 79]
Crystalline structure, T_{room}	Orthorhombic	[1]
Space group	Cmm2	[1, 9]

Table 5.5 *Continued.*

Property	Values	References
Parameters of the unit cell, T_{room} (Å)	$a = 17.59182$ $b = 17.62560$ $c = 7.98980$	[19, 18] (chapter 4, [6])
Volume of the unit cell (Å^3) T_{room}	1238.683	[34]
Curie temperature (°C) (depending on the composition)	552–585	[1, 5, 34, 35, 81–85]
Orthorhombic–tetragonal transition temperature (°C)	260	[1]
Spontaneous polarization (C m^{-2})	0.40	[34, 78, 65, 86]
Dielectric permittivity, T_{room}		
ε_x	238 ± 5	
ε_y	228 ± 5	[34]
ε_z	43 ± 2	
Coefficient of electromechanical coupling (K_t)	0.57–0.60	[14, 80, 83]
Domain of transparency (μm)	0.4–5	[34, 80]
Refractive indices		
$\lambda = 1.064$ μm		
n_x	2.2580	[34]
n_y	2.2567	
n_z	2.1700	
$\lambda = 0.532$ μm		
n_x	2.3676	[34]
n_y	2.3655	
n_z	2.2502	
Thermal coefficient of refractive indices (°C^{-1})		
dn_z/dT	8×10^{-5}	[34, 81]
dn_x/dT	-2.5×10^{-5}	
Angle between optical axes $2V$	13°	[34]
Absorption coefficient (cm^{-1})		
$\lambda = 1.064$ μm	10^{-4}–2×10^{-3}	[81, 87, 88]
$\lambda = 0.532$ μm	40×10^{-3}	[81]
Nonlinear optical coefficients relative to quartz d_{11}		
d_{31}	40 ± 2	[34]
d_{32}	40 ± 4	
d_{33}	55 ± 4	
d_{15}	40 ± 2	
d_{24}	38 ± 2	
Phase matching temperature (°C)		
T_{31}	90–113	[1, 5, 34, 80–82]
T_{32}	80–100	

Table 5.5 *Continued.*

Property	Values	References
Matching angle ('oo–e' interaction)		
($\lambda = 1.064\ \mu$m)		
θ_{31}	73°45′	[34]
θ_{32}	75°26′	
Linear electro-optic coefficients		
(10^{-12} m V^{-1})		
$\lambda = 0.633\ \mu$m		
r_{13}	15 ± 1	[34]
r_{23}	13 ± 1	
r_{33}	48 ± 2	
r_{42}	92 ± 4	
r_{15}	90 ± 4	
$r_{33} - (n_x/n_z)^3 r_{13}$	31.9 ± 0.4	
$r_{33} - (n_y/n_z)^3 r_{23}$	35.0 ± 0.4	
Half-wave voltage $V_{\lambda/2}$ (V)		
$\lambda = 0.633\ \mu$m	1830	[34]
Quadratic electro-optic coefficients		
(m^4 C^{-2})		
$g_{12} = \dfrac{r_{i3}}{2\varepsilon_0(\varepsilon_3 - 1)P_s}(i = 1)$	0.052 ± 0.006	[34]
$g_{11} = \dfrac{r_{i3}}{2\varepsilon_0(\varepsilon_3 - 1)P_s}(i = 2)$	0.045 ± 0.006	
$g_{44} = \dfrac{r_{ij}}{\varepsilon_0(\varepsilon_j - 1)P_s}\begin{array}{l}(i,j = 4; 2)\\ (i = 5; 1)\end{array}$	0.17 ± 0.01 0.11 ± 0.01	
Thermal expansion coefficient	10.4 ± 11.4	[3, 34]
(10^{-6} °C^{-1}) T_{room}		
Heat conductivity (W cm^{-1} grad^{-1})	0.035	[81]
Elastic constants (10^{11} H m^{-2})		
C_{11}	2.39	[83]
C_{12}	1.04	
C_{13}	0.50	
C_{22}	2.47	
C_{23}	0.52	
C_{33}	1.035	
C_{44}	0.65	
C_{55}	0.66	
C_{66}	0.76	
Elastic compliances (10^{-12} m^2 H^{-1})		
S_{12}	-1.98	[15, 88]
S_{13}	-1.20	
S_{23}	-1.25	
S_{22}	5.14	

Table 5.5 *Continued.*

Property	Values	References
S_{32}	8.33	
S_{44}	15.4	
S_{55}	15.4	
S_{66}	13.2	
Piezoelectric constants (C m^{-2})		
e_{15}	2.8	
e_{24}	3.4	
e_{31}	-0.4	
e_{32}	-0.3	
e_{33}	4.3	
Piezoelectric modules (10^{-11} C H^{-1})		
d_{15}	4.2	[15, 88]
d_{24}	5.2	
d_{31}	-0.7	
d_{32}	-0.6	
d_{33}	3.7	
Acoustic quality Q	$>1 \times 10^5$	[85]
Photoelastic constant P_{31}	0.2	[80]

6 Other Crystals with Tetragonal Potassium Tungsten Bronze-type Structure

In this chapter we again deal with alkaline and alkaline earth niobates and tantalates with tetragonal potassium tungsten bronze (TPTB)-type structure. As in the preceding chapters, we analyse their dielectric, optical, nonlinear optical and physico-chemical properties, as well as the phase diagrams and the methods of obtaining single crystals. Individual compounds are considered in a sequence corresponding to the occupation of pentagonal, tetragonal and trigonal vacancies in their crystal lattices by alkaline and alkaline earth ions. The general tendency is such that as the chemical composition of a compound becomes more complicated, there occur additional phase transitions, and the crystal composition differs more and more strongly from the stoichiometric composition of the melt. All this complicates the technology of obtaining single crystals.

6.1 Barium sodium potassium niobate crystals

One of the main defects of barium sodium niobate crystals is that they exhibit a non-ferroelectric phase transition near $260\,°C$, which is accompanied by crystal lattice twinning. This makes it difficult to manufacture elements of them for transforming radiation with $\lambda = 1.064\ \mu m$ into the second harmonic. Besides, for attaining a non-critical phase matching, crystals should be heated up to $80–100\,°C$, which also complicates their use for an intracavity SHG.

The system $K_x Na_{1-x} Ba_2 Nb_5 O_{15}$, in which sodium was partially replaced by potassium, was investigated in 1969 (chapter 5, [58], [59]; [1]). For a certain x value such a replacement eliminates crystal twinning and lowers the phase matching temperature up to room temperature. The dielectric, electro-optic and nonlinear optical properties of single crystals of this system were investigated later (chapter 5, [68]; [2–4]). In the tetragonal ferroelectric phase the crystals belong to the symmetry space group C_{4v}. The space group in the non-polar phase above the Curie temperature T_C remains unknown. It may be centrosymmetric if one takes into account the nature of the phase transition and the equality to zero of the nonlinear coefficient d_{31} (the latter fact was experimentally established).

6.1.1 Dielectric properties

$K_xNa_{1-x}Ba_2Nb_5O_{15}$ crystals were used to study the temperature dependences of the dielectric permittivity ε_c, from which their Curie temperature was then determined (figure 6.1). As is seen in the figure, it decreases monotonically from 576°C (pure barium sodium niobate) to 480°C for crystals obtained from a melt with $x = 0.6$. This is evidently connected with a slight decrease of the displacement of 'ferroactive' ions from the symmetric positions which they occupy in the paraelectric phase. This follows from the relation derived for other niobates (chapter 4, [32]; see also § 8.1):

$$T_C = (2.00 \pm 0.09) \times 10^4 (\Delta z)^2 \tag{6.1}$$

where Δz is the displacement of 'ferroactive' ions and the T_C values correspond to the Kelvin scale. Figure 6.1 also shows the temperature variation for the transition from the orthorhombic to the tetragonal phase, which determines the crystal detwinning temperature.

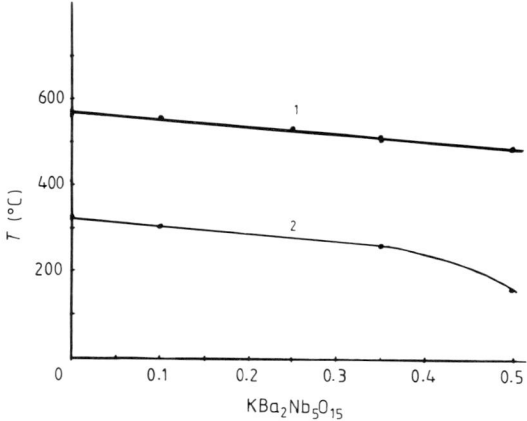

Figure 6.1 Dependences of the Curie temperature (1) and the temperature of phase transition 4mm → mm2 (2) of $K_xNa_{1-x}Ba_2Nb_5O_{15}$ crystals on potassium content in the initial melts [4].

The temperature dependences of ε_c for crystals of different compositions near the corresponding Curie temperatures are plotted in figure 6.2. Measurements were taken at a frequency of 1 kHz. On the basis of the $\varepsilon(T)$ graphs, the inverse dependences $1/\varepsilon$ were constructed; these are plotted in the same figure. The Curie–Weiss constants, calculated from these dependences, are respectively equal to $C_0 = 0.5 \times 10^5$ and $C_1 = 4 \times 10^5$ below and above the Curie point. From the data presented it follows that the replacement of Na ions by K ions does not induce a significant smearing of the ferroelectric phase transition.

This can be explained by the ordered distribution of potassium ions in crystallographic positions A1, because of the difference in the ion radii of sodium and potassium.

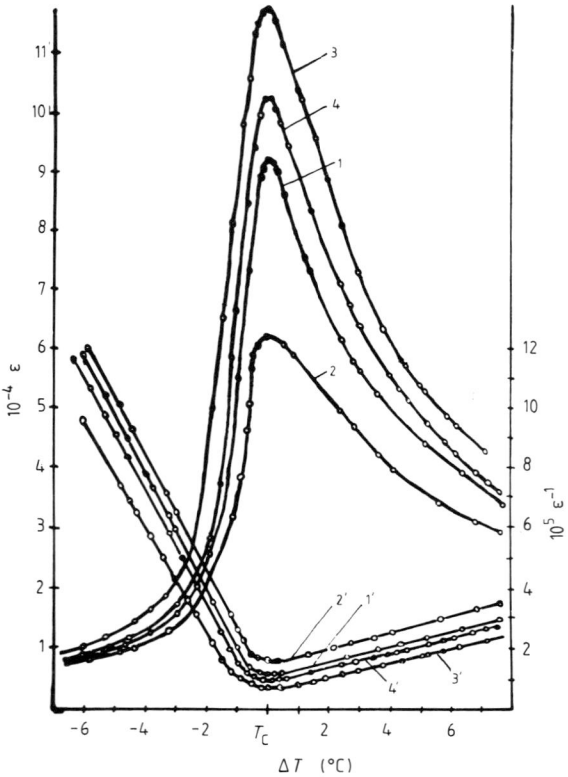

Figure 6.2 Temperature dependences of the dielectric permittivity ε_c (1–4) and $1/\varepsilon_c$ (1′–4′) in the vicinity of the Curie point of $K_x Na_{1-x} Ba_2 Nb_5 O_{15}$ single crystals of different compositions: $x = 0$ (1, 1′); 0.25 (2, 2′); 0.35 (3, 3′); 0.5 (4, 4′) [4]. The x value corresponds to the potassium content in the initial melt.

6.1.2 Optical properties

The refractive indices of the $K_{0.8}Na_{0.2}Ba_2Nb_5O_{15}$ crystal were measured by the least-deviation method using crystal prisms with a vertex angle of about 30° (figure 6.3). The prism was cut in such a way that the c axis of the crystal was parallel to the vertex edge. The peculiarity of the behaviour of the refractive indices lies in their deviation from the linear dependences (broken lines) at a temperature about 50 K above the Curie point (420°C). A continuous variation of refractive indices near T_C indicates a second-order ferroelectric transition, which agrees with the character of the hysteresis loop variations in this temperature interval (chapter 5, [68]). At 550°C, the refractive indices n_1 and n_3 acquire equal values, and the crystal becomes isotropic for light with $\lambda = 0.633$ μm.

Figure 6.4 illustrates the experimental dependences of the refractive indices of ordinary (n_1) and extraordinary (n_3) beams on the light wavelength for a $K_{0.8}Na_{0.2}Ba_2Nb_5O_{15}$ crystal, which are well described by the Sellmeier dispersion relation

$$n^2(\omega) = C_1 + C_2/(\omega_1^2 - \omega^2) \tag{6.2}$$

where ω_1 is the resonance frequency and C_1 and C_2 are the dispersion constants. Their values are listed in table 6.1.

6.1.3 Electro-optic properties

Crystals with $x \geqslant 0.7$ are optically uniaxial in both phases. In the polar phase there exists three independent linear electro-optic coefficients r_{13}, r_{33} and r_{15}. When a field is applied along the ferroelectric axis, the variation of the

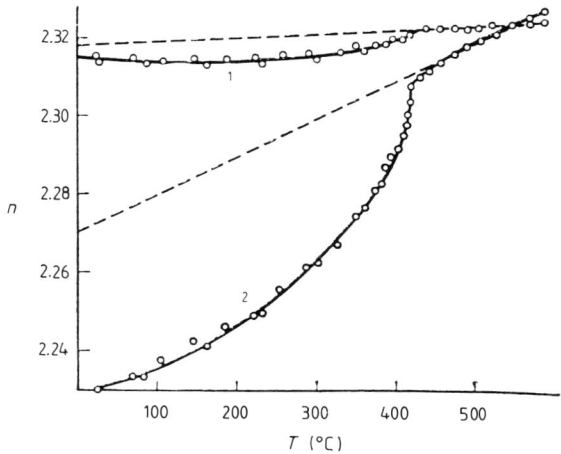

Figure 6.3 Temperature dependences of the refractive indices of (1) an ordinary and (2) an extraordinary beam for a $K_{0.8}Na_{0.2}Ba_2Nb_5O_{15}$ crystal for $\lambda = 6328$ Å [2].

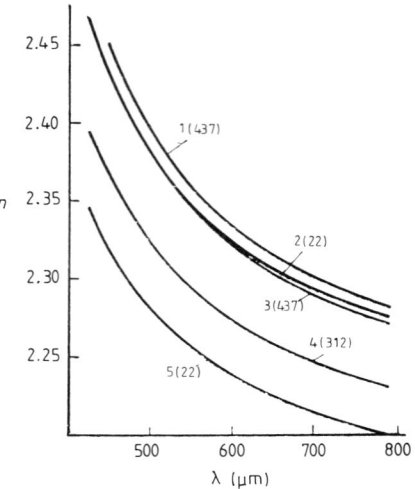

Figure 6.4 Dispersion of the refractive indices of ordinary $(1, 2)$ and extraordinary $(3–5)$ beams for a $K_{0.8}Na_{0.2}Ba_2Nb_5O_{15}$ crystal at different temperatures [2]. The temperature (in degrees Celsius) is given in brackets by the curves.

refractive indices is given by the expression

$$\Delta(1/n_i^2) = r_{13}\bar{E}_3. \tag{6.3}$$

The variation of birefringence of light transmitted along the a axis was measured in the electro-optic experiments [4]:

$$\Delta(n_3 - n_1) = \tfrac{1}{2}(n_3^2 r_{33} - n_1^3 r_{13})E_3. \tag{6.4}$$

The effective electro-optic coefficient

$$r_c = r_{33} - (n_1/n_3)^3 r_{13} \tag{6.5}$$

Table 6.1 Dispersion constants C_1 and C_2 of the refractive indices of the crystal $K_{0.8}Na_{0.2}Ba_2Nb_5O_{15}$ [2].

Refractive index	T (°C)	C_1	C_2 (eV²)	ω_1 (eV)
n_3	22	2.9198	46.737	5.1605
	312	3.2273	34.842	4.7109
	437	3.5539	26.075	4.3004
n_1	22	3.680	24.681	5.3004
	312	3.6307	25.357	4.3004
	437	3.5791	26.534	4.3004

is used for the amplitude modulation. This coefficient is related to the half-wave voltage $V_{\lambda/2}$ as $V_{\lambda/2} = \lambda/n_3^3 r_c$. The half-wave voltage $V_{\lambda/2}$ was measured using the dynamic method. A He–Ne laser with $\lambda = 0.63$ μm and a beam diameter of approximately 0.5 mm was used as a radiation source. The laser beam was directed along the a axis of the crystal. An electric field was applied along the ferroelectric axis c; the plane of polarization of light made an angle of 45° with this direction. The crystal was moved along the c axis, which made it possible to measure $V_{\lambda/2}$ at separate points of the crystal element (figure 6.5). An increase of the potassium content in the investigated crystals is followed by a decrease of the half-wave voltage caused by an increase of the electro-optic coefficients. The latter fact can be explained by an increase of the degree of bond ionicity in the crystal lattice upon replacement of sodium by potassium.

Under similar experimental conditions, Smith *et al* [2] measured an induced birefringence using a $\frac{1}{4}\lambda$ plate and a polarizer. The values of r_c and $V_{\lambda/2}$ obtained at room temperature for crystals of different compositions are tabulated in table 6.2. The table also presents the values of the dielectric permittivity ε_c and the ratio $r_c/\varepsilon_0\varepsilon_3$. For the majority of the indicated compositions, the values of this ratio are similar. It may therefore be concluded that differences in r_c and $V_{\lambda/2}$ values are due to the dependence of ε_c on the crystal composition.

From a comparison of the dispersion curves of the refractive indices of barium sodium niobate crystals (chapter 5, [22]) and potassium-containing crystals ($x = 0.8$) [2] it follows that an introduction of potassium ions produces a stronger effect upon the refractive index n_3 of an extraordinary beam than upon n_1 of an ordinary beam. The concentration variations of these refractive indices have opposite signs and are equal to $\Delta n_1/\Delta x = -0.9 \times 10^{-4}$ and $\Delta n_3/\Delta x = 2.0 \times 10^{-4}$, where x is the potassium ion

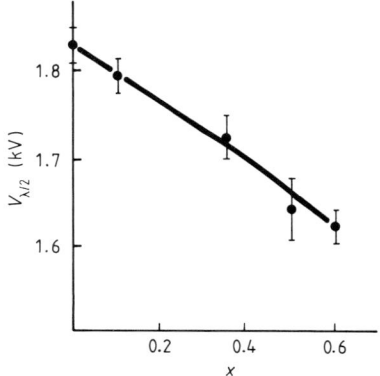

Figure 6.5 Dependence of the half-wave voltage $V_{\lambda/2}$ of $K_x Na_{1-x} Ba_2 Nb_5 O_{15}$ crystals on the composition for $\lambda = 0.628$ μm [4].

Table 6.2 Dielectric and electro-optic characteristics of $K_x Na_{1-x} Ba_2 Nb_5 O_{15}$ crystals (chapter 5, [68]) [2, 4].

x	$T\ (^\circ C)$	ε_3	$V_{\lambda/2}\ (V)$	$r_c\ (mV^{-1})$	$r_c/\varepsilon_0\varepsilon_3$ $(m^2\ C^{-1})$	$P_s\ (C\ m^{-2})$
0	560	51; 43	1830	36; 31.9	0.078	0.39
0.1	—	47	1795	32.1	0.077	0.37
0.35	519	47	1730	33.3	0.080	0.38
0.5	—	47	1660	34.4	0.082	0.39
0.7	445	47; 45	1830 ± 50	31 ± 1	0.078	0.39
0.8	420	55	1410 ± 40	38 ± 1	0.075	0.38
0.9	—	65	1090 ± 40	50 ± 2	0.090	0.45

concentration in mole fractions. The experimental values of the effective electro-optic coefficients for crystals of different compositions (table 6.2) were measured using the values of $\Delta n_1/\Delta x$ and $\Delta n_3/\Delta x$ [2].

The spontaneous polarization P_s for $K_x Na_{1-x} Ba_2 Nb_5 O_{15}$ crystals of different composition was estimated using the equation proposed by Di Domenico and Wemple for oxygen octahedral ferroelectrics:

$$r_{i3} = 2\varepsilon_0 \varepsilon_3 g_{i3} P_s. \tag{6.6}$$

The quantities $g_{13} = 0.052\ m^4\ C^{-2}$ and $g_{33} = 0.16\ m^4\ C^{-2}$, which had been obtained for the tungsten bronzes $KSr_2 Nb_5 O_{15}$ and other relative crystals [5–7], were used. The results of these calculations are presented in table 6.2. For pure barium sodium niobate the measured P_s values are in close agreement with the calculated ones.

The temperature dependence of $V_{\lambda/2}$ for a crystal with $x = 0.8$ (figure 6.6) was determined for small field amplitudes ($E < 10\ V\ cm^{-1}$) at a frequency of

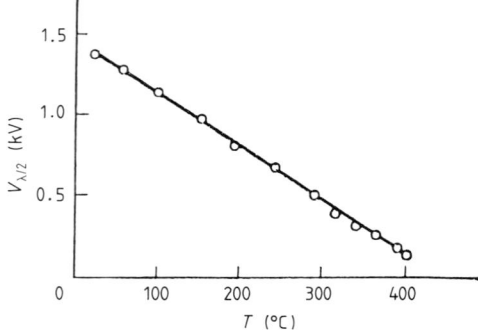

Figure 6.6 Temperature dependence of the half-wave voltage $V_{\lambda/2}$ of a $K_{0.8} Na_{0.2} Ba_2 Nb_5 O_{15}$ crystal [2].

3 kHz to avoid the conductivity effects at high temperatures. From the figure, it is seen that $V_{\lambda/2}$ decreases linearly up to 400°C with increasing temperature, and then falls sharply at T_C. A linear decrease of $V_{\lambda/2}$ was also observed for barium sodium niobate [8].

6.1.4 *Nonlinear optical properties*
The temperature dependence of the second harmonic output was measured [4], using the method proposed in [9]. A He–Ne laser with the wavelength $\lambda = 1.15\,\mu$m and a neodymium-doped garnet laser with the wavelength $\lambda = 1.06\,\mu$m were used. Their polarization was perpendicular to the optical axis of the crystal (figure 6.7). As is seen from the figure, as the content of potassium in crystals increases, the phase matching temperature decreases, i.e. it behaves like a Curie temperature. The results obtained show that barium sodium potassium niobate crystals may be used in devices for an effective transformation of Ne laser radiation into the second harmonic at room temperature.

The nonlinear coefficient d_{31} for the light transmitted perpendicular to the c axis was determined from the intensity of the second harmonic with $\lambda = 1.06\,\mu$m. Phase matching was observed at temperatures of -130°C for $x = 0.8$ and -80°C for $x = 0.7$ [2].

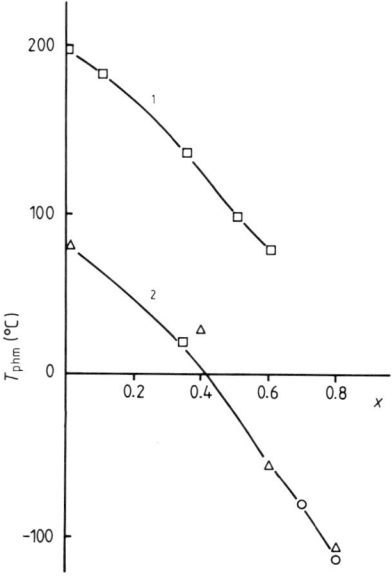

Figure 6.7 Dependences of SHG phase matching temperature on the composition of $K_xNa_{1-x}Ba_2Nb_5O_{15}$ crystals for (1) $\lambda = 1.15$, (2) $\lambda = 1.06\,\mu$m [2, 4]; $\triangle$, [2], $\square$, [4].

The observations of the second harmonic intensity $I^{2\omega}$ were made in the temperature interval ($T_{phm} - T_C$). As the temperature of a polydomain sample was increased, the second harmonic intensity first decreased in an irregular manner, and then fell sharply to zero at T_C. The T_C value thus obtained differed by several degrees from the T_C derived from the dielectric measurements. The peak of intensity $I^{2\omega}$ for single-domain crystals at temperatures far from T_{phm} was 0.01 of the peak of a polydomain sample (figure 6.8). This fact indicates that the domain size is comparable with the coherent length l_{coh}. The graph of the dependence has a typical cyclic nature due to the factor $(\sin^2 \psi)/\psi^2$ in the formula for the second harmonic intensity (see § 5.9). The broken line in figure 6.8 shows the envelope of the function of second harmonic output which is described by the equation [2]

$$I_n/I_0 = \{[d_{31}(n)/d_{31}(0)][(n+\tfrac{1}{2})\pi]^{-1}\}^2 \tag{6.7}$$

where $n \geqslant 1$. This expression was employed to determine the temperature dependence of d_{31} from the relation between the observed and calculated envelopes. It was established that d_{31} remains approximately constant up to a temperature $T = T_C - 200$ K and reaches zero at T_C, thus resembling the temperature dependence of P_s.

The values of the inverse temperature coefficient of the refractive indices $\Delta T/\Delta n = [d(n_3^{2\omega} - n_1^{\omega})/dT]^{-1}$, determined by the positions of the maxima of $I^{2\omega}$, are presented in figure 6.9. For comparison, the figure also shows the dependence of this coefficient for the wavelength $\lambda = 0.633$ μm obtained from the data of figure 6.3.

The nonlinear coefficients d_{31} for the compositions $x = 0.7$, 0.8 were determined with respect to d_{31} (LiNbO$_3$) through comparison of the intensities of the harmonics for corresponding T_{phm}. For both compositions $d_{31} = (2.5 \pm 0.3)d_{31}$(LiNbO$_3$), which is comparable with the value of d_{31} for a pure barium sodium niobate.

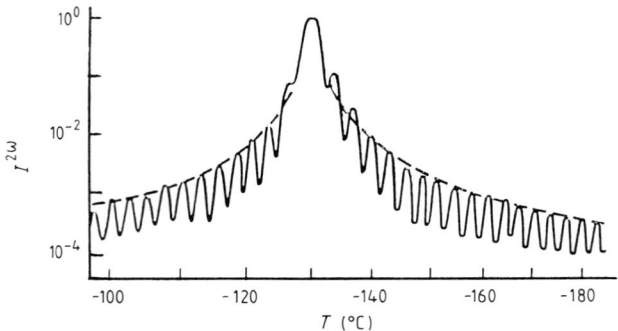

Figure 6.8   Temperature dependence of the intensity of SHG 1.06 → 0.53 μm for a K$_{0.8}$Na$_{0.2}$Ba$_2$Nb$_5$O$_{15}$ crystal [2].

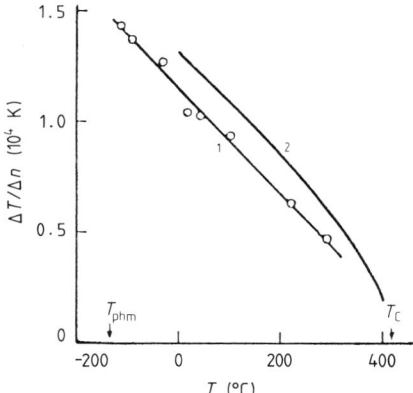

Figure 6.9 Temperature dependence of $\Delta T/\Delta n$ for a $K_{0.8}Na_{0.2}Ba_2$ Nb_5O_{15} crystal determined from the data of (1) figure 6.8 and (2) figure 6.3 for $\lambda = 0.633$ μm [2].

The main requirement imposed upon barium sodium niobate-type crystals is that, in intracavity second harmonic generation, the light of a second harmonic should not induce any changes in the refractive index of the crystal, as is the case with lithium niobate. The experiments [2] showed no degradation of harmonic output in a $K_{0.8}Na_{0.2}Ba_2Nb_5O_{15}$ crystal, even after many hours of work. But in dealing with this crystal, one finds it difficult to maintain the phase matching temperature ($-130°C$).

6.1.5 *Phase diagrams and growth of single crystals*
Giess *et al* (chapter 5, [58]) investigated the compositions of single crystals pulled from melts of the composition $(KNbO_3)_x(NaNbO_3)_y(BaNb_2O_6)_z$, for $x < 34$, $y < 34$, $z > 66$. These limits give an approximate localization of the ferroelectric phase with a tungsten bronze-type structure.

Crystals of the system $K_xNa_{1-x}Ba_2Nb_5O_{15}$ were grown by the Czochralski method using high-frequency heating in platinum crucibles. Thermal protection of the crucible was achieved with aluminium oxide ceramics. Oxygen was pumped through the growth chamber at a rate of $0.3\,l\,min^{-1}$ to avoid niobium reduction and the crystal turning blue. A crystal of $Ba_2NaNb_5O_{15}$ with the c axis aligned along the growth direction was used as a seed crystal. Crystal pulling rates were $2-6$ mm h^{-1}, rotation rates from 40 to 60 rpm. The temperature gradient above the melt surface was 18 K mm^{-1} for the first 5 mm of the growing crystal and 1 K mm^{-1} for the subsequent 50 mm. The crystal–melt system was usually cooled first by 300 K at a rate of 5 K h^{-1} and then the rate was increased to 15 K h^{-1} up to room temperature. In such a regime the crystal was well annealed and did not crack. Fast cooling induced cracking normal to the growth axis. The employed growth rates turned out to be optimal for establishing equilibrium conditions between the crucible and the growing crystal, although none of the investigated ternary compositions were congruently melting. The crystal compositions were always distinct from the melt compositions (figure 6.10 and table 6.3). It should be noted that the authors examined each time the first crystal grown from a given melt. The mass of the crystal was not more than 30 g, the mass of the melt being 400 g. The shift in the composition of the grown crystals relative to the melt composition was observed also in Bridgman crystallization in sealed tubes. The size of the grown crystals did not exceed 3 mm.

It was experimentally established that K_2O and Na_2O do not evaporate from the $K_xNa_{1-x}Ba_2Nb_5O_{15}$ melts with $x = 0.7$ and 0.8. Melt soaking from 20 to 430 h above the melting point did not produce any substantial effect upon the chemical composition or the Curie temperature of the grown crystals.

Crystals pulled from $K_{0.8}Na_{0.2}Ba_2Nb_5O_{15}$ melts were polarized by an applied electric field of $3\,kV\,cm^{-1}$ along the tetragonal c axis at a temperature of 100 °C.

The maximum diameter of Czochralski-grown crystals reached 20 mm; the mean diameter and length were 8 and 30 mm respectively. The grown crystals were chemically homogeneous. For a small crystal size with respect to the melt volume the chemical changes in the crystals were reduced to a minimum.

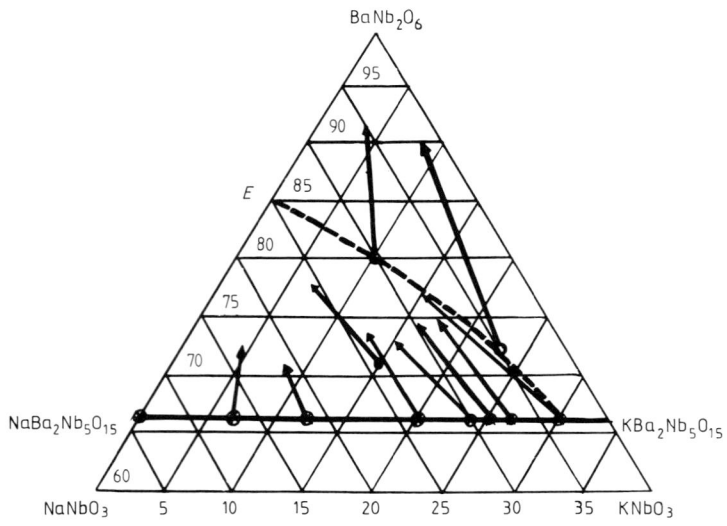

Figure 6.10   Part of the phase diagram $KNbO_3-NaNbO_3-BaNb_2O_6$ (chapter 5, [8]). Circles correspond to melt compositions; arrows indicate variation of the composition of the corresponding crystals. The broken line is an approximate boundary between $BaNb_2O_6$ and TPTB phases.

Table 6.3 Composition of crystals grown from $(KNbO_3)_x(NaNbO_3)_y(BaNb_2O_6)_z$ melts (chapter 5, [58]).

Melt (mol.%)			Crystal (mol.%)			T_{pht} (°C)	T_C (°C)	Lattice parameters (Å)	
x	y	z	x	y	z			a	c
15	9	76	7.9	13.0	79.1	247	548	12.469	4.001
6.7	26.6	66.7	3.5	22.6	73.9	217	538	12.471	4.001
11.7	21.6	66.7	6.6	20.9	72.5	175	519	12.473	4.003
10	15	75	5.6	16.1	78.3	148	492	12.492	4.006
20	13.3	66.7	11.1	14.3	74.6	70	469	12.499	4.011
16	11	73	8.4	14.5	77.1	—	457	12.502	4.011
23.3	10.0	66.7	13.6	13.1	73.3	—	453	12.508	4.012
15.5	8.7	75.7	9.9	11.2	78.9	—	447	12.502	4.011
14	11	75	8.1	14.4	77.5	—	441	12.499	4.008
25	8.3	66.7	14.2	10.6	75.2	—	432	12.514	4.015
26.7	6.6	66.7	16.3	8.3	75.4	—	420	12.514	4.017
30	3.3	66.7	14.8	9.1	76.2	—	—	12.532	4.021
10	10	80	3.3	4.8	91.2	—	—		
25	5	70	8.3	1.6	90.1	—	—		

The chemical homogeneity is confirmed by constancy in the Curie temperature along the crystal length as well as by the device use of these crystals for electro-optic purposes. After crystal pulling, three phases were usually discovered in the hardened melt: $BaNb_2O_6$, $K_xNa_{1-x}Ba_2Nb_5O_{15}$ and $KNbO_3$. Crystals grown from melts of the compositions $0.6 < x \leqslant 0.9$ at room temperature had a tetragonal tungsten bronze-type structure. Crystals of orthorhombic structure were grown from melts with $x \leqslant 0.6$.

A chemical analysis of the grown crystals shows that the melt is enriched with $KNbO_3$ in the process of crystal growth. Crystals with tetragonal tungsten bronze-type structure weighing less than 30 g were grown from melts, weighing 400 g, of the composition $(KNbO_3)_{15.5}(NaNbO_3)_{8.7}$ $(BaNb_2O_6)_{75.8}$. Crystals of greater weight had a $BaNa_2O_6$ phase. This indicates that the melt composition is somewhere near the boundary between the tungsten bronze and the $BaNa_2O_6$ phase.

It is seen from figure 6.10 that crystals grown from melts with a composition corresponding to tungsten bronzes contain more $BaNb_2O_6$ and less $KNbO_3$ than the melt, whereas the concentration of $NaNbO_3$ changes very little. All compositions containing more than 66 mol. % of $BaNb_2O_6$ have close melting points. The melting points of the grown crystals exceed the peritectic temperature of $KBa_2Nb_5O_{15}$ decomposition, 1391°C, and lie below the $BaNb_2O_6$ melting point, 1459°C. At room temperature the structure is tetragonal for $x > 0.6$ and orthorhombic for $x \leqslant 0.6$ (figure 6.11).

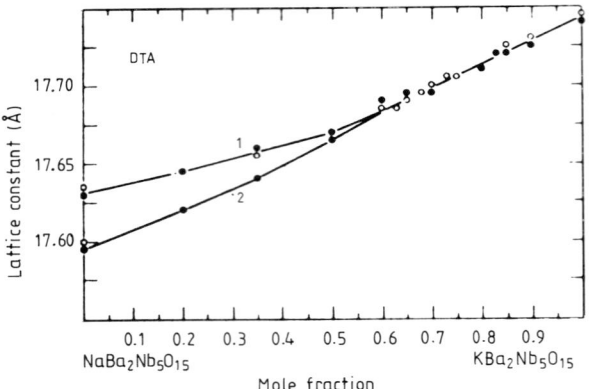

Figure 6.11   Dependence of the lattice parameters (1) a, (2) b on the composition of the cooled $K_x Na_{1-x} Ba_2 Nb_5 O_{15}$ melt (chapter 5, [58]).

X-ray phase analysis, along with the chemical analysis of the crystals, made it possible to establish an approximate existence region for the tetragonal tungsten bronze structure. An attempt to find in the ternary system $KNbO_3 - NaNbO_3 - BaNb_2O_6$ an optimal composition for obtaining homogeneous crystals was unsuccessful.

6.2   Crystals of barium lithium niobate

In 1970 Hirano *et al* [10] reported obtaining new crystals of $Ba_2LiNb_5O_{15}$ which possess properties favourable for practical use. The crystals are therefore uniaxial, and exhibit no room-temperature micro-twinning and no photo-induced inhomogeneities of the refractive index.

Single crystals of barium lithium niobate (BLN) were grown by the Czochralski method using the spontaneous crystallization method. They were clear pale yellow. X-ray and optical studies have shown that at room temperature the crystals possess the tetragonal tungsten bronze structure (point group 4mm) with lattice parameters $a = 12.44$ and $c = 3.998$ Å. Observations in polarized light did not reveal any micro-twins in them at room temperature.

6.2.1   *Dielectric properties*

The ferroelectric transition temperature T_C is approximately equal to 586 °C; the Curie–Weiss temperature $T_0 \simeq 545$ °C and the Curie–Weiss constant $C \simeq 1.88 \times 10^5$ K. The dielectric permittivity along the c axis at T_C is 4.6×10^3; at 30 °C it is 90. The run of the temperature dependence of the dielectric permittivity indicates that at T_C there occurs a first-order phase transition.

6.2.2 Optical properties

The optical properties of the crystals were measured at room temperature; a non-polarized light was transmitted along the c axis of the crystal. The quantity R was estimated in the wavelength range 0.3–0.6 μm using the expression for the intensity of light transmitted through an absorbing crystal [11]

$$I = (1 - R)^2 e^{-\alpha d}(1 + R^2 e^{-2\alpha d}) \tag{6.8}$$

where R is the reflection coefficient, d is the crystal thickness and α is the absorption coefficient (figure 6.12). The calculations show that the crystal is clear within the range 0.4 to 5 μm and has an absorption band near 3 μm, which is due to the presense of hydroxyl groups OH^- in the crystal.

The room-temperature refractive indices of $Ba_2LiNb_5O_{15}$ were determined by the least deviation method (figure 6.13) [12]. A xenon high-pressure lamp and a tungsten filament lamp served as radiation sources respectively in the visible and IR regions of the spectrum; the wavelengths were distinguished

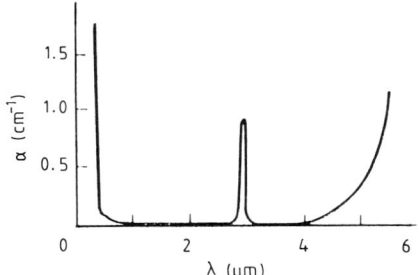

Figure 6.12 Room-temperature dispersion of the absorption coefficient α of $Ba_2LiNb_2O_{15}$ [12].

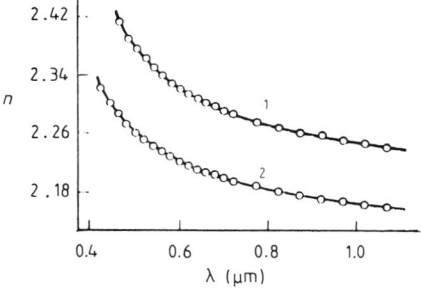

Figure 6.13 Room-temperature dispersion of the refractive indices of (1) an ordinary and (2) an extraordinary beam in barium lithium niobate [12].

using a grating monochromator. Measurements were carried out on two samples which gave similar results within experimental error ($\Delta n = \pm 0.0025$).

The temperature dependences of the refractive indices of ordinary (n_o) and extraordinary (n_e) rays were measured in the interval 20–80 °C on wavelengths of 514.5 nm (argon laser) and 1064 nm (YAG : Nd laser) [12]. The results of the measurements are presented in table 6.4.

The temperature dependence of the ratio n_o (1064 nm)$/n_e$ (532 nm), which is important for determining the angle θ_{phm} between the direction of the phase matching of the second harmonic and primary radiation of the YAG : Nd laser and the optical axis of the crystal, was investigated in [11]. The angle α, complementary to the angle θ_{phm}, is given by the formula

$$\tan \alpha = \left(\frac{(n_o^{\omega}/n_e^{2\omega})^2 - 1}{1 - (n_o^{\omega}/n_e^{2\omega})^2} \right)^{1/2}. \tag{6.9}$$

In second harmonic generation of YAG : Nd laser radiation in a $Ba_2LiNb_5O_{15}$ crystal, the fundamental wave (ω) is an ordinary beam and the second harmonic (2ω) is an extraordinary beam. The angle α measured at 24 °C was 13.2°; at 91 °C the angle α became zero. Taking into account all possible errors in the angle and temperature measurements and assuming the refractive indices to depend linearly on the temperature, Born and Wolf [11] calculated the temperature coefficient (K^{-1}):

$$\frac{d}{dT} \left(\frac{n_o^{\omega}}{n_e^{2\omega}} \right) = -(3.6 \pm 0.3) \times 10^{-5}. \tag{6.10}$$

In an earlier paper [10], the phase matching angle θ_{phm} was estimated to be equal to 70°, the light being transmitted along the *a* axis of the crystal and the polarization of the fundamental wave and of the second harmonic being respectively perpendicular to the transmission direction and parallel to the *c* axis.

The half-wave voltage $V_{\lambda/2}$ of this crystal was determined under the condition that an electric field is applied along the *c* axis, and a ray of a He–Ne laser with $\lambda = 6328$ Å polarized at an angle of 45° to this axis propagated along the *a* axis. The obtained value $V_{\lambda/2} \approx 1700$ V turned out to be a little lower than that of barium sodium niobate.

Table 6.4 Temperature coefficients of the refractive indices of a $Ba_2LiNb_5O_{15}$ crystal between 20 and 80 °C [12].

λ (nm)	$(1/n_o) \, dn_o/dT$ $(10^{-6} \, K^{-1})$	$(1/n_e) \, dn_e/dT$ $(10^{-6} \, K^{-1})$
514.5	± 7	34 ± 7
1064.0	± 7	23 ± 7

6.2.3 *Nonlinear properties*

The phase matching angle θ_{phm} and the temperature dependence of second harmonic output for a $Ba_2LiNb_5O_{15}$ crystal were measured on a cubic sample with an edge of 5 mm [12]. A continuous YAG:Nd laser radiating light with $\lambda = 1.064$ μm with a gaussian intensity distribution in the beam profile was used as a radiation source. The phase matching angle θ_{phm} at 24°C was 76.8°.

To measure the phase matching temperature under non-critical matching, a YAG:Nd laser beam was focused in the crystal volume. The temperature dependence of harmonic generation for a beam incident perpendicular to the optical axis is plotted in figure 6.14. The curve of the second harmonic intensity is asymmetric with respect to the maximum, which is similar to the dependence obtained on other crystals. The nonlinear coefficient d_{31} was estimated from the maximum value of the second harmonic intensity by the formula (chapter 5, [37])

$$d_{31}^2 = \frac{c\varepsilon_0^3 n^3 \lambda^2 P_{sh} \pi d_0^2}{2P_{fund}^2 8\pi^2 l^2 R} \tag{6.11}$$

where c is the velocity of light, ε_0 is the dielectric permittivity of the vacuum, n is the refractive index, λ is the wavelength of the characteristic radiation in the vacuum, P_{fund} and P_{sh} are the powers of the fundamental and generated harmonics, l is the optical path length in the medium, R is the attenuation factor at the phase matching angle and d is the laser beam diameter. As a

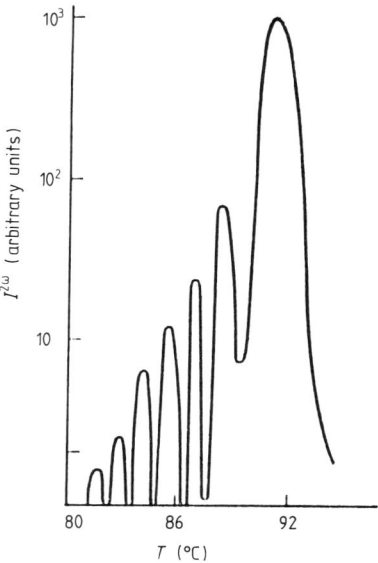

Figure 6.14 Temperature dependence of SH intensity for a barium lithium niobate crystal [12].

result, the following values of this coefficient were obtained ($m^2 V^{-2}$): 4.12×10^{-23} (from the phase matching angle) and 3.14×10^{-23} (from non-critical phase matching). The lower value of d_{31} obtained from the non-critical phase matching is explained by the fact that, when a light beam is transmitted perpendicularly to the crystal growth direction, a certain effect is produced by a slight difference in the refractive indices of regular striae. This effect is comparable with the losses of the light transmitted normally to the grating. Small variations in the refractive indices of the striae may lead to a certain disagreement between the measured d_{31} and θ_{phm} values and the true values. Because of this, the higher of the indicated d_{31} values must be the lower limit of the true value.

6.2.4   Optical damage

The optical damage of single crystals of $Ba_2LiNb_5O_{15}$ was studied using a single-mode YAG:Nd CW laser. The power density of the characteristic radiation with $\lambda = 1.064$ μm was 10^5 W cm^{-2}; on 1 cm of $Ba_2LiNb_5O_{15}$ crystal length the second harmonic power of 200 mW was obtained. Crystals of $Ba_2LiNb_5O_{15}$ were irradiated for an hour. Under such conditions they exhibited neither mechanical damage nor variations in the optical properties, provided their surfaces were free of scratches and absorbing particles. But repeated sharp second harmonic oscillations due to turn-off (on) of the Q-switch or turn-on (off) of the characteristic laser radiation induced irreversible changes in the refractive index in the illuminated region of the crystal. This affects the mode picture of the generation and leads to a decrease in the output power of the second harmonic.

It should be noted that an intense laser beam ($\approx 10^5$ W cm^{-2}) incident on a mechanically damaged crystal surface causes a rapid extension of the damaged area and thus complete crystal destruction. A typical crystal destruction period at a laser power of 5 W was 1 s. This effect was obviously due to crystal surface contamination which cannot be removed from the defective region.

6.2.5   Growth of single crystals

As has already been mentioned, the first crystals of $Ba_2LiNb_5O_{15}$ were grown by the Czochralski method [10]. To avoid crystal cracking, the authors applied a gradual cooling directly in the growth chamber in the first experiments. Crystal cracking may be a consequence of great elastic stresses that occur in the course of cooling. These questions were thoroughly investigated [13]. Single crystals were grown in a device where the pulling block was separated from the annealing block, and thus a crystal could be cooled under controlled conditions.

A platinum crucible of about 100 cm^3 was placed in a quartz vessel; the space between the crucible and the vessel walls was filled with Al_2O_3 powder to provide thermal insulation. A platinum screen placed above the crucible guaranteed a uniform temperature distribution in the space into which the crystal was pulled. A second, cone-shaped, platinum screen was placed directly above the melt. It served as a radiation screen and simultaneously formed a constant diameter of the growing crystal. The melting point of the compound $Ba_2LiNb_5O_{15}$, estimated using differential thermal analysis, was equal to $(1320 \pm 5)°C$. The seed crystal was aligned parallel to the c axis. Optimal pulling and rotation rates were respectively 1 mm h^{-1} and 10 rpm. At the end of the growth process, a crystal with a maximum length of 5 cm and diameter of 2 cm was moved to an annealing furnace, which had a temperature of 700°C and was controlled to an accuracy of ± 2 K. In this furnace, the crystal was cooled to room temperature with a constant cooling rate of 20 K h^{-1}. The temperature gradient along the crystal was less than 0.5 K.

The curves of the temperature expansion of the grown crystals turned out to be similar to those observed in $Ba_2NaNb_5O_{15}$ (see figure 5.4). X-ray topography has shown that a crystal of $Ba_2LiNb_5O_{15}$ is prone to the formation of mosaic structure and exhibits considerable mechanical stresses leading to friability. About a 30′ disorientation of neighbouring blocks is sufficient to induce cracking along the grain boundaries during the cooling process [13]. But, using a perfect seed crystal oriented to an accuracy of 2° parallel to the c axis and avoiding the formation of grain boundaries, a large crystal free of cracks was grown [13].

6.3 Crystals of potassium strontium niobate

Single crystals of $KSr_2Nb_5O_{15}$ (KSN) were grown by the Czochralski method and by the method of spontaneous crystallization from a large melt volume [14]. This compound melts at a temperature of $1470 \pm 10°C$ and has a tetragonal potassium tungsten bronze-type structure. The lattice constants $a = 12.47 \pm 0.02$ and $c = 3.942 \pm 0.004$ Å were determined from X-ray diffraction patterns using Si as indicator. The pycnometric density of the single crystal was equal to 5 g cm^{-3}, which corresponds to two structural units in a unit cell. It was established [14] that the cation ratio K/Sr (i.e. $KNbO_3 : SrNb_2O_6$) may vary over relatively wide limits within the tetragonal potassium tungsten bronze structure.

6.3.1 *Dielectric properties*
The values of the dielectric permittivity ε_c and spontaneous polarization P_s for KSN crystals of different compositions are listed in table 6.5. The spontaneous polarization was determined from the dielectric hysteresis loops and from integration of the pyroelectric charge; the dielectric permittivity ε_c was measured at a frequency of 15 kHz. As can be seen from the table, on moving away from the stoichiometric composition (11 mol.% of K_2O, 40 mol.% of SrO, 49 mol.% of Nb_2O_5), the T_C and P_s values decrease while

Table 6.5   Dielectric characteristics of KSN crystals at $22\,^\circ$C [15].

Composition (mol.%)				ε_c		
K$_2$O	SrO	Nb$_2$O$_5$	P_s (10^{-6} C cm^{-2})	polydomain	single domain	T_C ($^\circ$C)
11	40	49	29	900	490	155
5	44	51	25	1300	820	138
4	42	54	24	1500	1000	120

the ε_c value increases. Under the action of an applied field of 6 kV cm^{-1}, the room-temperature polarization leads to a considerable decrease in ε and tan δ.

The ε_c values vary widely, depending on the K/Sr ratio in the crystal (table 6.6). But it is seen from figure 6.15 that the Curie temperature depends weakly on the crystal composition.

For KSr$_2$Nb$_5$O$_{15}$ crystals above the ferroelectric phase transition temperature ($155\,^\circ$C), the Curie–Weiss law is obeyed and the C constant for the investigated samples lies within the range $C = (2-4) \times 10^5\,^\circ$C. At $T < 160\,^\circ$C, a linear dependence of P on E is observed. The coercive field E_c at room temperature is equal to 20 kV cm^{-1}. Thus these crystals may be rather easily polarized.

Crystals with tungsten bronze-type structures, e.g. BNN, exhibit proton diffusion during polarization in moist air at a high temperature [17]. As mentioned in the preceding chapter, the presence of hydrogen ions affects the dielectric properties of BNN crystals. The influence of protons upon the dielectric properties of KSN crystals was investigated through crystal annealing at a temperature of $550\,^\circ$C under a field of 500 V cm^{-1} for 16 h [16]. Annealing was carried out in a vacuum, in dry and moist oxygen, as well as in moist air. The results of these experiments have shown that, in contrast with the above mentioned niobates, KSN crystals exhibit no high-temperature proton diffusion. This conclusion was later confirmed by the absence of an OH absorption line in these crystals. Thus the variations in the dielectric

Table 6.6   Compositions of KSN crystals (mol.%) (to figure 6.15).

No	K$_2$O	SrO	Nb$_2$O$_5$	No	K$_2$O	SrO	Nb$_2$O$_5$
1	11.88	40.10	48.02	5	4.07	40.96	54.96
2	7.53	39.67	52.80	6	4.02	41.94	54.03
3	5.40	44.50	50.10	7†	5.00	45.00	50.00
4	5.15	40.65	54.20				

† Congruent composition K$_2$Sr$_9$Nb$_{20}$O$_{60}$.

properties of single-domain KSN crystals cannot be explained by diffusion of impurities into the crystal under the action of an applied electric field.

In KSN crystals, ferroelectric domains are aligned along the tetragonal c axis. If an electric field is applied along the c axis of a non-polar sample, then the domains antiparallel to the field direction change the polarization direction so that all the domains appear to be aligned parallel to the applied field. A reversal of the applied field, and therefore of the polarization direction, must not change the dielectric properties of the crystal. Experiments have shown, however, that the dielectric properties do not correspond to the initial ones if the crystal has been depolarized in a decreasing AC field, whereas under thermal depolarization (cooling through the Curie temperature) the dielectric properties of the crystal are completely restored. Besides, it was established that an application of an AC field to a polydomain crystal produces the same effect as a DC field does. This effect has not yet been explained, although one may assume that thermal depolarization of a crystal leads to an antiferroelectric state of extremely small ferroelectric domains if the field applied does not exceed 3 kV cm^{-1}.

Temperature dependences of the dielectric permittivity of KSN crystals (figure 6.15) usually have the form typical of relaxation-type ferroelectrics with a smeared phase transition. As has already been mentioned in chapters 3 and 4, complex oxides containing various cations in identical crystallographic positions have smeared phase transitions due to composition fluctuation. This

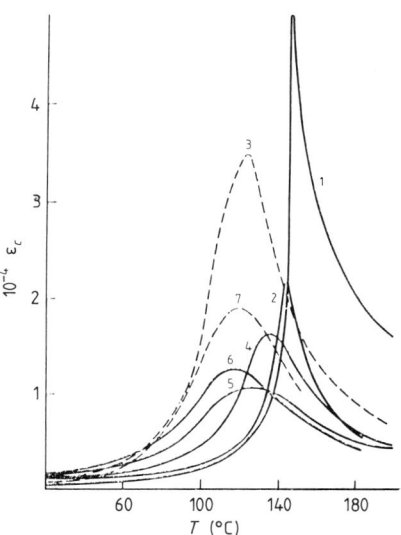

Figure 6.15 Temperature dependence of dielectric permittivity ε_c for different compositions of KSN crystals (see table 6.6) [16].

may manifest itself in single crystals whose compositions correspond to a partially occupied TPTB structure.

The composition $K_2Sr_9Nb_{20}O_{60}$ was assumed to be 'congruent' [16]. Crystals of this composition exhibit no striations. Table 6.7 presents several of their characteristics.

Figure 6.16 shows the dependence of the Curie temperature on the crystal composition: T_C varies for $KSr_2Nb_5O_{15}$ from 158 °C to 124 °C as the content of $SrNb_2O_6$ increases up to 84 mol.%. A more significant effect upon the value of the dielectric permittivity is observed at room temperature. This permittivity, for the same composition range, increases from 600 to 1400.

6.3.2 *Optical and electro-optic properties*

The effective electro-optic coefficient r_c, for a crystal to which a field E is applied along the c axis and light is transmitted along the a axis, is given by the expression

$$r_c = r_{33} - (n_o/n_e)^3 r_{13}$$

where n_o and n_e are the refractive indices of the ordinary and extraordinary beams respectively. The temperature dependence of this coefficient for $\lambda = 0.633$ μm ($n \approx 2.25$) is plotted in figure 6.17.

The field-induced birefringence was measured for the light with $\lambda = 6328$ Å using a $\frac{1}{4}\lambda$ wave plate and a polarizer. The relation between the measured optical phase delay Γ and r_c is given by

$$\Gamma = \pi n_e^3 r_c lE/\lambda. \tag{6.12}$$

At room temperature, $r_c = 1.3 \times 10^{-8}$ cm V^{-1}. This estimate is between the r_c values for $Ba_{0.25}Sr_{0.75}Nb_2O_6$ (14×10^{-8}) and $LiNbO_3$ (0.18×10^{-8}).

Table 6.7 Some characteristics of a $K_2Sr_9Nb_{20}O_6$ crystal [16].

Structure	Tetragonal potassium tungsten bronzes
Lattice parameters (Å)	$c = 7.832 \pm 0.004$;
	$a = 12.414 \pm 0.004$
Density (g cm^{-3})	5.058 (expt.), 5.069 (calc.)
Curie temperature T_C (°C)	120
Dielectric permittivity ε_c (21 °C, 1 kHz)	1600 (polydomain)
	800 (single domain)
Loss tangent tan δ ($v = 1$ kHz)	0.10 (polydomain)
	0.005 (single domain)
Half-wave voltage $V_{\lambda/2}$ (V)	260
Spontaneous polarization P_s	
(10^{-6} C cm^{-2})	24
$\gamma = dP_s/dT$ (10^{-8} C cm^{-2} K^{-1})	
($T = 20$–40 °C)	6.5

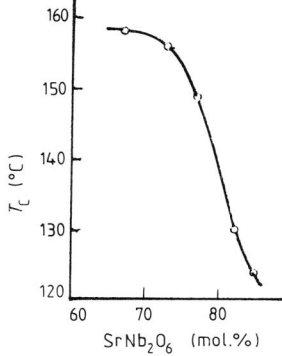

Figure 6.16 The Curie temperature of potassium strontium niobate single crystals as a function of composition [1].

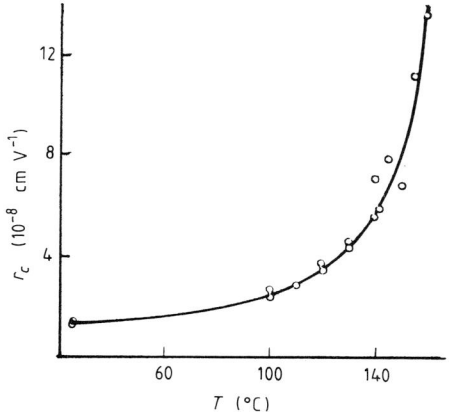

Figure 6.17 Temperature dependence of the effective electro-optic coefficient r_c of a $KSr_2Nb_5O_{15}$ crystal for $\lambda = 0.633\ \mu m$ [14].

The temperature dependences of $V_{\lambda/2}$ for potassium strontium niobate crystals of different composition were measured in [15]. The AC electric fields used in the experiments were much smaller than the coercive fields of the crystals. The obtained $V_{\lambda/2}$ values were compared with those calculated by the formula

$$V_{\lambda/2} \approx \frac{\lambda\eta^3}{2n^3\varepsilon_0 g}\left(\frac{1}{\varepsilon_c P}\right)$$

where g is the generalized electro-optic coefficient constant for oxygen octahedral compounds, η is the packing coefficient which for tungsten bronzes is close to unity, P is the polarization and ε_0 the dielectric permittivity of a vacuum. In the ferroelectric phase, induced polarization is usually disregarded and spontaneous polarization P_s alone is taken into account.

For KSN, the numerical value of $V_{\lambda/2}$ for $\lambda = 6328$ Å is determined by the relation

$$V_{\lambda/2} = 2.87 \times 10^6 / \varepsilon_c \bar{P}_s \qquad (6.13)$$

where P_s is measured in μC cm^{-2}.

Figure 6.18 shows that in the ferroelectric phase the calculated $V_{\lambda/2}$ values are lower than the experimental ones. This is perhaps a consequence of the dependence of the dielectric permittivity of non-stoichiometric crystals on the applied field. The dielectric permittivity of such crystals relaxes with the frequency of the AC field used to determine $V_{\lambda/2}$, which leads to an increase of measured values of the half-wave voltage. Additional experiments actually demonstrated that non-stoichiometric crystals have a frequency dependence of half-wave voltage for frequencies less than 100 Hz.

The dependence of $V_{\lambda/2}$ on the composition may be obtained using formula (6.13) by substituting into it the quantities P_s and ε for the single-domain crystals listed in table 6.5. The calculations show that crystals grown from melts strongly different from a stoichiometric one have low half-wave voltages. Such crystals show no striations, which indicates that the melt approaches a congruent one.

Due to remanent polarization, polydomain KSN crystals also possess a linear electro-optic effect. For them, $V_{\lambda/2} \approx 1.5$ kV, whereas for single-domain crystals it is reduced to 250 V.

6.3.3 *Phase diagrams, growth of single crystals*

In $KSr_2Nb_5O_{15}$, as in $Ba_2NaNb_5O_{15}$ (see chapter 5), alkaline and alkaline earth cations have an equally probable distribution in large A1 and A2 voids in the crystal lattice. If the electroneutrality conditions and the dimensional

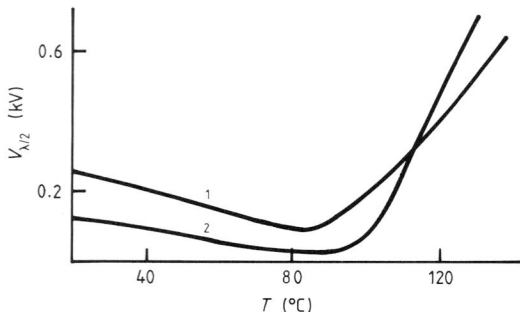

Figure 6.18 Experimental (1) and theoretical (2) temperature dependences of the half-wave voltage of a crystal of composition 0.04 (K_2O) 0.42 (SrO) 0.54 (Nb_2O_5) for $\lambda = 0.6328$ μm [15].

requirements on the structure formed of NbO_6 octahedra are met, then in the region of the phase diagram rich in an alkali metal the TPTB structure can be formed by way of occupation of C positions, and in the region rich in niobium by way of formation of voids in the A1 and A2 positions of the crystal lattice.

The phase diagram of the system $KNbO_3-SrNb_2O_6$, constructed on the basis of a differential thermal analysis and quenching of the melts, is plotted in figure 6.19. The solidus and liquidus lines, according to this analysis, are reproduced with an accuracy of ± 2 K, although an absolute accuracy of each measurement is ± 5 K.

The solid solution with a TPTB-type structure exists in the region located approximately from 59 (1100°C) to 87 mol.% of $SrNb_2O_6$ (1487°C). The phase boundary corresponding to 59 mol.% of $SrNb_2O_6$ is unexpected to some extent, since a solid solution containing less than 67 mol.% of $SrNb_2O_6$ must contain many K^+ or Sr^{2+} cations in C positions, and inasmuch as the Sr^{2+} ion radius (1.14 Å) is smaller than the K^+ radius (1.33 Å), the C vacancies might be expected to be occupied by Sr^{2+} ions. For 59 mol.% of $SrNb_2O_6$, the probability of the four vacant C positions, corresponding to one unit cell, being filled with Sr^{2+} ions is 0.0875. For 87 mol.% of

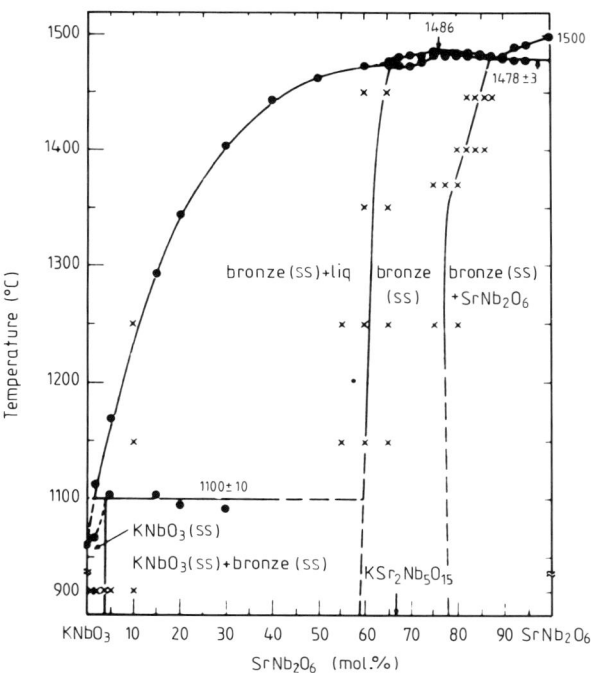

Figure 6.19 Phase diagram of the system $KNbO_3-SrNb_2O_6$ [1].

$SrNb_2O_6$, the probability of the six A1 and A2 positions being occupied by K^+ and Sr^{2+} ions is 0.892.

The distribution of alkaline earth ions over the crystallographic positions is confirmed by the variations in the crystal lattice parameters subject to the crystal composition [1]. When the melt contains more than 79 mol.% of $SrNb_2O_6$, the unit cell parameters become equal to or greater than the parameters of the compounds obtained by solid-phase synthesis. This suggests a narrowness of the TPTB field in which crystals contain less $SrNb_2O_6$ than the corresponding melt compositions.

The phase diagram implies that the composition $KSr_2Nb_5O_{15}$ is not congruent. Its variations can be reduced to a minimum if crystals are grown from a melt containing 78 mol.% of $SrNb_2O_6$, which has a congruent melting point at 1486°C. The presence of a plateau on the liquidus line and a high stability of the solid solution in the $SrNb_2O_6$-rich region are in complete analogy with the phase diagram of the system $NaNbO_3$–$BaNb_2O_6$ [18].

In crystals quenched rapidly from the crystallization temperature, a solid solution with a TPTB-type structure does not undergo decomposition. However, an annealing of a crystal containing 84 mol.% of $SrNb_2O_6$ for 40 h at a temperature of 1350°C has led to a decomposition; the powder patterns showed a mixture of $SrNb_2O_6$ and TPTB phases, the lattice parameters of the latter corresponding to a content of 80 mol.% of $SrNb_2O_6$.

The data of the chemical analysis of KSN crystals grown by the Czochralski method and by the melting method correspond, within experimental error ($\approx 1\%$), to the TPTB region of the $KNbO_3$–$SrNb_2O_6$ diagram. On this basis the investigated system was assumed to be truly binary.

The investigation of the phase diagram of $KNbO_3$–$SrNb_2O_6$ has made it possible to determine the congruent composition of the melt $KSr_9Nb_{20}O_{60}$, from which one can obtain striation-free crystals with a vertical acicular structure similar to the one observed in BSN crystals (see chapter 4). Crystals grown from a melt composition other than $K_2Sr_9Nb_{20}O_{60}$ are as a rule prone to striations. An X-ray analysis of crystals of such a composition [16] showed that the parameter c doubles, whereas no doubling was observed along the a axis.

6.4   Potassium lithium strontium and sodium lithium strontium niobates

6.4.1   *Dielectric and electro-optic properties*
Colourless single crystals of $Sr_4KLiNb_{10}O_{30}$ were grown by the Czochralski method [19, 20]. The dielectric permittivities measured in the directions parallel and perpendicular to the c axis at a frequency of 1.2 MHz at room temperature were 1200 and 900 respectively. In figure 6.20, the temperature dependence of ε_c for a $Sr_4KLiNb_{10}O_{30}$ crystal is presented, and an analogous dependence for a $Sr_2KNb_5O_{15}$ crystal is given for comparison. The ε_c

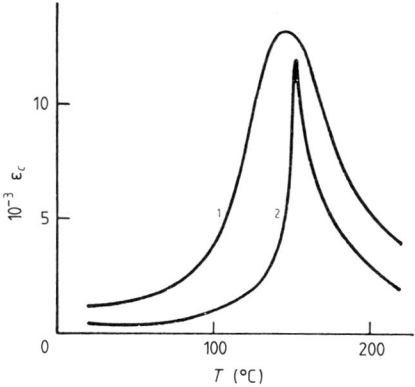

Figure 6.20 Temperature dependences of the dielectric constant ε_c of (1) $Sr_4KLiNb_{10}O_{30}$ and (2) $Sr_2KNb_5O_{15}$ [20].

maximum of the $Sr_4KLiNb_{10}O_{30}$ crystal is wider than that of a KSN crystal. The Curie temperature for both crystals is 145 °C. Spontaneous polarization of potassium lithium strontium niobate (PLSN) measured by the pulsed field method at room temperature was 0.28 C m^{-2}. The refractive indices n_o and n_e of this crystal for light with $\lambda = 0.633$ μm are 2.30 and 2.27 respectively. At about 140 °C, an optically negative crystal becomes optically positive. A conoscopic picture has shown that the crystal is biaxial but that the difference in the refractive indices n_x and n_y is small ($\Delta n \approx 10^{-5}$), and therefore we will proceed under the assumption of tetragonal symmetry. The lattice parameters of $Sr_4KLiNb_{10}O_{30}$ crystals are $a = 12.41 \pm 0.005$ Å and $c = 3.91 \pm 0.005$ Å.

If an applied DC electric field is parallel to the c axis of a $Sr_4KLiNb_{10}O_{30}$ crystal and the light beam is perpendicular, then $V_{\lambda/2} = 200$ V for $\lambda = 0.633$ μm. The electro-optic coefficient r_c calculated from the value of $V_{\lambda/2}$ is 2.6×10^{-10} m V^{-1} (figure 6.21). As has already been mentioned, the crystals are colourless and have a minimum optical absorption in the visible.

Above the Curie temperature, the crystal possesses a quadratic electro-optic effect, to which there corresponds an effective coefficient $g_c = g_{33} - (n_o/n_e)^3 g_{13} = 0.16$ m^4 C^{-2}. In the temperature interval from 140 to 200 °C, this value is close to $(g_{11} - g_{12}) = 0.17$ m^4 C^{-2} for KTN [7].

The crystals of $Sr_4NaLiNb_{10}O_{30}$ grown by Yano *et al* [21] differ from $Sr_4KLiNb_{10}O_{30}$, in that their electro-optic properties do not depend on temperature. Their room-temperature structure is assumed to be tetragonal. The lattice parameters $a = b = 12.33$ Å and $c = 3.88$ Å were determined from the powder patterns. But Ohta *et al* [22] discovered an orthorhombic superstructure with parameters $a' = 2\sqrt{2}a = 34.8$; $b' = \sqrt{2}a = 17.4$; $c' = 2c = 7.76$ Å. The crystal is optically biaxial and the refractive indices for $\lambda = 0.633$ μm along the orthorhombic axes at room temperature are $n'_a = 2.31$;

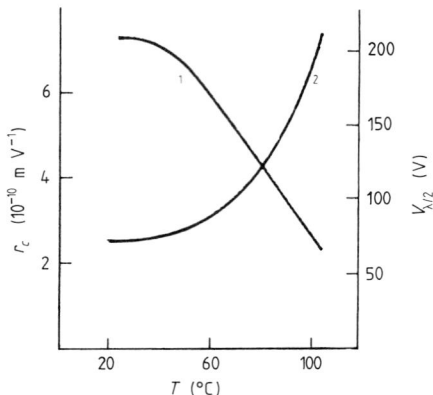

Figure 6.21   Temperature dependences of (1) the half-wave voltage $V_{\lambda/2}$ and (2) the electro-optic coefficient r_c of a $Sr_4KLiNb_{10}O_{30}$ crystal [20].

$n'_b = 2.30$; $n'_c = 2.27$. The angle between the optical axes is equal to $64°$. The domain structure of $Sr_4NaLiNb_{10}O_{30}$ at room temperature is not observed under a polarization microscope, and it is assumed therefore that at this temperature the crystal is single domain. The spontaneous polarization measured by the pulsed field method was equal to 18 $\mu C\ cm^{-2}$.

The temperature dependences of the dielectric permittivities in the direction of the ferroelectric axes of $Sr_4NaLiNb_{10}O_{30}$ and $Sr_{4.2}Na_{1.6}Nb_{10}O_{30}$ are plotted in figure 6.22(a). Both crystals exhibit two dielectric anomalies.

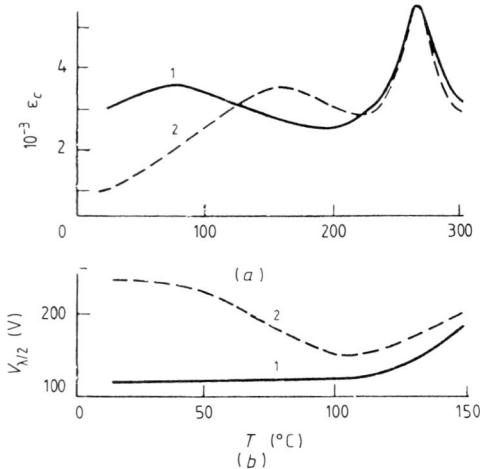

Figure 6.22   Temperature dependences of the dielectric constant ε_c for 1 kHz (*a*) and of the half-wave voltage $V_{\lambda/2}$ (*b*) of single-domain $Sr_4NaLiNb_{10}O_{30}$ (1) and $Sr_{4.2}Na_{1.6}Nb_{10}O_{30}$ crystals [21].

The first broad maximum of a $Sr_4NaLiNb_{10}O_{30}$ crystal is shifted towards low temperatures relative to the maximum of the other crystal, while the narrow maxima of both crystals coincide. At room temperature, for the $Sr_4NaLiNb_{10}O_{30}$ crystal $\varepsilon_c = 3000$, whereas for $Sr_{4.2}Na_{1.6}Nb_{10}O_{30}$ $\varepsilon_c \approx 1000$.

For an electric field applied along the c axis and a light beam with $\lambda = 0.633$ μm propagating along the orthorhombic a' or b' axes, $V_{\lambda/2} \approx 120$ V for a $Sr_4NaLiNb_{10}O_{30}$ crystal and $V_{\lambda/2} \approx 240$ V for a $Sr_{4.2}Na_{1.6}Nb_{10}O_{30}$ crystal. The value of $V_{\lambda/2}$ for the first crystal is temperature independent in the interval from room temperature to 120°C (figure 6.22(b)), which is advantageous for its use in electro-optics.

6.4.2 *Phase diagrams*

The phase diagrams of the system $SrNb_2O_6$–$KNbO_3$–$LiNbO_3$ (figure 6.23) were investigated on ceramics. The specimens were synthesized at temperatures from 100 to 1400°C subject to strontium content. The region separated by the full line is a single-phase region of compounds with TPTB-type structure. The chemical compounds $Sr_4KLiNb_{10}O_{30}$ and $Sr_{4.4}K_{1.1}Li_{0.5}Nb_{10}O_{30}$ with a maximum content of strontium in the unit cell were found to exist in this portion of the diagram. A similar situation takes place in the systems $SrNb_2O_6$–$KNbO$ [23, 24] and $BaNb_2O_6$–$NaNbO_3$ [25].

Colourless single crystals were obtained from melts of $Sr_4KLiNb_{10}O_{30}$ and $Sr_{4.4}K_{1.1}Li_{0.5}Nb_{10}O_{30}$ by the Czochralski method. The melting temperature of these compounds is about 1430°C. An approximate chemical formula of a $Sr_{4.4}K_{1.0}Li_{0.2}Nb_{10}O_{30}$ single crystal grown from the second of the indicated melts was established through comparison between the lattice parameters of the crystal and ceramic specimens of the indicated composition, and also using micro-roentgen analysis.

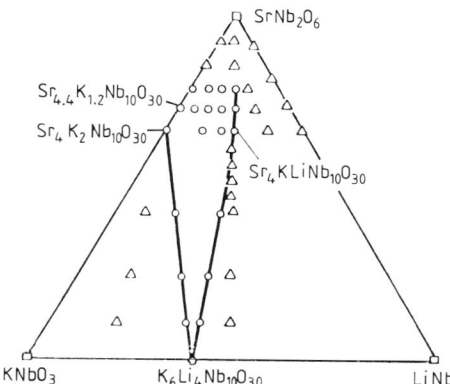

Figure 6.23 Phase diagram of the system $SrNb_2O_6$–$KNbO_3$–$LiNbO_3$ obtained in the study of ceramic samples [21]. Circles are located in the single-phase and triangles in the two-phase region.

6.5   Alkali rare earth niobates

6.5.1   *Ferroelectric properties*

Burns *et al* [26] modified the compound $KSr_2Nb_5O_{15}$ by introducing lanthanum into it. Thus they enlarged the room-temperature dielectric permittivity and increased the electro-optic coefficients r_{ij}, since r_{ij} and ε_i are directly proportional to each other. The modified crystals were grown by the Czochralski method from melts of the compositions $(KNbO_3)_{0.45}(SrNb_2O_6)_{0.55}$ with additions of 0, 3, 6 and 9 mol.% of $LaNb_3O_9$. The Curie temperature in this system decreased with increasing content of $LaNb_3O_9$ in the melt, and ε increased at room temperature, as expected.

For single crystals of this system, the coefficients of the Devonshire–Ginzburg equation were determined:

$$F(P, T) - F(0, T) = A(T - T_0)\bar{P}^2 - B\bar{P}^4 + C\bar{P}^6 + \dots . \quad (6.14)$$

The quantities A, B and T_0 were estimated from three experimental dependences described by the expressions

$$\varepsilon = \frac{2\pi/A}{T - T_0} \quad (6.15)$$

$$\frac{B}{2C} = \left(\frac{1}{A}\frac{dE}{dT_C}\right)^2 \quad (6.16)$$

$$\frac{B^2}{4C} = A(T_C - T_0) \quad (6.17)$$

as well as from the relations $\bar{E} = (\partial F/\partial P)$ and $4\pi/\varepsilon = (\partial \bar{E}/\partial \bar{P})$. Equation (6.15) is the expression for the Curie–Weiss law above the transition temperature, from which the coefficient A and the Curie–Weiss temperature T_0 are determined. The right-hand side of equation (6.16) is connected with the shift of the phase transition temperature in the electric field. The right-hand side of equation (6.17) involves the difference between the transition temperature and the Curie–Weiss temperature.

Figure 6.24 illustrates the temperature dependences of the dielectric permittivity ε_c for different electric fields applied to the crystal grown from a melt with 6 mol.% of $LaNb_3O_9$.

The quantities A, B and C for single crystals of the indicated system and several dielectric constants calculated from experiment are listed in table 6.8. The A, B and C values are of the same order of magnitude as for displacive-type ferroelectrics, e.g. $BaTiO_3$ and $KNbO_3$. It is known that for order–disorder-type ferroelectrics the quantities A, B and C are larger by at least a factor of 10^2, 10^3 and 10^7 respectively. Thus the obtained crystals may be regarded as displacive-type ferroelectrics.

It should be noted that a random distribution causes variations in the crystal properties (see table 6.8). For some of these compositions, the dielectric

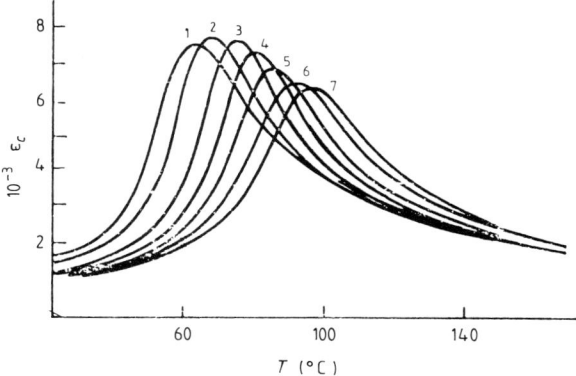

Figure 6.24 Temperature dependences of the dielectric constant ε_c for a crystal of composition $(KNbO_3)_{0.45}(SrNb_2O_6)_{0.55}$ with an addition of 6 mol.% of $LaNb_3O_9$ for different applied fields [26]. E is equal to: (1) 0; (2) 1725; (3) 3450; (4) 5175; (5) 6900; (6) 8625; (7) 10 350 V.

permittivity is rather high at room temperature, and therefore one may expect high values of the linear electro-optic coefficients.

Scott *et al* [27] studied the crystal chemistry and the dielectric properties of tetragonal bronzes $K_2RENb_5O_{15}$, where RE are trivalent rare earth ions. The K^+ and RE^{3+} ions are assumed to take A1 and A2 positions in the crystal structure. Alkali rare earth niobates exhibit structural transformations, but ferroelectric properties are obviously absent from these materials.

For the appearance of ferroelectricity a strongly polarizable ion, e.g. Pb^{2+}, should necessarily be present. Trivalent rare earth ions obviously do not possess high polarizability, and therefore alkali rare earth niobates possess no ferroelectric properties. In this connection, it is of interest to note that there exists a ferroelectric $KBiNb_5O_{15}$ with a Curie temperature of $\approx 160\,°C$ [28]. At room temperature it has an orthorhombic structure with the lattice parameters $a = 17.75$, $b = 17.90$ and $c = 7.84$ Å. The ion radius of Bi^{3+}, equal to 0.93 Å, is close to the ion radii of rare earth cations but, as distinct from rare earth cations, Bi^{3+} has a high polarizability.

Single crystals of $La_{0.1}Li_{0.1}NaBa_{1.8}Nb_5O_{15}$ were grown [29] by the Czochralski method. The temperature gradient above the melt surface was approximately 50 K cm^{-1}. The pulling and rotation rates were 2 mm h^{-1} and 60 rpm. An interesting peculiarity of these crystals is that the composition of a grown crystal obviously corresponds to the melt composition at a temperature near 1450 °C. The grown crystals were yellow, but after annealing at 1000 °C they became colourless. The crystals had perfect cleavage planes (001) and were biaxial. Comparison of $La_{0.1}Li_{0.1}NaBa_{1.8}Nb_5O_{15}$ and $Ba_2NaNb_5O_{15}$ showed that at room temperature both crystals belong to the orthorhombic tungsten bronze structure (symmetry mm2). The lattice

Table 6.8 Dielectric characteristics and coefficients of the Devonshire–Ginzburg equation for single crystals of compositions $(KNbO_3)_{0.45}(SrNb_2O_6)_{0.55}(LaNb_3O_9)_x$ at room temperature [26].

x (mol.%)	ε	T_C (°C)	T_0 (°C)	$\dfrac{dE/dT_C}{(V\,cm^{-1}\,K^{-1})}$	A (10^{-5} °C)	B (10^{-13} CGS units)	C (10^{-22} CGS units)
0	430	156	145	240	1.82	2.0	0.49
3	380	119	91	300	2.79	11	4.3
6	1650	59	40	300	3.1	12	5.8
9	5300	37	−8	235	2.0	14	5.4

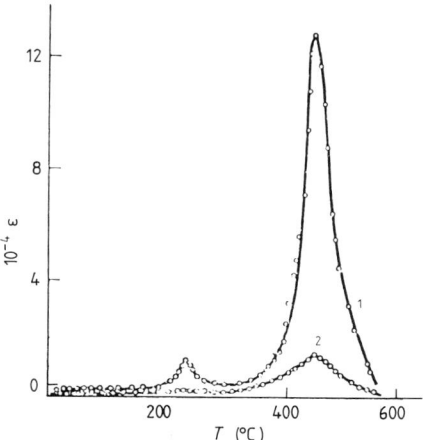

Figure 6.25 Temperature dependences of dielectric permittivities (1) ε_c and (2) ε_a of a $La_{0.1}Li_{0.1}NaBa_{1.8}Nb_5O_{15}$ crystal [29].

parameters of the former crystal are $a = 17.763$, $b = 17.215$ and $c = 7.853$ Å. Figure 6.25 illustrates the temperature dependences of the dielectric permittivity of this crystal along the c and a axes at a frequency of 10 kHz. In the region of 250 °C a dielectric anomaly is observed, which corresponds to the phase transition from the orthorhombic to the tetragonal phase; $T_C = 440$ °C, and the ε_c value at this point reaches 1.28×10^5, which exceeds by several times the one for barium sodium niobate. The room-temperature dielectric permittivities are $\varepsilon_c = 710$ and $\varepsilon_a = 350$. The coercive field E_c determined by the loops of the dielectric hysteresis at the same temperature is equal to 5.5 kV cm^{-1}. It decreases linearly with increasing temperature; at 140 °C $E_c \approx 2.1$ kV cm^{-1}. The remanent polarization is equal to 29.1 μC cm^{-2}. The crystal of $La_{0.1}Li_{0.1}NaBa_{1.8}Nb_5O_{15}$ is of interest for second harmonic generation YAG : Nd lasers.

6.6 Crystals of potassium lithium barium niobates

The electro-optic crystals with room-temperature TPTB structure were grown from the ternary system $KNbO_3$–$LiNbO_3$–$BaNb_2O_6$ (chapter 5, [68]).

6.6.1 *Dielectric and electro-optic properties*
The Curie temperatures T_C of single crystals of TPTB solid solutions of the systems $KNbO_3$–$LiNbO_3$–$BaNb_2O_6$, $KSr_2Nb_5O_{15}$–$KBa_2Nb_5O_{15}$ and $KBa_2Nb_5O_{15}$–$KPb_2Nb_5O_{15}$ are plotted in figure 6.26 as a function of the normalized parameter $\eta = \sqrt{10c/a}$. The parabolic dependence of T_C on η,

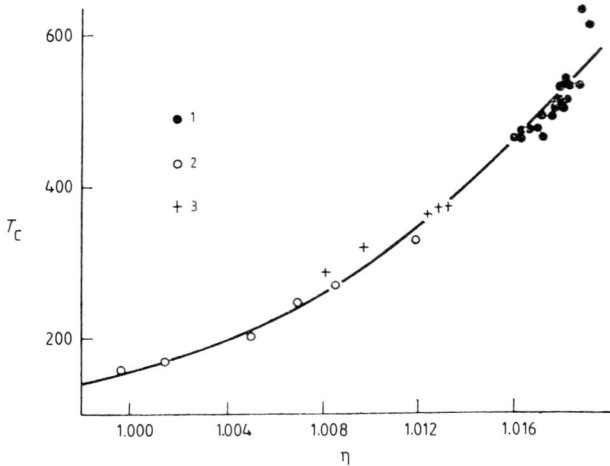

Figure 6.26 The Curie temperatures of single crystals of solid solutions of the systems (1) $KNbO_3-LiNbO_3-BaNb_2O_6$, (2) $KSr_2Nb_5O_{15}-KBa_2Nb_5O_{15}$ and (3) $KBa_2Nb_5O_{15}-KPb_2Nb_5O_{15}$ as functions of normalized lattice parameter $\eta = \sqrt{10c/a}$ (chapter 5, [27, 68]).

implied by the Devonshire–Ginzburg equation, becomes more obvious when applied to these three systems.

The relatively high Curie temperatures of $KNbO_3-LiNbO_3-BaNb_2O_6$ crystals may be due to the presence of lithium ions. The dielectric characteristics of these crystals point to a first-order phase transition over a comparatively wide range of solid solution compositions, such as $KSr_2Nb_5O_{15}$ [26], whereas $NaBa_2Nb_5O_{15}$ has a second-order phase transition. The grown unannealed crystals have very high dielectric losses which lessen substantially after annealing. In these crystals, the non-critical phase matching temperature T_{phm} for SHG with $\lambda = 1.06$ μm depends on their compositions above the Curie temperature (table 6.9). This tendency agrees with the results reported in [1], where crystals of the system $K_2O-Li_2O-Nb_2O_5$ with a TPTB-type structure are shown to have very high T_{phm} for SHG in YAG:Nd lasers.

6.6.2 *Phase diagrams, growth of single crystals*

The system $KNbO_3-LiNbO_3-BaNb_2O_6$ was investigated using X-ray diffraction, differential thermal and chemical analyses of the grown crystals (chapter 5, [68]) (figure 6.27). An incongruent composition with a melting temperature of 1354°C (± 5) and peritectic at 55 mol.% of $BaNb_2O_6$ was found on the side of $LiNbO_3-BaNb_2O_6$. The solid solution $LiNbO_3-BaNb_2O_6$ with a TPTB-type structure exists over the range 67 to 78 mol.% of $BaNb_2O_6$. Large crystals of good optical quality, as can be judged by the SHG efficiency,

Table 6.9   Curie point T_C and phase matching temperature T_{phm} for SHG ($\lambda = 1.06 \ \mu m$) for crystals of the system $(KNbO_3)_x-(LiNbO_3)_y-(BaNb_2O_6)_z$ (chapter 5, [68]).

Composition (mol.%)				
x	y	z	T_C (°C)	T_{phm} (°C)
14.5	8.1	77.4	466	−95
31.3	9.3	59.4	457	−23
32.9	8.9	58.2	471	60
33.5	7.9	58.6	461	80
32.0	16.0	52.0	480	103
19.0	22.7	58.3	493	188
0.0	22.6	77.4	612	150
0.0	32.5	67.5	630	

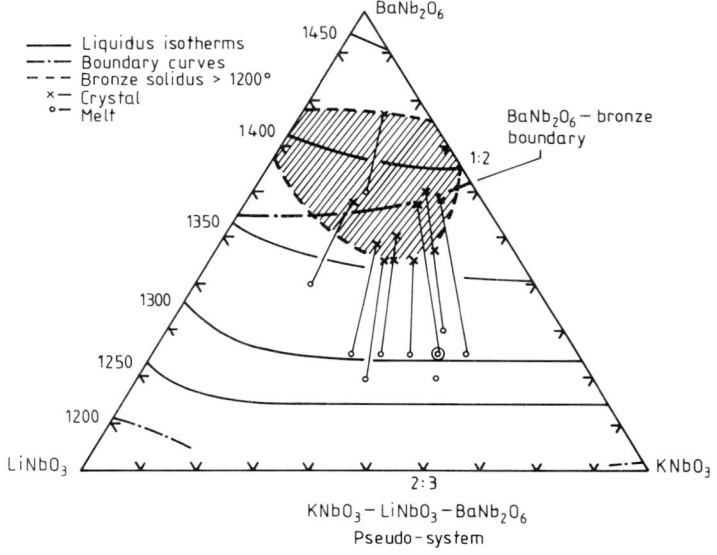

Figure 6.27   Ternary phase diagram of the pseudo-system $KNbO_3-LiNbO_3-BaNb_2O_6$ (chapter 5, [68]). Full lines are liquidus isotherms, chain lines are boundaries between phases, the TPTB region is enclosed by broken lines. Arrows indicate composition variations occurring in the course of crystal growth.

were obtained from a melt of the composition $50KNbO_3-25LiNbO_3-25BaNb_2O_6$ (point ◎ in figure 6.27).

It should be noted that mixed K–Li–Ba bronzes are produced from melts with excess alkali metal, and therefore the cation C positions with the coordinate number 9 are occupied in this structure, as distinct from the

Table 6.10 Some characteristics of TPTB-type single crystals.

Single crystals	ρ (g cm^{-3})	T_C (°C)	ε_c	$V_{\lambda/2}$ (V) ($\lambda = 0.63\ \mu m$)	δ†	T_{phm} (°C)	References
$Sr_2NaNb_5O_{15}$	5.0	270	2000	200	—	—	(chapter 5, [9])
$Ba_2NaNb_5O_{15}$	5.39	560	50	1400	20	80–100	(chapter 5, [9])
$K_3Li_2Nb_5O_{15}$	4.3	420	100	130	10	60–100	(chapter 5, [9])
$Na_3Li_2Nb_5O_{15}$	—	>800	100	—	—	—	(chapter 5, [9])
$K_2NaLi_2Nb_5O_{15}$	—	—	500	—	—	—	(chapter 5, [9])
$Ba_{1.6}Na_{1.4}Li_{0.4}Nb_5O_{15}$	—	≈560	50	1500	—	—	(chapter 5, [9])
$Ba_{0.5}Sr_{0.5}Nb_2O_6$	—	130	500	250	—	—	[23]
$Sr_2KNb_5O_{15}$	—	156	430	500	—	—	[24]
$Sr_4KLiNb_{10}O_{30}$	—	145	1200	200	—	—	[20]
$Sr_4NaLiNb_{10}O_{30}$	—	260	3000	120	—	—	[21]
$Ba_2LiNb_5O_{15}$	—	586	90	1700	—	28	[18]

† δ are nonlinear Miller coefficients expressed relative to KDP.

systems $NaNbO_3$–$BaNb_2O_6$ and $KNbO_3$–$BaNb_2O_6$ considered earlier. The TPTB solidus line in the system is rather gently sloping, and the crystal composition changes significantly with temperature. As a result, crystals grown from one and the same melt may differ significantly in the composition. However, the inhomogeneity of the crystals does not produce a noticeable effect upon the efficiency of SHG in YAG:Nd laser radiation because the growth striae lie perpendicularly to the c axis, and the laser beam measuring 0.1 mm in diameter propagates parallel to the a axis of the crystal. The radiation transformation efficiency for crystals of the composition marked in the phase diagram by point A is twice as high as that for lithium niobate.

Thus, the pseudo-system $KNbO_3$–$LiNbO_3$–$BaNb_2O_6$ has a wide range of solid solutions with tetragonal tungsten bronze-type structure and a phase matching temperature higher than room temperature, which do not require detwinning. The dielectric, electro-optic and nonlinear optic properties of several single crystals with tetragonal tungsten bronze-type structure are listed in table 6.10.

6.7 The ferroelectric $Ba_6Ti_2Nb_8O_{30}$

The ferroelectric $Ba_6Ti_2Nb_8O_{30}$ is the only crystal with a TPTB-type structure containing Ti ions in the places occupied by Nb, which was obtained as a comparatively large single crystal [30]. The ferroelectric properties of this crystal, grown by the Verneuil method, were established by Fang *et al* [31]; the crystal structure was investigated by Stephenson [32].

6.7.1 *Dielectric properties*
The room-temperature dielectric permittivities ε_c and ε_a at a frequency of 10 kHz amount to 209 ± 5 and 193 ± 5 respectively (figure 6.28). The dielectric permittivity ε_c has a marked anomaly at the transition point, 245°C. The linear variation of $1/\varepsilon_c$ takes place above and below the Curie temperature; the ratio of the slopes of the curves is 5.5. The inverse dielectric permittivity has a discontinuity at the Curie point. The Curie constant calculated from the slope in the paraelectric phase is equal to 2.8×10^5 K, which is comparable with the Curie constant of $BaTiO_3$ and other oxygen octahedral ferroelectrics. By analogy with barium titanate, one may assume that $Ba_6Ti_2Nb_8O_{30}$ has a first-order phase transition.

The spontaneous polarization P_s was measured by integration of the pyroelectric currents as well as the dielectric hysteresis loops. Below 150°C hysteresis loops are not observed, due to a high coercive field of this crystal. The room-temperature spontaneous polarization P_s corrected from the results of both these methods is equal to 0.22 ± 0.001 C m^{-2}, which is less than the value estimated from the Abrahams relation (chapter 4, [52]).

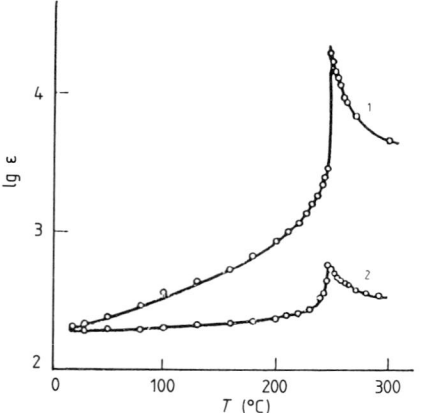

Figure 6.28 Temperature dependences of the dielectric constant of a $Ba_6Ti_2Nb_8O_{30}$ single crystal along (1) the c axis and (2) the a axis [30].

6.7.2 Optical and electro-optic properties

The refractive indices of $Ba_6Ti_2Nb_8O_{30}$ were measured using the least-beam deviation method on crystalline prisms in the wavelength range 4358–6438 Å (table 6.11). The dispersion of the refractive indices is described with a high degree of approximation by the one-term Sellmeier expression (§ 9.3). The values of the parameters S_0 and λ_0 for the two principal refractive indices are presented in table 9.3. The temperature dependences of the refractive indices n_o and n_e of the crystal for $\lambda = 0.5893$ μm were investigated in the interval from room temperature to 350 °C (figure 6.29). The refractive index n_o of an ordinary beam increases weakly with increasing temperature up to the Curie point, and above the Curie point it becomes almost temperature-independent. The refractive index n_e of an extraordinary beam increases monotonically with temperature. Below T_C, the behaviour of n_e is typical of ferroelectrics with tungsten bronze-type structure: the slope of $\delta n_e/\delta T$ gradually increases with temperature.

For the point symmetry group C_{4v}, to which $Ba_6Ti_2Nb_8O_{30}$ crystals belong, the indicatrix of the refractive indices under the action of an applied

Table 6.11 Refractive indices n_o and n_e of $Ba_6Ti_2Nb_8O_{30}$ crystals [30].

λ (Å)	n_e	n_o	λ (Å)	n_e	n_o
6438	2.276	2.314	5086	2.333	2.376
5893	2.293	2.332	4800	2.351	2.399
5791	2.297	2.337	4358	2.396	2.448
5461	2.311	2.349			

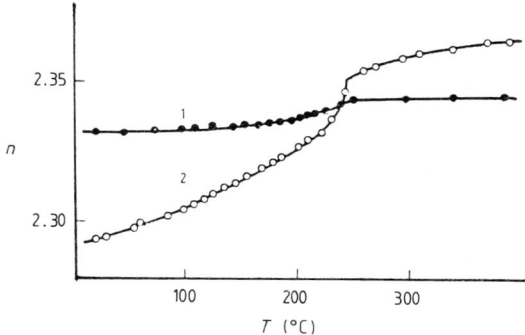

Figure 6.29 Temperature dependences of the refractive indices n_o (1) and n_e (2) of a $Ba_6Ti_2Nb_8O_{30}$ crystal ($\lambda = 5893$ Å) [30].

electric field E is described by the equation

$$(1/n_o^2 + r_{13}E_3)x^2 + (1/n_o^2 + r_{13}E_3)y^2 + (1/n_e^2 + r_{33}E_3)z^2$$
$$+ 2r_{51}E_2yz + 2r_{51}E_1xz = 1.$$

Crystals with the symmetry group C_{4v} have three independent non-zero electro-optic coefficients r_{13}, r_{33} and r_{51}. Two of them, r_{13} and r_{33}, were measured using an interferometer at the wavelength $\lambda = 6328$ Å. Their numerical values were $r_{33} = (1.17 \pm 0.02) \times 10^{-10}$, $r_{33} = (0.42 \pm 0.01) \times 10^{-10}$ m V^{-1}.

The effective half-wave voltage calculated from the expression

$$V_{\lambda/2} = \lambda / |n_e^3 r_{33} - n_o^3 r_{13}| \tag{6.18}$$

taking account of the measured electro-optic coefficients and the refractive indices $n_e = 2.279$, $n_o = 2.318$, was equal to 734 V. The $V_{\lambda/2}$ measurements by the dynamic method gave a value of 735 ± 5 V, which is in close agreement with the calculated one. At a high frequency of 200 MHz, the effective electro-optic coefficient $r_c^s = (0.37 \pm 0.03) \times 10^{-10}$ m V^{-1} and the corresponding half-wave voltage $V_{\lambda/2} = 1.4 \pm 0.2$ kV, which is twice as high as the one at low frequencies. The difference between r_c^s and r_c is due to the difference in the dielectric permittivities at low and high frequencies. Thus, in the investigated crystals the clamping effect manifests itself rather strongly at high frequencies.

In the general case, the linear electro-optic effect in ferroelectric crystals is interpreted as a quadratic electro-optic effect based on spontaneous polarization [6]. The quadratic electro-optic constants are given by the expressions

$$g_{33} = \frac{r_{33}}{2\varepsilon_0(\varepsilon_3 - 1)P_s} \qquad g_{13} = \frac{r_{13}}{2\varepsilon_0(\varepsilon_3 - 1)P_s} \tag{6.19}$$

where ε_0 is the dielectric permittivity of the vacuum. For $\varepsilon_0 = 8.85 \times 10^{-12}$, $\varepsilon_3 = 209$ and $P_s = 0.22\ \mathrm{C\ m^{-2}}$ one obtains the quadratic electro-optic constants $g_{33} = 0.15$, $g_{13} = 0.05\ \mathrm{m^4\ C^{-2}}$, which are comparable in magnitude with analogous coefficients of other octahedral ferroelectrics.

6.7.3   Nonlinear optical properties

From symmetry conditions, $\mathrm{Ba_6Ti_2Nb_8O_{30}}$ has only two independent nonlinear optical coefficients, d_{33} and d_{31}, which were measured by Maker's method [33, 34] since the phase matching conditions for this crystal were unknown. The values of the coefficients were determined with respect to lithium niobate: $d_{33} = (0.37 \pm 0.05)d_{33}(\mathrm{LiNbO_3})$ and $d_{31} = (0.27 \pm 0.05)d_{33}$ $(\mathrm{LiNbO_3})$. In absolute values $d_{33} = (15.1 \pm 2.0)$ and $d_{31} = (11.0 \pm 2.0) \times 10^{-12}\ \mathrm{m\ V^{-1}}$ for $d_{33}(\mathrm{LiNbO_3}) = 40.7 \times 10^{-12}\ \mathrm{m\ V^{-1}}$ (chapter 4, [14]).

The Miller coefficients δ, which are defined by the relation (see chapter 9)

$$\delta_{ij} = d_{ij}/\varepsilon_0(n_\omega^2 - 1)^2(n_{2\omega}^2 - 1) \tag{6.20}$$

have the following values: $\delta_{33} = (2.5 \pm 0.2) \times 10^{-2}$ and $\delta_{31} = (1.7 \pm 0.2) \times 10^{-2}\ \mathrm{m^2\ C^{-1}}$. The relative signs of the coefficients d_{33} and d_{31} were determined by the Miller method [35]. These coefficients were found to be of the same sign, as in the case of $\mathrm{Ba_2NaNb_5O_{15}}$.

Knowing the Miller coefficients, one could evaluate the spontaneous polarization of the crystal from the Jerphagnon relation [36]:

$$|\delta_{33} + 2\delta_{31}| = 0.2P_s.$$

Substituting the above δ_{ij} values, we obtain $P_s = (0.29 \pm 0.03)\ \mathrm{C\ m^{-2}}$, which is somewhat greater than the experimental value $P_s = (0.22 \pm 0.01)\ \mathrm{C\ m^{-2}}$.

The coefficients d_{33} and d_{31}, which were determined from Maker's interference peaks, decrease with increasing temperature, tending to zero at the Curie point (figure 6.30). The coherence lengths l_{33} and l_{31} also decrease

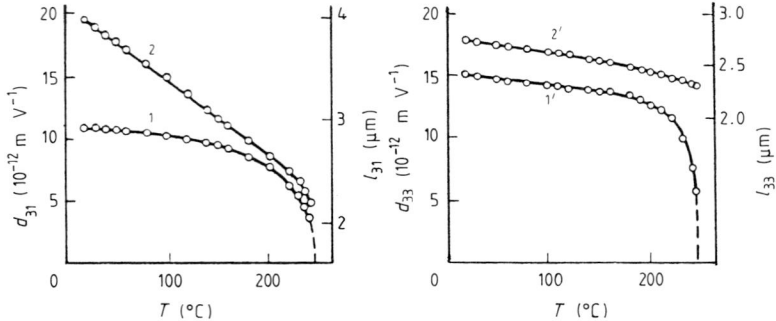

Figure 6.30   Temperature dependences of nonlinear coefficients (1) d_{31}, (1') d_{33} and of coherence lengths (2) l_{31}, (2') l_{33} of a $\mathrm{Ba_6Ti_2Nb_8O_{30}}$ single crystal.

monotonically with increasing temperature, but the decrease rate of l_{31} is higher.

To obtain the temperature dependences of δ_{33} and δ_{31}, it was assumed that n_{e} for $\lambda = 0.53\ \mu\mathrm{m}$ varies with temperature in the same manner as for $\lambda = 0.59\ \mu\mathrm{m}$, for which the temperature dependence was found experimentally. Such an assumption is admissible since the dispersion of quadratic electro-optic coefficients g_{33} over the wavelength range $(0.53-0.59)\ \mu\mathrm{m}$ is not large. Using the temperature dependences of n_{o}, n_{e}, d_{31} and d_{33}, Itok and Iwasahi [30] obtained the temperature dependences of δ_{33}, δ_{31} and P_{s} normalized to the room-temperature values (figure 6.31). The figure shows that δ_{33} and δ_{31} are proportional to P_{s}.

6.7.4 *Growth of single crystals*

The starting materials were prepared by mixing high-purity reagents $BaCO_3$, Nb_2O_5 and TiO_2 in stoichiometric proportions. A solid-phase synthesis of $Ba_6Ti_2Nb_8O_{30}$ was carried out at $1000\,^{\circ}\mathrm{C}$ for several hours.

Single crystals were grown by the Czochralski technique in a high-frequency heating chamber. Crystals grown in a nitrogen atmosphere were black, probably due to the oxygen losses. When grown in an oxygen atmosphere, they had plenty of micro-cracks and were non-transparent. The black crystals were bleached by annealing in oxygen at a temperature of $700-800\,^{\circ}\mathrm{C}$ for $20-30$ h.

In the crystalline plates (001), striation was observed which disappeared under poling. One may assume that the observed striation corresponds to a 180° domain pattern formed in the growth process. The crystal was poled by heating to $20\,^{\circ}\mathrm{C}$ above the Curie temperature and slow cooling under the

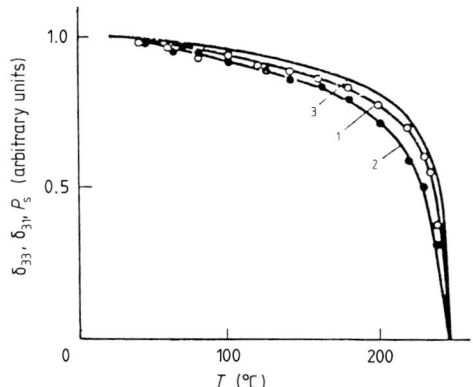

Figure 6.31 Temperature dependences of Miller coefficients (1) δ_{33}, (2) δ_{31} and (3) of spontaneous polarization P_{s} of a $Ba_6Ti_2Nb_8O_{30}$ crystal, normalized to the values obtained at room temperature [30].

field in a silicone liquid. The degree of polarization was checked by chemical etching of the crystals in boiling hydrofluoric acid.

6.8   Crystals of $Ba_{4+x}Na_{2-x}Nb_{10-x}Ti_xO_{30}$

A substitution of Ti for Nb in $Ba_2NaNb_5O_{15}$ crystals makes it possible to obtain a number of solid solutions from BNN (see chapter 5) to $Ba_6Ti_2Nb_8O_{30}$ (see chapter 8). An analysis of ceramic samples of the $Ba_{4+x}Na_{2-x}Nb_{10-x}Ti_xO_{30}$ system has revealed that an increase of x causes lowering of the Curie temperature and a decrease of the room-temperature orthorhombic lattice deformation which hampers a wide use of BNN single crystals.

6.8.1   *Dielectric properties*

The temperature dependences of the dielectric permittivity ε_c of the grown crystals have the form typical of ferroelectrics; the Curie–Weiss constant is $\approx 3.2 \times 10^5$ K. The dielectric permittivity ε_a is essentially temperature-independent (the distinction between ε_a and ε_b was not established) [37]. The magnitude of the spontaneous polarization was estimated from the pyroelectric current measurements. The values of ε_c, ε_a and P_s at room temperature for the investigated samples are presented in table 6.12. The temperature dependences of ε_c and ε_a for several compositions are shown in figure 6.32.

Table 6.12   Ferroelectric and electro-optic properties of $Ba_{4+x}Na_{2-x}Nb_{10-x}Ti_xO_{30}$ crystals [37].

x	Symmetry at T_{room}	ε_c	ε_a	P_s (C m^{-2})	n_c
0 (BNN)	mm2	43.0	244	0.40	2.218
0.23 (A)	mm2	45.1	216	—	2.200
0.35 (B)	mm2	47.0	206	0.34	2.237
0.78 (C)	mm2	63.8	187	—	2.268
1.30 (D)	4mm	111	336	0.26	2.2685

x	$n_b \approx n_a$	$2V$† (grad)	$V_{\lambda/2}$ (V)	r_c (10^{-12} m V^{-1})	$1/\varepsilon_c V^2_{\lambda/2}$ (10^{-9} V^{-2})
0 (BNN)	2.322	14.25	1830	33.0	7.0
0.23 (A)	2.297	13.5	1660	35.7	8.1
0.35 (B)	2.321	11.5	1560	36.3	8.7
0.78 (C)	2.347	9.75	1170	46.4	11.4
1.30 (D)	2.3275	—	960	57	9.8

† $2V$ is the angle between the optical axes of the crystal.

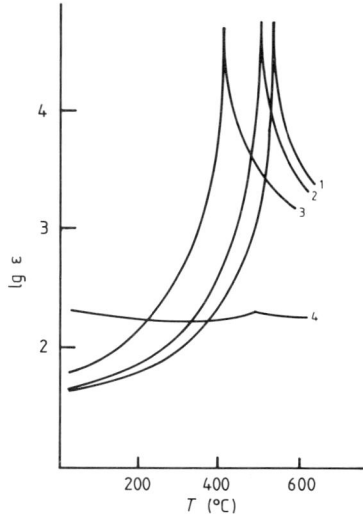

Figure 6.32 Temperature dependences of dielectric permittivities (1–3) ε_c, (4) ε_a of $Ba_{4+x}Na_{2-x}Nb_{10-x}Ti_xO_{30}$ single crystals of compositions (1) 0.23, (2, 4) 0.35, (3) 0.75 [38].

6.8.2 *Optical and electro-optic properties*

Crystals of the compositions A, B and C (see table 6.12) are biaxial and optically negative; crystal D is uniaxial and optically negative. The value of the birefringence $\Delta n = n_a - n_b$ and the angle $2V$ between the optical axes were determined at room temperature by the conoscopic picture (figure 6.33). It should be noted that the birefringence Δn_{ab} is small in the absolute value and decreases with increasing x. Therefore, in the study of the temperature

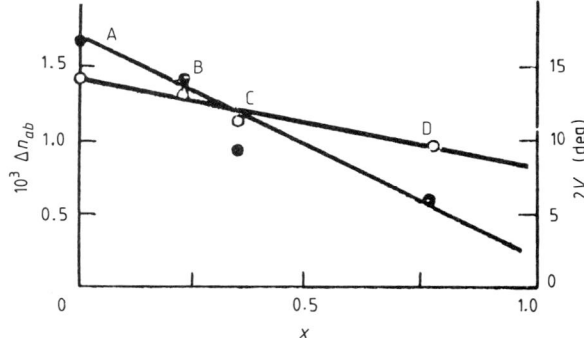

Figure 6.33 Birefringence Δn_{ab} (points) and angle $2V$ between optical axes (circles) as functions of the composition of a $Ba_{4+x}Na_{2-x}Nb_{10-x}Ti_xO_{30}$ crystal. Crystal compositions are presented in table 6.12 [38].

dependence of the refractive indices and electro-optic effect, the crystals A, B and C may be regarded as uniaxial, i.e. one may assume $n_a = n_b$, $r_{13} = r_{23}$. The dependence of Δn_{ac} on T is plotted in figure 6.34.

The values of the half-wave voltage $V_{\lambda/2}$ were determined by modulation methods. The light transmitted along the a axis was modulated by an electric field applied along the c axis. A plot of the effective electro-optic coefficient r_c, calculated from the half-wave voltage by the formula $r_c = \lambda/n^3 V_{\lambda/2}$, versus temperature is shown in figure 6.35(a).

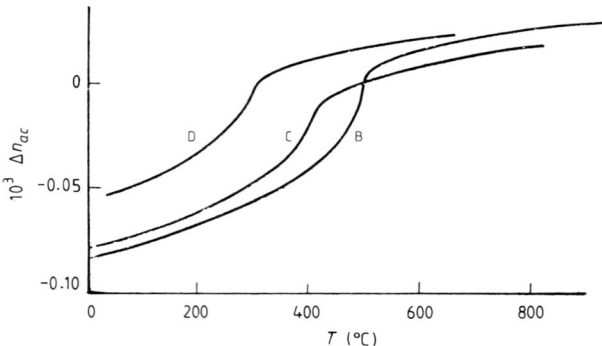

Figure 6.34   Temperature dependences of birefringence Δn_{ac} of crystals B, C, D (see table 6.12) [38].

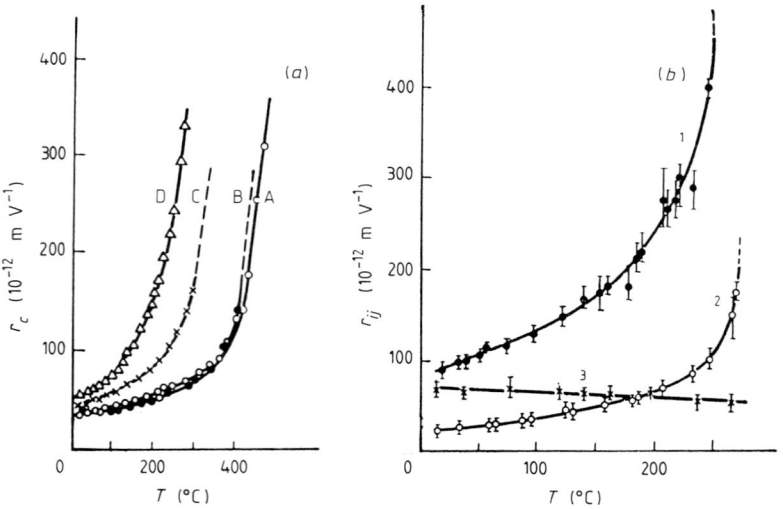

Figure 6.35   Temperature dependences of the effective electro-optic coefficient r_c for crystals A, B, C, D (a) and of the electro-optic coefficients (1) r_{33}, (2) r_{31}, (3) r_{42} for crystal D (b). Crystal compositions are presented in table 6.12 [38].

The electro-optic coefficients r_{ij} were determined separately using a Mach–Zehnder interferometer. For room temperature the obtained values were $r_{13} = 26$, $r_{33} = 92$ and $r_{42} = 68 \times 10^{-12}$ m V^{-1}. The corresponding temperature dependences are presented in figure 6.35(b).

As has already been mentioned in the preceding sections, the criterion of applicability of electro-optic crystals may be the quantity $1/\varepsilon_c V_{\lambda/2}^2$, which is a kind of figure of merit. The corresponding values of this characteristic for the obtained crystals are listed in table 6.12. As x increases, the figure of merit first rises and then falls.

6.8.3   *Growth of single crystals and the structure properties*

Crystals were grown by the Czochralski method in a high-frequency heating chamber. The growth conditions were typical of this method: the temperature gradient above the melt was 100 K cm^{-1}, the pulling rate reached approximately 5 mm h^{-1}, the seed rotation rate was approximately equal to 30 rpm, the cooling rate of the crystal in the postcrystallization period reached approximately 15 K h^{-1}.

The compositions of the melts and of the grown single crystals, the melting temperatures and the Curie points are presented in table 6.13. The crystal composition was determined from the composition–Curie temperature phase diagram constructed for this system from an analysis of the dielectric properties of ceramic samples.

The samples A, B and C at room temperature are in the orthorhombic phase. The temperature T_0 of transition from the orthorhombic to the tetragonal phase was determined from the maximum elastic compliance of the crystal bar cut out at an angle of 45° to the c axis. The variation of this temperature as a function of the composition is plotted in figure 6.36. Below T_0 the crystal symmetry is mm2(C_{2v}); between T_0 and T_C the symmetry is 4mm(C_{4v}). The crystal D, possessing a room-temperature tetragonal structure, has the composition $Ba_{5.37}Na_{0.66}Nb_{8.88}Ti_{1.05}O_{30}$. It should be noted that

Table 6.13 Lattice parameters, Curie temperatures and melting temperatures of $Ba_{4+x}Na_{2-x}Nb_{10-x}Ti_xO_{30}$ crystals [38].

	x	Lattice parameters† (Å)				
Melt	Crystal	a	b	c	T_C (°C)	T_{mel} (°C)
0 (BNN)	0	17.59	17.62	7.99	560	1450
0.25	0.23 (A)		17.63	7.995	510	1440
0.30	0.35 (B)		17.64	8.00	495	1420
1.00	0.78 (C)		17.65	8.01	405	1370
1.50	1.30 (D)		12.520	4.014	300	1360

† Lattice parameters of specimens A, B and C correspond to $T > T_0$.

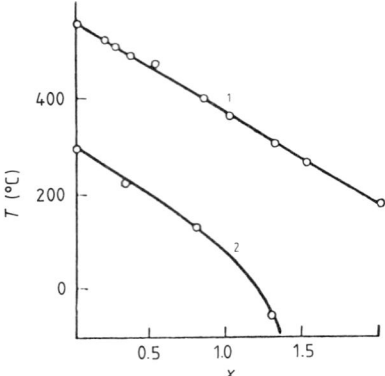

Figure 6.36 Dependences of (1) the Curie temperature T_C and (2) the temperature T_0 of transition from orthorhombic to tetragonal phase on the composition of a $Ba_{4+x}Na_{2-x}Nb_{10-x}Ti_xO_{30}$ crystal [38].

the composition D deviates from the nominal composition of the indicated system, although the general formula almost preserves the form $A_6B_{10}O_{30}$. Strictly speaking, crystals D cannot be regarded as solid solution in the pseudo-binary $Ba_2NaNb_5O_{15}-Ba_6Ti_2Nb_8O_{30}$ system. Such a system requires a more thorough analysis of phase equilibria.

In a tetragonal crystal D, micro-twins do not occur, and a lower melting temperature as compared with other crystals facilitates the growth process. Thus, this crystal has some advantages over BNN in electro-optic device use. But attempts to grow large crystals [38, 39] were unsuccessful. This is primarily connected with the difficulty of keeping the crystal from lateral growth. For crystals grown from a congruent melt these difficulties will be overcome. To establish the congruent composition one should investigate the phase diagram of $BaO-Na_2O-Nb_2O_5-TiO_2$.

6.9   Single crystals of potassium lithium niobate

As has already been mentioned, the tetragonal potassium tungsten bronze structure can be characterized by the general chemical formula $(Al)_2(A2)_4(C)_4(Bl)_2(B2)_8O_{30}$, where positions B1 and B2 are occupied by the cations forming the lattice and A1, A2 and C by the cations providing electroneutrality. In ferroelectrics with the occupied tungsten bronze structure, all A1 and A2 positions are occupied by cations, and at room temperature they do not exhibit an optically induced variation of the refractive index, which is observed in lithium niobate and in other materials.

Potassium lithium niobate (PLN) crystals, $K_3Li_2Nb_5O_{15}$, have a completely occupied structure, i.e. C positions are also filled with lithium ions, and consequently they are not subject to optical damage. PLN crystals do not exhibit twinning typical of BNN. The melting temperature of PLN is approximately 1000°C, which is somewhat lower than that of the majority of tungsten bronze-type ferroelectrics. This circumstance facilitates the growth of single crystals by the Czochralski method.

6.9.1 *Dielectric properties*

For crystals $(K_2O)_{0.30}(Li_2O)_{0.7-x}(Nb_2O_5)_x$ the maximum of the dielectric permittivity ε_c widens with increasing x, which hampers the determination of the phase transition temperature (figure 6.37). The dielectric permittivity ε_a does not show any anomaly, but decreases gradually with increasing temperature.

The temperature dependence of P_s (figure 6.38(a)) was determined from the pyroelectric currents. Smith *et al* [41] noticed the influence of the crystal surface treatment upon the magnitude of pyroelectric currents. Due to this fact, the reproducibility of the results of measurements did not exceed 10%.

6.9.2 *Optical and electro-optic properties*

The temperature dependence of the refractive indices of PLN single crystals was determined by the method of least deviation in a prism made of the investigated material (figure 6.39). The room-temperature birefringence of the crystal, equal to 0.14, is the highest value observed in ferroelectric niobates.

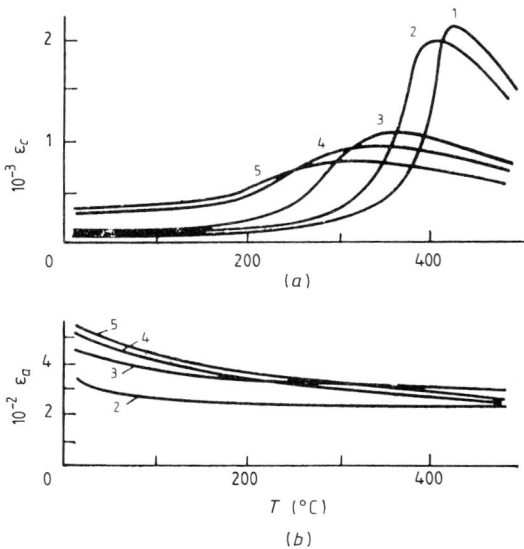

Figure 6.37 Temperature dependences of dielectric permittivities ε_c (a) and ε_a (b) of potassium lithium niobate crystals with x equal to (1) 0.530, (2) 0.534, (3) 0.542, (4) 0.547, (5) 0.550 [40].

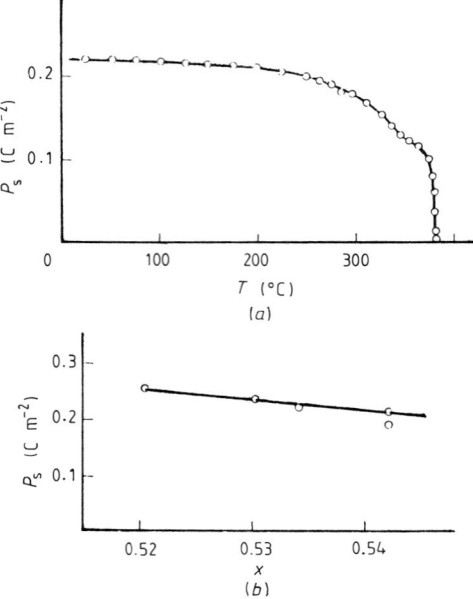

Figure 6.38 Temperature dependence of spontaneous polarization of a potassium lithium niobate crystal with $x = 0.534$ (a) and the dependence of P_s on concentration at room temperature (b) [40].

The sharp decrease of n_e below T_C is due to the spontaneous polarization of the crystal. The continuous variation of the refractive indices at the phase transition temperature is indicative of a second-order phase transition. Above T_C, the crystal lattice vibrations contribute to the temperature dependence $n(T)$, and therefore the extrapolation of $n(T)$ to the low-temperature region is not sufficiently correct (broken lines in figure 6.39). The high value of the birefringence offers an opportunity of reaching phase matching for light with $\lambda = 0.90 \ \mu m$, generated by a GaAs laser, which enables this laser to be used in optical communication.

The investigated PLN crystals of all compositions are optically negative ($n_o > n_e$). For the compositions $x = 0.534$ and $x = 0.547$, the temperature variations of n_o and n_e somewhat differ from those shown in figure 6.39. The refractive index of the extraordinary beam increases linearly with increasing niobium content in the crystal, whereas the refractive index of the ordinary beam does not depend on the concentration (figure 6.40).

Crystals of PLN are transparent over the range 0.4 to 5 μm. The refractive indices of the $K_3Li_2Nb_5O_{15}$ crystal, suitable for electro-optic use and for SHG, are presented in table 6.14.

A laser beam with a wavelength $\lambda = 0.488 \ \mu m$ and a power of 300 mW focused inside a $K_3Li_2Nb_5O_{15}$ crystal did not induce variations in the

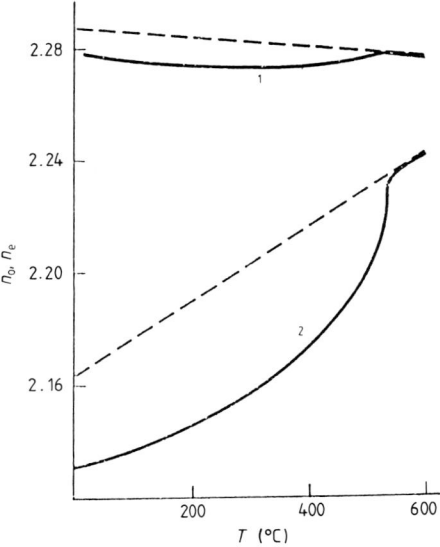

Figure 6.39 Temperature dependences of refractive indices of (1) an ordinary beam n_o and (2) an extraordinary beam n_e in a potassium lithium niobate crystal with $x = 0.52$ for $\lambda = 0.633$ μm [41]. Broken lines show extrapolation of $n(T)$ into the low-temperature region.

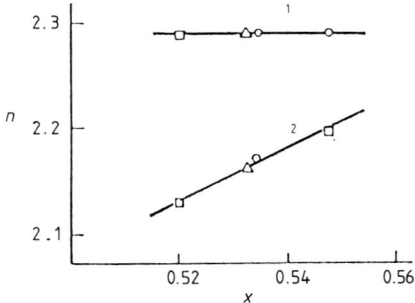

Figure 6.40 Dependences of refractive indices (1) n_o and (2) n_e on the composition of potassium lithium niobate crystals [40]. Room temperature, $\lambda = 0.633$ μm.

refractive index. Under identical irradiation conditions, a variation of the refractive index in $LiNbO_3$ was induced by a laser beam power less than 1 mW.

The matrix of the linear electro-optic coefficients of a PLN crystal, which belongs to the point symmetry group 4mm, contains three independent non-zero coefficients r_{13}, r_{15} and r_{33}. For an electric field applied along the ferroelectric c axis of a $K_3Li_2Nb_5O_{15}$ crystal and for light propagating

Table 6.14 Refractive indices of a $K_3Li_2Nb_5O_{15}$ crystal for several wavelengths [40].

λ (μm)	n_e	n_o
0.628	2.163	2.277
0.532	2.197	2.326
1.064	2.112	2.208

normally to this axis and polarized at an angle of $45°$, the half-wave voltage for $\lambda = 0.633\ \mu$m is 930 V [40]. This gives $r_c = 6.8 \times 10^{-9}$ cm V^{-1}. For a beam polarized parallel to the c axis, it was found from the phase delay that $|n_e^3 r_{33}| = 7.9 \times 10^{-8}$ cm V^{-1}; for a beam polarized perpendicularly to the c axis $|n_o^3 r_{13}| = 1.05 \times 10^{-8}$ cm V^{-1}. From a comparison of the results of various measurements, r_{13} and r_{33} were found to be of the same sign.

For PLN crystals with x ranging from 0.525 to 0.55 the half-wave voltage at $25°$C is equal to 1350 V; to this half-wave voltage there corresponds an effective electro-optic coefficient $r_c = 4.8 \times 10^{-9}$ cm V^{-1} [41]. For the dielectric permittivity $\varepsilon_c \approx 60$ the electro-optic constant $(r_c/\varepsilon_c\varepsilon_0)$ amounts to 0.09 m^2 C^{-1}, which is a little higher than for relative ferroelectric niobates.

6.9.3 Nonlinear optical properties

For the point symmetry group 4mm, to which PLN crystals belong, the nonlinear polarization quadratic with respect to the field is expressed in terms of the amplitudes of the electric field E and nonlinear coefficients:

$$\bar{P}_x = 2d_{15}\bar{E}_x\bar{E}_y \qquad \bar{P}_y = 2d_{15}\bar{E}_y\bar{E}_z$$
$$\bar{P}_z = d_{13}(\bar{E}_x^2 + \bar{E}_y^2) + d_{33}\bar{E}_z^2.$$

According to Kleinman's symmetry conditions [42], the equality $d_{13} = d_{15}$ holds.

Second harmonic generation in PLN crystals was investigated using a single-mode YAG:Nd laser (1.06 μm) [41]. From figure 6.41 it is seen that T_{phm} embraces a wide temperature interval from 140 to $450°$C, which is a consequence of the large birefringence (see figure 6.39).

The nonlinear coefficient d_{31} was determined relative to d_{31} of LiNbO$_3$ by way of comparing the second harmonic intensities at corresponding T_{phm}. For crystals of $(K_2O)_{0.3}(Li_2O)_{0.7-x}(Nb_2O_5)_x$, where x ranges between 0.525 and 0.535, the nonlinear coefficient $d_{31} = (1.9 \pm 0.3)d_{31}(LiNbO_3)$.

The high phase matching temperatures for $\lambda = 1.06\ \mu$m suggest the possibility of phase matching of radiation of the injection GaAs laser with $\lambda = 0.90\ \mu$m for a beam directed perpendicular to the optical axis. This was later seen in PLN crystals of the composition $x = 0.525$, which for the fundamental wave $\lambda = 0.90\ \mu$m have $T_{phm} = 110°$C. The corresponding

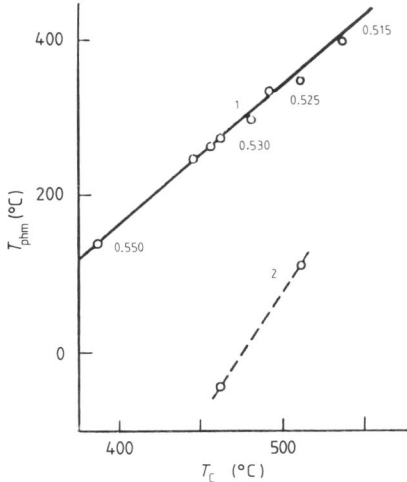

Figure 6.41 Dependences of the phase matching temperature on the Curie temperature for $(K_2O)_{0.3}(Li_2O)_{0.7-x}(Nb_2O_5)_x$ crystals of different compositions for λ_0 equal to (1) 1.06, (2) 0.9 μm [41]. The x values are given at the points. Room temperature.

temperature for $\lambda = 1.06$ μm is equal to 350°C. BNN crystals have no phase matching for the fundamental wave $\lambda = 0.90$ μm above room temperature.

For crystals of the composition $K_3Li_2Nb_5O_{15}$ the nonlinear coefficients d_{ij} were estimated relative to d_{11}^Q of quartz (chapter 5, [4]); $d_{33} = (18 \pm 2)d_{11}^Q$ and $d_{31} = (14 \pm 2)d_{11}^Q$. Similar values were reported in [43]. Phase matching for this crystal can be attained without birefringence for temperatures from 60 to 100°C.

6.9.4 *Phase diagrams and growth of single crystals*
As has been shown above, potassium lithium niobates exhibit a substantial variation of their properties depending on the growth conditions and melt composition.

To establish reproducible growth conditions for single crystals and to study the influence of the composition upon the chemical and physical properties of the crystals, Scott *et al* [44] examined the ternary phase diagram of the system K_2O–Li_2O–Nb_2O_5 (figure 6.42). The circles in the diagram indicate compositions of the samples in the region of solid solutions of tungsten bronzes and along 50 mol.% Nb_2O_5 and 30 mol.% K_2O isopleths, which were thoroughly investigated. The compositions and melting temperatures of the compounds in the systems K_2O–Nb_2O_5 and Li_2O–Nb_2O_5 were borrowed from [86, 45]. The cross-hatched area of the phase diagram stretching from 51 to 68 mol.% Nb_2O_5 along the 30 mol.% K_2O isopleth corresponds

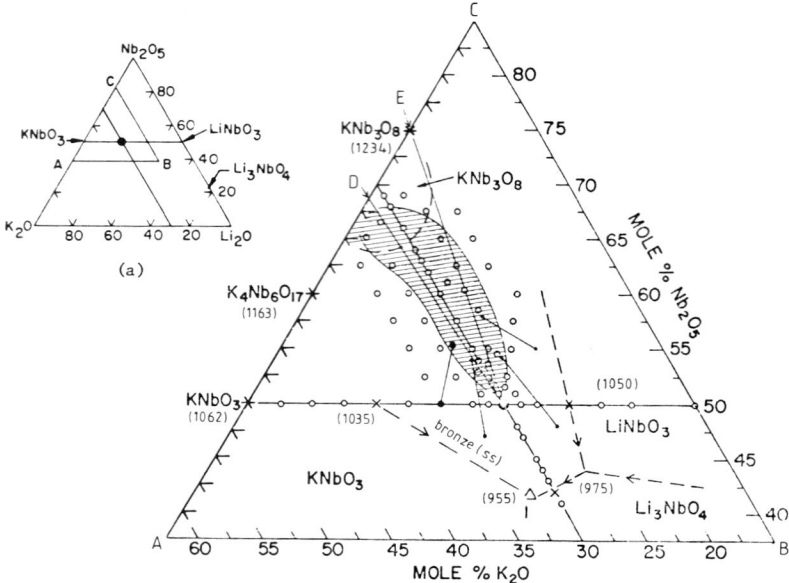

Figure 6.42   Ternary phase diagram of the system $K_2O–Li_2O–Nb_2O_5$ [44]. Arrows indicate variations of crystal composition relative to melt compositions.

to the single phase of tungsten bronze solid solution. As is seen from the diagram, the composition $30K_2O–20Li_2O–50Nb_2O_5(K_3Li_2Nb_5O_{15})$ does not get into this region.

Thus, a 'completely occupied' tungsten bronze structure, in the unit cell of which all the ten cation positions are occupied, is not stable, and the solid solution exists only in the niobium-rich region. The approximate composition of the eutectics at 955°C, denoted by a triangle, is $32.0K_2O–26.3Li_2O–41.7Nb_2O_5$, which was established through a chemical analysis (± 1 mol.%) of the eutectic matrix. The chemical composition of the peritectic point (975°C) has not been precisely established.

The cross section along the 30 mol.% K_2O isopleth (figure 6.43) is the main one in the region of solid solutions of tungsten bronzes and passes approximately through the centre of this system (see figure 6.42). The melt crystallization in this cross section proceeds along the liquidus line from 51 to 41 mol.% of Nb_2O_5 and ends with the formation of eutectics at a temperature of 955°C. The solidus line gradually ascends from the boundary of the solid solution, from 51 to 63 mol.% of Nb_2O_5, where it intersects the liquidus line at the congruent point. But due to a relatively high error in the determination of the temperature (± 2 K), the existence of congruence in this cross section has not been precisely established. Solid solution crystals grown from a melt containing 30 mol.% of K_2O contain as a rule less potassium

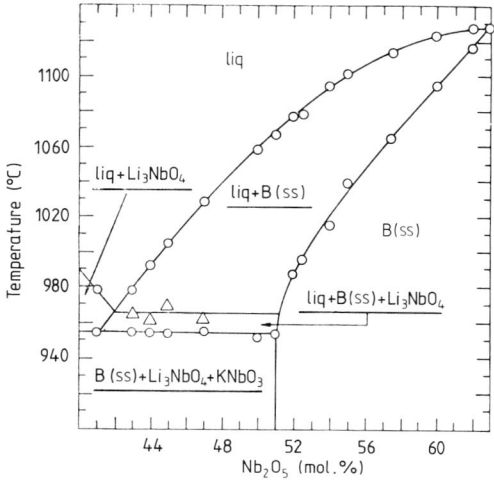

Figure 6.43 Cross section along a 30 mol.% K_2O ternary phase diagram of the system $K_2O-Li_2O-Nb_2O_5$ up to 63 mol.% of Nb_2O_5 [44].

oxide, and therefore the liquidus and solidus lines in this cross section are not determined exactly.

The results of the chemical analysis of the first crystals grown from the melt are tabulated in table 6.15. Although these results can be interpreted within an accuracy of ± 1 mol.%, the difference in the amount of K_2O in the melt and crystal apparently decreases as the niobium concentration increases. These data, as well as figure 6.43, may give grounds for assuming that the maximum stable cross section in the system of solid solutions stretches along the lines of 30 mol.% K_2O isopleth from about the composition $29K_2O-19.5Li_2O-51.5Nb_2O_5$ to 63 mol.% of Nb_2O_5. Without exactly

Table 6.15 Compositions of melts and of $K_2O-Li_2O-Nb_2O_5$ crystals grown from these melts [44].

Melt composition (mol.%)			Crystal composition (mol.%)		
K_2O	Li_2O	Na_2O_5	K_2O	Li_2O	Na_2O_5
30.0	26.0	44.0	29.5	18.3	52.2
30.0	24.0	46.0	29.3	17.6	53.1
30.0	22.0	48.0	29.5	16.8	53.7
30.0	20.0	50.0	29.9	14.2	55.9
30.0	15.0	55.0	29.8	11.4	58.8
30.0	10.0	60.0	30.1	9.2	60.7

determining this cross section, table 6.15 shows that the gap between the liquidus and solidus lines lessens as the niobium content increases up to 60 mol.%, and the congruent point is above this composition. The table also implies that the phase diagram of figure 6.43 is reliable enough as a reference point for expected compositions of crystals grown along the cross section of 30 mol.% K_2O of a ternary system.

An X-ray diffraction analysis of crystals of PLN solid solutions has shown that they have a tetragonal cell with parameters $a \approx 12.6$ and $c \approx 4$ Å only below approximately 55 mol.% of Nb_2O_5. For higher niobium concentration, the crystals have orthorhombic symmetry. This was established optically, since due to a small orthorhombic distortion a powder X-ray analysis of crystals and ceramics of such compositions could fix only tetragonal structure. The superstructure of the crystal lattice, observed in other complex niobates, has not been discovered in orthorhombic PLN crystals.

The lattice parameters measured on ceramic samples of different compositions are shown in figure 6.44. The appearance of the orthorhombic phase for 55 mol.% of Nb_2O_5, for which the ratio $\sqrt{10}c/a = 1$ (figure 6.44), was also

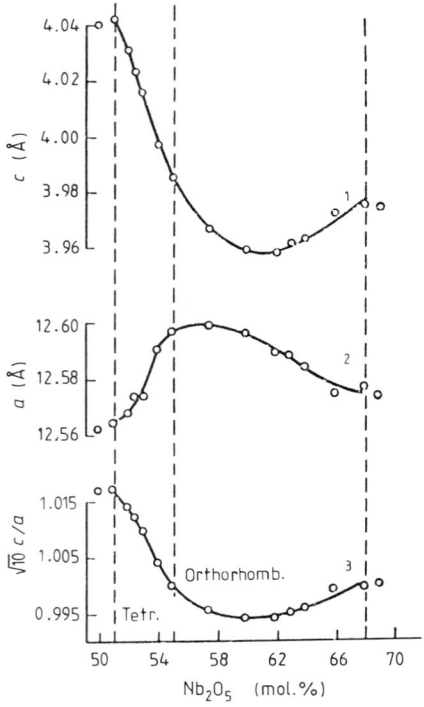

Figure 6.44 Dependence of potassium lithium niobate crystal lattice parameters: (1) c, (2) a, (3) the ratio $\sqrt{10}c/a$ on the content of Nb_2O_5 [46].

reported by Carruthers and Grasso (chapter 5, [27]). But orthorhombic potassium lithium niobates are not ferroelectrics. The complicated variation of the lattice parameters c and a is explained by replacement of K^+ ions in the central positions by Nb^{5+} ions. The charge is compensated by the formation of surroundings and the appearance of lithium vacancies in positions with the coordination number 9 (C position). Another location of ions in the PLN structure is also possible [45].

The temperature dependences of the lattice parameters of $K_3Li_2Nb_5O_{15}$ were investigated on ceramic samples [45]. The tetragonal phase was shown to be retained in the transition through the Curie point ($\approx 425°C$). This crystal exhibited anisotropy of the temperature variation of the lattice parameters: the parameter c gradually decreases with increasing temperature, whereas the parameter a continuously increases.

The first $K_3Li_2Nb_5O_{15}$ crystals were grown from a melt with an excess of potassium or lithium (chapter 5, [41]). Crystals accessible for measurements were obtained by cooling a melt of the composition $0.525K_2CO_3$, $0.255Li_2CO_3$ and $0.72Nb_2O_5$ moles. The melt was placed in a second 100 cm^3 platinum crucible. The crucible was placed in a muffle furnace at $1250°C$ and was cooled at a rate of 1 K h^{-1} down to $800°C$; the furnace was then turned off and cooled naturally. Later, large crystals were grown from a melt whose volume was 100 times as large as the one reported in the above mentioned paper. The equipment for melt crystallization in such a volume was described in [47]. In a crucible with a diameter and height up to 20 cm, columnar crystals were grown with a length up to several centimetres (along the c axis) and a diameter up to two centimetres. The crystals were cracking perpendicular to the c axis, but this was not an obstacle for their application, since the preferable direction for device use is the a axis.

Crystals of PLN were also grown by the Czochralski method. They exhibited less cracking if pulled along the a axis. The growth temperature was about $1000°C$. The pulling and rotation rates were about 6.5 mm h^{-1} and 50 rpm. Crystals grown along the a axis had a rectangular cross section whose longer side coincided with the c axis.

Single crystals of $K_3Li_2Nb_5O_{15}$, like single crystals of $KTa_xNb_{1-x}O_3$ (see chapter 2), were also grown by the Kyropoulos method [46]. The starting oxides were mixed with excess K_2CO_3 and Li_2CO_2 (30–33 mol.% and 20–22 mol.% respectively). A growing crystal was raised at regular intervals at a rate of $1-2 \text{ mm h}^{-1}$; the melt was cooled at a rate of about 1 K h^{-1}. The crystals were pulled along the a and c axes, or in the [110] direction. The seed crystal was rotated at a rate of 60 rpm. Good crystals were grown only in the [110] direction; along other directions they were prone to cracking. The crystals were pale yellow and were transparent in the wavelength range from 0.4 to 5 μm. In the process of growth it was established that the crystallization rate is much greater along the c axis than along other directions. The majority of $K_3Li_2Nb_5O_{15}$ crystals cracked parallel to the junction plane (001) in the cooling process in the growth chamber. Three reasons for this cracking can be pointed out. (i) The (001) plane in these crystals has a high reticular density, and interatomic bonds in this plane are strong. (ii) Imperfections in the crystal may be due to a high growth rate along the [001] direction. (iii) Anisotropic thermal expansion may take place along the c and a axes, which increases the strain in the crystal. Growth of PLN single crystals by vertical zone melting is described in [48].

6.10   Ferroelectric $K_3Li_2(Ta_xNb_{1-x})_5O_{15}$ crystals

Single crystals of the solid solutions $K_3Li_2(Ta_xNb_{1-x})_5O_{15}$ for $0 \leqslant x \leqslant 1$ were grown by way of partial replacement of Nb by Ta in a $K_3Li_2Nb_5O_{15}$ crystal. Crystals of PLTN have a tungsten bronze-type structure and better electro-optic properties than PLN [49].

The dependence of the Curie temperature on the PLTN composition decreases rapidly as the Ta content in the crystal increases and becomes negative (on the Celsius scale) for $x = 0.5$ (figure 6.45). This composition is intermediate between tetragonal and orthorhombic phases. The completely replaced $K_3Li_2Ta_5O_{15}$ has the Curie temperature $T_C = -266°C$.

The room-temperature dielectric permittivity increases from ≈ 100 for $x = 0$ up to 356 for $x = 0.46$, whereas the ε_c maximum decreases with increasing Ta concentration. The half-wave voltages of this series of crystals were measured on He–Ne laser radiation with $\lambda = 0.633$ μm. The $V_{\lambda/2}$ values, as well as the figure of merit $1/\varepsilon V_{\lambda/2}^2$ of crystals of different composition, are presented in table 6.16. The half-wave voltage and the figure of merit reach their minimum and maximum (respectively) when $x = 0.33$. The maximum

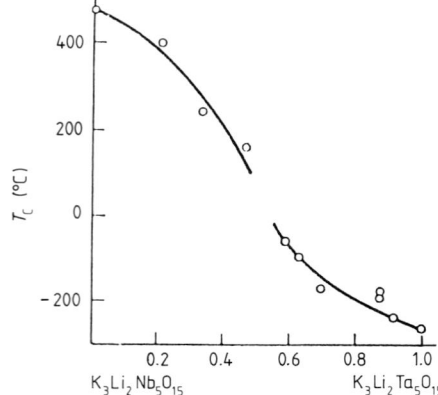

Figure 6.45   Dependence of the Curie temperature on the composition of single crystals of solid solutions of $K_3Li_2Nb_{5(1-x)}Ta_{5x}O_{15}$ [49].

Table 6.16 Dielectric and electro-optic properties of $K_3Li_2(Ta_xNb_{1-x})_5O_{15}$ [49].

x	ε_c	$V_{\lambda/2}$ (V)	$1/(\varepsilon V_{\lambda/2}^2)$ $(10^{-8}\ V^{-2})$
0	100	960	1.1
0.33	207	450	2.4
0.46	356	980	0.29

value of the figure of merit is $2.4 \times 10^{-8}\ V^{-2}$, which is twice as high as that for $K_3Li_2Nb_5O_{15}$ and three times as high as that for BNN.

An examination of PLTN under a polarization microscope has shown that the interference patterns for $x \leqslant 0.5$ are typical of uniaxial crystals, but crystals become biaxial if $x > 0.5$. Micro-twins similar to those observed in BNN crystals occur in crystals with $x \gtrsim 0.7$.

It was revealed in X-ray studies that PLTN crystals with $x \lesssim 0.5$ have a tetragonal structure below T_C, but for $x \gtrsim 0.5$ the crystal structure at T_C transforms from unknown to orthorhombic. High-temperature X-ray studies have shown that, for $x \lesssim 0.5$, the tetragonal distortion of a perovskite-type subcell is due to the appearance of ferroelectric phase. Pure $K_3Li_2Ta_5O_{15}$ ($x = 1$) is crystallized in an orthorhombic tungsten bronze-type structure with the lattice constants $a = 17.78$, $b = 17.83$, $c = 3.931$ Å (at 25°C). It is highly probable that this crystal belongs to the space symmetry group Cmmm.

Single crystals of PLTN solid solutions, which were grown in the direction perpendicular to the c axis, exhibited no optical inhomogeneity in the direction of this axis but had irregularities in the direction of growth due to variation of the Nb/Ta ratio.

Single crystals of PLTN were grown by the Kyropoulos method. The starting oxides were taken with an excess of K_2CO_3 and Li_2CO_3; the melting temperature was varied from 1000 to 1250°C depending on Ta content. During growth the crystal was raised at regular intervals at a rate of 1 mm h^{-1}. The melt was cooled at a rate of about 1 K h^{-1}. The rotation rate was 30–60 rpm. Chemical analysis of PLTN crystals has shown that they are enriched with tantalum due to the large difference between the solidus and liquidus lines in the system.

7 Nonlinear Optical Crystals with Lamellar Structure

In recent years crystals with effective nonlinear optical properties have appeared. They have a structure different from that of potassium tungsten bronzes. These crystals are the β phase of barium borate, β BaB_2O_4 (BBO), and potassium titanate phosphate $KTiOPO_4$ (KTP). They possess a number of useful properties, of which the most important are a high threshold of optical damage formation, a wide temperature and angular matching area, and harmonic generation in the UV region of the spectrum.

These crystals widen the potentialities of nonlinear optics. At the end of the chapter we analyse the efficiency and the spectral regions of applicability of nonlinear optical crystals.

7.1 Barium metaborate BaB_2O_4 (BBO)

The new nonlinear optical material, β barium metaborate, attracts researchers' attention since, according to the available results, it enables laser radiation to be transformed from the UV to the middle IR region of the spectrum [1]. The material is highly transparent from 0.19 to 3.5 μm, and it has a high birefringence and low dispersion, which permits phase matching for harmonic generation between 0.2 and 1.5 μm. The effective coefficient of transformation of radiation with $\lambda = 1.06$ μm into the second harmonic is six times higher than that for KDP crystals, the crystal damage threshold reaching about 13.5 GW cm^{-2} for 1.06 μm and 7.0 GW cm^{-2} for 0.53 μm for pulse durations of 1 ns and 250 ps respectively. Apart from the other unique properties, the material is characterized by a high optical homogeneity ($\Delta n = 10^{-6}$ cm^{-1}), a high temperature stability (temperature width is equal to about 55°C) and good mechanical properties [2, 3].

260

7.1.1 *Crystal structure*

There exists two polymorphic modifications of barium metaborate: the low-temperature β form and the high-temperature α form, the phase transition proceeding at $925 \pm 5°C$. α barium metaborate has tetragonal syngony, space group $R\bar{3}C$ (D_{3d}^6), unit cell parameters $a = 7.2351$ Å, $c = 39.192$ Å ($c_0/a_0 = 5.4169$); each unit cell contains 18 formula units [4]. The data encountered in the literature on β modification of BaB_2O_4 refer this form either to monoclinic syngony (space group $C2/c$ with lattice parameters $a = 11.135$ Å, $b = 12.67$ Å, $c = 8.381$ Å, $c/a = 0.6615$, $\beta = 100°2'$, unit cell contains 12 formula units) [5] or to trigonal syngony (space group $R3$, lattice parameters $a = b = 12.532$ Å, $c = 12.717$ Å, the hexagonal unit cell contains six formula units) [6].

The compound $BaB_2O_4/Ba_3(B_3O_6)_2$ has a lamellar structure of alternating Ba^{2+} ions and $(B_3O_6)^{-3}$ rings. The unit cell consists of 12 metaborate rings arranged in three groups with 18 Ba^{2+} atoms among them. Each Ba^{2+} atom is surrounded by seven oxygen atoms. Ring-shaped anion groups $(B_3O_6)^{3-}$ are practically planar. The three-fold axis passes through the centre of the ring perpendicular to the plane of the groups. The ring-shaped and exocyclic B–O bonds are respectively ~ 1.40 Å and ~ 1.32 Å long. The angles between B–O–B and O–B–O bonds are approximately equal to $122°$ and $118°$ respectively. Distortion of an anion ring results from the variations in the bond length and angles up to $\sim 2\%$, and from the shift of exocyclic oxygen atoms from the plane to 0.11 Å (figure 7.1) [1]. The position of the Ba^{2+} ion in the structure determines its symmetry which is acentric at low temperatures and centrosymmetric above the temperature of the $\beta \to \alpha$ transition. The interaction between barium and anion metaborate groups is weak. Barium cations are assumed not to make a direct contribution to the nonlinear effect. Nonlinear optical properties of $\beta\, BaB_2O_4$ are associated with charge transfer from oxygen to boron in the metaborate anion $(B_3O_6)^{3-}$. Quadratic nonlinear optical coefficients were calculated semi-empirically [1] on the

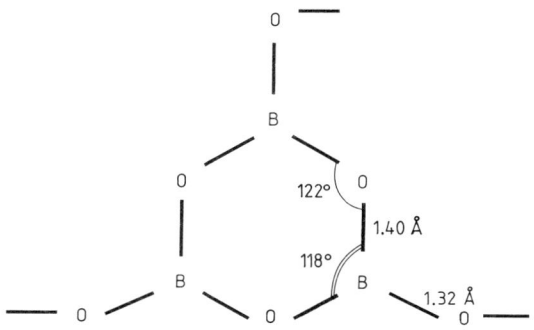

Figure 7.1 Structure of metaborate ion $(B_3O_6)^{3-}$ [1].

basis of the 'oriented' gas model. The results are in close agreement with experiment.

Both polymorphic modifications of BaB_2O_4 were obtained and investigated in [7]. Polycrystalline samples of both the modifications were obtained using solid-phase synthesis.

7.1.2 Synthesis and phase transitions

Barium metaborate was synthesized from barium and boron oxides as well as from $BaCO_3$ and boric acid. Synthesis proceeded in a resistance furnace with temperature varying from 700 to 1000°C, a soaking time from 2.5 to 7 h and a subsequent cooling down to room temperature at a rate of inertial cooling. At low temperatures, the starting material badly 'smokes' due to boric acid decomposition and may contain effluent of the reaction products and entail violation of stoichiometry of the resultant product. It is preferable to carry out a step-by-step synthesis with soaking at 150 and 700°C.

According to the differential thermal analysis (DTA) data (figure 7.2), the dehydration of orthoboric acid in the temperature range 120–140°C proceeds at a high rate, which leads at the beginning of the synthesis process to H_3BO_3 foaming above the reacting mass. In the temperature range between 300 and 400°C, the differential curve shows up a smeared exo-effect which is apparently due to the phase transition in the boron oxide. The solid-phase synthesis is most intense in the temperature range between 650 and 800°C; it is accompanied by large weight losses and leads to the formation of a dense BaB_2O_4 cake. For BaB_2O_4 synthesized at a temperature of 800°C

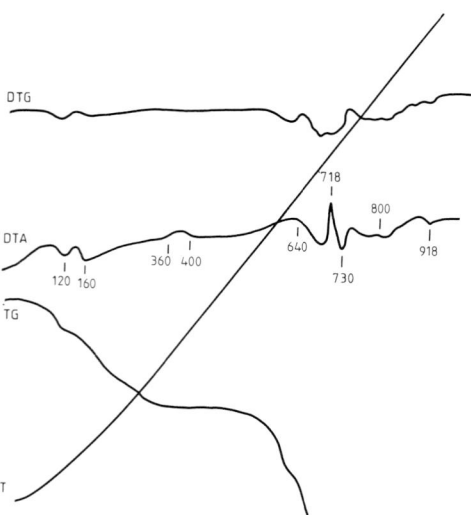

Figure 7.2 Thermogram of BaB_2O_4 synthesis from HBO_3 and $BaCO_3$ [7].

over 5 h, the DTA curves show no anomalies up to 900°C. The endo-effect registered at a temperature of 918°C is connected with the transition of the low-temperature β form of BaB_2O_4 into the high-temperature α modification.

X-ray analysis shows that single-phase β BaB_2O_4 can be obtained at a temperature of 800°C within 5 h of soaking. The first manifestations of α phase appear at a temperature of 830°C. An increase of temperature up to 880°C leads to violation of the single-phase state. Although at this temperature the β modification is predominant, some part of the material acquires the α form, which is seen from the appearance in the diffractogram of two basic reflexes typical of the α phase. A further temperature increase up to 930–980°C results in a subsequent increase in the α phase content (along with the main peaks on the X-ray patterns there appear peaks of lower intensity of the α phase). The solid-phase synthesis at a temperature of 1000°C within 5–6 h of soaking leads to the formation of the single α phase state (figure 7.3).

The DTA data on the synthesis from barium and boron oxides indicate that the formation of BaB_2O_4 ends at a temperature of about 1050°C. The melting starts at 1106°C. A BaB_2O_4 melt shows a high degree of overcooling, and the melt is crystallized rapidly at a temperature near 900°C, i.e. overcooling reaches about 200°. Examinations show that crystallization of an overcooled melt leads to the formation of the β phase irrespective of the phase composition of the starting material.

According to X-ray phase analysis, in the region of angles 2θ between 10° and 60° up to 60 diffraction peaks are observed, which for the β phase can be found both in monoclinic and in trigonal syngonies. The calculation of the mean deviation of experimental and calculated interplane distances Δd for trigonal and monoclinic syngonies allows the β form to be referred to trigonal syngony.

7.1.3 *Methods of obtaining* BaB_2O_4

Large transparent crystals of β BaB_2O_4 about 30×10 mm in size have been grown by a high-temperature solution top-seeding method. β BaB_2O_4 has a melting point of 1095°C and a transition temperature of 925 ± 5°C. As it is grown under a high temperature, the crystal is characterized by anti-deliquescence and good mechanical properties for easy cutting and polishing. Obviously, it is far superior in physical and chemical properties to the solution-grown crystals.

The phase diagram of the pseudo-binary BaB_2O_4–$Na_2B_2O_4$ system is shown in figure 7.4. It can be seen from this diagram that there is no such problem as the compositional heterogeneity induced by temperature fluctuations during the crystal growth. Thus, no growth striae occur in β BaB_2O_4 as is often the case with niobate and tantalate crystals. Therefore, β BaB_2O_4 has excellent optical properties.

Single crystals of α BaB_2O_4 were grown by the Czochralski and Stepanov methods [8]. The specificities of crystallization of α BaB_2O_4, which provide

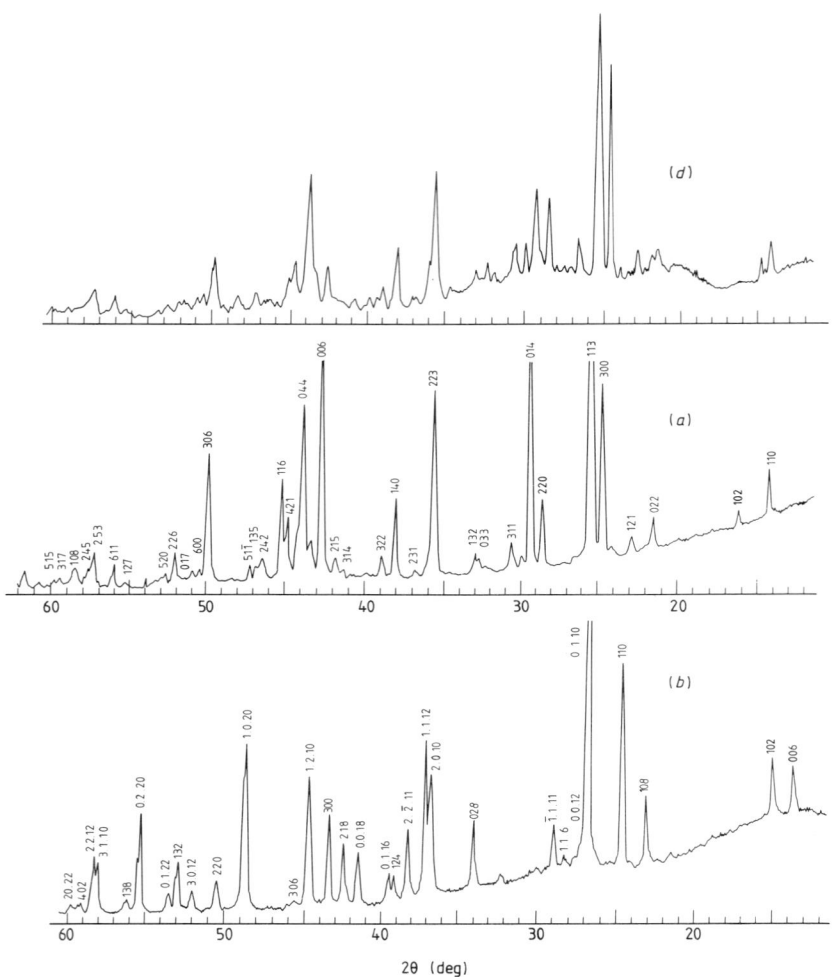

Figure 7.3   X-ray pattern of BaB_2O_4 synthesized at 800°C (a, β BaB_2O_4), 880°C (b, β $BaB_2O_4 + \alpha$ BaB_2O_4) and 1000°C (d, α BaB_2O_4) [7].

optically perfect crystals, require relatively low pulling rates (no more than 2–3 mm h^{-1}) and a very delicate melt temperature regulation in the course of growth. The process is optimal when seeding and growth of a crystal up to a given diameter proceeds at a constant temperature, which makes it possible to avoid entrapment of bubbles and inclusions of second phase by the growing crystal. Faceting observed in Czochralski-grown crystals makes it difficult to obtain a constant diameter. α BaB_2O_4 crystals have a perfect cleavage along the (100) plane.

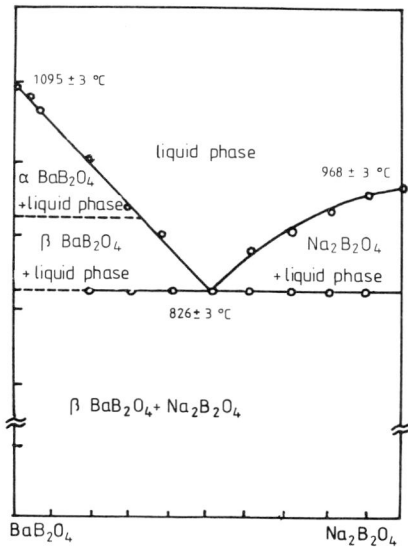

Figure 7.4 Phase diagram of pseudo-binary BaB_2O_4–$Na_2B_2O_4$ system [10].

The density of BBO single crystals $\rho_{(20)}$ measured using the hydrostatic weighing method is equal to 3.643 g cm^{-3}, which is somewhat smaller than the value $\rho = 3.751$ calculated from the data of X-ray diffraction analysis. This is connected with the defect structure of the real crystal.

7.1.4 *Mechanical properties*

The analysis of micro-hardness made it possible to establish the anisotropy of the mechanical properties of BBO crystals and their dependence on the conditions of surface preparation. Micro-hardness for cuts parallel and perpendicular to the triad axis differs approximately by a factor of two. The crystal is characterized by a high degree of embrittlement. For cuts perpendicular to the trigonal axis, total image destruction is observed under a load of 20 g, and under a load of 10 g 50% of images have chips. Crystal cutting by diamond discs and a subsequent lapping and polishing using diamond-containing paste result in the appearance of micro-chips, which substantially degrade the mechanical properties of the surface. On such a surface, the imprint shape is strongly distorted and has the shape of a double pyramid. The best results have been obtained after surface treatment by polarite. The micro-hardness measured on the growth face {100} which was not mechanically treated was ~ 750 kg mm^{-2}. BBO crystals are not hygroscopic.

7.1.5   Dielectric properties

The temperature dependences of the dielectric permittivity (ε) and loss tangent (tan δ) for single crystals of barium borate have been investigated in the temperature interval 4–1200 K (figure 7.5). The ε value at 273 K is about 10 and remains constant up to 600 K. Within the range 600–1200 K, the dielectric permittivity has temperature anomalies at 700 K, 880 K and 1150 K. The maximum value of the dielectric permittivity $\varepsilon_{max} = 80$ corresponds to a phase transition point at 1153 K. Anisotropy of the dielectric permittivity is expressed weakly. At room temperature the loss tangent is 0.003, but increases rapidly with increasing temperature. The high values of tan δ are due to the high conductivity of single crystals of barium borate. The conductivity of samples in the direction of the c axis is $\sim 1 \times 10^{-9}\,\mathrm{S\,m}^{-1}$ at room temperature and $1.5 \times 10^{-5}\,\mathrm{S\,m}^{-1}$ at 850 K. Figure 7.6 shows the dependence of the logarithm of electric conductivity on the inverse temperature for BBO crystals at a frequency of 1 kHz. On the curve $\ln \sigma = f(1/T)$ one can distinguish three regions described by the Arrhenius formula $\sigma_i = \sigma_0 \exp(E_a/kT)$. The activation energies of charge carriers calculated from this dependence are 0.36 eV for 500–700 K, 1.85 for 700–820 K and 0.95 eV for 811–1150 K.

7.1.6   Optical properties

The region of optical transparency stretches from 0.2 to 3.3 μm. The transmission spectra of monocrystalline barium niobate are presented in figure 7.7. A small absorption band near 2.9 μm is possibly connected with the absorption of OH groups present in the crystal, as was shown above for lithium and barium sodium niobate crystals. The calculated width of the

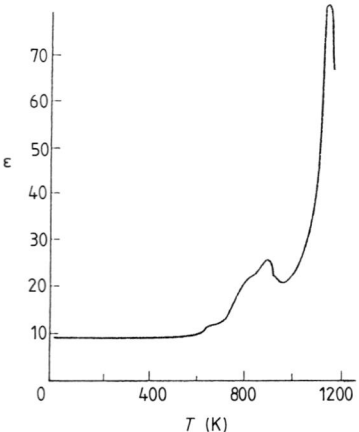

Figure 7.5   Temperature dependence of dielectric permittivity of BaB_2O_4 at a frequency of 1 kHz [8].

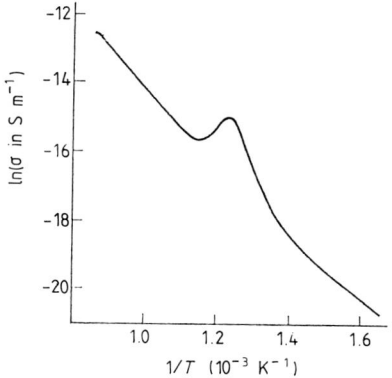

Figure 7.6 Electric conductivity of BBO as a function of temperature (frequency 1 kHz) [8].

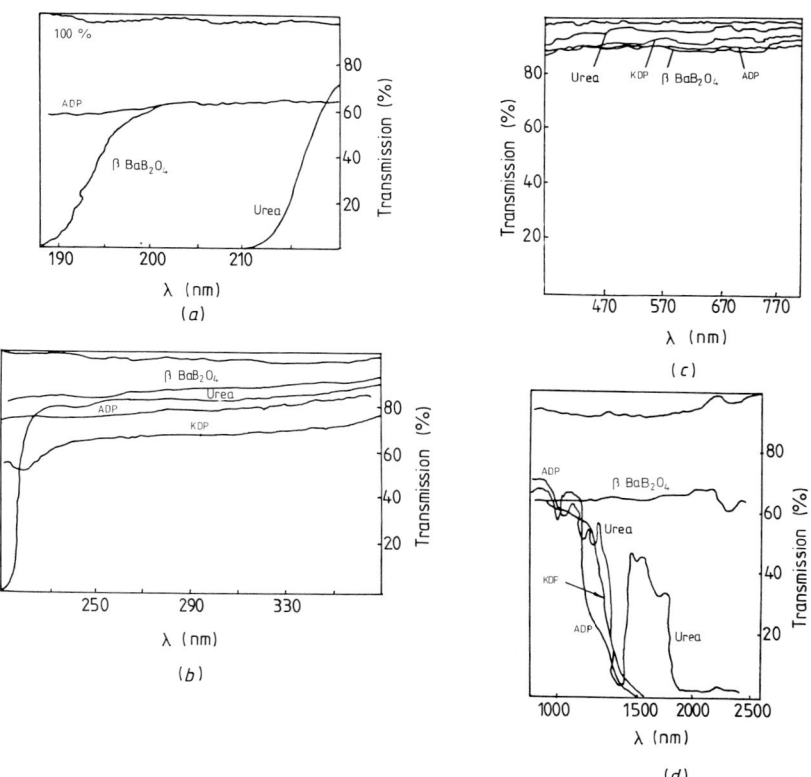

Figure 7.7 The transmission of β BaB$_2$O$_4$, urea and ADP crystals at (a) 190–200 μm, (b) 210–370 μm, (c) 370–800 μm and (d) 800–2500 μm [10].

forbidden band is $\Delta E \approx 6.2$ eV. β BaB$_2$O$_4$ is a negative uniaxial material with large birefringence and relatively small dispersion. The refractive indices have been measured at 12 wavelengths from 0.2288 μm to 1.079 μm, by the method of minimum deviation [9]. A Sellmeier equation of the form

$$n^2 = A + B/(1 - C/\lambda^2) \tag{7.1}$$

(λ is the free space wavelength in μm) was fitted to these data by a least-squares technique with the result

$$
\begin{array}{llll}
A = 1.2365 & B = 1.4878 & C = 0.014\,51 & \text{for } n_o \\
A = -0.3061 & B = 2.6622 & C = 0.006\,35 & \text{for } n_e.
\end{array}
$$

Standard deviations between the calculated indices and measured values are $\Delta n_o = 0.000\,84$ and $\Delta n_e = 0.000\,80$ [3].

Figure 7.8 shows the measured birefringence of the β BaB$_2$O$_4$ crystal as well as the theoretical curve derived from the dispersion equation of the refractive index. Owing to the larger birefringence and lower dispersive power, the β BaB$_2$O$_4$ crystal has a much wider phase matchable region.

As a nonlinear optical (NLO) crystal, BBO offers [11]:

(i) a broad phase-matchable range from 409.6 nm to 3500 nm;

(ii) a wide transmission region from 200 nm to 3500 nm;

(iii) a large effective second harmonic generation (SHG) coefficient about three to six times greater than that of potassium dihydrogen phosphate (KDP), depending upon the wavelength;

(iv) a high damage threshold of 10 GW cm^{-2} for 100 ps pulsewidths and 5 GW cm^{-2} for 10 ns pulsewidths, measured at the Nd:YAG wavelength of 1064 nm;

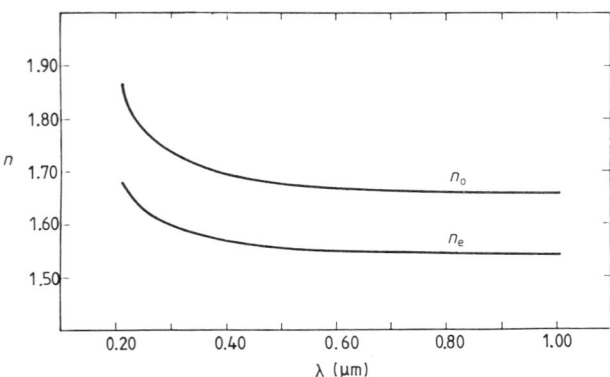

Figure 7.8 Dispersion curves of the refractive index of BBO crystals [10].

(v)   a low thermal birefringence computed with the refractive indices n_o (ordinary ray) and n_e (extraordinary ray) of

$$\Delta n_o / \Delta T = -15.9 \times 10^{-6} \, {}^{\circ}C^{-1}$$

$$\Delta n_e / \Delta T = -5.17 \times 10^{-6} \, {}^{\circ}C^{-1}$$

(vi)   a high optical homogeneity computed with n_o of $\Delta n_o = 1 \times 10^{-6} \, cm^{-1}$.

7.1.7   Nonlinear phase matching properties

The simplest coherent nonlinear process is SHG. This is the conversion of unfocused plane wave radiation with a frequency ω to its second harmonic frequency 2ω at the phase matched (PM) condition. SHG is also popularly known as frequency doubling. The efficiency η of SHG expressed as the ratio of the output intensity $P(2\omega)$ to the input intensity $P(\omega)$ is given by

$$\eta = \frac{P(2\omega)}{P(\omega)} = \tanh^2(x) \tag{7.2}$$

where x is proportional to the crystal length, the pump intensity and the effective nonlinear coefficient d_{eff} defined at the PM angle. The effective nonlinear coefficient d_{eff} depends upon the Kleinman symmetry of the crystal.

The two techniques for phase matching discussed here are denoted type I and type II. These two processes are defined according to whether the fundamental frequency ω and the second harmonic frequency 2ω have the same or different planes of polarization. Knowing the refractive indices n_o and n_e, the PM angles θ_m for type I and type II processes can be calculated. For a negative uniaxial crystal, such as BBO, where $n_e < n_o$, the PM angle for type I phase matching is given by

$$\sin^2 \theta_m(I) = \frac{n_o(\omega)^{-2} - n_o(2\omega)^{-2}}{n_e(2\omega)^{-2} - n_o(2\omega)^{-2}}. \tag{7.3}$$

For type II phase matching, the PM angle is given by

$$\left(\frac{\cos^2 \theta_m(II)}{n_o(2\omega)^2} + \frac{\sin^2 \theta_m(II)}{n_e(2\omega)^2} \right)^{-1/2}$$

$$= \frac{1}{2} \left[n_o(\omega) + \left(\frac{\cos^2 \theta_m(II)}{n_o(\omega)^2} + \frac{\sin^2 \theta_m(II)}{n_e(\omega)^2} \right)^{-1/2} \right]. \tag{7.4}$$

In the SHG process, it is important to note that it is the intensity of the incident radiation, expressed in MW cm^{-2}, rather than the power that is of utmost concern. Therefore, high-energy short pulses focused into a small area are most efficiently converted, as compared to long pulses with the same given average power.

For pulses with high peak power, it may not be necessary or desirable to focus the laser radiation onto a smaller area. Focusing tightly to provide higher efficiencies may thermally detune the crystal or produce permanent damage. For short pulses with low peak power energies, or for continuous wave laser radiation, it is often necessary to focus the radiation to a small beam waist.

The BBO crystal has a small acceptance angle and may not necessarily produce higher efficiencies than other NLO crystals. Fortunately, the damage threshold of BBO is very high, and the problems of thermal detuning are small due to the small thermal birefringence. The experimental values of phase matchable angles at various wavelengths, together with the calculated ones derived from the dispersion equation, are shown in table 7.1 and figure 7.9 respectively. It is shown that the β BaB$_2$O$_4$ crystal can be phase matchable in the 1500–200 nm (second harmonic) region.

7.1.8   *The SHG coefficients*

The reduced second-order susceptibility tensor for point group 3 is of the form

$$\begin{pmatrix} d_{11} & -d_{11} & 0 & d_{14} & d_{15} & -d_{22} \\ -d_{22} & d_{22} & 0 & d_{15} & -d_{14} & -d_{11} \\ d_{31} & d_{31} & d_{33} & 0 & 0 & 0 \end{pmatrix}. \tag{7.5}$$

Applying Kleinman symmetry conditions, we have $d_{31} = d_{15}$ and $d_{14} = 0$. The expressions for the effective nonlinear coefficients are as follows: for two o rays and one e ray

$$d_{eff} = d_{15} \sin\theta + \cos\theta(d_{11}\cos 3\varphi - d_{22}\sin 3\varphi) \tag{7.6}$$

and for one o ray and two e rays

$$d_{eff} = \cos^2\theta(d_{11}\sin 3\varphi + d_{22}\cos 3\varphi). \tag{7.7}$$

Since the birefringence of β BaB$_2$O$_4$ is negative uniaxial, equation (7.6) applies to type I phase matching, and equation (7.7) applies to type II phase matching. The nonlinear coefficients have been measured by the Maker fringe and second harmonic techniques [9]. In [9] a coordinate system with X parallel to a lattice mirror plane is used. Here the coordinate system is rotated by 90° about Z to give Y parallel to a lattice mirror plane. With this convention the measured nonlinear optical coefficients are

$$d_{22} = \pm(4.10 \pm 0.20)d_{36}(\text{KDP})$$

$$d_{31} = \pm(0.07 \pm 0.03)d_{22}(\beta \text{ BaB}_2\text{O}_4)$$

$$|d_{11}| = 0.05|d_{22}(\beta \text{ BaB}_2\text{O}_4)|.$$

If we now approximate d_{11} to zero, the effective nonlinear coefficients are the same as for point group 3m with the IRE convention of $m \perp X$. For this

Table 7.1 Phase matching angles for β BaB$_2$O$_4$ [10].

Laser system	Nd:YAG			Ruby	Dye	Cu ion	Ar$^+$ ion
Wavelength (nm)	1064–532	532–266	1064–212	694.3–347.2	580–290	510–255	528.7 – 454.4 → 264.4–227.3
θ_{pm} (Expt.)	21° ± 1°	48° ± 1°	55° ± 1.5°	36° ± 1°	41° ± 1°	50° ± 1°	50° – 65°
θ_{pm} (Calc.)	21.12°	48.14°	53.5°	35.82°	41.03°	50.27°	49.06° – 62.94°

$d_{11} = 4.1 \, d_{36}$ (KDP)

$d_{31} = 0.07 \, d_{11}$

$d_{eff} = d_{11} \sin\theta_{pm} + \cos\theta_{pm}(-d_{11}\cos 3\varphi + d_{22}\sin 3\varphi) \approx d_{31}\sin\theta_{pm} - d_{11}\cos\theta_{pm}\cos 3\varphi.$

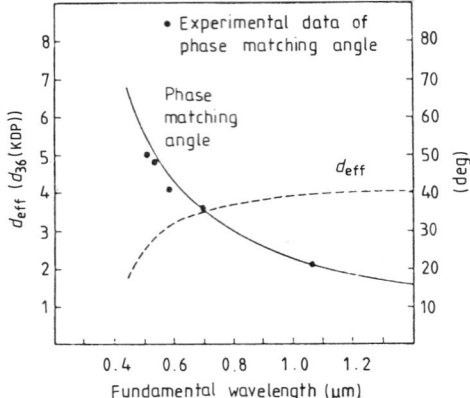

Figure 7.9   The effective SHG coefficients and phase matchable angles at various wavelengths [10].

case, the effective nonlinear coefficient for two o rays and one e ray is

$$d_{eff} = d_{15} \sin\theta - d_{22} \cos\theta \sin 3\varphi \qquad \text{(type I)} \qquad (7.8)$$

and for two e rays and one o ray it is

$$d_{eff} = d_{22} \cos^2\theta \cos 3\varphi \qquad \text{(type II)}. \qquad (7.9)$$

Some additional properties of β BaB$_2$O$_4$ relevant to nonlinear frequency conversion are listed in table 7.2. A comparison is made with properties of other nonlinear materials commonly used in the visible and ultraviolet spectral regions. Phase matching angles and angular acceptances are calculated from dispersion equations for the materials. The accuracy of the nonlinear coefficients of the materials is not always precisely known; it has been pointed out that many conflicting values exist in the literature [12]. Also, damage thresholds are being improved greatly for a number of materials, and older published values may no longer be applicable.

7.1.9   *Second harmonic generation*

The second harmonic measurements for 1064 to 532 nm conversion with the stable resonator Q switch laser pump source are shown in figure 7.10. In the first of these observations, the performance of a 6.8 mm long β BaB$_2$O$_4$ type I crystal is compared to a 24 mm long type II KD*P crystal. The harmonic conversion efficiency increases linearly as pump energy is increased for both materials, and the conversion in β BaB$_2$O$_4$ is approximately 1.8 times that in KD*P. This ratio is consistent with the parameters listed for the two materials in table 7.2. Conversion in both crystals is well within the limit of weak focusing [13] $n\lambda l/(2\pi w_0^2) \ll 1$, where l is the crystal length, n is the index of refraction, λ is the free space wavelength of the fundamental and w_0

Table 7.2 Selected second harmonic generation properties for some nonlinear crystals [3].

Crystal	Transmission range (μm)	Refractive indices λ (μm)	n_o	n_e	Nonlinear coefficient (d_{36} (KDP))	Phase matching λ_{SHG} (μm)	Damage threshold (GW cm^{-2})	$\partial\Delta k/\partial\theta$ (cm^{-1} mrad^{-1})	Walk-off angle	Temperature acceptance (°C)
β BaB$_2$O$_4$	0.19–3.5	1.06	1.657	1.539	$d_{11}=4.1$	0.2–1.5	13.5	10.6 (type I)	2.7°	55
		0.53	1.674	1.554				6.8 (type II)		
KD$_2$PO$_4$	0.20–1.5	1.06	1.49	1.46	$d_{36}=0.76$	0.26–0.75	6	4.6 (type I)	1.4°	7
		0.53	1.51	1.47				2.1 (type II)	1.3°	
Urea	0.21–1.4	1.06	1.477	1.583	$d_{14}=2.98$	0.24–0.7	5	8.1 (type II)	3.0°	?
		0.53	1.490	1.596						
KTP	0.35–4.5	λ (μm)	n_a n_b n_c		$d_{31}=13.5$ $d_{32}=10.4$ $d_{33}=28.4$	0.53–2.2	3		1.0°	25
		1.00	1.740 1.749 1.831							
		0.50	1.787 1.797 1.898							

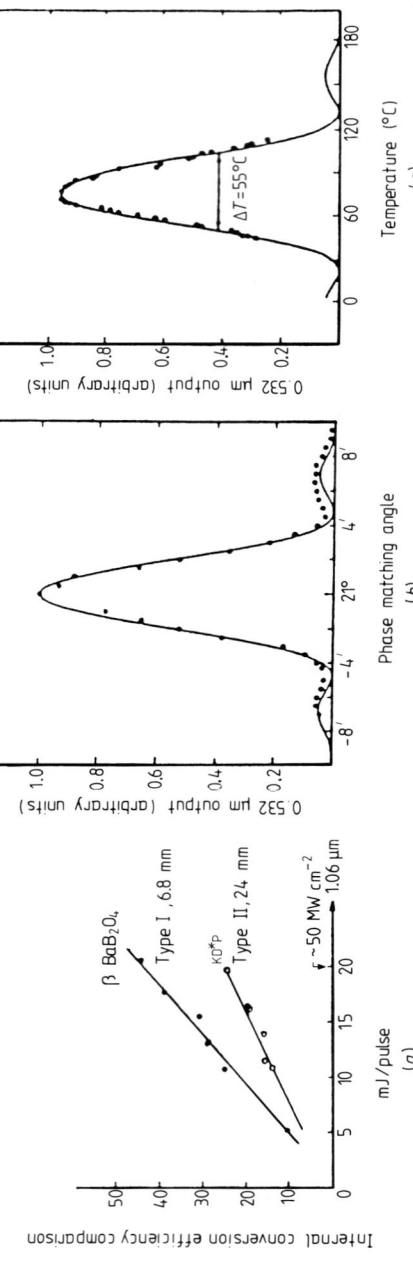

Figure 7.10 Second harmonic measurements using a 1064 nm 24 ns Q switched laser pulse. The beam had a gaussian-like transverse distribution of 2 mm diameter with nearly diffraction-limited divergence. (a) Comparison of harmonic generation as a function of fundamental pulse energy in β BaB$_2$O$_4$ and KDP crystals. (b) Angle-tuned, and (c) temperature-tuned phase matching curves for β BaB$_2$O$_4$ [3]. (a) Type I, 21° cut, 6.8 mm length. (b) TEM$_{00}$ mode, $t = 24$ ns, beam radius $= 1$ mm, $\Delta\theta < 0.3$ mrad, $\theta = 21°$.

is the beam amplitude radius. In this case the monochromatic plane-wave approximation for second harmonic generation

$$I_2(l) = (I_1(0))^2 \left(\frac{\omega_1 d_{\text{eff}}}{n_1 c} \right)^2 \frac{[\mu_0/\varepsilon_0]^{1/2}}{n_1} l^2 \, \text{sinc}^2(\Delta k l/2) \qquad (7.10)$$

can be used to analyse the conversion. Here $I_1(0)$ is the input fundamental intensity, $I_2(l)$ is the harmonic intensity at crystal length l, ω_1 is the angular frequency of the fundamental radiation, c is the velocity of light and $[\mu_0/\varepsilon_0]^{1/2} = 377 \, \Omega$. Restricting the comparison to the case of perfect phase matching, $\Delta k = k_2 - 2k_1 = 0$, the ratio of harmonic conversion will be the ratio $(l d_{\text{eff}})^2/n_1^3$ for the two crystals. Using values from table 7.2 and noting that KD*P of point group 42m has an effective nonlinear coefficient $d_{\text{eff}} = \sin(2\theta_{\text{m}})d_{36}$ for type II phase matching, the predicted ratio is 1.96, assuming the two coefficients d_{11} and d_{15} for β BaB$_2$O$_4$ can be added. The predicted ratio is 1.78 if they are subtracted. Both of these ratios agree with the observed value within the accuracy to which the nonlinear coefficients are known. The observed tuning curve for second harmonic generation in β BaB$_2$O$_4$ (figure 7.10(b)) matches closely a sinc^2 curve. This agreement indicates that the condition of weak focusing is satisfied, and that the optical uniformity of the crystalline material is very good.

The relative potential of the two materials was not accurately compared in the above measurements because the harmonic conversion process was not optimized for either crystal. The allowed crystal length will be determined in part by phase matching limitations. A comparison that considers phase matching limits is obtained from the ratio $(l_{\text{max}} d_{\text{eff}})^2/n_1^3$ for the two materials, where l_{max} is the maximum interaction length. In the case of diffraction-limited gaussian beams, this length is determined by the birefringent walk-off angle ρ; $l_{\text{max}} \sim 1/\rho$. When beam divergence limits conversion, the interaction length depends on angular acceptance, which in this case is determined by the change in wavevector mismatch Δk with respect to the change in angle of propagation θ: $l_{\text{max}} \sim (\partial \Delta k/\partial \theta)^{-1}$. This comparison indicates that β BaB$_2$O$_4$ would perform 4.7 times better than KD*P for walk-off angle limitation and 1.3 times better for divergence limitation. However, in phase modulation or broad spectral bandwidth limit conversion, the group velocity mismatch becomes important and β BaB$_2$O$_4$ could offer only 0.11 times the conversion of KD*P. This comparison assumes that crystals of arbitrary length are available and beam parameters do not change. There are many other factors that could affect harmonic conversion.

For high conversion levels of fundamental radiation with high average intensity, it is necessary to consider the sensitivity of phase matching to temperature change. Small amounts of the fundamental and harmonic radiation will be absorbed, resulting in a heating of the crystal. Also of concern are the absorption coefficients at these two wavelengths, and the degree to which other effects (such as two-photon absorption, strain induced birefringence and optical damage) will limit the conversion process. The absorption coefficients of β BaB$_2$O$_4$ were so low that they were not measured here. However, the temperature dependence of phase matching was observed by heating the β BaB$_2$O$_4$ sample at second harmonic generation in an oven. The observed temperature phase matching width when the crystal was held in a fixed angular orientation was 55°C (figure 7.10(c)), a large value compared to that of other nonlinear crystals. The large temperature width and high damage threshold of β BaB$_2$O$_4$ suggest that this material is a good choice for applications of high average intensity.

The highest levels of harmonic conversion and the highest intensities reported in [3] were achieved with the mode-locked laser system. An internal energy conversion efficiency of 84% was observed at a 2 GW cm^{-2} fundamental-pulse peak intensity. In the measurement of internal conversion efficiency, the ratio of total harmonic energy to input fundamental energy is calculated using corrections to remove the Fresnel reflections at the crystal faces. In principle, it should be possible to make these reflections very small with anti-reflection coatings; however, the crystals used in the measurements were not coated. A simple model was used to calculate harmonic conversion. A gaussian beam of spherical wavefront was assumed, with no absorption in the material. The harmonic generation at high depletion levels was described by an elliptic function [14] and averaged over the pulse in time and space. The agreement between experiment and this simple model (figure 7.11) is reasonable. Conversion is limited by some phase mismatch, but the assumption that the pulse is described by a gaussian distribution in

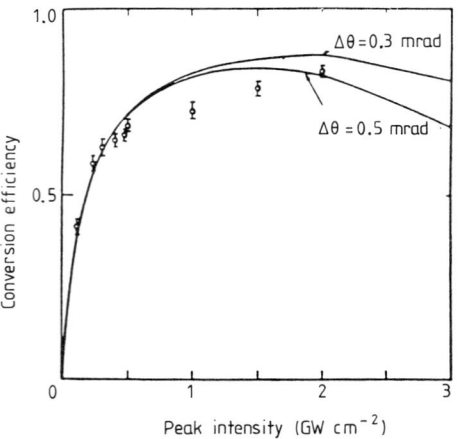

Figure 7.11 Second harmonic internal energy conversion efficiency for a 1 ns pulse generated in a mode-locked Nd:YAG laser (1064–532 nm, $L = 6.8$ mm). Conversion is shown as a function of fundamental pulse peak intensity [3]. Full curve shows calculated results.

time and space is an over-simplification. The level of conversion achieved here, not optimized or extended to the damage limit of the material, compares favourably with the highest conversion efficiency reported elsewhere; 92% in KDP [15]. An important conclusion is that β BaB$_2$O$_4$ can operate at high peak intensities at both fundamental and harmonic wavelengths without serious two-photon absorption or other nonlinear limitations.

The unstable-resonator Q switched laser system provided the highest peak powers and highest average powers reported in [3]. Average fundamental intensities were as high as 27 W cm^{-2}. The 0.5 mrad divergence was approximately 2.5 times the diffraction limit for the 6.5 mm beam diameter. This laser system had a harmonic conversion apparatus consisting of KD*P crystals for second harmonic generation and summing to third harmonic, and an ADP crystal for cascaded generation to fourth harmonic. The performance of the 6.8 mm β BaB$_2$O$_4$ harmonic generation crystal was compared directly with that of the second harmonic crystal supplied with the laser (figure 7.12). The conversion efficiency of the β BaB$_2$O$_4$ crystal was between 1.3 and 1.7 times better than that of the KD*P crystal throughout the operating range of the laser system. The beam characteristic of this laser should place the conversion process in the divergence limited regime, rather than in the regime of beam walk-off limitation expected for beams of gaussian transverse distribution.

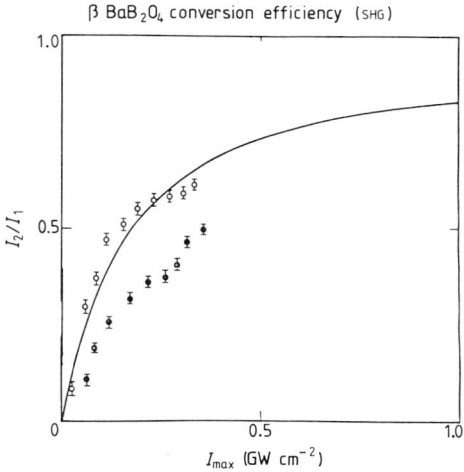

Figure 7.12 Second harmonic overall conversion efficiency for an unstable-resonator Q switched Nd:YAG laser pulse with 0.5 mrad beam divergence ($\tau = 8$ ns, $L = 6.8$ mm, 1064–532 nm). Conversion is shown as a function of peak fundamental intensity [3]. $\bigcirc$ β BaB$_2$O$_4$; $\bullet$ KD*P; ——— calculated.

7.1.10   *Higher harmonics*

It was shown that for certain conditions β BaB_2O_4 offered an advantage over KD*P. Possibly β BaB_2O_4 would offer significant advantages for higher average intensity applications, but this has not been confirmed experimentally. One application in which it has been shown that β BaB_2O_4 offers a significant advantage over other materials is in short-wavelength nonlinear frequency conversion. Wavelengths as short as 205 nm have been generated in β BaB_2O_4 by second harmonic conversion of dye laser radiation [16]. Measurements of third, fourth and fifth harmonic generation of 1064 nm neodymium laser radiation are presented [3].

A comparison was made between β BaB_2O_4 and the harmonic generation apparatus supplied with the laser system described above. The 6.8 mm, 21° β BaB_2O_4 crystal was used for second harmonic generation. The sum generation process of the third harmonic was performed in a 5.5 mm, 64° type II β BaB_2O_4 crystal by combining the fundamental and second harmonic. Cascaded fourth harmonic generation from the second harmonic was performed in a 5 mm, 48° type I crystal. The results obtained with the unstable-resonator Q switched laser pump source are presented in table 7.3. Again the direct substitution of β BaB_2O_4 provided an improvement in performance.

Measurements of fourth harmonic generation in β BaB_2O_4 were also made using the mode-locked laser system. Figure 7.13 shows the observed experimental conversion efficiency and a calculated conversion efficiency curve for the 5 mm 48° crystal in 532 to 266 nm conversion. A conversion efficiency of 52% was consistently achieved with a peak intensity of 160 MW cm^{-2} at 532 nm. A beam diameter of 4 mm was used for these measurements. The agreement between the experimental data and the theoretical curve is quite good, and indicates that any effects relating to absorption are small in the cascaded harmonic conversion.

Fifth harmonic generation was performed by summing the fundamental and fourth harmonic in a 6.4 mm, 53° β BaB_2O_4 crystal. Previously KB5 [17] was the only crystalline material that had been used for nonlinear frequency

Table 7.3   Harmonic generation using β BaB_2O_4 and KD*P with unstable-resonator Q switched laser system [3].

Crystal		Pulse energy (mJ)		
	Fundamental	2nd harmonic	3rd harmonic	4th harmonic
β BaB_2O_4	220	105	39	18.5
	600	350	140	70.0
KD*P	600	270	113	45.0

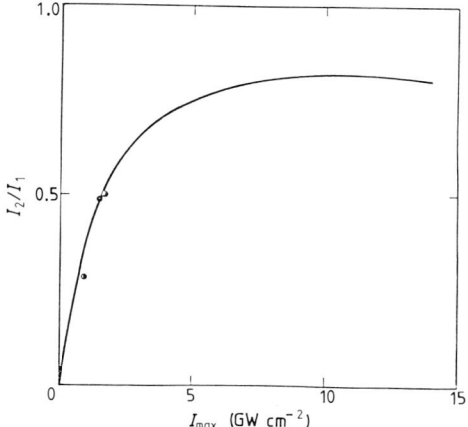

Figure 7.13 Harmonic energy conversion efficiency for generation of the 266 nm fourth harmonic of mode-locked Nd:YAG laser pulses. Conversion efficiency is shown as a function of peak 532 nm intensity incident on the barium borate crystal [3]. $\tau = 1$ s; $L = 5$ mm. Full curve shows calculation.

conversion at this wavelength. However, the conversion efficiency in KB5 was very low because of the small effective nonlinear coefficient: only $0.11d_{36}$(KDP). The effective nonlinear coefficient of β BaB$_2$O$_4$ is $2.8d_{36}$(KDP) for this sum generation process. The arrangement used for fifth harmonic generation is shown schematically in figure 7.14. The second and fourth harmonics were generated in β BaB$_2$O$_4$ crystals of 6.8 mm and 5 mm lengths respectively. All crystals were cut for type I phase matching. Again, both the mode-locked laser and the unstable-resonator Q switched laser were used as pump sources. The experimental data are presented in table 7.4. An overall conversion efficiency of 1064 nm fundamental energy to 213 nm fifth harmonic energy

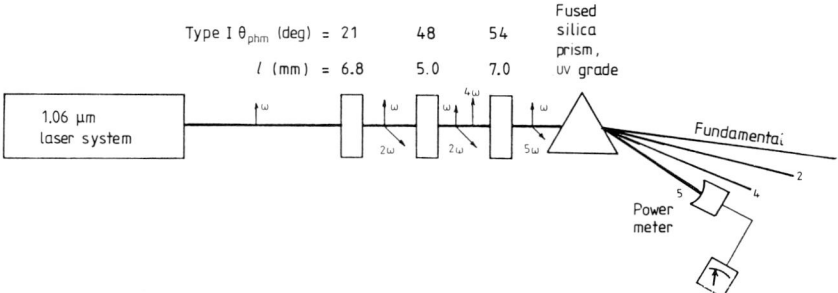

Figure 7.14 Schematic representation of experimental set-up used for fifth harmonic generation in barium borate [3].

Table 7.4   5th harmonic generation of 212 nm in β BaB$_2$O$_4$ [3].

1.06 μm laser system	Fundamental (mJ)	2HG (mJ)	4HG (mJ)	5HG (mJ)
Mode-locked† 1 ns, 4 mm dia. 0.3–0.5 mrad	34.0	23.31	10.0	5.0
Quanta ray	220	105	18.5	5
DCR laser	540	300	40	20

† 2HG/total = 68.6%; 4HG/2HG = 42.9%; 5HG/4HG = 50%; 5HG/total = 11.4%.

of 11% was achieved with the mode-locked laser. Lower conversion efficiency, but higher fifth harmonic energy, was obtained with the unstable-resonator Q switched laser. At a pump energy of 550 mJ, 20 mJ of fifth harmonic energy was generated. At higher pump energy, the fifth harmonic energy did not increase further. It is thought that this effect is due to strong depletion at the peak of the fundamental pulse. It should be possible to optimize further the fifth harmonic generation process to avoid this problem and generate greater amounts of 213 nm radiation. Nevertheless, 11% overall conversion and 20 mJ of fifth harmonic generation, to the best of our knowledge, are the highest reported values for this conversion process.

7.1.11   *Optical parametric oscillation*

The nonlinear generation of tunable coherent radiation in the ultraviolet is also of great interest. The authors of [18] have achieved tunable optical parametric oscillation (OPO) in β BaB$_2$O$_4$. A 532 nm pump wavelength was used for this initial demonstration; however, the optical properties of the material are well suited for shorter pump wavelengths. The stable-resonator Q switched system was the laser pump source for this experiment. A Nd:YAG amplifier was added to increase 1064 nm energy to 60 mJ, and the 6.8 mm β BaB$_2$O$_4$ crystal was used to generate 25 mJ of second harmonic energy which in turn was used to pump the OPO. The experimental apparatus is shown schematically in figure 7.15.

The 24 ns laser pulse was shortened to 18 ns in the second harmonic generation process. Before introduction into the OPO cavity, the pump beam was focused to a beam amplitude radius of $w_p = 0.25$ mm. Two flat mirrors spaced by 2.5 cm were used to form the parametric oscillator cavity. The mirror reflectivities were 90% and 76% at the signal wavelength. The OPO threshold was 3 mJ near the degeneracy point. When pumped at 10 mJ near degeneracy, the OPO output energy was 3 mJ, and the output pulse duration was 12 ns. The tuning range extended from 940 to 1220 nm (figure 7.16(*a*)); the spectral extent of tuning was limited by the reflectivity of the OPO cavity mirrors.

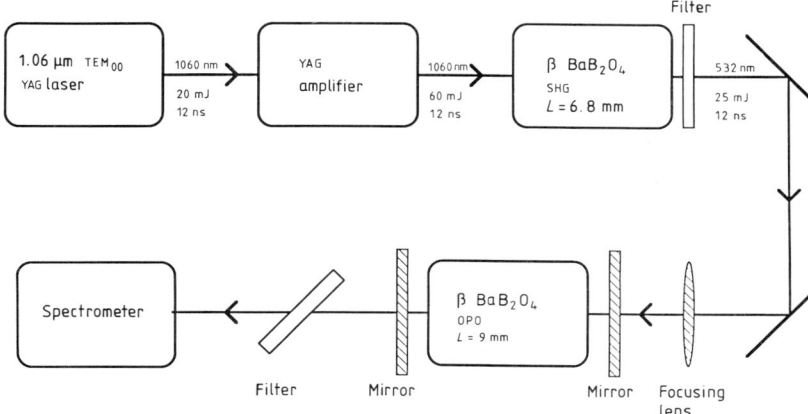

Figure 7.15 Schematic drawing of barium borate optical parametric oscillator (OPO) set-up [3].

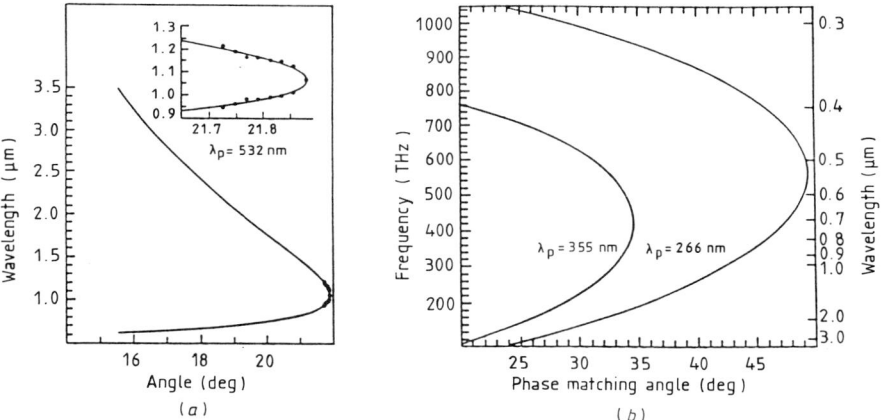

Figure 7.16 Angle tuning curves for barium borate OPO. (*a*) Observed and calculated tuning curves for a 532 nm pumped barium borate OPO; the inset shows on an expanded scale the range over which parametric oscillation was observed; (*b*) calculated tuning curves for 355 nm and 266 nm pump wavelengths [3].

The large birefringence and small dispersion of β BaB$_2$O$_4$ allow angle tuning of phase matching for the parametric process over wide spectral ranges. The phase matching range for the 532 nm pumped process extends continuously from 630 nm to 3.5 μm with a rotation of the internal propagation direction between 15° and 21°. Calculated tuning curves for 266 nm and 355 nm pump wavelengths are shown in figure 7.16(*b*). Two different crystals would probably be required to obtain the 25° tuning range needed to achieve wavelength tuning between 290 nm and 3.5 μm calculated for the 266 nm pump. A 355 nm pumped OPO would tune from degeneracy near 34° to a 410 nm and 3.5 μm signal and idler pair at 19°.

7.2   Continuously tunable generation from the ultraviolet to the near infrared in β barium borate

For experiments, two β BaB$_2$O$_4$ crystals were used [19]. Sample 1 is cut for third harmonic generation (THG) of a Nd:YAG laser ($\lambda = 1064$ nm) in a type II process $\omega(e) + 2\omega(o) \rightarrow 3\omega(e)$. Sample 2 is cut for fourth harmonic generation of a Nd:YAG laser in a type I process $2\omega(o) + 2\omega(o) \rightarrow 4\omega(e)$. In preliminary experiments, using a weakly focused beam from a Q switched Nd:YAG laser, the authors checked for both samples the crystal cut in a SHG process (type II SHG of 1064 nm for sample 1, type I SHG of 532 nm for sample 2). For both samples the crystal cut was found to be exact to within 1°. Type II SHG deficiency at 1064 nm in sample 1 was measured relative to a type II SHG process in a KTP crystal. The latter turned out to be 11.2 times more efficient than the former, thus yielding for β BaB$_2$O$_4$ a value for d_{22} which is well within the above stated error bars. The authors of [19] used d_{eff}(KTP) $= 7.42$ pm V^{-1}, according to [20] and [21]. The measured 50% transmission range of the samples (not corrected for Fresnel reflection losses) was 200 nm–2.6 μm.

In an optical parametric emission/optical parametric amplifier (OPE/OPA) configuration, β BaB$_2$O$_4$ can most usefully be pumped by either the second, third or fourth harmonics of a picosecond-pulsed Nd laser system (YAG, glass etc) or by the fundamental or second harmonic of a short pulse rhodamine dye laser–amplifier combination. Moreover, for each of these pumps, both type I (e$\rightarrow$o + o) and type II (e$\rightarrow$o + e) interactions can be phase matched over a broad range of signal/idler wavelengths. Figure 7.17 (figure 7.18) represents the calculated angle tuning curve for a type I (type II) process pumped by the third harmonic of a Nd:YAG laser ($\lambda_p = 355$ nm). Regardless of the particular type of interaction involved, it is clear than an extremely broad (20 000 cm^{-1}) tuning range can be achieved with one crystal. This feature makes β BaB$_2$O$_4$ particularly attractive as an OPE/OPA, especially since such a device does not require any external dielectric mirrors whose bandwidths inevitably would limit the tunability. The limit of the tuning range can be even further extended into the UV by using a second β BaB$_2$O$_4$ crystal for SHG of the OPE/OPA signal. This is shown schematically in figures 7.17 and 7.18 by the upper curves. In this way, continuous tuning between 205 nm and 2.6 μm becomes possible for the first time ever.

To verify the tuning characteristics, Vanherzeele *et al* [19] have experimentally investigated OPE in β BaB$_2$O$_4$ samples. Frequency measurements were performed with a 0.64 m Czerny–Turner grating monochromator. The

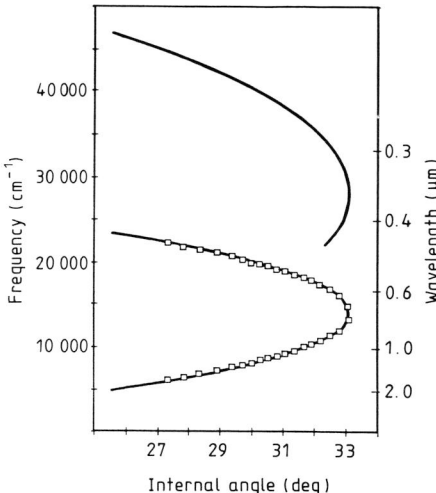

Figure 7.17 Angle tuning curve for a β BaB$_2$O$_4$ type I OPE/OPA pumped by the third harmonic of a Nd:YAG laser ($\lambda_p = 355$ nm); the upper curve refers to the OPE/OPA–SHG combination. All angles refer to internal propagation in the OPE/OPA crystal. The dots represent the experimental data [19].

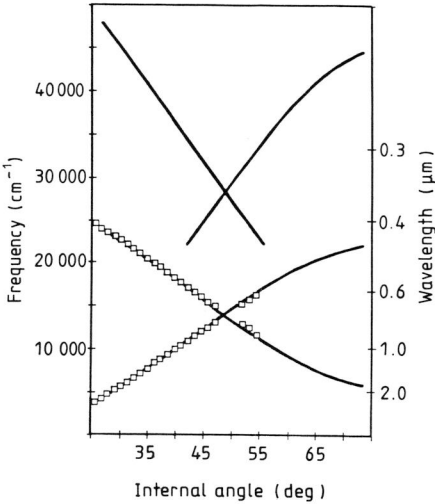

Figure 7.18 The same as figure 7.17, but for a type II OPE/OPA [19].

β BaB$_2$O$_4$ crystals were mounted on top of a rotation stage, which allowed angular positioning with an accuracy of better than 0.5°. The wavelength of the parametric spontaneous emission in both type I and type II processes was measured as a function of incidence angle of the pump beam. For the latter, the second, third and fourth harmonics of a Q switched Nd:YAG laser weakly focused inside the sample were used. For all three pump wavelengths, good agreement between prediction and experimental data is obtained, thus validating the accuracy of the Sellmeier equations [16]. For illustration, the experimental results obtained when pumping with the third harmonic of YAG are superimposed on the predicted curves shown in figures 7.17 and 7.18.

An OPE/OPA is an ideal candidate for generation of short pulses (picosecond or even sub-picosecond), because of its inherent pulse shortening mechanism in sufficiently short crystals [22]. For spectral investigations the bandwidth of the parametric picosecond pulses is important. Several processes contribute to the spectral width of the signal or idler pulses [23]. We consider the following major contributions: (i) the effect of the pump beam divergence; (ii) off-axis parametric amplification within the beam diameter of the pump. As a result of the finite beam divergence of the pump, parametric pulses are generated in different directions with different centre frequencies. The frequency width Δv of the signal pulses due to this effect can be estimated from $\Delta v(\varphi) \approx (\mathrm{d}v/\mathrm{d}\theta)\varphi/n_p$, where φ is the divergence of the pump beam (outside the crystal), and n_p is the index of refraction of the crystal at the frequency of the pump. The factor in the brackets represents the slope of the tuning curve. From a glance at figures 7.17 and 7.18 it becomes obvious that this contribution to Δv will generally be more important for a type I process than for a type II process. In a short crystal (length L), parametric emission occurs over a relatively large divergence ($\approx w/L$) where w represents the pump beam diameter. This can result in a wide frequency band due to off-axis phase matching. For many applications, nearly diffraction- and bandwidth-limited radiation is highly desirable. Therefore an OPE is usually followed by another OPA with a suitable distance D between them, as first proposed by Kung [24]. (In practice one can conveniently use the same crystal for both the OPE and the OPA in a double-pass configuration.) In this way, the divergence can be limited to $\approx w/D$. In a type II configuration, the frequency variation as a function of the angle of emission in a plane parallel to the principal plane (defined by the pump wavevector and the optical axis) is given by [25, 26]

$$\Delta v(\alpha) = v(\alpha) - v_0 = \frac{1}{2\pi bc}\left(\frac{k_s k_p}{2k_i}\alpha^2 + \frac{k_s}{k_i}\frac{\partial k_i}{\partial \theta_i}\alpha\right)$$

$$b = \frac{1}{u_s} - \frac{1}{u_i}$$

(7.11)

where Δv is in cm^{-1}. In equations (7.11), α is the angle between the pump and signal wave directions, θ_i is the angle between the idler wave and the optical axis, u denotes the group velocity, k is the wavevector, and the subscripts s, i and p refer to the signal, idler and pump respectively. In equations (7.11) all angles are taken inside the crystal. (A negative angle means a decreasing angle with respect to the optical axis.) Notice the asymmetric dependence of $\Delta v(\alpha)$ with respect to $\alpha = 0$, due to the linear term in the first equation. This term vanishes both for type I phase matching and for type II phase matching if α is taken perpendicular to the principal plane. By using the dispersion data in [16], the quantities $\Delta v(\varphi)$ and $\Delta v(\alpha)$ have been calculated and measured for an arbitrarily chosen signal $v_0 = 19.327 \, cm^{-1}$ (pump wavelength 355 nm). The calculated results for $\Delta v(\alpha)$, both for type I and for type II phase matching, are shown in figure 7.19 for external angles α. The experimental data for type II phase matching, also shown in this figure, clearly fit the theory relatively well. The data were obtained by simply moving a narrow slit across the signal beam and measuring the signal wavelength accordingly. Notice that no fitting procedure whatsoever was involved to determine the dispersion parameters in equations (7.11). The slight deviation of the experimental data from the calculated results is due to the contribution of $\Delta v(\varphi) \approx 77 \, cm^{-1}$ in this type II process (pump beam divergence $\varphi = 5$ mrad). For the corresponding type I process, the effect of the divergence of the pump beam is the major contribution to the frequency width of the parametric signal. The calculated $\Delta v(\varphi) \approx 187 \, cm^{-1}$, in 20% agreement with the measured value, dominates the off-axis phase matching effect $\Delta v(\alpha)$ in this case. From the above data, one can deduce the necessary divergence decrease to obtain a given linewidth narrowing in a subsequent OPA.

Since in β BaB_2O_4 the birefringence is much larger than the dispersion (hence the large tuning range), the degree of Poynting vector walk-off will be quite substantial. For the type I process shown in figure 7.19 the walk-off angle between signal/idler and pump is 72 mrad. However, this amount can be greatly reduced by using a type II process where both the signal (or idler) and the pump are extraordinary beams: for the particular case shown in figure 7.19, the walk-off between idler and pump is only 2.6 mrad. Therefore, a type II OPA process in β BaB_2O_4 will be in general more efficient than a type I interaction, despite the fact that the effective nonlinear coefficient is somewhat larger for the latter. This feature has been qualitatively demonstrated by using β BaB_2O_4 sample 1 for amplification of a 532 nm signal. A collimated but weak 1.064 μm idler (derived from the Nd:YAG) was injected into the crystal, to generate a strong green signal by parametric difference mixing of the 355 nm pump with the idler in a type II configuration. The measured efficiency was in qualitative agreement with the predicted one, assuming parametric interaction over the full crystal length. The authors of [19] have

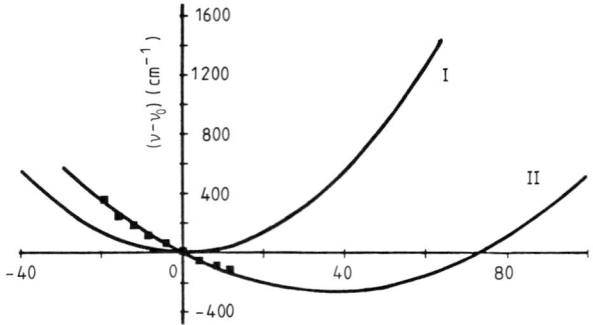

External angle, α (mrad)

Figure 7.19   Frequency deviation versus external angle of emission due to non-collinear phase matching in a 355 nm pumped β BaB$_2$O$_4$ OPE ($v_0 = 19.327$ cm^{-1}). The squares represent the experimental data [19].

demonstrated single-pass spontaneous parametric emission and amplification in relatively short β BaB$_2$O$_4$ crystals. The observed angle tuning curves, in close agreement with the calculated ones, reveal a very wide (20 000 cm^{-1}) tuning range. For subsequent amplification, the spectral width of the parametric emission was also studied. Again a relatively good agreement between theory and experiment has been obtained. From these data, an OPE/OPA device can be constructed which fulfils the particular spectral needs of a given application. With the output of the OPE/OPA frequency doubled in a subsequent β BaB$_2$O$_4$ crystal, the entire wavelength region between 205 nm and 2.6 μm can be covered using the third harmonic of a Nd laser system as the pump source. Damage problems with previous similar devices, such as LiNbO$_3$ or LiIO$_3$, present no difficulty in the case of β BaB$_2$O$_4$.

7.2.1   *Features of the BBO crystal and its applications*
The salient features of β BaB$_2$O$_4$ (BBO) are summarized as follows:

   (i)   Very wide transparency range and phase matching angle tuning 190–3000 nm;
   (ii)   High damage threshold (up to 13 GW cm^{-2} for YAG laser at 1 ns) and wide temperature acceptance width (55°C, about 10 times better than KDP and twice as good as KTP);
   (iii)   High nonlinear $d_{22} = 4.1d_{36}$ (KDP), with an effective figure of merit 3–14 times that of KDP (type I SHG of 1064 nm) (see figure 7.40);
   (iv)   Chemical stability and near freedom from moisture.

   BBO crystal has been identified as the best crystal for frequency doubling of dye lasers for generating deep-UV sources. Efficient and tunable sources (201–310 nm) have been reported for phase matching at room temperature

Table 7.5 Applications of type I BBO crystals [48].

Angle cut	Applications†
23°	SHG of Nd:YAG; OPO (0.6–3 μm)
31°	THG of 1064 nm; SHG of Alexandrite laser; OPO (0.5–1.2 μm)
40°	SFM of 1064 and 532 nm; SHG of dye laser (560–650 nm)
48°	4HG of 1064; SHG of dye laser (500–560 nm)
51°	5HG of 1064; SHG of dye laser (480–530 nm); Doubling of Ar laser, Cu vapour laser
60°	SHG of dye laser (440–490 nm)
65–90°	SHG of dye laser (400–440 nm)

† Phase matching by angle tuning of $\pm 8°$ (external) for applications of OPO and UV sources from doubling dye lasers, where a crystal dimension (in the tuning direction of n_e) of 5–8 mm is required for a pumping laser beam diameter of 1–2 mm. ([48], p 262).

[16]. More recently, a 195 nm source has been achieved using BBO operated at low temperature from the sum frequency mixing of a dye laser. A VUV source at 71 nm has been recently reported in the third harmonic generation of a 213 nm pumping beam (generated from 5HG of Nd:YAG) in neon gas. An output peak power of a 1–2 W of this 5HG (at 71 nm) has been achieved.

A wide range of frequency conversion using BBO crystals for various pumping sources is shown in table 7.5, where a specific angle cut may be used for various applications. The efficiencies of BBO crystals used for the high-order harmonics of Nd:YAG are shown in tables 7.3 and 7.4, where the results from DKDP crystals are compared.

7.3 Crystals of potassium titanate phosphate (KTiOPO₄) and their analogues

A new nonlinear material, $KTiOPO_4$ (KTP), is receiving rare reviews from experimenters who have evaluated it as a frequency doubler for neodymium lasers. The crystal has a nonlinear coefficient that appears to be nearly as large as that of barium sodium niobate, the most-nonlinear material previously available. Unlike BSN, however, KTP has a high damage threshold, close to that of the most damage-resistant nonlinear crystals. With a crystal 1.4 millimetres long, S E Harris obtained a second harmonic conversion efficiency of about 0.5% at a power density of 57 MW cm⁻² for a Q switched Nd:YAG laser in TEM_{00} mode. At 400 MW cm⁻² the conversion efficiency increased to about 22% but the crystal may have been damaged. The conversion efficiency is about 40 times greater than that for caesium dihydrogen arsenate of the same length. KTP also has a large acceptance angle: nearly 2° for the 1.4 mm crystal.

D Decker compared a 40 cm potassium dideuterium phosphate crystal with a 3 mm KTP sample. At comparable pulsed input power levels, he observed 38% conversion efficiency with the KDP and 30% with the KTP.

R Rice placed a small KTP sample inside the cavity of a continuous Nd:YAG laser and found its nonlinearity to be only slightly less than that of BNN. Rice also measured the thermal conductivity of KTP as about 0.13 W cm^{-1} °C^{-1}: twice that of BNN. This high thermal conductivity, combined with KTP's low optical absorption (roughly 0.0005 per centimetre at 1.06 μm) and the insensitivity of the crystal's indexes to temperature, makes KTP impervious to thermally induced degradation of phase matching in high-power lasers: a common problem with BNN.

In a separate experiment, Rice bonded an acousto-optic transducer to a KTP crystal and found the material's diffraction efficiency to be very low. McDonnell–Douglas had been interested in fabricating an acousto-optic modulator and a frequency doubler from the same crystal.

D Hon reported 50 W of average power from a continuously pumped Q switched laser. The KTP crystal was placed inside the resonator and the harmonic output, which was stable at the 50 W level for only 10 or 15 s, was extracted in two separate beams. The infrared power density at the nonlinear crystal was between 50 and 100 MW cm^{-2}, well below the damage threshold which Hon measured to be about 400 MW cm^{-2}. Zumsteg *et al* [27] investigated the system $K_x Rb_{1-x} TiOPO_4$, where $0.0 \leqslant x \leqslant 1.0$.

7.3.1 *Crystal structure*

$K_x Rb_{1-x} TiOPO_4$ exists as a solid solution for $0.0 \leqslant x \leqslant 1.0$. All members of this family are orthorhombic and belong to the acentric point group mm2 (space group P*na*2). Each cell contains eight formula units and has the lattice parameters shown in figure 7.20. The structure [28] of this family is characterized by chains of TiO_6 octahedra linked at two corners with alternating long and short bonds. As we will see later, it is these Ti–O chains which are partially responsible for both the large nonlinear optical effects as well as some of the ferroelectric properties.

Structure studies of $K_x Rb_{1-x} TiOPO_4$ imply that its parent symmetry is mmm. No experimental confirmation of this can be made since dissociation occurs before the transition. In making the transition from mmm to mm2, the Rb(K) ions are displaced 0.116 Å (0.143 Å) along the c axis, and the Ti atoms are displaced in such a way as to create chains of alternating long and short Ti–O bonds. The resulting polarization state is extremely stable, since the temporary formation of two short bonds on the Ti necessary to reverse the sense of polarization is energetically very unfavourable.

The absorption band at 2.8 μm (figure 7.21) most likely represents an OH stretching band, indicating that H$^+$ has been incorporated into the structure. The stretching frequency corresponds to an O–H–O distance of approximately 3.1 Å. The structural analysis shows four O–O distances of 3.13 Å and four

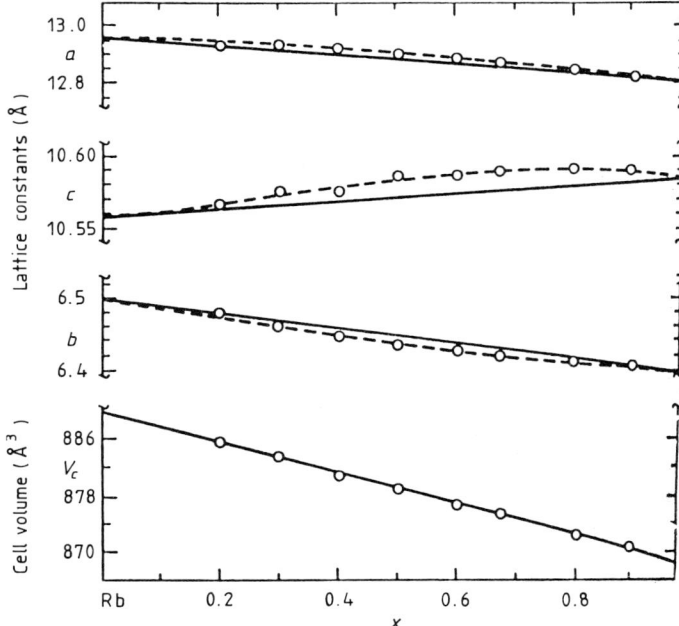

Figure 7.20 Lattice parameters of $K_xRb_{1-x}TiOPO_4$ [27].

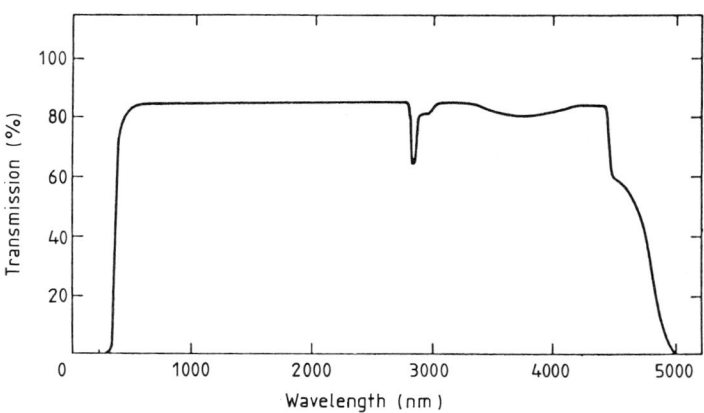

Figure 7.21 Optical transmission of $RbTiOPO_4$ [27].

O–O distances of 3.23 Å per unit cell. The closest other O–O distances to 3 Å are 3.45 and 2.89 Å, both sufficiently far from 3.1 Å to remove them as possible proton sites. This position corresponds to a site midway between two phosphate groups. Although this is not a Rb^+ site, charge neutrality requires that H^+ replace Rb^+.

The broad absorption at 3.8 μm is most likely due to H$_2$O, and the absorption beginning at 4.5 μm is due to molecular (PO$_4$ and TiO$_6$) absorption bands.

7.3.2   *Ferroelectric properties*

It should be pointed out that, although Zumsteg *et al* [27] consider this material to be ferroelectric, it has not been shown to be ferroelectric in the conventional sense of reversing the direction of polarization with an electric field. It is hypothetically ferroelectric because the symmetry relation between super- and sub-point group permits the existence of a ferroelectric stage, and X-ray measurements show a net polarization. Reversal is not possible because of the high coercive force of the Ti–O chains and because ionic conductivity prevents the application of large electric fields at high temperatures.

Examination of the crystal structure of KTP [28] indicates that domain walls parallel to (100) crystal planes are probable, since for this configuration domain reversal results in only slight rotation of the PO$_4$ tetrahedra and the TiO$_6$ octahedra. Of course, the Ti^{4+} ions are displaced in the TiO$_6$ octahedra (average $\Delta Z \approx 0.30$ Å) and K$^+$ undergoes a fairly large displacement ($\Delta Z \approx 1.6$ Å). Also, since a portion of the Z directed polarization consists of contributions from Ti–O chains oriented along (011) and (0$\bar{1}$1) directions, domain walls parallel to (100) planes which do not intersect these chains will be preferred. The lattice distortions and domain wall energies that would result for domain walls parallel to any other plane are considerably higher than for (100) planes and hence domain walls along these planes will be less common.

Voronkova *et al* [30] investigated the dielectric permittivity of KTP, RTP and TlTiOPO$_4$ (TTP) crystals, and also their electrical conductivity and temperature dependence of nonlinear optical susceptibility. The temperature dependence of SHG was measured using a powder technique. Crystals were ground to grains of size 1–2 μm, which is of course less than the expected coherence length. A pulsed laser with $\lambda = 1.06$ μm and a heater were used, which enabled the specimen temperature to reach 1000°C.

The dielectric permittivity and electrical conductivity were measured at frequencies of 1 and 9 MHz in the temperature range 20 to 1000°C.

Figure 7.22 illustrates the temperature dependences of SHG (in the units of SHG intensity of a standard lithium niobate specimen) and of the relative dielectric permittivity of KTiOPO$_4$ crystals measured at a frequency of 1 MHz in the direction of the polar Z axis. One can see that the dielectric permittivity of these crystals has a sharp anomaly at 934 $\pm$ 2°C typical of ferroelectric phase transitions. Measurements along the X and Y axes show that the dielectric permittivity gradually increases from 19 $\pm$ 2 at 20°C to about 100 at 1000°C, exhibiting only a slight anomaly ($\varepsilon_{max} = 140$–160) at the transition point. These data indicate that KTP crystals belong to uniaxial ferroelectrics with spontaneous polarization only in the Z direction and with

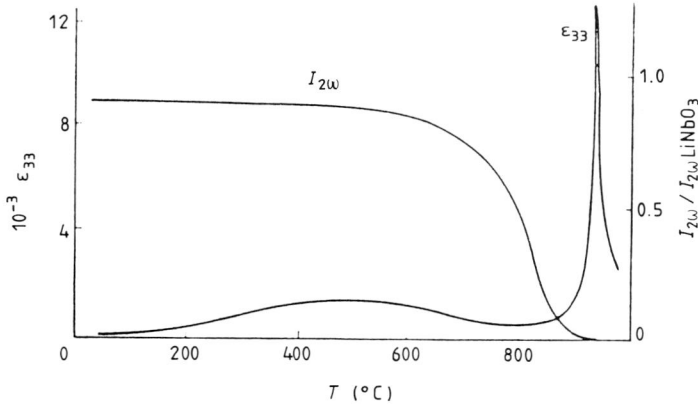

Figure 7.22 Temperature dependences of SHG intensity $I_{2\omega}$ and of dielectric permittivity ε_{33} of KTiOPO$_4$ crystals [30].

a probable symmetry variation, Pna2$_1$ $\rightleftarrows$ Pnan, under phase transition [31, 32]. As can be judged by the Curie constant, $(1.0 \pm 0.4) \times 10^5$ K, these crystals undergo a displacive-type phase transition. Thermal differential analysis has shown that this phase transition is not accompanied by a significant thermal effect and is a second-order transition, which is also indicated by the form of anomaly in the dielectric permittivity. The broad maximum of ε_{33} in the interval 200–800°C is of relaxation origin [33, 34].

As can be seen from figure 7.22, the temperature dependence of SHG intensity is in close agreement with the results of dielectric measurements. Up to about 600°C the SHG intensity remains rather high and practically constant. However, with a further temperature increase $I_{2\omega}$ starts decreasing rapidly and drops to zero near the phase transition point, which corresponds to a transition from a polar into a non-polar state.

The temperature dependences of ε_{33} and $I_{2\omega}$ for RbTiOPO$_4$ and TlTiOPO$_4$ crystals have shown that the ferroelectric phase transition is observed at 789 ± 2°C and 581 ± 2°C respectively [35]. The Curie–Weiss law for the temperature dependence of the dielectric permittivity ε_{33} is obeyed in this case both above and below the transition point, and the 'law of two' holds.

Room-temperature dielectric properties were measured from 10 Hz to 1 GHz [36].

The dielectric frequency spectra for ε_{22} and ε_{33} are shown in figure 7.23 (ε_{11} is nearly identical to ε_{22}). The dielectric constant ε_{33} is characterized by a large dispersion at low frequency, piezoelectric resonances from ~ 500 kHz to 10 MHz and an essentially flat response from 10 MHz to 1 GHz. The loss tangent ($\varepsilon''/\varepsilon'$) is very large at low frequencies, reducing to the 0.01–0.001 region at high frequencies. The low-frequency dispersion in ε_{22} is much

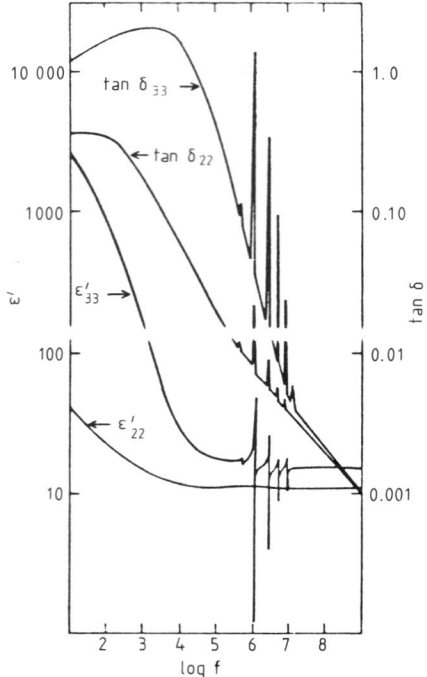

Figure 7.23   Frequency dependence of dielectric constants ε_{22} and ε_{33} for KTP [36].

smaller than that for ε_{33} and is probably due to slight sample misorientation and fringing fields.

The dielectric properties of KTP are typical for ionic conductors where the susceptibility follows the general Debye form

$$\varepsilon(\omega) - \varepsilon_{\infty} \sim 1/[1 + (\mathrm{i}\omega/\omega_p)^{1-\beta}] \tag{7.12}$$

where ω_p is a characteristic frequency that is associated with the dielectric loss peak. For KTP, $\beta \approx \frac{1}{3}$ and $\omega_p < 10$ Hz. ω_p is identified with an ion hopping rate, and is assumed to be thermally activated with an activation energy E_a. This activation energy was measured by Kalesnikas *et al* [33] to be ≈ 0.35 eV and was attributed to potassium ion conduction.

The concrete cause of ferroelectric transitions is obviously displacement of titanium ions in $\mathrm{TiO_6}$ octahedra linked by their vertices into chains along the Z axis. In this case there form alternating short and long titanium–oxygen bonds [28]. At the same time it is assumed that it is just such a non-equivalence of Ti–O bonds that is responsible for the high optical nonlinearity [27]. Thus, nonlinear optical properties of KTP crystals and their analogues are fully explained by their ferroelectric properties. The basic physical properties of KTP and their analogues are listed in table 7.6.

Another important aspect of the relations between ferroelectric and nonlinear optical properties is the possibility to divide a KTP crystal into electric domains. For symmetry reasons, these domains can be only 180° ones and, therefore, are undetectable by usual optical methods. But they may strongly influence SHG by violating the coherence conditions in the crystal.

The electro-optic and pyroelectric techniques have been used to correlate second harmonic generation (SHG) properties at 1.06 μm with domains and to monitor the effects of poling [37]. Generally a single-domain KTP crystal shows high SHG conversion efficiency and a single SHG maximum as a function of temperature (temperature tuning) or of angle (angle tuning). Alternatively, multidomain crystals show lower conversion efficiencies and/or multiple SHG maxima. Note that the domain wall is at an angle to the crystal faces since for 1.06 μm the SHG propagation direction for phase matching is about 24° from the *a* axis (*x* direction). The domain related multiple-peaked and single-peaked angle tuning curves agree with theory [38] and are similar to those discussed in reference to $Ba_2NaNb_5O_{15}$ (chapter 5).

For many ferroelectric materials single-domain crystals can also be obtained by poling, which is usually accomplished by subjecting the crystal to a high electric field while cooling through the Curie temperature. Since the reported Curie temperature of KTP, 934°C, is quite close to its decomposition temperature, 1050–1100°C, the upper poling temperatures were limited to about 500°C. Despite this relatively low temperature, poling was successful in a majority of the multidomain crystals where poling was attempted.

The poling process consists of applying a series of DC current pulses along the polar Z axis at gradually increasing temperatures. Evaporated aluminium electrodes were used with a thin layer of platinum paste to make contact with the poling source. Starting at 200°C, and at every 50°C thereafter, approximately 5 mA cm^{-2} of current is applied to the crystal for 15 s at each temperature, until the maximum temperature of 500°C is reached. Because of the relatively high ionic conductivity of KTP, the voltage needed to generate this current density varies from about 700 V at 200°C to about 25 V at 500°C. The currents and pulse lengths are limited since high current and/or prolonged application of current causes crystal damage. As a final step the crystal is cooled at about 50°C h^{-1} in the presence of a residual voltage of about 1 V. Angle tuning curves before and after poling are shown for a typical crystal in figure 7.24. It is evident from this figure that after poling a single domain-like peak with a full width half-maximum of about 2° (external) appears between and replaces the two original peaks. Furthermore, regardless of the conversion efficiency of the double peaked crystals, after poling the crystals yield similar amounts of green output under our test conditions (~ 40 MW cm^{-2}). This translates to roughly 35% conversion efficiency for

Table 7.6 Basic physical properties of KTiOPO$_4$ crystals and their analogues [30].

| Crystal | Lattice parameters (nm) | | | Density (g cm^{-3}) | Melting point (°C) | Curie point T_C (°C) | Curie constant (10^3 K) | Dielectric permittivity, ε_{33} | | Electric conductivity at 300°C (10^{-5} Ω^{-1} cm^{-1}) | Refractive indices for $\lambda = 530$ nm | | |
	a	b	c					T_C	25°C		n_g	n_m	n_p
KTiOPO$_4$	1.283	0.641	1.061	3.02	1148	934	100	14 000	19	200	1.889	1.791	1.779
RbTiOPO$_4$	1.303	0.653	1.057	3.64	1070	789	64	1 700	23	5	1.885	1.800	1.785
TlTiOPO$_4$	1.289	0.649	1.060	5.41	1035	581	65	6 500	20	10	—	—	—

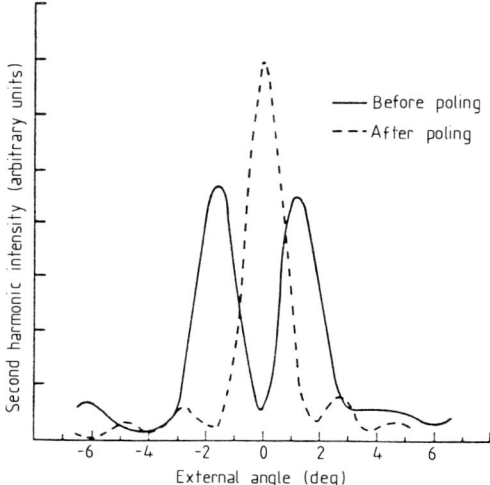

Figure 7.24 SHG angle tuning curves for a multidomain KTP crystal before and after poling [37].

a 5.5 mm length poled crystal, which is similar to the single-domain crystal value. Pyroelectric maps of the crystals show at least two regions of different polarity before poling and for most crystals a single region after poling.

Some crystals failed to pole even after several attempts. Examination of the pyroelectric maps of nearly all these crystals indicates a more complicated domain configuration, such as the zig-zag walls characteristic of head-to-head domain orientations shown in figure 7.25. These crystals contain higher-energy walls and will probably require higher temperatures and/or fields to pole.

7.3.3 *Conductivity*

In ion materials, the conductivity is the sum of the electron σ_e and ion σ_i conductivities, $\sigma = \sigma_e + \sigma_i$. If $\sigma_e \ll \sigma_i$, the ion conductivity is expressed by the

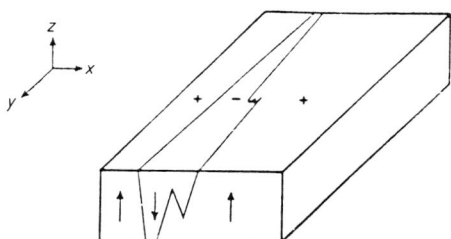

Figure 7.25 Typical high-energy domain wall configuration observed in KTP [37].

well known Arrhenius formula:

$$\sigma_i = (\sigma_0/T) \exp(-E_a/kT). \tag{7.13}$$

Figure 7.26 shows the dependence $\lg(\sigma_{33}T) = f(10^3 T^{-1})$ at a frequency of 1 kHz. The activation energy E_a calculated from the dependence (7.13) in the region $T = 280$ K changes slightly—from 0.31 to 0.36 eV. For another crystal at a temperature of 295 K and a frequency of 1 kHz, the electrical conductivity $\sigma_{33} = (1.5–3.5) \times 10^{-6}$ S m^{-1}.

Many ion–covalent crystals exhibit high ion conductivity, due to a high ion mobility of silver, copper, potassium, lithium, sodium, caesium, rubidium, fluorine etc [38]. A characteristic feature of such compounds is 'friability' of their crystal structure, which allows a large number of equivalent intersites inside a unit cell, into which ions may migrate from their normal stable position. Another characteristic feature is the existence of 'windows' for the motion of these ions with a low activation energy. The crystal structure of KTP comes well within this description.

Mobile K^+ ions in KTP occupy part of the voids in the crystal lattice, the voids along the 2 axis remaining vacant. These voids are obviously those 'channels' along which K^+ ions move under a temperature effect upon the crystal lattice of $KTiOPO_4$, and are responsible for the high ion conductivity.

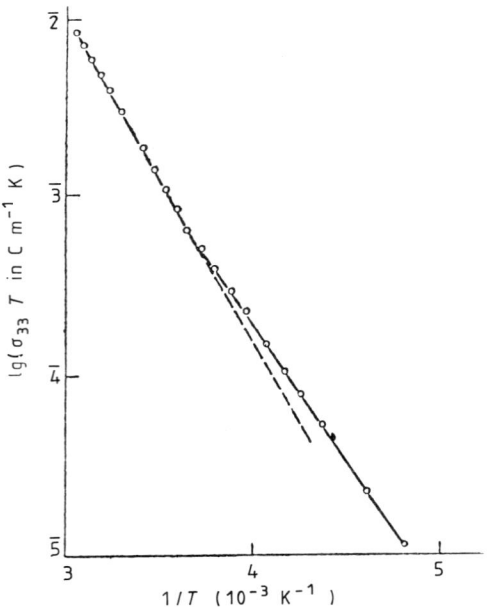

Figure 7.26 Temperature dependence of electric conductivity at a frequency of 1 kHz along the c axis [33].

The disorder of the K sublattice may be accompanied by a decrease of $\varepsilon(0)$. Each cation transition from a site to an intersite creates a dipole moment P, polarizes the medium and has an energy gain $P E$. If the number of intersites in the lattice is large, the dipole can have an arbitrary direction and can change it in the external electric field, thus creating relaxation polarization.

After the relaxation polarization is switched off, ε and $\tan \delta$ in the direction of the c axis are not large and the conductivity for the frequency $v = 9 \times 10^9$ Hz at 295 K is equal to $\sigma_{33}(v) = 0.11 \, \text{S m}^{-1}$. The low-frequency conductivity therefore increases by a large factor, which confirms the jump-like mechanism of conductivity. The anomalies of the microwave dielectric permittivity disappear, as in the case of the usual structure order–disorder phase transition.

In RTP and TTP crystals with large rubidium and thallium cations, the conductivity is lower by 1–2 orders of magnitude.

The high electrical conductivity of KTP crystals does not prevent their use as frequency doublers, parametric oscillators etc. Furthermore, this characterisic of KTP crystals is a favourable factor, since a high conductivity near the phase transition temperatures decreases the probability of formation of ferroelectric domains due to neutralization of the depolarizing field. The high electrical conductivity may also explain the stability of these crystals to optical damage.

7.3.4   *Optical properties*

The transmission characteristics of $RbTiOPO_4$ are shown in figure 7.21. As can be seen from the figure, the transparency range extends from 0.35 μm to 4.5 μm. No attempt has been made to correct for reflections. The lack of any noticeable colour and the constant transmission from 0.35 to 2.7 μm indicate that the absorption is very small in this region [27].

For a material to be useful as a nonlinear optical material, it must be possible to have the fundamental radiation stay in phase with that generated at the second harmonic, i.e. the phase velocity must be the same for both the second harmonic and the fundamental waves. Thus if $n_1(\theta, \phi, \omega)$ and $n_2(\theta, \phi, \omega)$ are the two refractive indices associated with a particular frequency ω and propagation direction θ and ϕ, type I phase matching occurs when $n_1(\theta, \phi, 2\omega) = n_2(\theta, \phi, \omega)$ and type II phase matching occurs when $n_1(\theta, \phi, 2\omega) = \frac{1}{2}[n_1(\theta, \phi, \omega) + n_2(\theta, \phi, \omega)]$. The most favourable situation, that of non-critical phase matching, occurs when the propagation direction is identical to one of the principal directions of the index ellipsoid. For this special case, alignment and divergence considerations are less critical, since the phase mismatch ΔK varies quadratically for small deviations from the propagation direction, rather than linearly. An additional advantage is that the beam walk-off angle is zero.

While the phase matching conditions are an intrinsic property of any given material, it is possible to alter them by changing the temperature. Alternatively, in cases such as this, where a range of compositions exists with nearly identical nonlinear optical properties, the composition can be varied to alter the phase matching conditions.

To determine phase matching conditions, refractive index measurements were made as a function of temperature, wavelength and composition using the method of minimum deviation. Values of n_x and n_z were determined using the prism formed by the intersection of the natural (110) crystal faces, while n_y and n_z were determined using the (201) surfaces. The characteristic natural prism angle of about 63° limits measurement to refractive indices of less than 1.91. For larger values of n, prisms with an angle of about 45° were made from suitably oriented crystals.

A complete dispersion curve for KTP was made at room temperature (figure 7.27). Then, to determine the wavelength range for non-critical phase matching that could be achieved by temperature tuning, n was measured as a function of temperature at several wavelengths for $x = 0.0$ and 1.0. An example of the results of such a run is shown in figure 7.28. As can be seen, n is a very slowly varying function of λ and, more importantly, the birefringence is nearly independent of the temperature. As a result, the wavelength for phase matching is much less sensitive to temperature change than other nonlinear optical materials. Using the measured temperature coefficients of the indices of refraction, the width of the phase matched peak of a 1 cm crystal was calculated as approximately 7.5°C.

It should be noted that the calculation of the wavelength λ_{ncpm}, where non-critical phase matching occurs, is very uncertain. This is primarily due to

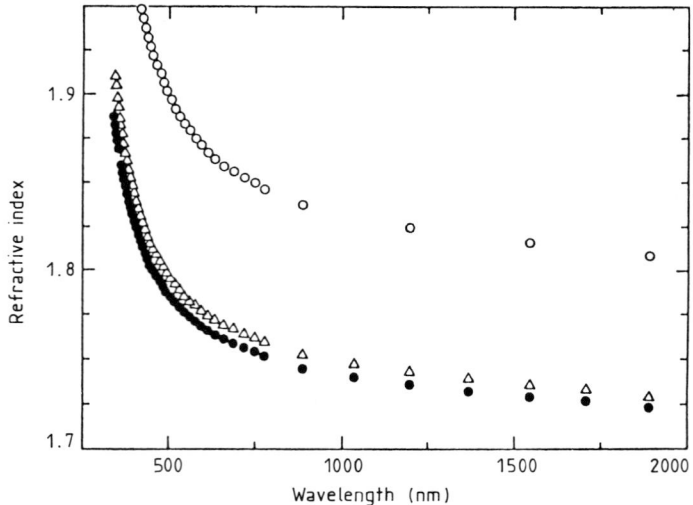

Figure 7.27 Refractive index versus wavelength of KTiOPO$_4$. Full circles are n_x; triangles n_y and open circles n_z [27].

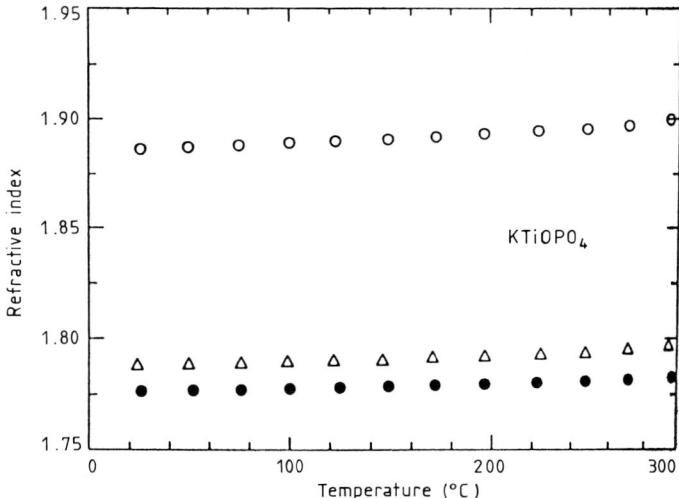

Figure 7.28 Temperature dependences of the refractive indices of KTiOPO$_4$. Full circles show n_x, triangles n_y, and open circles n_z [27].

the relative flatness of the n versus λ curve in the infrared and the uncertainties in the experimental data. To illustrate the effects of temperature on this wavelength, n versus λ was measured at several different temperatures and then fitted to a single-term Sellmeier relation. While the fit is not perfect, it does allow a systematic comparison of λ_{ncpm} at each temperature. The absolute value, however, is uncertain to ± 250 Å. Of special importance is the question of whether or not any member of this family can be non-critically phase matched at 1.06 μm. The results of these measurements indicate that, while λ_{ncpm} is close to 1.06 μm, it cannot be made to match for any value of x or for any practical temperature.

Phase matching conditions for the 1.06 μm Nd line were calculated using room-temperature data at 1.06 and 0.53 μm for $x = 1.0$, 0.68 and 0.0. The results of this calculation are shown in figure 7.29. Measurements of the phase matching angle for propagation in the principal planes compare favourably with these calculations.

It should be noted that, for second harmonic generation of light, not only is it necessary to have phase matching; it must also be possible for the polarization induced by the fundamental wave to couple to that of the second harmonic. The general solution to this problem is rather complicated, but for propagation in the principal planes, one can show that for $K_x Rb_{1-x} TiOPO_4$ there will be coupling for type II phase matching in the xy and yz plane and for type I phase matching in the xz plane.

Although the solution for the phase matched propagation direction for arbitrary wavelength must be done numerically, one can use the refractive

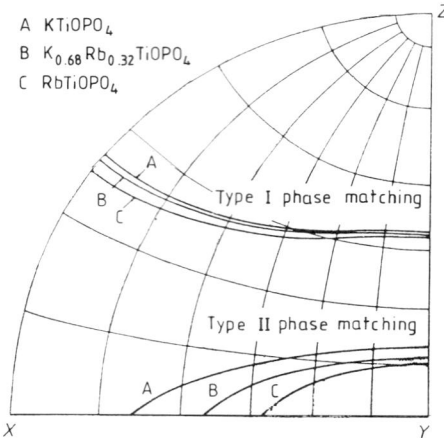

Figure 7.29   Phase matching directions in $K_xRb_{1-x}TiOPO_4$ for the doubling of 1.06 μm [27].

index data to determine the wavelength range over which the different types of phase matching are possible. In this case, type I phase matching is possible over the entire transparency range of the crystal and type II phase matching is possible for wavelengths larger than approximately 1 μm.

Stolzenberger [39] has carried out a detailed study of nonlinear optical properties of KTP crystals.

Figure 7.30 illustrates the basic experimental arrangement. A Q switched Nd:YAG laser was used for temperature acceptance width, angular acceptance width, efficiency, and uniformity testing. The cavity was fitted with an aperture for TEM_{00} operation, and a beam profile showed a near-gaussian intensity distribution. A half-wave plate was used to rotate the polarization of the input beam to the 45° plane necessary for type II frequency doubling. A 3:1 Galilean telescope was used to reduce the beam diameter. After passing

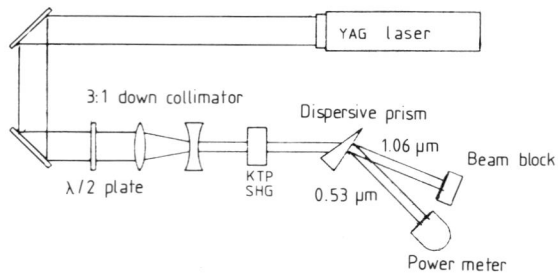

Figure 7.30   Basic experimental testing set-up [39].

through the KTP sample, the IR and visible radiation were separated by a dispersive prism. The green light entered the prism at the Brewster angle and, therefore, suffered negligible loss. A volume absorbing calorimeter was used to measure radiant flux (laser power).

For the temperature acceptance width measurements, an Inrad heater/cooler served as a test chamber. The samples were mounted under dry nitrogen onto a massive temperature controlled heat sink and oriented for maximum doubling at room temperature. Relative efficiency was determined by comparing the input and output flux.

To measure the angular acceptance width, samples were mounted on a goniometer/rotation stage and oriented to reflect the laser source back on itself. From this reference direction the crystals were rotated in θ and φ (see figure 7.31), while the doubling efficiency was recorded. Absolute efficiency was determined by dividing the measured green radiant flux by the IR flux incident on the sample. The samples were not anti-reflection coated to eliminate Fresnel reflection losses. The actual efficiency, therefore, is somewhat higher than that reported.

The optimum propagation direction for type II SHG at 1.064 μm was found to be $\theta = 90°$, $\varphi = 23°19'$. The angle (or temperature) acceptance width–length product provides an indication of crystal quality. Imperfections in the crystal caused a broadening of the angular acceptance width–length product, which ranged from 13 mrad cm for an optically perfect crystal to 43 mrad cm for severely non-uniform crystals. A temperature acceptance width–length product of 20°C cm was found. Efficiency testing was done at a power density of 100 MW cm^{-2}. The highest energy conversion efficiency measured was 50% for an uncoated crystal 1 cm in length [39].

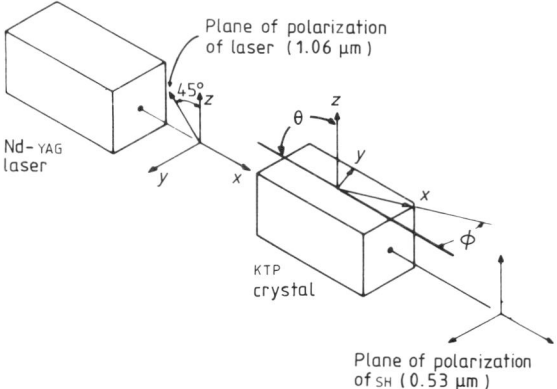

Figure 7.31 Type II phase matching in KTP, x, y and z corresponding to the crystallographic axes a, b and c, respectively [44].

7.3.5 *Beam walk-off*

The double refraction of acentric birefringence crystals can often prevent full use of its NLO properties. Double refraction will cause the fundamental beam to separate physically from the generated beam. As a result, the interaction length is limited to

$$l_a = W_0 \sqrt{\pi/\rho} \tag{7.14}$$

where ρ is the double refraction angle and W_0 is the beam radius.

In general, calculation of the double refraction angle is complicated, and one must consider the walk-off of both orthogonally polarized extraordinary waves. For propagation in one of the principal planes, however, only the extraordinary wave departs from the original propagation direction. Ideally, one would like to have propagation along one of the principal axes of the index ellipsoid. There are two advantages to using that direction. There is no walk-off and the effects of beam divergence of misalignment are minimized. As pointed out before, $90°$ phase matching is not possible in these materials at 1060 nm. However, as can be seen from table 7.7, beam walk-off is minimized for the phase matched condition, with propagation in the xy plane.

7.3.6 *Nonlinear optical properties*

If a NLO material is exposed to light with electric field components $E(\omega)$, then for the mm2 point group, the induced second harmonic polarization $P_i(2\omega)$ is given by

$$P_x = 2d_{15}E_x E_z$$
$$P_y = 2d_{24}E_y E_z \tag{7.15}$$
$$P_z = d_{31}E_x^2 + d_{32}E_y^2 + d_{33}E_z^2$$

where d_{ij} is the nonlinear optical coefficient given the usual reduced subscript notation. As can be seen, different polarizations and propagation directions are necessary to determine all of the components. This was accomplished by

Table 7.7 Walk-off angles for KTiOPO$_4$ [27].

Position	λ (μm)	Walk-off angle
Type II phase matching in xy plane	0.53	0.077°
	1.06	0.057°
Type II phase matching in yz plane	0.53	0.74°
	1.06	0.64°
Type I phase matching in yz plane	no coupling possible	
Type I phase matching in xz plane	0.53	3.12°
	1.06	2.67°

Table 7.8 Nonlinear optical coefficients of $KTiOPO_4$ [27].

ij	d_{ij} $(10^{-12}\,\mathrm{m\,V^{-1}})$	$\Delta_{ij}/\Delta_{11}^{quartz}$
31	6.5	3.9
32	5.0	3.0
33	13.7	6.2
24	7.6	4.5
15	6.1	3.8

making measurements on x and y plates cut from oriented crystals of $K_xRb_{1-x}TiOPO_4$. The relative magnitudes of the coefficients were determined using the Maker fringe method. These results were then calibrated by direct comparison with d_{36} of a crystal of KDP.

The results of these measurements are summarized in table 7.8. Measurements were made for samples with $x = 0.0$, 0.68 and 1.00, but there was no measurable difference in d_{ij} as a function of x.

Miller [40] has shown that, while the nonlinear optical coefficients vary by as much as four orders of magnitude, a parameter Δ_{ijk} can be defined which relates the nonlinear and linear optical susceptibilities and is constant to within an order of magnitude. This constant, defined in the equation

$$\Delta_{ijk} = \frac{d_{ijk}}{\varepsilon_0[n_i^2(2\omega)-1][n_j^2(\omega)-1][n_k^2(\omega)-1]} \tag{7.16}$$

is representative of the intrinsic nonlinearity of the material. The ratio of Δ_{ijk} of $KTiOPO_4$ to that of Δ_{111} of quartz is shown in table 7.8. To put these values in perspective, one should note that the largest $\Delta_{ij}/\Delta_{11}^{quartz}$ for $Ba_2NaNb_5O_{15}$ is 2.5, while $\Delta_{33}^{KTiOPO_4}/\Delta_{11}^{quartz}$ is more than twice this value.

7.3.7 *Optical damage*

Many nonlinear materials are limited in their applications by damage caused by the large-amplitude electric fields needed in nonlinear optical applications. $LiNbO_3$ is a well known example of this problem, and only material grown under the most careful conditions can be made with acceptable threshold levels. $Ba_2NaNb_5O_{15}$ has been reported to damage at power densities of $3\,\mathrm{MW\,cm^{-2}}$. The damage is not only structural; in some instances it can also be seen as changes in the index of refraction.

No optical damage to $K_xRb_{1-x}TiOPO_4$ has been detected, either after several hours' exposure to the highly focused output of a 5 W CW Nd:YAG laser or with repeated pulses of $150\,\mathrm{MW\,cm^{-2}}$ from a Q switched Nd:YAG laser. The lack of damage was ascertained by examining interferograms of the crystals before and after exposure.

7.3.8   *Electro-optic properties*

The space group of KTP is $Pna2_1$; thus the crystallographic directions a, b, c correspond to the optic axes x, y, z, c being the polar axis.

In determining the electro-optic coefficients from DC to 13 MHz a sinusoidal driving voltage was applied across the crystal and the modulated light detected with a photodiode [36]. With this technique the measured electro-optic signal is related to the electro-optic coefficients by

$$r = 2\Delta I\lambda/I_0\pi ln^3 E \qquad (7.17)$$

where ΔI is the maximum modulated light intensity, $I_0 = I_{max} - I_{min}$, λ is the wavelength (6328 Å), l is the optical path length under the applied field E, and n is the refractive index ($n_1 = 1.7634$, $n_2 = 1.7717$, $n_3 = 1.8639$). The directions of light propagation and applied field and the appropriate refractive indices needed to determine the various r coefficients are given in table 7.9.

Using a technique with 100% modulation, the electro-optic coefficients are related to the detected modulated light intensity by

$$r = \sqrt{32\Delta I/I_0}(\lambda/\pi ln^3 E) \qquad (7.18)$$

where ΔI is the detected signal amplitude when $I = I_{min}$ and E is the amplitude of the RF field.

The measured coefficients r are summarized in table 7.10. The results reported in table 7.10 represent measurements averaged over many samples.

A more fundamental measure of the electro-optic properties of a material is the polarization optic coefficient f which is related to the r coefficients by $r_{ij} = \varepsilon_0 f_{ij}(\varepsilon_{jj} - 1)$. For KTP the coefficient f_{33} is $0.28 \text{ m}^2\text{C}^{-1}$, which is approximately twice higher than for any other inorganic material, indicating a large degree of acentricity in the crystal structure. This large f coefficient correlates well with the large Miller Δ_{33} coefficient obtained from SHG measurements.

It is interesting to use the measured r coefficients and the quadratic electro-optic g coefficients, which have been shown to be remarkably constant for a variety of oxygen octahedral ferroelectrics, to predict the spontaneous

Table 7.9   Orientational parameters for equations (7.17) and (7.18) [36].

Coefficient	Propagation direction	E field direction	n
r_{13}	y	z	n_1
r_{23}	x	z	n_2
r_{33}	x, y plane	z	n_3
r_{51}	x, z plane 45° from x	x	$n_5 = \sqrt{2n_1 n_3}(n_1^2 + n_3^2)^{-1/2}$
r_{42}	y, z plane 45° from y	y	$n_4 = \sqrt{2n_2 n_3}(n_2^2 + n_3^2)^{-1/2}$
r_{c1}	y	z	n_3
r_{c2}	x	z	n_3

Table 7.10 Electro-optic constants of KTP [36].

	Low-frequency $(10^{-12} \text{ m V}^{-1})$	High-frequency $(10^{-12} \text{ m V}^{-1})$
r_{13}	$+9.5 \pm 0.5$	$+8.8 \pm 0.8$
r_{23}	$+15.7 \pm 0.8$	$+13.8 \pm 1.4$
r_{33}	$+36.3 \pm 1.8$	$+35.0 \pm 3.5$
r_{51}	7.3 ± 0.7	6.9 ± 1.4
r_{42}	9.3 ± 0.9	8.8 ± 1.8
r_{c1}	$+28.6 \pm 1.4$	$+27.0 \pm 2.7$
r_{c2}	$+22.2 \pm 1.1$	$+21.5 \pm 2.2$

polarization P_s of KTP. The relationships between r, g and P_s are

$$r_{i3} = 2g_{12}\varepsilon_0(\varepsilon_{33}-1)P_s/\eta^3 \qquad i=1,2$$

$$r_{33} = 2g_{11}\varepsilon_0(\varepsilon_{33}-1)P_s/\eta^3 \qquad\qquad (7.19)$$

$$r_{ij} = g_{44}\varepsilon_0(\varepsilon_{ij}-1)P_s/\eta^3 \qquad ij = 42, 51$$

where η is the octahedral packing density of KTP relative to that in the perovskites and, for oxygen octahedral ferroelectrics, $g_{12} \approx 0.045 \text{ m}^2\text{ C}^{-2}$, $g_{11} \approx 0.17 \text{ m}^4\text{C}^{-2}$ and $g_{44} \approx 0.11 \text{ m}^4\text{C}^{-2}$. The average value of P_s obtained using (7.19), with $\eta = 0.59$, is 0.17 C m^{-2}. P_s can also be predicted for displacive ferroelectrics from a point charge model, the most simple being

$$P_s = \frac{1}{V}\sum_i Z_i \Delta z_i \qquad\qquad (7.20)$$

where V is the unit cell volume, Z_i are the effective charges and Δz_i the atomic displacements along the polar axis from a position giving a zero net polarization. For KTP, $\Delta z(\text{Ti}) = 0.150 \text{ Å}$, $\Delta z(\text{P}) = 0.037 \text{ Å}$ and $\Delta z(\text{K}) = 0.831 \text{ Å}$, and hence from (7.20), P_s is 0.24 C m^{-2}, which is fairly close to the value determined from the r coefficients, considering the simple models used. Initial attempts to measure P_s directly using high-voltage pulsed fields have not been successful due to the high ionic conductivity in the polar direction.

To compare KTP with other materials for electro-optic modulator applications, it is also important to know the temperature dependence of the electro-optic coefficients and the thermal retardation coefficient k. The coefficient r_{c1} was found to be nearly independent of temperature (to within $\pm 10\%$) from ~ 100 to ~ 900 K. The other r coefficients are assumed to be similarly temperature independent. These coefficients are expected to decrease to zero at 1238 K, which is the recently measured Curie temperature [29]. The coefficients $k = (\text{d}l\Delta n)/l\text{d}T$ and $\text{d}ln/l\text{d}T$ for amplitude modulation using r_{c1} and phase modulation using r_{33}, respectively, where l is the optical path length, n the refractive index, and Δn the birefringence, were measured at 6328 Å in KTP and are given in table 7.11. Table 7.11 also compares various

Table 7.11 Electro-optic modulator materials [36].

			Phase				Amplitude		
Crystal	ε	n	r (pm V^{-1})	k ($10^{-6}\,^\circ$C^{-1})	$n^7 r^2$ (pm V^{-1})$^{-2}$	r (pm V^{-1})	k ($10^{-6}\,^\circ$C^{-1})	$n^7 r^2$ (pm V^{-1})$^{-2}$
KTP	15.4	1.86	35.0	31	6130	27.0	11.7	3650
KD*P	48	1.47	24	9	178	24	8	178
LiIO$_3$	5.9	1.74	6.4	24	335	1.23	15	124
LiNbO$_3$	27.9	2.20	28.8	82	7410	20.1	42	3500

materials with KTP for bulk electro-optic modulator applications. In this table ε, n, r, and k are appropriate high-frequency parameters for the particular modulator configuration listed and $n^7 r^2/\varepsilon$ is a figure of merit which is related to a bandwidth/driving power ratio. Table 7.11 clearly shows KTP to be an attractive material for both phase and amplitude electro-optic modulator applications.

Nonlinear optical properties of KTP crystals are summarized in table 7.12.

7.3.9 Intra-cavity second harmonic generation

Liu *et al* [41] demonstrated the high efficiency of KTP crystals for intra-cavity second harmonic generation in an acousto-optically Q switched Nd:YAG laser oscillator.

The crystal was cut and oriented for type II interaction at 1.06 μm with one face anti-reflection coated at 1.06 μm and the other face anti-reflection coated at both 1.06 and 0.53 μm. This configuration gives an effective nonlinear constant of 1.9×10^{-11} m V^{-1}.

The crystal was placed inside a Nd:YAG laser oscillator with the YAG optically pumped with a CW krypton lamp contained in an elliptical cavity. The Nd:YAG oscillator was acousto-optically Q switched at 5 kHz. Both oscillator mirrors, separated by 120 cm, were flat and were dielectrically coated for maximum reflection at 1.06 μm. The output mirror was dichroic coated to allow maximum transmission at 532 nm. An oscillator is shown schematically in figure 7.32. No additional optics were introduced inside the

Table 7.12 Some properties of KTiOPO$_4$ [41].

Structure	Orthorhombic
Point group	mm2
Refractive indices	at 530 nm
(Positive biaxial)	$n_x = 1.7787$
	$n_y = 1.7924$
	$n_z = 1.8873$
	at 1.06 μm
	$n_x = 1.7400$
	$n_y = 1.7469$
	$n_z = 1.8304$
Transmission	350 nm–4 μm
Density	2.945 g cm^{-3}
Nonlinear coefficient	$d_{31} = 6.5$
(10^{12} m V^{-1})	$d_{32} = 5.0$
	$d_{33} = 13.7$
	$d_{24} = 7.6$
	$d_{15} = 6.1$
Phase matching angle	$26°$
(Type II at 1.06 μm)	

Figure 7.32 Experimental arrangement for intra-cavity SHG in acousto-optically Q switched Nd:YAG oscillator using KTiOPO$_4$. M1, dichroic-coated output mirror with maximum transmission for 532 nm and maximum reflection for 1.06 μm; WP, half-wave plate for 1.06 μm; A0, modulator; M2, rear mirror with maximum reflection at 1.06 μm; YAG, Nd:YAG laser (3 mm diameter and 79 mm long); KTP, the doubling crystal (5.8 mm × 4.1 mm in cross section and 3.5 mm in length) angle-tuned for type II interaction [41].

cavity for focusing. A half-wave plate at 1.06 μm was inserted inside the cavity. The second harmonic output was optimized by rotating the wave plate. The wave plate was slightly tilted with respect to the optical axis of the resonator to avoid the étalon effect. The one-way second harmonic output measured at the dichroic output mirror was plotted as a function of the lamp current and is shown in figure 7.33. An average power of 5.6 W was measured at a lamp current of 39 A at 5 kHz. The pulse width measured was about 200 ns. Above 28 A, the output increased almost linearly proportional to the lamp current. No saturation was observed up to 39 A, which was the limit of the power-supply current available. For comparison, by replacing the KTiOPO$_4$ crystal with 10 mm anti-reflection coated LiIO$_3$ and repeating

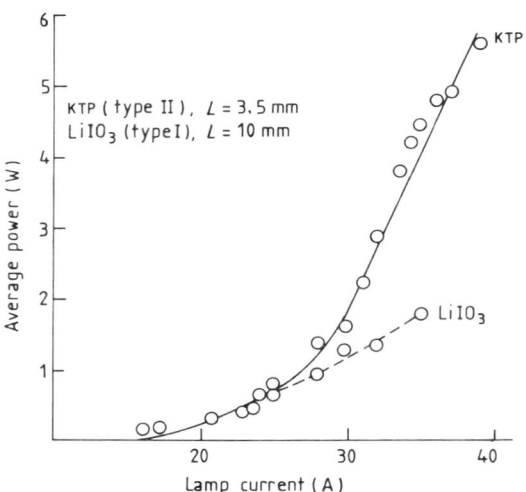

Figure 7.33 Intra-cavity SHG in KTP versus LiIO$_3$ in the acousto-optically Q switched Nd:YAG oscillator at 5 kHz. Second harmonic outputs for KTiOPO$_4$ and for LiIO$_3$ plotted as functions of the lamp current [41].

the experiments under similar conditions, a maximum average power of less than 2 W was achieved in this case and was limited by the damage of the $LiIO_3$ crystal used in this configuration.

Although it has long been recognized that intra-cavity doubling provides the most efficient means for converting energy from the fundamental to its second harmonic, this assumed advantage has never been fully verified because of the lack of a nonlinear material having the required properties, including a high nonlinear coefficient, a high damage threshold and favourable phase matching characteristics. Internal doubling has been applied only to low-power or CW laser oscillators, and little progress has been made for high-power applications. Consequently, previous analysis of intra-cavity second harmonic generation has been limited to the small-signal case [42]. These discussions, however, are not applicable to a description of the present results. One noticeable observation in the present study was the absence of the pulse stretching mentioned in earlier work [43]. For a Q switched laser, the inversion density ΔN and the fundamental and the second harmonic photon flux densities N_{ph}, normalized respectively to the threshold population density ΔN_{th}, can be expressed as

$$\frac{dn}{dt} = -\sigma_c \varphi_1 n \qquad (7.21)$$

$$\frac{d\varphi_1}{dt} = -\frac{\varphi_1}{\tau_c}\frac{\varphi_1}{\tau_c} - \frac{\varphi_1}{\tau_1} \tanh^2(\beta\varphi_1\tau_1/\tau_c)^{1/2} \qquad (7.22)$$

$$\frac{d\varphi_2}{dt} = \frac{\varphi_1}{\tau_1} \tanh^2(\beta\varphi_1\tau_1/\tau_c)^{1/2} \qquad (7.23)$$

where

$$n = \Delta N/\Delta N_{th}$$

$$\varphi_{1,2} = N_{ph}/\Delta N_{th} \qquad \text{for } \omega_1 \text{ and } \omega_2$$

where τ_c is the cavity photon lifetime, τ_1 is the single-pass cavity transit time, σ is the stimulated transition cross section and β is a characteristic coupling parameter that depends on the nonlinear optical properties of the material, the gain of the cavity and other cavity parameters. It can be shown that, under the condition that $\beta\varphi_1\tau_1/\tau_c < 1$, pulse stretching would occur as a result of overcoupling. Here we are in a region where $\beta\varphi_1\tau_1/\tau_c > 1$, and efficient doubling regulated the intra-cavity circulating power and provided a condition under which the total photon flux change inside the cavity ($\varphi_1 + 2\varphi_2$) is almost equal to the change of population inversion density. In the present experiment, the maximum two-way second harmonic power was approximately equal to the maximum fundamental power output under the optimum coupling condition: i.e. almost 100% conversion efficiency was achieved.

7.4 Growth of crystal KTP

Crystals of $Rb_{1-x}TiOPO_4$ which are optically clear and suitable for characterization were prepared by two different methods. The preferred method consists of slow-cooling a homogeneous solution of TiO_2 in an aqueous flux of the appropriate mixture of potassium and rubidium phosphate. This solution is cooled from 850 to 600°C over a period of one week and then quenched. During the crystal growth, the collapsible gold tube containing the reagents is maintained at a constant pressure of approximately 3000 atm.

The second method also involves hydrothermal techniques. In it, the TiO_2 nutrient is transported through an aqueous rubidium potassium phosphate flux held at a constant temperature gradient of about 10°C in^{-1}. The cold end of the gold tube, where nucleation and growth occur, is usually held at 500–700°C. Uncontrolled spontaneous nucleation makes this method less desirable than the first [27].

There are few papers devoted to the technology of KTP crystals. In this respect, there is an ideal paper by Bordui *et al* [44], which gives a sufficiently complete presentation of the material. We shall present some points from this paper, to provide an insight into the specificities of the technology of KTP crystals.

Herein are the results of an extended study into the growth of $KTiOPO_4$ crystals from high-temperature solution. Specifically, a process was developed for crystal growth of $KTiOPO_4$, from its solution in $K_6P_4O_{13}$, using a seeded, slow-cooling technique. Throughout this process, 10 day growth runs routinely produce flaw-free crystals large enough to yield plates $10 \times 10 \times 7$ mm^3, oriented in the optimal direction for second harmonic generation, having properties measured to be at least as good as $KTiOPO_4$ grown by other methods.

The liquid composition $K_6P_4O_{13}$ is a viable high-temperature solvent for crystal growth of $KTiOPO_4$. Notable advantages of this solvent composition are that it is relatively non-volatile and that it contains no species foreign to $KTiOPO_4$ itself. Liquid solutions of $KTiOPO_4$ in $K_6P_4O_{13}$ were made by *in situ* reaction of TiO_2.

The solubility of $KTiOPO_4$ in $K_6P_4O_{13}$ over the temperature range 900 to 1000°C is shown in figure 7.34. These data were gathered over a large number of growth experiments and may be considered accurate within 1%.

A new furnace system was designed and constructed specifically for the growth of $KTiOPO_4$ from its solution in $K_6P_4O_{13}$.

The central component of the furnace system constructed following these design considerations is a vertically oriented, resistance heated tube furnace, shown schematically in figure 7.35. The key element in the furnace is a sodium filled Inconel heat pipe which, lined with a protective silica glass tube, defines the furnace cavity. The highly efficient heat transfer within the hollow annular walls of the heat pipe effects a high degree of spatial temperature uniformity

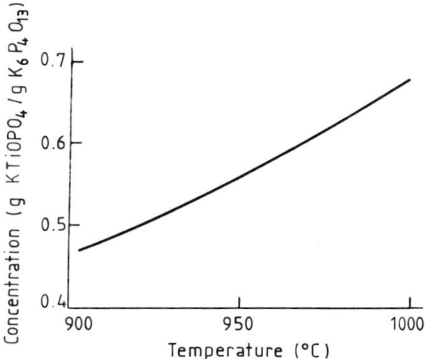

Figure 7.34 Solubility versus temperature for $KTiOPO_4$ in $K_6P_4O_{13}$ [44].

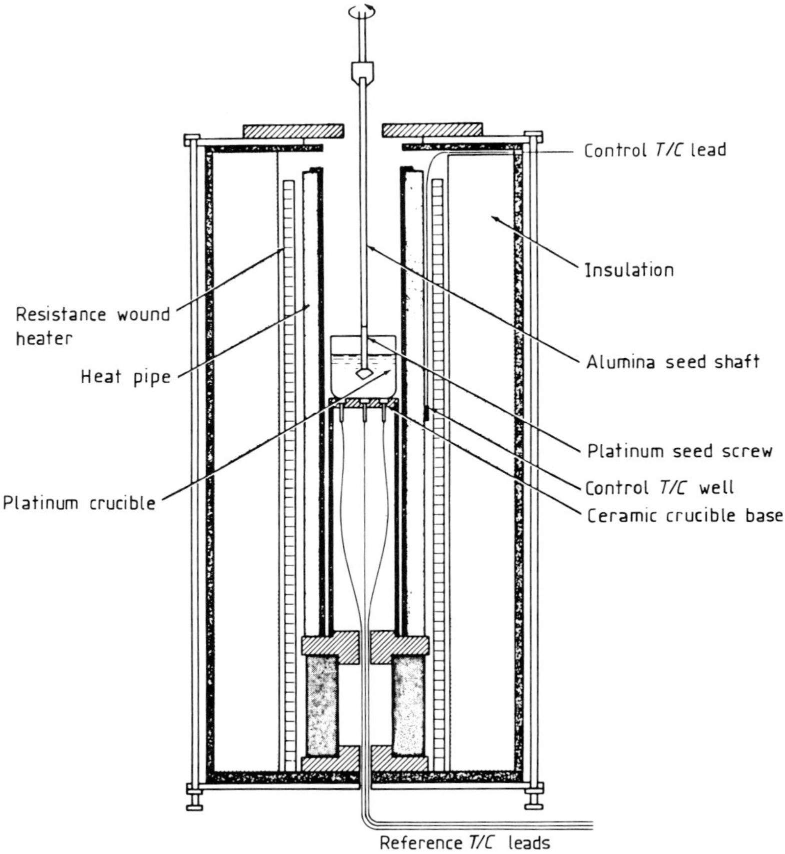

Figure 7.35 Heat pipe based furnace system, schematic representation [44].

and stability in the central region of the furnace cavity. The furnace accommodates a platinum growth crucible 7 cm in height by 8 cm in diameter with a capacity of roughly 230 cm^3 of growth solution.

A tubular steel frame structure provides support for both the furnace and a puller unit mounted above. The furnace rests directly on a heavy duty adjustable $X-Y$ table mounted to the bottom cross members of the frame. The puller unit is supported by the top cross members of the frame. The puller features a controllable rotation rate and rotation reversal periodicity and can be equipped with a variety of platinum seed rods and stir paddles to facilitate a range of stirring and seeding functions.

Power to the furnace's resistance heaters is provided by an analogue temperature controller. The control thermocouple is positioned in a small Inconel well mounted roughly halfway along the outer wall of the heat pipe, as shown. Additional reference thermocouples are positioned in the crucible support plate, as shown. Temperature programming and data acquisition are coordinated by a desktop computer. Crystal growth runs were performed in the temperature range 970 to 900°C using cooling rates ranging from 0.5 to 10°C per day.

Thermal characterization of the furnace was performed involving direct profiling of an unstirred growth solution with a calibrated, platinum sheathed thermocouple assembly, combined with heat transfer analyses to establish the validity and accuracy of the thermocouple measurements. Such characterization indicated spatial temperature uniformity within ± 1.9°C throughout the entire growth solution volume. Temperature stability over several days at any single location in the solution was better than ± 0.3°C. Run-to-run repeatability of the time averaged temperature at a given location in the solution for a given furnace set-point was within ± 0.1°C.

Bordui *et al* [44] give very important advice concerning the preparation of the seed crystal and the seeding process itself which promotes the growth of high-quality crystals. This advice is to a certain extent general for the given crystal growth method.

First, the preferred seed attachment technique involved drilling ultrasonically and threading a small $KTiOPO_4$ crystal, securing it onto the end of a threaded platinum seed rod. There was no evidence of any problem associated with the consequent advancement of a growing crystal over and around the rod.

Second, within the investigated ranges of operating parameters, the achievement of initially uniform and flaw-free growth of a seed crystal was determined to require the controlled dissolution of the original seed surfaces prior to initiation of supersaturation and consequent growth. It is assumed that this necessity has as its basis a surface degradation phenomenon that occurs during the brief interval the seed crystal is exposed to the air inside the heated furnace cavity prior to its immersion in solution. The controlled surface dissolution in turn requires temperature controllability and repeatability in the solution to within ± 0.5°C, and a seed mounting secure enough to ensure integrity in the face of dissolution.

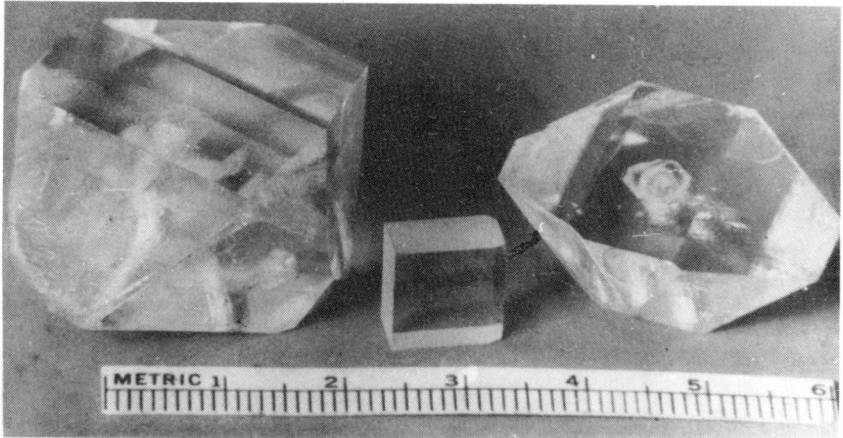

Figure 7.36 $KTiOPO_4$ inclusion-free crystals and fabricated plate [44].

Both naturally faceted crystals produced through induced nucleation and fabricated parallelepipeds cut from larger boules were used as seeds. In all cases, the transition from initial surface dissolution to faceted, stable growth was marked by a thin 'veil' of solution inclusion, in a manner identical to that observed under similar conditions in many crystal growth operations from aqueous solution. The preferred configuration for avoiding solution inclusion flaws involves having a growing crystal mounted on the end of a platinum seed rod with a 90° bend roughly 12 mm from its end. The crystal is submerged roughly midway between the solution surface and the crucible bottom and rotates with periodic reversal, thereby traversing through the solution at some radius from the crucible centreline.

Figure 7.36 shows crystals grown in the heat pipe furnace system using the surface dissolution seeding process and the 90° crystal orientation configuration. The overall process takes about 10 days.

7.5   Material selection of high-efficiency nonlinear crystals

With regard to practical applications, a crystal possessing a large nonlinear optical effect is a necessary but not a sufficient condition. In order to serve a practical purpose, a good nonlinear optical crystal should possess other requisite qualities such as high chemical stability, good mechanical strength, phase matching capability, excellent optical homogeneity, high damage threshold and wide transparency range (preferably in ultraviolet regions). These additional requirements call for the design of a complete set of experimental procedures to search for a new type of nonlinear optical crystal.

The flow diagram of this procedure is shown in figure 7.37. The procedure mainly consists of the following four parts.

(i) Search for powder specimen. Generally speaking, the selection of effective nonlinear optical crystals from a wide variety of available crystalline substances is by no means easy, since the growth of good-quality single crystals is, in most cases, a long and tedious process. It is therefore preferable to start with a good understanding of the crystal in its polycrystalline powdered form. Kurtz and Perry [45] have introduced a powder test technique which permits a rapid evaluation of the nonlinear optical properties of the material under consideration. This technique allows an estimation of the magnitude of this effect and gives us information about phase matchable conditions.

In recent years, the introduction of tunable dye lasers has enabled us to determine exactly the phase matchable region of the specimen, thus greatly enhancing the efficiency of research. Table 9.2 illustrates the application of this method for the estimation of the SHG efficiency of various nonlinear optical materials. The experimental data available are in fairly good agreement within one order of magnitude with those obtained from the corresponding single crystals.

(ii) Crystal growth and its quality improvement. After the powdered specimen has been tested and shown to possess a large SHG effect and phase matchable property, the next step is to grow sufficiently large crystals with optimal optical homogeneity. The work to be carried out in this step includes phase diagram determination, choice of growth method, purification of specimen, observation of defect and identification, improvement of crystal quality etc.

(iii) An overall measurement and checkup of various properties of the crystal. This includes determination of crystal structure, SHG coefficients,

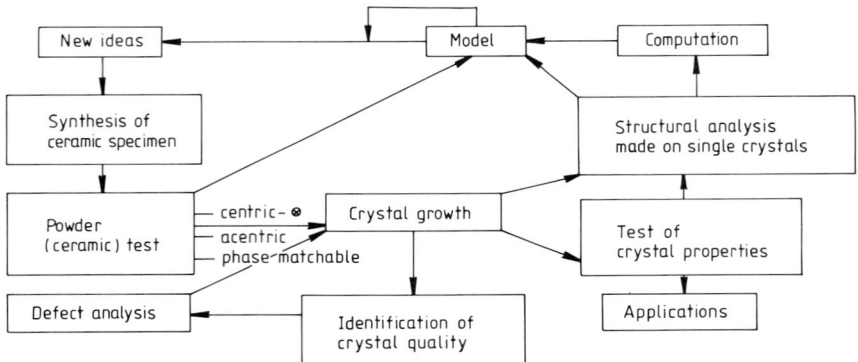

Figure 7.37   Flow diagram of the procedures in search for new-type nonlinear optical crystals [10].

absorption spectra, birefringence, optical homogeneity, damage threshold, harmonic conversion efficiency and various other electrical properties. These will enable us to assess the practical usefulness of the crystal as a nonlinear optical material.

(iv) Theoretical analysis. The above mentioned experimental result makes it possible to calculate once again the SHG coefficients of this new type of nonlinear optical material so as to ascertain whether or not the relation between the SHG effect and the microscopic structure is compatible with the original theoretical expectation. Under these circumstances, we are now in a position to improve further our theoretical working model and to attain a higher degree of perfection [46]. The practical application of nonlinear optical crystals in different spectral regions has been considered in more detail [47].

7.5.1 *Material selection rule*
The key issues of material selection for frequency conversion may be summarized as follows [48]:

(1) high conversion efficiency;
(2) high damage threshold;
(3) wide phase match and transparency range;
(4) large size with good optical homogeneity;
(5) low cost and easy fabrication;
(6) chemical (lack of moisture) and mechanical stability.

The first issue of high conversion efficiency is governed by both the laser behaviour and the crystal condition.

The widely used KDP or DKDP covers the above issues of 3, 4 and 5 but not 1, 2 or 6. $LiIO_3$ is good for issues 1, 3 and 5, but not 2, 4 or 6. DCDA meets 4 and 5 but not 1, 2, 3 or 6. Most of the organic crystals such as m-NA, POM and MNA have very high nonlinear coefficients; the transparency ranges are, however, cut off at about 450 nm. Urea is an exception which covers issues 1, 2, 3 and 4; it does not fulfil the criteria of 5 or 6. The recently developed crystal KTP satisfies 1, 2, 3 and 6, but it is too expensive and seriously limited by issue 4. The new crystal of β BaB_2O_4 (BBO) may be the only one which satisfies most of these issues. Another new crystal called LAP may be as good as BBO and is particularly attractive due to issues 2 and 4.

7.5.2 *Tunable laser sources ($0.2-12 \mu m$)*
Tunable laser sources ranging from UV to IR may be achieved by frequency conversion of fixed or tunable lasers, using nonlinear crystals. Second harmonic generation (SHG), third harmonic generation (THG) and sum frequency mixing (SFM) have been widely used for frequency up-conversion. Optical parametric oscillation (OPO) provides a unique method for frequency down-conversion with continuous angle- or temperature-tuned output.

Shorter-wavelength (100–200 nm) and mid-IR laser sources may also be obtained by further Raman-shifting of the nonlinear crystal converted output.

Using the direct output of commercial lasers or their harmonics as the pumping sources, OPO in nonlinear crystals provides a very wide tuning spectrum, from far IR to VUV (see figure 7.38).

The major properties of nonlinear crystals, which are transparent in the spectral ranges 0.2–5 μm and 0.5–30 μm, are summarized in tables 7.13 and 7.14 respectively. The new crystals of $MgO:LiNbO_3$, KTP, $KNbO_3$ and BBO, and m-NA, POM, COANP, DAN and structure-modulated $LiNbO_3$ have been included in table 7.13.

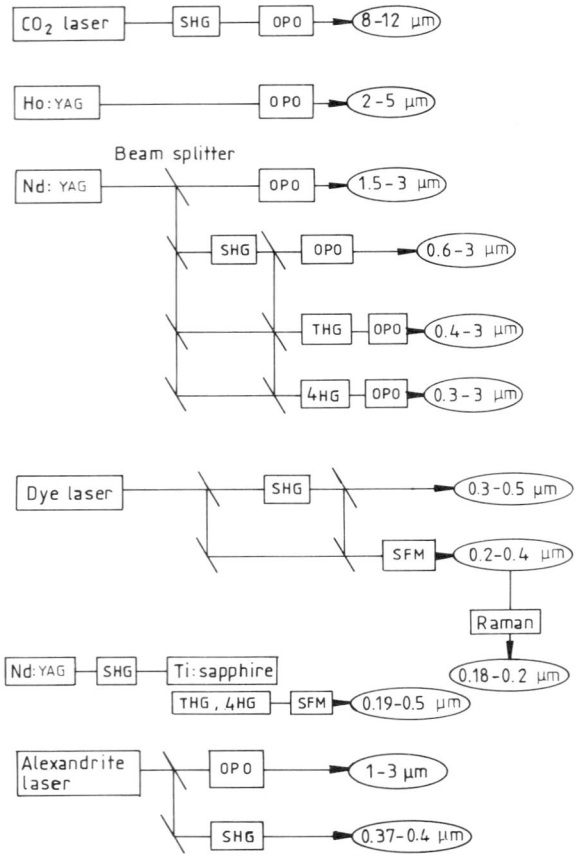

Figure 7.38 Schematics of tunable sources via nonlinear crystals pumped by various commercial lasers. The appropriate crystals for these processes are characterized by their transmission spectra (referred to tables 7.12 and 7.13) [47].

Table 7.13 Nonlinear crystals with transparency ranges $0.2-5$ μm [47].

Material	Transparency range (nm)	Phase matching range (type)	Damage threshold (GW cm^{-2})	Relative figure of merit†
KDP-KH$_2$PO$_4$	200–1500	$\begin{cases} 517-1500(\text{I}) \\ 732-1500(\text{II}) \end{cases}$	0.2	1.0 1.8
DCDA-CsD$_2$AsO$_4$	2700–1660	1034–1600(I)	0.5	1.7
Urea ((NH$_2$)CO)	210–1400	$\begin{cases} 473-1400(\text{I}) \\ 600-1400(\text{II}) \end{cases}$	1.5	6.1 10.6
BBO (β BaB$_2$O$_4$)	198–3300	$\begin{cases} 400-3300(\text{I}) \\ 526-3300(\text{II}) \end{cases}$	5	26 15
LAP	220–1950	440–1950(I)	10	40
LiIO$_3$	300–5500	570–5500(I)	0.5	50
m-NA	500–2000	1000–2000(I)	0.2	60
MgO:LiNbO$_3$	400–5000	800–5000(I)	0.05	105
KTP (KTiOPO$_4$)	350–4500	1000–2500(II)	1	215
POM	414–2000	830–2000(I)	2	350
MAP	472–2000	900–2000(I)	3	1600
KNbO$_3$	410–5000	840–1065(I)	0.35	$\begin{cases} 1460 \\ 1755\ddagger \end{cases}$
COANP	480–2000	960–2000	—	4690
DAN	430–2000	860–2000	—	5090
Structure LiNbO$_3$	400–5000	800–5000	0.05	$2460N^2$§

†The relative figures of merit are based on the value of KDP (type I phase match for SHG of 1.064 μm), all at room temperature except for DCDA, MgO:LiNbO$_3$ and KNbO$_3$, where high-temperature non-critical phase match is operated. The absolute value of KDP is calculated at 0.025 (pm V^{-1})^2.

‡ For KNbO$_3$ crystals, room-temperature 90° phase match at 860 nm (using d_{32}) and at 986 nm (using d_{31}), where $d_{32} = 1.33 \times d_{31} = 20.4$ (pm V^{-1}).

§Structure LiNbO$_3$ grown in quasi-phase matching conditions, where N is the number of periodic layers of the structure-grown crystal.

LAP, L-arginine phosphate monohydrate
m-NA, meta-nitroaniline
POM, 3-methyl-4-nitropyridine N-oxide
MAP, methyl (2,4-dinitrophenyl) aminopropanoate
COANP, 2N-cyclooctylamino-5-nitropyridine
DAN, 3-acetamido-4-dimethylaminonitrobenzene.

7.5.3 High-efficiency crystals for laser diode applications

As shown in figure 7.39, visible coherent sources may be achieved via either: (i) direct doubling of laser diode output frequency, where 830–860 nm, 904 and 1064 nm are the commercially available output wavelengths and may be converted into visible sources via nonlinear crystals; or (ii) frequency doubling of diode-pumped solid state lasers such as Nd:YAG (output of

Table 7.14   Nonlinear crystals with transparency ranges 0.5–30 μm [47].

Material (reference)	Transparency range (μm)	Absorption (cm^{-1}) (at 10.6 μm)	Damage threshold $(MW\ cm^{-2})$	Relative figure of merit
AgGaS$_2$	0.5–13	0.09	15	1.0†
CdSe	0.75–20	0.016	50	1.6
AgGaSe$_2$	0.71–18	0.05	12	6.3
TAS	1.26–17	0.04	16	6.5
CdGeAs$_2$	2.4–18	0.23	40	9.2
ZnGeP$_2$	0.74–12	0.9	3	14.0
Te	3.8–32	0.96	45	270

† Figure of merit (evaluated at phase matching for SHG of 10.6 μm) of AgGaS$_2$ 14.0 $(pm\ V^{-1})^{-2}$ is chosen as one unit.

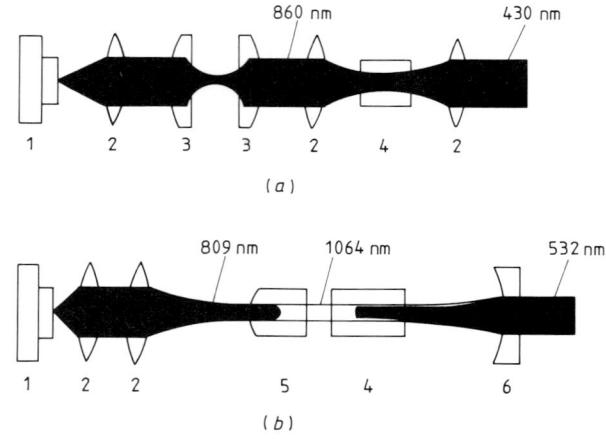

Figure 7.39   Visible lasers generated from direct doubling of laser diode (*a*) and SHG of diode-pumped solid state lasers (*b*) [47]. 1 Laser diode, 2 focusing lens, 3 cylindrical lens, 4 nonlinear crystal, 5 Nd:YAG crystal, 6 concave lens.

1064 nm), Nd:YLF (1047 nm), Ho:YLF (2060 nm), Nd:YAG (1079 nm) or Nd glass (1053 nm).

For application (i), only crystals with very high nonlinear coefficients are suitable for efficient conversion due to the intrinsic nature of the laser diode output. This output ranges from 20 mW (single mode) to 2 W (multimode) and the beam quality (divergence and spectral width) is rather poor. The existing candidate for this application is the KNbO$_3$ crystal which may be operated in the non-critical phase matching condition. Potential candidates must include the organic crystals of POM, MAP, COANP and DAN. The structure-grown $LiNbO_3$ which is still in the laboratory research stage may be another good candidate. Fibres made from highly efficient nonlinear crystals such as $LiNbO_3$ and BBO may also provide potential sources for this application.

For application (ii), intra-cavity doubling is usually required, where KIP and $MgO:LiNbO_3$ (operated at high temperature) fulfil that requirement and are in several commercial products. Potential candidates must also include those described in application (i); however, low absorption at the fundamental wavelength (e.g. 1064 or 1079 nm) is critical for the intra-cavity SHG. External cavity SHG of diode-pumped lasers may also be achieved for the pulsed operation.

7.5.4 *Key parameters for frequency conversion*

Given a good pumping source (narrow band, small divergence and good temporal and spatial profiles), high conversion efficiency may be achieved only when the crystal meets the following conditions: (i) low absorption for both the fundamental and the harmonics; (ii) it is phase matchable within large acceptance widths of angle, spectrum and temperature (or high thermal conductivity); (iii) large aperture length or small walk-off angle; (iv) large figure of merit (E number).

A large nonlinear coefficient d is a necessary but not sufficient condition for a crystal to give a high efficiency. In [48] an E number is introduced which can actually reflect the conversion efficiency and take into account the effects of phase matching angle (or crystal symmetry), crystal refractive index and the wavelength of the pump source. Given the E number, the conversion efficiency for SHG (plane wave, phase matched case) or THG (under appropriate conditions) is given by [48]

$$\eta = \tanh^2(E^{1/2}) \tag{7.24a}$$

where

$$E \sim (d_{\text{eff}}^2/n_2 n_1^2)^{1/2} IL/\lambda_p^2 \tag{7.24b}$$

where L is the crystal length; I and λ_p are the intensity and wavelength of the pump; n_1 and n_2 are the refractive indices of the crystal evaluated at the fundamental and harmonic wavelength; and finally d_{eff} is the effective nonlinear coefficient of the crystal evaluated at the phase matching angle. The computations of the phase matching angles have been based upon the measured refractive indices, and the E number is shown in figure 7.40 for pump wavelengths ranging from 400 to 1500 nm. Figure 7.41 provides the guidance for materials selection for high efficiency. The important features are summarized as follows: (1) for KDP, type I is more efficient than type II for $\lambda_p < 800$ nm; (2) for SHG of dye lasers (550–750 nm), urea and BBO are the best candidates; (3) $LiIO_3$ is excellent for SHG around 600 nm, but with

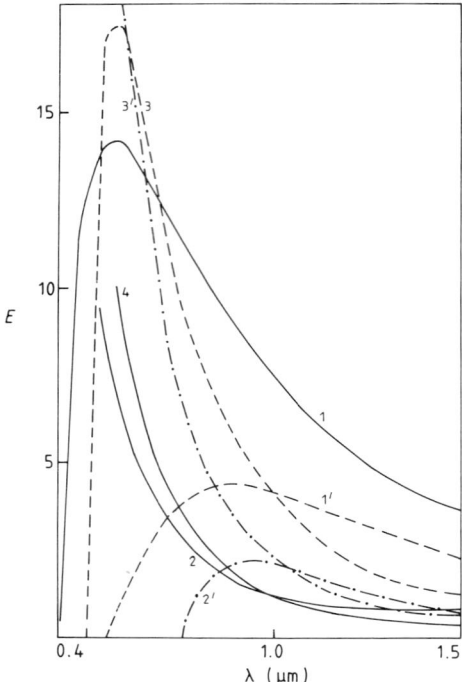

Figure 7.40 Dependence of the figure of merit E of different crystals on the wavelength of fundamental radiation [47]. 1, 1' β BaB$_2$O$_4$; 2, 2' KDP; 3, 3' urea; 4, LiIO$_3$. Full curves show type I generation; broken curves show type II generation. The figure of merit E is normalized by KDP (type I generation, SHG of 1.064 μm).

a rather low damage threshold (50 MW cm^{-2}). It should also be noted that $E = 1$ for KDP (I) at 1064 nm, and E(BBO) is 7–14 times that of KDP(I) for $\lambda_p = 1000$–600 nm, where BBO is an excellent crystal for SHG of dye lasers. The above described results are valid for the ideal situation in which the pump beam quality is extremely good, and the acceptance widths (angular and spectral) of the crystal are much larger than that of the pump beam. In many cases, type II crystals are actually more efficient than type I, due to the fact that most type II crystals have wider angular and spectral acceptance widths. We shall discuss this issue as follows.

7.5.5 *Effects of beam quality*

The plane wave, phase matched expression of the SHG efficiency is valid for the ideal system where the crystal acceptance widths are much wider than that of the laser beam. A better expression for the first-order approximation

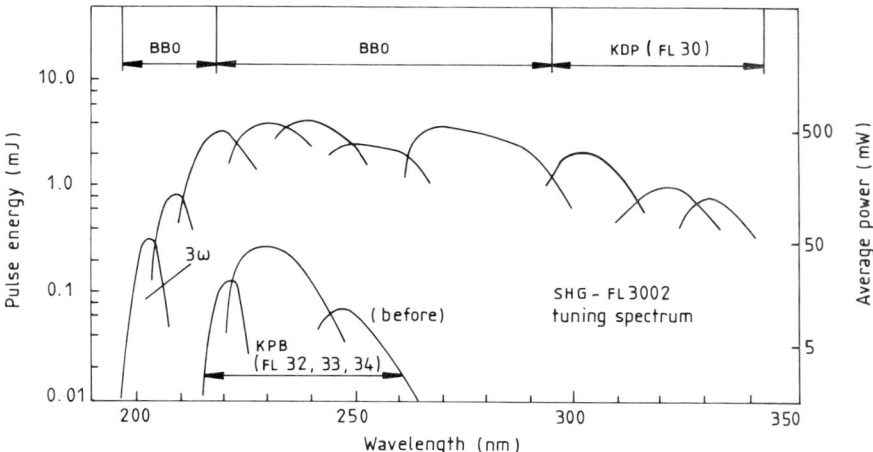

Figure 7.41 The performance of a BBO crystal for tunable dye lasers in UV (195–300 nm), where a BBO crystal length of 7 mm was used for typical conversion efficiencies of 10 to 15%, which is 5–10 times better than that of KPB crystals [47].

may be

$$\eta^{(1)} = \eta^{(0)} - (\Delta S/4)^2 [\tfrac{1}{2} \sinh(2E^{1/2}) - E^{1/2}] \tag{7.25}$$

where $\eta^{(0)}$ is given by equation (7.24a) for the phase matched case whose value is reduced by the second term given by the phase mismatching factor ΔS, which may be expanded in a Taylor series up to first order as

$$\Delta S = K_1(\Delta\theta L) + K_2(\Delta T L) + K_3(\Delta\lambda L) \tag{7.26}$$

where $(\Delta\theta L)$, $(\Delta T L)$ and $(\Delta\lambda L)$ are respectively defined as the angular, temperature and spectral acceptance widths of the crystal. Therefore the conversion efficiency is a sensitive function of the ratio between the divergence angle of the pump beam and $(\Delta\theta L)$ and similarly for the spectral bandwidth of the pump.

Examples of the effects of beam divergence of the pump on the conversion efficiency for the case of SHG of a Nd:YAG laser (1064 nm) are analysed as follows. Defining an effective figure of merit (F_{eff}) including the effects of beam divergence, one can calculate the F_{eff} for both type I and II crystals. Calculations for the values of $F_{eff} = E(\Delta\theta L)^2$ show that, for SHG of 1064 nm: for KDP, type II (eo–e) is 2.3 larger than type I (oo–e); for urea, type II (oe–o) is 3 times better than type I (ee–o); and for BBO, type II is 1.5 times better than type I. However, type II KDP SHG phase matching is cut off around 730 nm and type II BBO around 525 nm. It may therefore be concluded that for the application of SHG of dye lasers (400–750 nm), type I BBO is better than type II, where the calculations carried out by Lin show that for pumping wavelengths shorter than 750 nm F_{eff} (type II) < F_{eff} (type I).

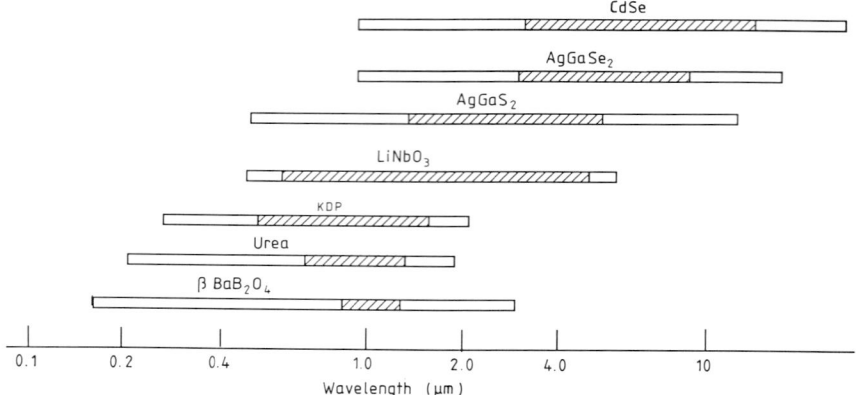

Figure 7.42   Parametric oscillators [47].

Nonlinear crystals play an essential role for the generation of coherent sources in the new spectral regimes where the direct output from existing lasers is impossible or difficult to achieve. An ideal crystal should include all the features of wide transparent and phase matching spectral range, wide angular, temperature and spectral acceptance widths, high damage threshold and nonlinear coefficient, high quality at large size, and chemical and mechanical stability.

For high-power laser applications, such as laser fusion, a high-damage crystal with large size (up to 20–40 cm) will be the goal for the development of crystals such as KDP, LAP and BBO.

For low-power laser applications, such as diode-pumped lasers and integrated optics devices, crystals with very high nonlinear coefficients (100–10 000 times that of KDP) and/or very large aperture length would be required.

In conclusion, figure 7.42 represents the regions of the spectrum overlapped by optical parametric generators on various crystals.

PART III

SOME PHYSICAL ASPECTS OF OXYGEN OCTAHEDRAL FERROELECTRICS

This part includes two chapters. Chapter 8 is devoted to the interaction of the laser radiation in the visible range with ferroelectric crystals, for the most part with niobates and tantalates of alkali and alkaline earth metals. Analysis of these problems is of great scientific and practical interest, because most of these materials are applied as modulators and transducers of radiation of quantum generators. We deal here with the interaction between radiation and crystals that induces inhomogeneities of the refractive index (the photorefractive effect). This effect results in an increase of the residual light transmission and of the interval of temperature matching of second harmonic generation.

All these questions are of interest from two points of view: from the point of view of an increase of crystal stability under laser radiation; and on the contrary, an increase of sensitivity for the purpose of using these crystals as a medium for recording holographic information. Several physical models of the mechanism of interaction between laser radiation and ferroelectric crystals are proposed. The operation of one or another mechanism depends on the structure of the energy levels in the forbidden band of a crystal as well as on the type of the effect produced (a spot or a holographic grating). The structure of the energy bands of a crystal can be affected by the introduction of impurities, reduction of cations or ion diffusion into the crystal lattice. Thus the crystal sensitivity to laser radiation can be controlled.

In chapter 9 we describe the attempts to look for regularities common to all the compounds considered throughout the book. The oxygen octahedra containing Nb, Ta and Ti ions which compose the crystal structure of the compounds of interest are such a common factor.

Di Domenico and Wemple have formulated a unified theory of the optical properties of oxygen octahedral ferroelectrics based on the structure of the electron energy bands. The theory emphasizes the fundamental importance of the BO_6 octahedron as the basic structure element.

From the crystal lattice dynamics, it is known that the static dielectric behaviour of these materials is mainly determined by BO_6 octahedra. In the case of the optical properties, they determine the lower boundary of the conduction band and the upper boundary of the valence band. These bands are similar for all the oxygen octahedral ferroelectrics because the d orbitals of B cations and the 2p orbitals of O anions joined in each octahedron make a considerable contribution to the energy bands. Other ions contribute to the higher-lying conduction bands. But their contribution is insignificant under the condition of low electron polarizability of ions. The model proposed by Di Domenico and Wemple is therefore inapplicable to ferroelectrics containing Pb and Bi ions.

To apply ferroelectric materials, it is of importance to know exactly the dependence of the material properties on the temperature, and their variation from material to material at a given temperature. The Devonshire–Ginzburg phenomenological theory of ferroelectricity predicts the temperature dependence of the ferroelectric properties of some materials, but it cannot predict the dependence of the properties on the composition of a material. Given the criterion of materials technology, one can choose a better material for a specific application and look for materials with high nonlinear optical characteristics.

Another aspect of this problem is a correct choice of nonlinear crystals for concrete applications. To this end one can use the figures of merit which include, along with the nonlinear optical coefficients, the ferroelectric constants of a material. The outlined problems are the subject of chapter 9.

8 Photorefractive Properties of Oxygen Octahedral Ferroelectrics

A wide practical use of optical quantum generators has revealed a whole number of new effects due to the interaction between intense radiation and matter. One such effect is a local reversible variation of birefringence in ferroelectric crystals under laser irradiation—the so-called optical distortion effect (photorefraction). The photorefractive effect on the one hand limits the application of nonlinear crystals in quantum electronics, and on the other hand gives new possibilities for employing these crystals as a medium for: reversible holographic information recording with a high diffraction efficiency; creating light-controlled space–time modulators; and for amplification of weak light fluxes. In this chapter we describe several model representations of the photorefractive effect and the application of oxygen octahedral ferroelectric crystals in holography.

8.1 Experimental observation of light-induced variation of refractive index

A light-induced variation of the refractive index in ferroelectric crystals of $LiNbO_3$ and $LiTaO_3$ was first observed by Ashkin et al [1] and Chen et al [2]. Under the action of a focused gas laser beam with a power of several mW, these crystals readily showed optical inhomogeneity. It occurs along the laser beam mainly due to variation of the refractive index of an extraordinary beam n_e.

The inhomogeneity is a region with a lowered n_e, which stretches from the beam axis and is surrounded by a small area with a periodic variation of this refractive index. Using the interference of ordinary and extraordinary rays in a lithium niobate crystal, it was found that $\Delta n > 2 \times 10^{-4}$.

Figure 8.1 presents a series of interference inhomogeneity patterns occurring

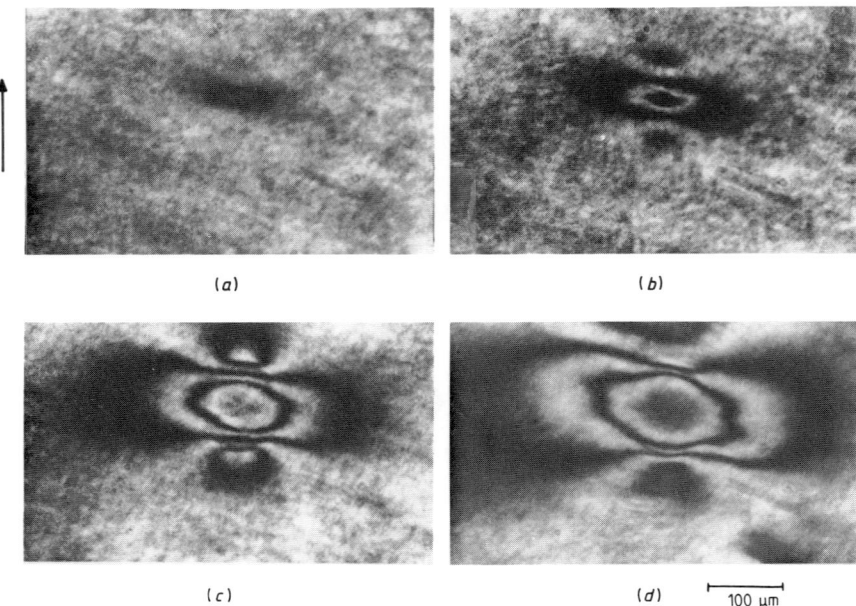

Figure 8.1 Refractive index inhomogeneities in a $LiNbO_3$:Fe crystal induced by laser radiation with $\lambda = 0.633\ \mu m$ for different exposures: 5 s (*a*), 20 s (*b*), 200 s(*c*), 1200 s (*d*). Radiation power 20 mW, beam diameter 70 μm. The arrow indicates the direction of spontaneous polarization in the crystal plate.

in a $LiNbO_3$: Fe crystal under the action of a constant-power He–Ne laser with $\lambda = 0.633\ \mu m$ for different exposures. The patterns were examined under a polarization microscope.

Inhomogeneities of the refractive index were also observed when light with $\lambda = 0.5147\ \mu m$ was transmitted along the *a* axis of a single-domain crystal of $BaTiO_3$ [2]. If such an irradiated crystal is examined under a polarization microscope, the region near the beam centre, which is single domain, appears to be surrounded by plenty of small antiparallel domains. The crystal may then again be made as a single-domain one.

Such a phenomenon, associated with refractive index inhomogeneity, was also observed in a paraelectric crystal of potassium tantalate niobate [3] under the action of an applied electric field of several kV cm^{-1}. The effect remained noticeable even after the field was switched off, although in the absence of an electric field the effect does not occur. Visually, the picture was different from that observed in $LiNbO_3$, apparently due to the different symmetry of the crystals.

Further development of these studies is due to Chen [3]. Experiments on optically induced variations of a refractive index were carried out at room temperature on polarized single crystals of LiNbO$_3$ and LiTaO$_3$. When a sample was exposed to a linearly polarized light with the polarization plane making an angle of 45° with the optic axis of the crystal, the beam passing through the sample (about 1 cm thick) split into three beams: a central unbiased ordinary beam and two extraordinary ones. Thus an induced Δn for an extraordinary beam is larger than for an ordinary one. Spatial variations of Δn along the c and b axes for different exposures are represented in figure 8.2. It should be noted that along the c axis Δn reverses sign on the boundary of the beam cross section, while along the b axis the sign remains negative. As the exposure time increases, the region of optical damage becomes asymmetric. It was established that induced variations Δn are largely limited to the diameter of the beam. Figure 8.3 shows the variation of $\Delta(n_e - n_o)$ exposure for different levels of laser radiation power. An Ar laser was used whose light with a power of 20 mW was focused at a spot of 10^{-2} cm in diameter. The magnitude of Δn reached saturation for 1 s at a radiation power of 20 mW. In a linear region, Δn is proportional to the product of the beam power and the exposure time.

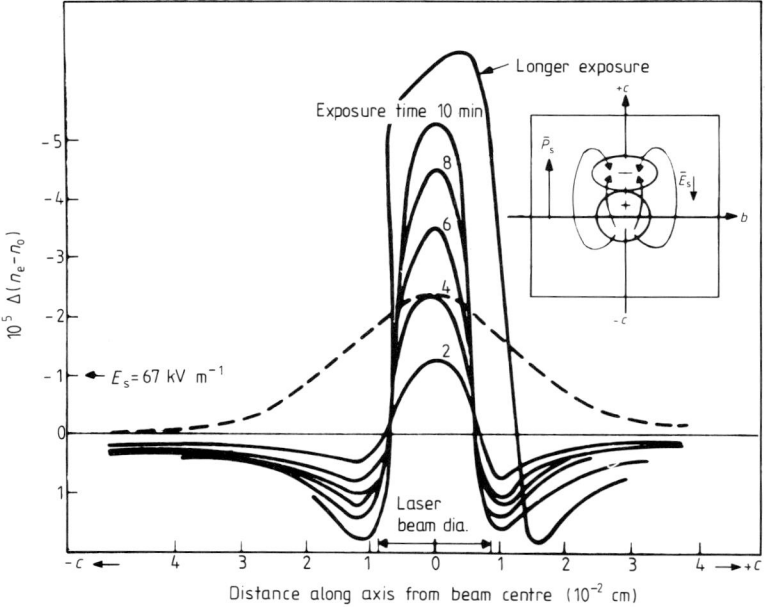

Figure 8.2 Photo-induced variations of the refractive index of LiNbO$_3$ along c axes (full curves) and b axis (broken curve) for different exposure times (given on the curves in minutes) to a laser beam propagating along the a axis [3]. d is the laser beam diameter. The distribution of the space charge field E_s is given in the top right-hand corner.

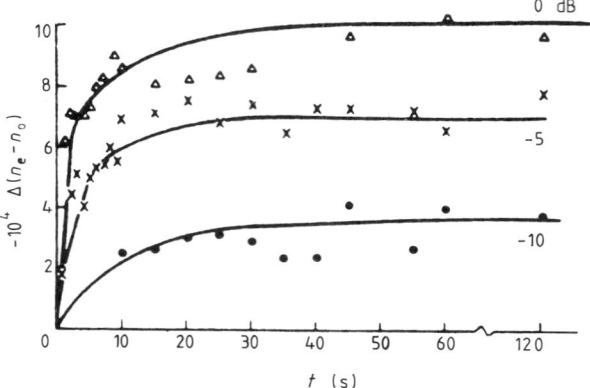

Figure 8.3   The variation $\Delta(n_e - n_o)$ in a $LiNbO_3$ sample depending on the time of exposure for different levels of Ar laser power (numbers on the curves are the attenuations) [3].

In the analysis of the mechanism of formation of an optical distortion, one essential question is its dependence on the wavelength of the irradiating light. The dependence of Δn on λ makes it possible to judge the position of the energy levels from which the electron goes to the transition band. This dependence for $LiNbO_3:Fe^{3+}$ crystals between the wavelengths of 0.4 and 0.8 μm was analysed [4] by measuring the difference between the maximum negative and maximum positive induced variations of the refractive index (figure 8.4).

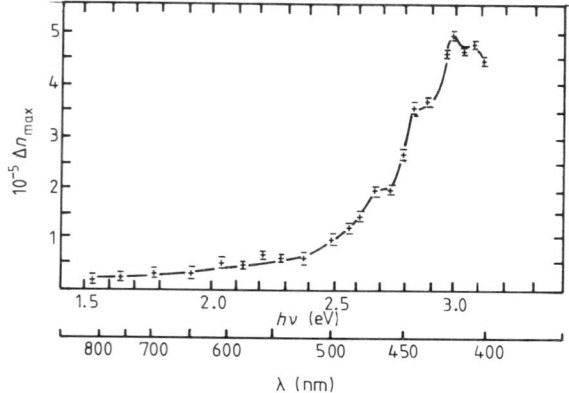

Figure 8.4   The maximum value of induced variation of the refractive index in $LiNbO_3:Fe$ as a function of the wavelength of irradiating light. The photon density is about 2×10^{17} photons s^{-1} cm^{-2} [4].

On the basis of this dependence, one may draw the following conclusions. (i) Optical inhomogeneities start appearing for a wavelength of 0.8 μm and increase significantly for a wavelength less than 0.5 μm. (ii) The peak of damage near 0.415 μm is approximately three times as large as that at $\lambda = 0.488$ μm, five times as large as that at $\lambda = 0.5145$ μm and ten times as large as that at $\lambda = 0.6328$ μm. The maxima of inhomogeneities near the wavelengths of 0.470, 0.400 and 0.415 μm are ascribed by the authors to excitations of electrons (or holes) from local levels in a forbidden band.

8.2 Physical models of the photorefractive effect

8.2.1 *The Chen model*
Chen [3] has proposed a model which explains qualitatively the observed experimental facts. The model is based on the assumption that in a crystal there exist electron traps, some of which are filled with electrons, the rest remaining unoccupied until irradiation. Besides, it is assumed that in a crystal there exists an internal electric field E_i directed opposite to the vector of the spontaneous polarization P_s.

Photo-excited electrons drift towards the positive pole of P_s, leaving behind the positively charged ionized traps. They drift until they are entrapped by vacant traps. Since photo-excitation is absent outside the limits of the irradiated region and since for deep traps the temperature appears to be insufficient for a secondary excitation, charge separation occurs and the electric field of the spatial charge E_s appears.

As a result of the electro-optic effect inherent in the indicated crystals, there occurs an induced birefringence $\Delta(n_e - n_o)$. It is induced mainly by the variation of n_e due to a considerable difference in electro-optic coefficients of lithium niobate [5, 6]. To obtain the quantities $\Delta n \approx 10^{-3}$ observed in experiment, the external electric field applied to the crystal should be equal to $E = 6.7 \times 10^4$ V cm^{-1}. The spatial distribution of the entrapped electrons and positive ionized centres, which leads to the observed distribution of the refractive indices, is shown in the upper portion of figure 8.2. The laser beam irradiated region is shown by a circle centred at the origin (the b and c axes) and the laser beam propagates normally to them.

The mathematical aspect of the problem [3, 7] is developed by analogy with the theory of photoconductivity of semiconductors [8]. It is assumed that in the space between the bands there exist the energy levels of the electron traps with initial density N_t (cm^{-3}), of which N'_t are filled and $N_t - N'_t = \Delta N$ are vacant. The velocity of electron motion from the irradiated region is n/t_n, where n is the free electron density (cm^{-3}) and t_n is the mean time of an electron passage through the cross section of a light beam of diameter d, i.e.

$$t_n = d/(2\mu^* E) \tag{8.1}$$

where μ^* is the effective mobility, which is equivalent to the microscopic mobility μ of free carriers if the electron entrapment is disregarded. In the case where the probability of repeated entrapment is high, μ^* will be less than μ in accordance with the ratio of the time of electron stay in the conduction band to the time of electron stay in the trap. In equation (8.1) E is the field in which electrons drift. It can be expressed as

$$E = E_i - E_s \tag{8.2}$$

where E_i is the internal crystal field and E_s is the laser-induced space charge field responsible for the refractive index variation.

The velocity of photo-excitation f for a weak absorption can be expressed as

$$f = \beta I \alpha n_t \equiv A I n_t \tag{8.3}$$

where β is the number of electrons released from the traps per unit energy of the absorbed radiation, I is the density of the incident radiation flux (W cm^{-2}), α is the absorption coefficient and n_t the concentration of filled traps. The kinetic equations describing the variations in the free electron concentration n (cm^{-3}) inside the irradiated region can therefore be written in the form

$$dn/dt = A I n_t - n(N_t - n_t)B - n/t_n \tag{8.4}$$

$$dn_t/dt = - A I n_t + n(N_t - n_t)B \tag{8.5}$$

where $B = vS$, v is the thermal velocity of free electrons and S is the cross section of the entrapment of free electrons. Equation (8.4) establishes that the rate of variation of free carriers in an exposed region is composed of the photo-excitation velocity, with a subtraction of the velocity of a repeated entrapment, and the velocity of electron motion away from the exposed region.

The number of electrons passing through 1 cm^2 during motion from an exposed region within the exposure time t_e will be given by

$$N(t_e) = \int_0^{t_e} (nd/t_n)dt \tag{8.6}$$

where t_e is the exposure time, the rest of the notation remaining as before. If the repeated entrapment is disregarded and the electron drift time t_n is assumed to be constant, then with the assumption that $AI \ll 1/t_n$ and $t_e > t_n$, equations (8.4)–(8.6) give

$$\Delta(n_e - n_o) \sim N(t_e) \approx N_t' d[1 - \exp(- A I t_e)] \approx N_t' d A I t_e. \tag{8.7}$$

Equation (8.7) shows that the induced variations of the refractive index below saturation are proportional to the energy density of the incident radiation, which is proved by experimental evidence.

The amount of ionized traps, and therefore the probability of repeated entrapment of carriers, increase with increasing exposure time, which slows down the rate of variation of Δn. When the velocity of a repeated entrapment exceeds the velocity of the electron drift from the exposure region, the relation

$$B(N_t - n_t) \gtrsim 1/t_n = C[I/(N_t - n_t)] \tag{8.8}$$

holds, where $C \equiv \mu E \beta \alpha/(dvS)$ [8]. A thermal repeated excitation of entrapped electrons, which has been disregarded in the previous equations, lowers the value of $\Delta(n_e - n_o)$.

8.2.2 *The polarization mechanism*

Johnston [9] developed a theory, distinct from Chen's theory, of a photo-induced variation of the refractive index in ABO_3-type pyroelectrics which include $LiNbO_3$ and $LiTaO_3$. He proceeded from the standpoint that a pyroelectric crystal is allowed by the symmetry variation of the macroscopic polarization (the density of electric dipole moments), caused by ionization or by occupation of certain traps, as well as by lattice polarization by the space charge field [10]. The resultant density of the polarization charge $\rho_p = -\nabla P_s$ [11] acts as an electric field charge under the action of which the photo-excited electrons diffuse from an exposed to an unexposed region.

An induced optical inhomogeneity is also observed when the trap density on the beam periphery is sufficient to entrap photo-excited electrons, and recombination on the donor levels is absent. In the stationary state there exists equilibrium between the field related to the space charge and the field of the polarized charge ρ_p. The macroscopic electric field E_s induced by the space charge ρ_s changes P_s by the value $\Delta P_s = (\varepsilon - 1)E_s$, where ε is the static dielectric permittivity. In the stationary state, the total macroscopic electric field of all the sources is equal to zero. This condition will be written in the form

$$\nabla(P_- - P_+) = (\varepsilon - 1)\rho_s \tag{8.9}$$

(P_- and P_+ are the contributions to P_s of occupied traps and ionized donors respectively).

Let us consolidate the Johnston model in application to $LiNbO_3$. To this end we assume that in the crystal there exist defects D^+, D^- and D^0, which are distinct in the charge state and form donor and acceptor levels. A normal state of the defect is neutral, D^0, and its ionization leads D^0 to the state D^+ with an increase of the unit cell polarization p_i of the crystal. The ionization energy is ~ 1.8 eV. The entrapment of an electron leads to the formation of the defect D^-, the energy of this process being 0.5–0.8 eV. The defect D^- causes a decrease of p_i. Such a scheme of defect formation in $LiNbO_3$ agrees with the data reported in [12].

The relation between the refractive index variations and the polarization is given by

$$\Delta n_e = -(n_e^3/2)[r_{33}\Delta P_3/(\varepsilon_{33} - 1)] \tag{8.10a}$$

$$\Delta n_o = -(n_o^3/2)[r_{13}\Delta P_3/(\varepsilon_{33} - 1)] \tag{8.10b}$$

where r_{ij} are macroscopic electro-optic coefficients. Using equations (8.10(a)) and (8.10(b)), as well as the experimental values for LiNbO$_3$, $P_s = 0.71$ C m^{-2} and $\Delta(n_e - n_o) = 3 \times 10^{-3}$ [3], we obtain $\Delta P_s/P_s = 3 \times 10^{-3}$. The sign of the change of birefringence shows that the defects must lie in the cation sublattice, and thus the defect ionization increases p_i.

8.2.3 The diffusion mechanism

Amodei and Staebler [13] expressed another view on the nature of the internal field under the action of which electrons drift from an exposed region of a crystal. Owing to the temperature dependence of ion conductivity and the high pyroelectric coefficient, an internal electric field is formed in these crystals upon cyclic heating and cooling [14]. When the temperature exceeds 100°C, the conductivity of lithium niobate becomes high enough, and the resultant pyroelectric field relaxes within several minutes. Upon crystal cooling, the pyroelectric effect changes sign, and due to a rapid decrease of electric conductivity a significant part of the induced charge is retained for many weeks if the crystal is not exposed to light. The field which remains after the crystal is cooled may be precisely the internal field postulated by Chen.

Two mechanisms leading to the electron motion from the exposed part of a crystal—the drift under the action of the internal field (as suggested by Chen) and the electron diffusion—were discussed by Amodei [15]. Both mechanisms may be responsible for charge separation and for the refractive index variation in a ferroelectric crystal.

Amodei and Staebler [13] considered the relation between these mechanisms on an example of recording an interference lattice which is formed in a crystal by superposition of two plane waves. Under the assumptions that the lifetime of free electrons is small and that the number of electrons excited into the conduction band is insignificant compared with the total number of entrapped electrons, one can write an expression similar to the light intensity distribution upon interference of two plane waves:

$$n(x) \approx a(1 + m \cos Kx) \tag{8.11}$$

where a is proportional to the maximum light intensity I_0, to the entrapment time and to the absorption coefficient; $K = 2\pi/L$ is the spatial frequency; and m is the modulation ratio equal to $2(I_1 I_2)^{1/2}/(I_1 + I_2)$, where I_1 and I_2 are the intensities of the object and reference beams.

Under the influence of an electric field and due to thermodiffusion, the ionized free carriers generate a spatially inhomogeneous current whose density is given by the equation

$$J(x) = eD_n \frac{dn}{dx} + en\mu_n E(x) \tag{8.12}$$

where e is the electron charge, D_n is the diffusion coefficient, μ_n is the electron mobility and $E(x)$ is the total electric field in the x direction. The current passing through the crystal leads to the formation of a space charge whose density is determined by the relation

$$\Delta\rho/\Delta t = \nabla J. \tag{8.13}$$

Integrating this expression, one can calculate the space charge.

As mentioned above, two mechanisms of charge motion were considered by Amodei and Staebler: the drift induced by the internal field and thermodiffusion. The critical value of the field E_{cr}, for which both mechanisms make equal contributions to the conductivity, is given by the expression [16]

$$E_{cr} = (kT/e)K \tag{8.14}$$

(k is Boltzmann's constant; T is the temperature). At room temperature, equation (8.14) can be written in the form $E_{cr} > 0.16/L$ V cm^{-1}, where L is the grating constant.

Thus, if an applied external or internal field in a crystal is substantially greater than E_{cr}, the diffusion may be neglected and one may calculate the space charge field including the drift component alone. If there are no fields, the diffusion is the only mechanism operating.

For the applied fields and exposures for which the appearing space charge field will be much greater than E_{cr}, the model discussed above predicts a linear increase of the space charge field, which is proportional to the light intensity distribution

$$E_s(x) \sim -E_m \cos Kx \tag{8.15}$$

where $E_m = e\mu_n E_0 m/\varepsilon$. It is of importance to note that the maximum of the field coincides with the minimum of the light intensity of the created interference pattern, and the magnitude of the field is independent of the grating constant L. This feature characterizes the drift-induced charge distribution in the electric field, the diffusion being neglected.

For the above picture of the space charge, the resultant field becomes nonlinear and is determined by the expression

$$E(x) = \frac{J_0}{e\mu_n n_0(1 + m \cos Kx)} \tag{8.16}$$

where J_0 is the current density in equilibrium. This expression is derived for a 100% modulation on the assumptions of small diffusion length and high trap concentration. A small diffusion length leads to an effect equivalent to a decrease of the modulation ratio m, and the finite trap concentration restricts the maximum field for a given spatial frequency.

In the case where external or internal fields are absent, the diffusion mechanism determines the formation of space charge fields. It may be shown [16] that the equilibrium value of the space charge field under these conditions is given by

$$E_\rho(x) = \left(\frac{kT}{e}\right)\left(\frac{Km \sin Kx}{1 + m \cos Kx}\right). \tag{8.17}$$

If we expand this expression into a Fourier series and restrict ourselves to the first term, which gives the main contribution to a holographic synthesis, we come to

$$E_1(x) = \left[\frac{kTKm}{(1 - m^2)^{1/2}}\right] \sin Kx. \tag{8.18}$$

The field strength maxima obtained from this expression are shifted in phase by 90° relative to the maxima formed by the charge carrier drift. This phase shift leads to different behaviours of the object and reference beams interacting during recording.

8.2.4 Optical excitation

The variation of the dipole moments of impurity ions in ferroelectrics under the action of a light pulse was investigated [17, 18].

Cu- and Cr-doped $LiNbO_3$ and $LiTaO_3$ crystals exhibit, besides the pyroelectric effect, a variation of the macroscopic polarization due to the difference in the dipole moments of the ground and excited states of the impurity ions. The acentric polar lattice polarizes impurity ions. Excitation by a light pulse leads to variation of the dipole moment of the ligands and the surrounding ions of the lattice due to redistribution of the impurity ion charge. This phenomenon is the basis of the mechanism of optical mixing and rectification for frequencies higher than those of the pyroelectric effect.

In a polar crystal lattice, the variations of the dipole moments Δp of excited Cr^{3+} ions sum up to give a macroscopic variation of polarization ΔP. In a non-polar lattice this effect is not observed, since Δp does not have a unique direction, and they are mutually compensated in the entire macroscopic volume. Taking into account all of these points, the crystal polarization may be expressed by the equation

$$P(t) = (\gamma W \Delta p / hv_1 d) \exp(-t/\tau) + P_1(t) \tag{8.19}$$

where γ is that part of the incident power which transforms into heat, τ is the electron–photon relaxation time, W is the total incident power and $P_1(t)$ is the polarization variation due to the pyroelectric effect.

From the ratio of experimental values P_{max}/P_∞, where P_{max} is the maximum polarization variation near $t = 0$ and P_∞ is the asymptote of the $P(t)$ curve for a large t, one can obtain not only the excited state relaxation time τ but also the value of the dipole moments of chromium ions Δp in the excited state, because

$$\frac{P_{\max}}{P_\infty} = \left[\frac{\gamma}{c_p}\left(\frac{v_1}{v_2} - 1 \right) + \frac{\Delta p}{hv_2}\right]\left[\frac{\lambda}{c_p}\left(\frac{v_1}{v_2} - \varphi Q \right) \right]^{-1} \tag{8.20}$$

where v_1 and v_2 are the frequencies corresponding to the optical transitions $^4T_1$, $^4T_2$ of Cr^{3+} ions in the crystal lattice; c_p is the heat capacity of the crystal; γ is the pyroelectric coefficient; Q is the quantum efficiency of radiation from the excited state of chromium ion $^4T_2$; and φ is the correction to the Stokes shift for Cr ions in the lattice of lithium niobate.

From figure 8.5 it is seen that Δp depends weakly on temperature and amounts to 0.6×10^{-18} CGS units. For $LiTaO_3$: Cr, $\Delta p = 0.8 \times 10^{-18}$ CGS units at low temperatures; at room temperature it increases up to 1.0×10^{-18} CGS units. To confirm the fact that the observed effects are due to excitation of the dipole moments of chromium ions, the experiments were reproduced with unalloyed $LiNbO_3$ and $LiTaO_3$ crystals. In this case, the measured $P(t)$ were small and corresponded to a small pyroelectric effect due to a residual radiation absorption by the crystals.

The dipole moment variation Δp is the result of several independent processes: charge redistribution on excited chromium orbitals, Cr^{3+} ion displacement due to the formation of bonds between the excited state of Cr^{3+} with odd modes of vibration of the lattice and the dipole induced by lattice polarization. The contributions of these effects to the resultant change of the

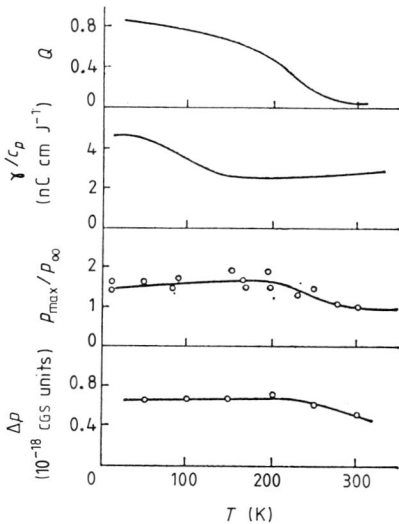

Figure 8.5 Temperature dependences of quantum efficiency Q, of the ratios χ/c_p, $P_{\max}/P_\infty$ and of dipole moment variation Δp of Cr ions under excitation in $LiNbO_3$:Cr [18].

dipole moment Δp were not separated by the authors of [18]. It should be noted that the time of the electric response of a pyroelectric crystal to the momentum of incident radiation is determined only by the probability of electronic transition.

8.2.5   Photovoltaic effect

Fe- or Cu-doped $LiNbO_3$ crystals irradiated by an argon laser with $\lambda = 0.5145$ μm showed a photovoltaic effect [19–22]. The authors believe that the effect describes qualitatively a photo-induced refractive index variation, which can thus be explained without involving any other models.

For all the crystals, the photoelectric current density J along the polar axis c, during irradiation of the entire crystal volume, was proportional to the absorbed power density

$$J = K_1 \alpha I \tag{8.21}$$

where I is the intensity, α is the linear absorption coefficient, and K_1 is a constant which depends on the nature of the absorbing centre and on the wavelength and is independent of the crystal geometry, electrode configuration and impurity concentration. No photoelectric current was observed in the direction perpendicular to the polar axis. For Fe-doped $LiNbO_3$ crystals K_1 was found to be equal to 3×10^{-9} A cm W^{-1}; for Cu-doped crystals $K_1 = 5.5 \times 10^{-10}$ A cm W^{-1}. The constant K_1 increased as the wavelength decreased. For a crystal thickness of 0.033 cm, the photo-electromotive force of saturation was about 1600 V for a radiation intensity of about 150 mW cm^{-2}, which corresponds to a saturating field of the order of 0.5×10^5 V cm^{-1}. As soon as the intensity of incident light exceeded 250 mW cm^{-2}, there occurred an electrical breakdown of the crystal. The photoconductivity of a crystal containing 0.2 wt. % of Fe_2O_3 and possessing an absorption coefficient $\alpha = 38$ cm^{-1} for an external field $E \approx 10^4$ V cm^{-1} was determined by the expression

$$\sigma = 1.5 \times 10^{-14} + 1.4 \times 10^{-12} I \; (\Omega \, \text{cm})^{-1}. \tag{8.22}$$

The optically induced variation of the refractive index in these crystals was measured by holographic and compensational methods.

For a short pulse, the variation of the refractive index is described by the formula

$$\Delta n = K_2 \alpha \int_0^t I \, dt = K_2 \alpha \Delta W \tag{8.23}$$

where ΔW is the energy density of light incident on the crystal. The photorefractive sensitivity K_2 (the variation of the refractive index per photo-excited electron) turned out to be independent of the impurity concentration and the incident light intensity, but was determined by the nature of the absorption centres. So, for Fe- and Cu-doped crystals, K_2 was estimated to be 1.4×10^{-5} and 2×10^{-6} cm^3 J^{-1} respectively.

After the light was turned off, the refractive index variation and the photo-signal fell to zero at the rate of the dielectric relaxation of the crystal. This is especially visible in strongly Fe-doped crystals. For these crystals, the relaxation time was 200 s, which corresponded to a resistivity of 7×10^{13} Ω cm.

For LiNbO$_3$, the refractive index variation due to the electro-optic effect is $\Delta n_e = 1.7 \times 10^{-8} E$, where the electric field is defined as

$$E = (1/\varepsilon\varepsilon_0) \int_0^t J \, dt. \tag{8.24}$$

Here

$$J = K_1 \alpha I + \sigma E + eD \, dn/dz. \tag{8.25}$$

For a small E and a small dn/dz, the relation

$$E = (K_1 \alpha/\varepsilon\varepsilon_0) \Delta W \tag{8.26}$$

holds. Thus the photorefractive sensitivity K_2 from equations (8.23)–(8.25) is given by

$$K_2 = \Delta n/(\alpha\Delta W) = 1.7 \times 10^{-8} K_1/\varepsilon\varepsilon_0. \tag{8.27}$$

For the photocurrents of Fe- and Cu-doped crystals, it was respectively found that

$$[\Delta n/(\alpha\Delta W)]_{Fe} = 2.3 \times 10^{-5} \text{ cm}^3 \text{ J}^{-1}$$

$$[\Delta n/(\alpha\Delta W)]_{Cu} = 4 \times 10^{-6} \text{ cm}^3 \text{ J}^{-1}$$

which is in close agreement with the above mentioned results of measurements.

The saturation field can be obtained from the expression (8.25) for $J = 0$:

$$E_{sat} = K_1 \alpha I/\sigma \tag{8.28}$$

where K_1 does not depend on E.

From experimental data on the photoconductivity of Fe-doped crystals of LiNbO$_3$, the saturation field E_{sat} was estimated to be 0.9×10^5 V cm^{-1}, which corresponds to a refractive index variation $\Delta n = 1.5 \times 10^{-3}$ and is in satisfactory agreement with experiment.

The volume photovoltaic effect, which is one of the basic properties of ferroelectric crystals, is explained by the local asymmetry of the position of the impurity ion in a ferroelectric matrix. Because the distance $Nb^{5+} - Fe^{2+}$ is different in the $+c$ and $-c$ directions, the probabilities of charge transfer in these directions (p_+ and p_-) under photoexcitation into the conduction band are different. The conduction band of a LiNbO$_3$ crystal is formed by d orbitals of Nb ions [23]. Photoexcitation induces a current running in a preferable direction of P_s. The probability of thermal recombination in the $\pm c$ direction will also be distinct. The currents $J_\pm$ in the $\pm c$ directions are given by the expression

$$\frac{\mathrm{d}J_\pm}{\mathrm{d}t} + \frac{1}{\tau_\pm}J_\pm = \frac{e\alpha I}{h\nu}v_\pm p_\pm \qquad (8.29)$$

where $\tau_\pm$ is the lifetime determined by the probability of thermal recombination; $v_\pm$ are the velocities of the carriers. The resultant current in the stationary state is expressed by

$$J = J_+ - J_- = (\alpha I e/h\nu)(l_+ p_+ - l_- p_-) \qquad (8.30)$$

where $l_\pm = v_\pm \tau_\pm$. The excitation of impurities in the localized state, caused by the difference of the dipole moments in the ground and excited states, leads to the appearance of the internal transition current. Thus the recombination process does not yield equal and oppositely directed currents. In the case of thermal excitation we have $J = 0$, since the probabilities of excitation and recombination in the $\pm c$ directions are equal.

The model of the anomalous photovoltaic effect proposed by Glass was developed further in [24–28]. Belincher *et al* [24] considered stationary photo-induced currents in ferroelectrics, taking account not only of impurity excitation but also of the recombination carriers, their scattering on impurities and photons. The authors restrict themselves to a simple model of the energetic structure of the crystal, when in the forbidden band there exists an impurity level.

Transition currents associated with the laser-induced polarization reversal were observed in barium strontium niobate crystals [29–39] and in alloyed crystals of lithium niobate and barium sodium niobate [40, 41]. The photorefraction mechanisms in the above mentioned crystals were established. It should be noted that the photorefraction mechanism is determined, not only by the nature of the crystal and impurities, but also by the manner of irradiation by light.

8.2.6 *Other mechanisms*
Levanyuk and Osipov [42] considered the appearance of optical distortion as a result of excitation of donor–acceptor pairs by light. Broad-band dielectrics contain approximately the same number of donors and acceptors. The illumination of such a crystal by short-wavelength light induces ionization of a negatively charged acceptor. The electron that passes into the conduction band is captured by a positively charged donor, i.e. in spite of two charged defects two neutral ones appear. The electron polarizability of the defects and their contribution to the refractive index of the medium change essentially. Thus, from this point of view, a light-induced refractive index variation Δn may occur in any high-resistance crystals.

The magnitude of the induced variation Δn will be proportional to the concentration of the excited donors (acceptors) N_D^0. After the light is turned off, the variation Δn associated with light excitation of impurities will remain unchanged for a time which is determined by the probability of electron transition w_{DA} from donor to acceptor. The transition probability $w_{DA} = w_0 \exp(-2R/a)$, where $R = (N_D^0)^{1/3}$ is the mean distance between excited impurities, and a is the dimension of neutral defects. Levanyuk and Osipov derived the system of equations for determining N_D^0; its solution leads to the consequences which were reported earlier by Chen [3].

Examining the mechanism of the appearance of optical distortion connected with the electro-optic effect, Levanyuk and Osipov observed [43] that a light-induced impurity centre affects not only the dipole moment but also the polarizability. As a result, under the action of light the spontaneous polarization changes by the value

$$\Delta P_{imp} \approx \alpha f P_s N_D^0 \qquad (8.31)$$

where $f P_s$ is the effective microscopic field which acts on the impurity.

In the above arguments, the polarization reversal of the crystal lattice upon impurity excitation is disregarded, which is justified for 'rigid' ferroelectrics. Levanyuk and Osipov associate the spontaneous polarization reversal with the change in the polarizability, and not with the charge state of the impurity centre as suggested by Johnston [9].

8.3 Photo-induced distortion of the crystal structure in lithium niobate

Lithium niobate (LN) is one of the most thoroughly investigated ferroelectrics possessing the photorefractive effect. But the action of light induces not only a refractive index variation in this crystal but also a photoelectric current. Light scattering induced by either one beam or two intersecting beams (in hologram recording) has been revealed in a LN:Fe crystal [44]. Jumps of birefringence of the light-induced focused laser radiation have been reported [45]. Electric noise occurring in irradiated LN:Fe crystals has been reported in [46, 47]. The authors arrive at the conclusion that the noise is due to the partial reversal of spontaneous polarization inside an exposed region. These data suggest that the action of light induces variation, not only of the space distribution of an electric charge, but also of the crystal structure of LN. To clarify this point, attempts were made [48–52] to investigate the action of light upon a LN:Fe crystal using the X-ray diffraction method.

To study the variation in the crystal structure, two types of experiments were performed: (i) changing the half-width of the diffraction maximum oscillation curve; (ii) obtaining X-ray topograms. In both cases, a two-crystal spectrometer and Cu Kα radiation were used. Plates of pure and Fe-doped (0.05 and 0.15 wt. % of Fe) lithium niobate with thicknesses of 120 and 190 μm

($\mu t = 5.1$ and 8.1 respectively) were used as samples. Plates were cut normally to the $[1\bar{1}0]$ direction, which made it possible to obtain topograms of crystals in highly intense reflexes (006) and (110). The plates were lapped (ground), polished and etched in a mixture of HF and HNO_3.

The prepared crystal plates were preliminarily examined for the sake of choosing the highest-quality samples on which all further measurements were made. The degree of perfection was estimated through an analysis of the topograms obtained in a reflected beam. The figures of merit for the crystals were the minimum number of dislocations, the absence of striations or small electric domains.

The action of light upon the crystals was investigated as follows. First of all, before the action of light, the oscillation curves were recorded and X-ray topograms taken. Then the crystals were irradiated. To this end, a He–Ne laser with $\lambda = 0.63$ μm, a He–Cd laser with $\lambda = 0.44$ μm and a mercury lamp were used. During the illumination, the intensity of transmitted X-ray beams and the oscillation curves were recorded. After the illumination, the X-ray topograms on reflexes (006) and (110) were taken.

In both types of experiments, under a local illumination of a crystal, the anomalous transmission of X-rays in an exposed region decreased, which indicates distortion of the crystal structure. The distortions observed on the topograms were caused by the deformation of (00h) planes perpendicular to the ferroelectric axis. The deformation of the (hh0) planes parallel to the ferroelectric axis was substantially weaker.

Figure 8.6(a) presents a topogram of LN:Fe crystals in a reflected beam before the illumination. The light lines on the topogram are due to crystal

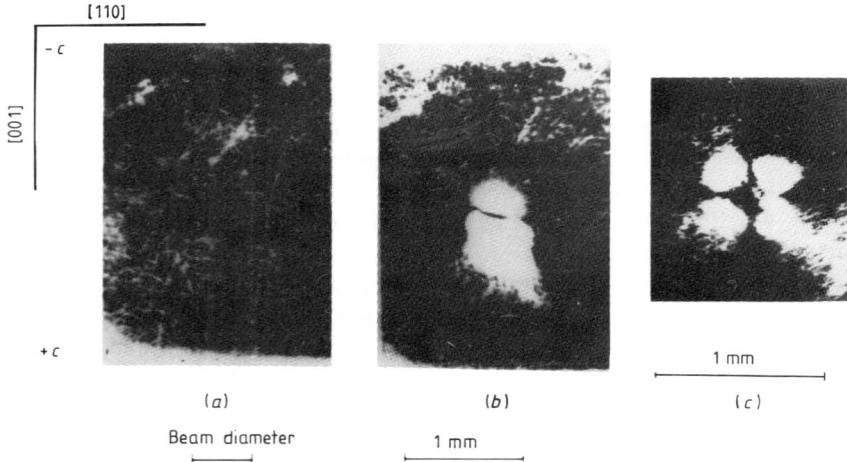

Figure 8.6 X-ray topograms of a $LiNbO_3$:Fe crystal before (a) and after (b, c) exposure [48]. Reflexes (006) (a, b) and (110) (c). Enlargement $\times 20$.

lattice distortion on dislocations whose density, estimated from the topogram, is approximately $10^3 \, \text{cm}^{-2}$, which is evidence of the high quality of the samples.

The topogram shown in figure 8.6(b) was obtained after the crystal was exposed to a focused He–Ne laser beam. The region of distortion in figure 8.6(b) is two semicircles in which the effect of anomalous X-ray transmission is absent. Between the semicircles, a narrow strip is seen where the Borman effect is observed. On the topogram taken on the reflex (110), the disappearance of the Borman effect is observed in the four sectors separated by strips, parallel and perpendicular to the ferroelectric axis of the crystal·and intersecting in the centre of the exposed region (figure 8.6(c)). On topograms, optical damage is localized mainly in the region of a light spot. In the case of reflex (006), one also observes a 'tail' of distortions in the $+c$ direction from a light spot.

When He–Ne and He–Cd lasers and a mercury lamp with a light filter were used as radiation sources, no qualitative distinctions in the observed picture of distortions were found. The effect was weaker in unalloyed crystals.

The traces of the effect of light upon the crystal structure relax very slowly with time. After a specimen had been kept in the dark for a month, it still showed those traces. The distortions were completely erased when the whole specimen was exposed to the light of a mercury lamp for several minutes or heated up to 170°C. The original topographical picture was fully reconstructed.

When a crystal is irradiated, ferroelectric macro-domains appear in the vicinity of the $+c$ face, whereas near the $-c$ face they are not observed. These macro-domains are unstable and soon vanish. So no domains were seen on a topogram taken seven days after exposure. The conclusion that the diffraction contrasts observed near the $+c$ face are ferroelectric domains was drawn on the basis of the results reported in [53, 54], where the defect structure of LN was investigated by the X-ray topology method.

To establish the kinetics of the formation of light-induced defects in LN crystals, a series of experiments was performed where, along with the measurements of the maximum Laue reflection intensity I, the half-width of the oscillation curve β and the diffraction angle 2θ were measured (figure 8.7). At an early stage of crystal irradiation, the angle 2θ slowly decreases, which corresponds to an increase of the lattice constant, the oscillation curve width exhibiting no noticeable variation at this stage. After the angle 2θ reaches its minimum value corresponding to the variation $\Delta c = 4 \times 10^{-3} \, \text{Å}$, for which the relative lattice deformation is 0.03%, it starts increasing rapidly up to a value close to the initial one. The starting moment of the increase of the angle 2θ coincides, within experimental error, with the beginning of the fall of the reflection intensity and the beginning of the broadening of the oscillation curve. The latter testifies to a significant violation of the crystal structure. An increase of the radiation intensity up to 16 W cm^{-2}

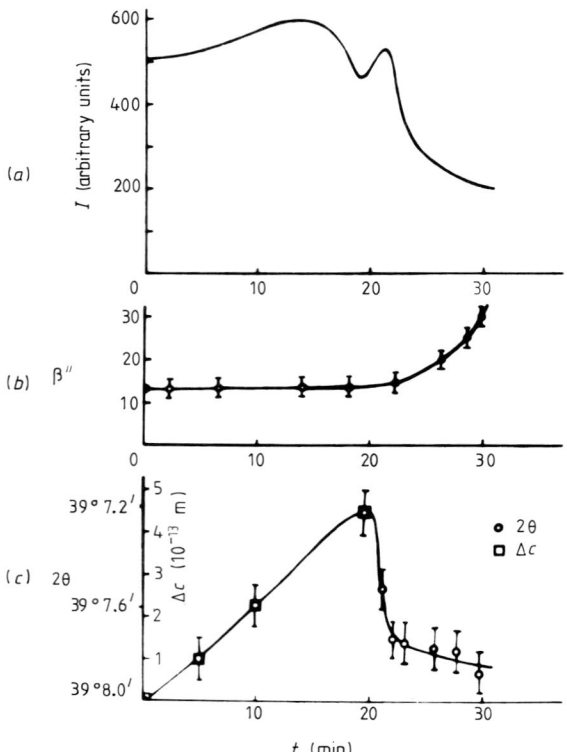

Figure 8.7   Diffraction maximum intensity I (a), width of oscillation curve β of this maximum (b) and lattice constant variations Δc (c) as functions of exposure to laser radiation for a power density of 0.2 W cm^{-2}, $\lambda = 0.44$ μm [48].

causes the oscillation curve broadening from 14″ to 140″, which corresponds to the presence of defects. Within the first five hours after the light is turned off, the oscillation curve width relaxes stronger than within the subsequent 150 hours.

To find out the character of the processes that induce broadening of diffraction maxima, the half-widths of X-ray reflexes (006) and (00.12) of irradiated and non-irradiated samples were measured on a diffractometer. The processing of these data showed that the true broadening of the X-ray lines lies, to an accuracy of 25%, on the curve $\beta \sim \tan \theta$, which testifies to a main contribution from the micro-deformation of the lattice to the broadening. Separation of disorientation and micro-deformation effects has shown that the disorientation $\Delta \theta = 30''$ and the relative micro-deformation $\Delta d/d = 6 \times 10^{-4}$.

To determine the mechanism of photo-induced distortion of the crystal structure, a polarization optical picture of the irradiated region was examined (figure 8.8). The optic axis of the crystal makes an angle of 45° with the polarizer and analyser axes. The photorefraction was induced by radiation of a He–Cd laser with an intensity of $100 \, \text{mW cm}^{-2}$, which was incident on the crystal through a 0.3 mm diaphragm placed near the crystal. The boundary of the light spot coincides with the central circle, but the region of the birefringence variation exceeds significantly the dimensions of the light spot, especially in the direction perpendicular to the c axis. Figure 8.9(a) represents the birefringence variation $\Delta(n_e - n_o)$ along the [001] direction, which was obtained from the data of figure 8.8 using the formula [55]

$$I = I_0 \sin^2 [2\pi\Delta(n_e - n_o)l/\lambda] \qquad (8.32)$$

where l is the crystal thickness along the beam, I and I_0 are the intensities of the transmitted and incident light, and the rest of the notation is the same as before.

The variation of birefringence in a crystal can be described in terms of the variation of the internal electric field or of the variation of spontaneous polarization P_s. But the light-induced P_s variation may proceed only in an exposed region, since the electric field of a space charge spreads over a substantially larger region of the crystal. The experimentally observed variation of birefringence exceeds greatly the exposed region, especially in the direction perpendicular to a ferroelectric axis c, and therefore one may assume the birefringence variation to be associated with the electric field of the space charge. The distribution of the electric field E_z along the c axis in a crystal coincides with the distribution of Δn plotted in figure 8.9. The values

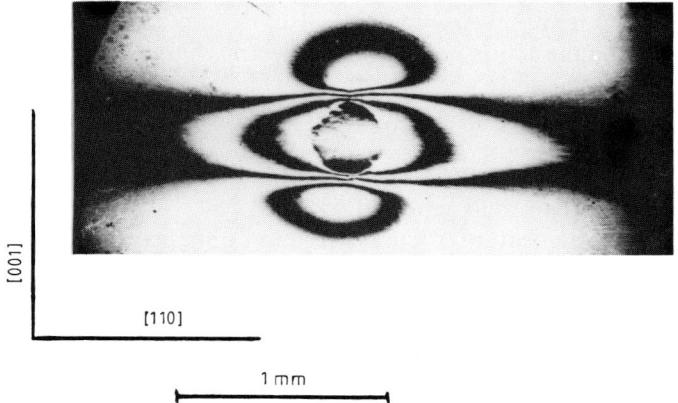

[001]

[110]

1 mm

Figure 8.8 Microphotograph of a $LiNbO_3$:Fe crystal (0.15 wt.%) in crossed polarizers after exposure to laser radiation [50]. Enlargement × 50.

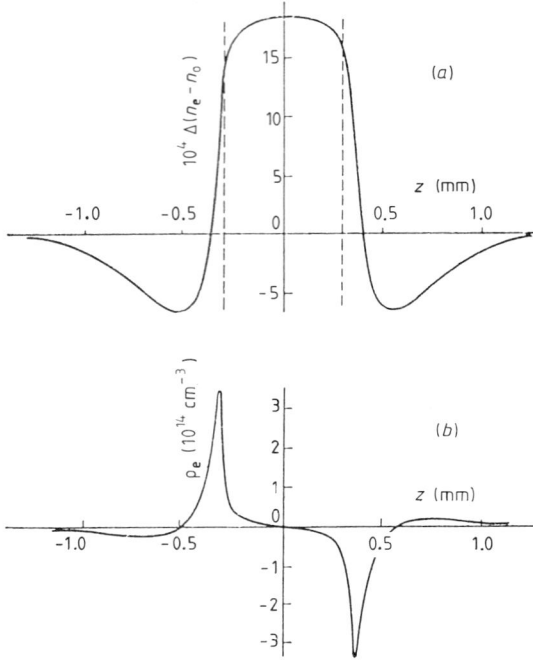

Figure 8.9 Variation of birefringence $\Delta(n_e - n_o)$ (*a*) and of the space charge density ρ_e (*b*) along the ferroelectric axis [001] [49]. Broken lines show the boundary of the light spot on the crystal.

of the electro-optic coefficients and the refractive indices for the computation of the distribution are taken from [56]. Taking into account the symmetry of the crystal and the action, we obtain the electric charge density distribution in a plate

$$\varepsilon_{33}(\partial E_z/\partial z) = \rho(z) \tag{8.33}$$

which is plotted in figure 8.9(*b*). The maximum value of $\rho = 3.5 \times 10^{14}$ cm^{-3}, which is close to the results reported by Chen [3].

The characteristic feature of the distribution obtained is the symmetry of the positive and negative charge densities relative to the centre of the exposed region, in contrast to Chen's scheme constructed on the assumption that photo-excited electrons escape from the exposed region, leaving it charged positively. From figure 8.9(*b*) it is seen that the charge density in the centre of the light spot is equal to zero.

Photo-induced distortion of the crystal structure may be explained as follows. The increase of the crystal lattice constant *c* observed at the initial stage of exposure is connected with macro-deformation of the central part of the exposed region. One of the reasons for the observed macro-deformation may be the inverse piezoelectric effect, which is due to the electric field induced by the photoelectric effect [19]. The maximum field of the photoelectric effect in LN:Fe has the value $\sim 10^5$ V cm^{-1}, which as a result of the piezo-effect creates the deformation $\Delta c/c \approx 10^{-4}$ close to the experimentally observed value.

Comparing figures 8.6(b) and 8.9(b), one can conclude that the vanishing of the Borman effect is caused by the crystal lattice deformation induced by micro-deformations around centres which change their charge in an exposed region of the crystal.

As shown by Amodei and Staebler [13], one may conclude that, when LN:Fe crystals are exposed to light, Fe ions change charge. The electrons are photo-excited with Fe^{2+}, are transported due to the photoelectric effect in the P_s direction and are captured by Fe^{3+} ions. Charged point defects in a piezoelectric medium cause deformations that decrease by the law $1/r^2$ [57], which is in turn responsible for the experimentally observed broadening of the diffraction maxima. Neutral defects, according to [57], do not lead to a broadening of the diffraction maxima.

Displacement of atoms in an electric field of the centres with a changed charge proceeds mainly along the c axis of the crystal, which is connected with the relation of the tensor components of the piezoelectric coefficients: $d_{33} \gg d_{13}$ [56].

Figure 8.6(b) shows that in the [001] direction from the light spot there exists a substantial structure distortion which is, perhaps, due to the formation in the crystal of micro-domains of opposite signs [58, 59]. But the authors were wrong to assume division into domains inside an exposed region. Indeed, according to [60], inside an exposed region the field of space charges E_s is directed along P_s, whereas beyond the limits of this region E_s and P_s have different directions. Therefore in these regions there may appear needle-like micro-domains which are formed in the fields below threshold ones [61]. In the cross section they are equal to $\lesssim 1$ μm and cannot therefore be resolved on a topogram. On the topogram, one can see only the region of stresses caused by the existence of 180° domain walls.

8.4 Recording of volume phase holograms

A scheme, one version of which is presented in figure 8.10, is used in experiments on the study of optical memory in materials [62]. The recording is performed using a beam of an argon ($\lambda = 0.488$ μm) or helium–cadmium ($\lambda = 0.440$ μm) laser. At these wavelengths, crystals are very sensitive to optical distortion. Reading is carried out by a He–Ne laser beam with $\lambda = 0.633$ μm. For this wavelength, the sensitivity of the crystal is small, and therefore reading does not lead to hologram erasure. Sometimes, recording and reading are realized at one and the same wavelength.

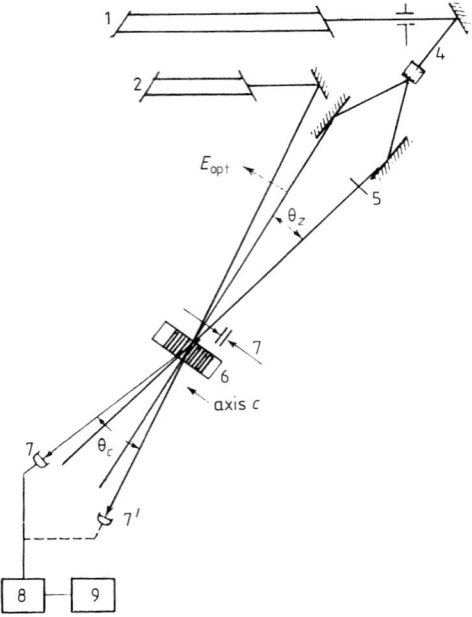

Figure 8.10 Set-up for hologram recording in crystals at a wavelength $\lambda = 0.488$ μm with a simultaneous measurement of diffraction efficiency at $\lambda = 0.633$ μm [62]; (1) the recording laser, $\lambda = 0.488$ μm; (2) the reading laser, $\lambda = 0.633$ μm; (3) diaphragm; (4) Wollaston prism; (5) $\lambda/4$ plate, crystal with a recorded holographic grating with a period L; (7, 7′) radiation detectors; (8) amplifier; (9) recorder.

As a result of superposition of two plane waves, there appears an interference pattern in the form of light and dark strips. In light strips, the crystal birefringence is varied, i.e. a sinusoidal diffraction grating with the constant L is recorded:

$$L = \frac{\lambda_{rec}}{2\,\sin(\theta_{rec}/2)} \qquad (8.34)$$

where λ_{rec} is the wavelength of the recording light and θ_{rec} is the angle between the recording beams.

In the reading of volume holograms ($h \gg \lambda$ and $h \gg L$, where h is the hologram thickness) at a wavelength λ_{read}, the angle between the incident and the deviated beams is determined by the Bregg conditions

$$2\,\sin(\theta_{read}/2) = \lambda_{read}/L \qquad (8.35)$$

where $\theta_{read}/2$ is the angle between the incident beam and the normal to the plate. The angles θ_{rec} and θ_{read} are related by the equation

$$\frac{\sin(\theta_{rec}/2)}{\sin(\theta_{read}/2)} = \frac{\lambda_{rec}}{\lambda_{read}}. \tag{8.36}$$

For small angles, one can write

$$\theta_{rec}/\theta_{read} = \lambda_{rec}/\lambda_{read}. \tag{8.37}$$

From the expressions (8.34) and (8.35), it is seen that in the general case the Wulf–Bregg condition for $\lambda_{rec} \neq \lambda_{read}$ cannot be satisfied for all the gratings arising in the recording of a complex object whose hologram is a set of gratings with different constants.

The diffraction efficiency of a sinusoidal grating in the reading by means of an extraordinary beam with the wavelength λ is given by Kogelnik's formula [63]

$$\eta = \sin^2\left[\frac{\pi d \Delta n_e}{\lambda \cos(\theta/2)}\right] \tag{8.38}$$

where Δn_e is the amplitude of modulation of the refractive index of an extraordinary beam and d is the crystal thickness. Experimentally, the diffraction efficiency is determined as the ratio of the intensity of the diffracting reading beam to the intensity of the beam transmitted by the crystal when a hologram in the crystal is not recorded. Thus, in an experimental determination of the diffraction efficiency, one can disregard the reflection and scattering losses, which makes it possible to compare different media for hologram recording.

To estimate the sensitivity of the material to optical recording, one introduces the parameter S—the sensitivity defined as the inverse density of the incident light energy necessary for attaining a given diffraction efficiency. The use of the parameter S in the majority of papers is justified by the simplicity of its determination, but it is inconvenient for comparing the results of different experiments. From equation (8.38) it is seen that η depends on the sample thickness and on the geometry of recording (the angle θ_{rec}). Another way of estimating the susceptibility of a material consists of the determination of the sensitivity as the ratio of the refractive index variation to the density of the light energy dn/dW, which can be measured using the scheme presented in figure 8.11.

The possibility of using a material for holographic recording is also determined by the resolution measured by the number of lines per cm of crystal length. For $LiNbO_3$, this value is estimated to be 1500 lines/mm [62]. Figure 8.12 presents a graph characterizing the sensitivity to recording and the resolution of the various materials studied by Tufte and Chen [64] on the basis of the results reported in [62]. As is seen from the graph, the ferroelectric crystals, the recording in which is based on a light-induced

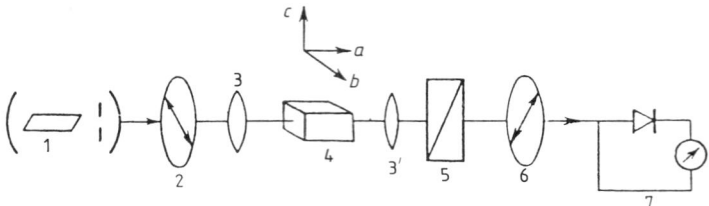

Figure 8.11 Experimental set-up for observing optically induced variation of the refractive index [3]: (1) He–Ne laser ($\lambda = 0.633\ \mu$m); (2) polarizer; (3, 3′) lenses; (4) LiTaO$_3$ crystal; (5) Senarmon recorder; (6) analyser; (7) radiation detector.

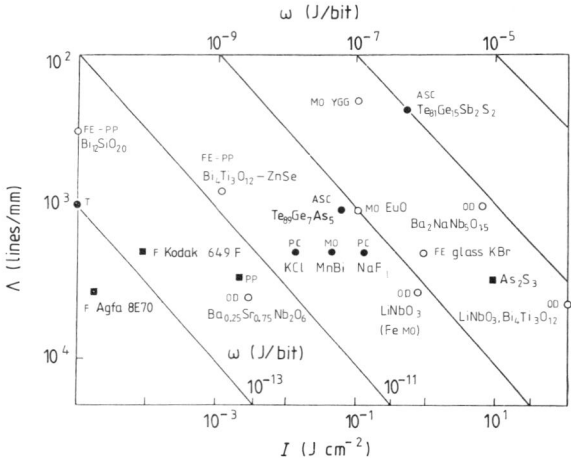

Figure 8.12 Recording sensitivity (energetic I and informational ω) versus resolving power Λ for various media used for holographic information storage [64]. Notation: OD, material subject to optical distortion; PP, photo-polymer; MO, magneto-optic film; FE, ferroelectric; F, film; ASC, amorphous semiconductor; PC, photochromic material; T, thermoplastic material; open circles, reversible recording; full circles, reversible recording with a limited number of cycles; squares, irreversible recording.

refractive index variation, have a resolution comparable with that of photo-plates and a photosensitivity lower than that of photo-plates. They are also fit for repeated recording and possess a number of other useful properties which we shall discuss below.

8.4.1 *Sensitivity to recording*

The sensitivity to an optically induced refractive index variation can be expressed in terms of the parameters of the material as follows:

$$\frac{dn_e}{dJ} = el\left(\frac{n_e^3 \; r_{33}}{2 \; \varepsilon_0 \varepsilon_{33}}\right)\frac{\alpha\beta}{\hbar\omega} \tag{8.39}$$

where el is the change in the electric dipole moment, whose mechanism may differ for different materials and different experimental conditions. For the most widespread case of the drift of photo-excited electrons in an internal or external electric field, the drift length of the electron $l = \mu\tau E$. The expression in brackets describes the electro-optic effect. Although the electro-optic coefficients may change noticeably, the value in brackets does not change considerably for different crystals. The ratio $\alpha\beta/\hbar\omega$ gives the number of excited centres or the number of photoelectrons in the conduction band.

Pure ferroelectric crystals do not usually possess a high sensitivity to light, which makes them applicable for modulation of laser radiation. But alloying by transition metals, violation of stoichiometry, annealing in reducing atmosphere, irradiation by γ-rays and application of an electric field make them much more photosensitive.

The effect of alloying upon sensitivity of $LiNbO_3$ and $LiTaO_3$ crystals is described in detail in [44, 65]. The use of transition metals as alloying additions is due to their ability to give d electrons reversibly to the conduction band under the action of light. Measurements of the diffraction efficiency in recording at wavelengths $\lambda = 0.488$ μm and $\lambda = 0.633$ μm showed a considerable increase in the sensitivity of crystals alloyed with Mn, Fe and Cu, whereas Cr, Co and Ni did not affect essentially the diffraction efficiency. An increase in the diffraction efficiency in the course of hologram recording in a pure $LiNbO_3$ crystal and in crystals with various additions is demonstrated in figure 8.13 [44]. Alloying with Fe increases the sensitivity by a factor of 500. There exists a relation between the absorption spectra and the sensitivity to recording in crystals with different additions. In Cr, Co and Ni, optical absorption is due to intra-band transitions, whereas for Mn, Fe and Cu the absorption is due to ionization.

The sensitivity of Fe-doped crystals to recording is determined by the concentration of Fe^{2+} ions, which in the $LiNbO_3$ lattice have a wide absorption band with a maximum near 0.480 μm. It is assumed that under photoexcitation an Fe^{2+} ion gives one photoelectron to the conduction band; then in the course of diffusion this electron is captured by an Fe^{3+} ion in an unexposed region. In a completely oxidized crystal, the valent state of iron is Fe^{3+}. But as a rule a real crystal contains Fe^{2+} ions whose content increases with increasing concentration of the doping addition, or upon annealing in a reducing atmosphere of H_2, and also in a vacuum or in an inert gas. An increase of the Fe^{2+} concentration leads to an increase of absorption at the recording wavelength (figure 8.14).

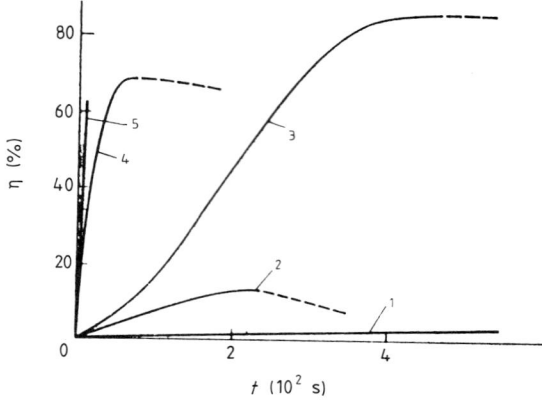

Figure 8.13   Diffraction efficiencies under recording (full lines) and erasure (broken lines) for LiNbO$_3$ crystals: (1) undoped; (2) Ni-doped; (3) Mn-doped; (4) Cu-doped; (5) Fe-doped [44]. Radiation power density 0.4 W cm^{-2}, $\lambda = 0.488$ μm.

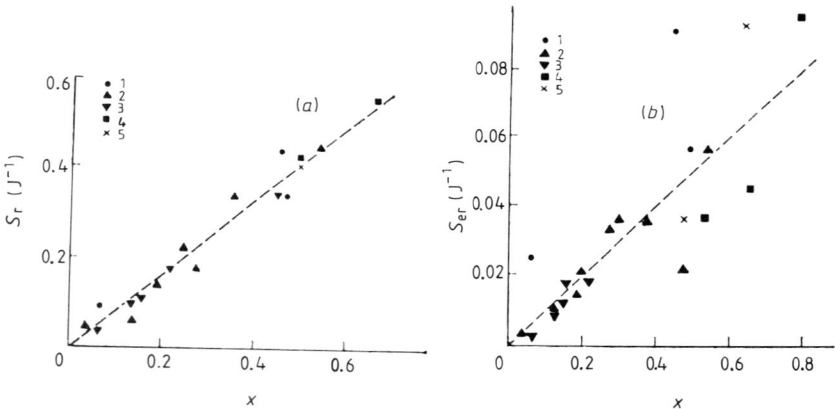

Figure 8.14   Dependence of sensitivity (*a*) to recording and (*b*) to erasure at a wavelength $\lambda = 0.488$ μm on the fraction x of the absorbed energy in Fe-doped LiNbO$_3$ crystals with different degrees of reduction [44]. Fe concentrations (wt.%) are: (1) 0.015; (2) 0.03; (3) 0.03; (4) 0.05; (5) 0.1. (1), (2), (4) and (5) are congruent compositions; (3) is a stoichiometric composition.

Micheron and Besmuth [66] studied the influence of impurities upon BaTiO$_3$ sensitivity to recording. Barium titanate crystals were doped with Fe, Ni and Nb in the process of growing. When these crystals are doped with 0.24 wt.% of Fe, the diffraction efficiency increases by almost two orders of magnitude as compared with undoped crystals under the same recording conditions. The impurity of 0.65 wt.% of Fe produces a smaller effect, whereas doping with 0.1 wt.% of Nb_2O_5 lowers the sensitivity compared with the undoped crystal. The difference in the influence of impurities upon the sensitivity to recording is explained [66] by the fact that in the $BaTiO_3$ grating Fe^{2+} and Ni^{2+} ions work as donors, whereas Nb^{5+} ions create an acceptor level and thus lower the photosensitivity. The properties of doped crystals of $KNbO_3$, a ferroelectric analogue of $BaTiO_3$, have been described in [67]. In this crystal, the highest effect is achieved by simultaneous doping with donor and acceptor impurities ($Fe + Mo$).

Along with doping, the sensitivity of a crystal to light can be controlled by varying its stoichiometry. Angert *et al* [68] obtained the dependence of a maximum variation of the refractive index Δn on the ratio Li/Nb in $LiNbO_3$ and Li/Ta in $LiTaO_3$. As this ratio increases, the sensitivity of the crystal to light also increases (figure 8.15), which is evidently due to the absence of oxygen vacancies.

A substantial increase in the sensitivity is achieved through crystal exposure to γ-rays [13]. An application of a DC electric field also increases the sensitivity to recording. This is especially so in materials with a small coercive field, such as PLZT ceramics, $Ba_{0.25}Sr_{0.75}Nb_2O_6$ and $BaTiO_3$. In crystals with a large coercive field, e.g. $LiNbO_3$, strong electric fields up to $E = 20 \text{ kV cm}^{-1}$ are required to obtain a considerable effect.

8.4.2 *Hologram fixation*

The storage time of recorded holograms depends on the nature of the material, storage and read-out conditions and can be controlled by different types of exposure. In undoped crystals of $BaTiO_3$, the storage time at room temperature is less than 1 s [66]; in pure $LiNbO_3$ crystals the storage time

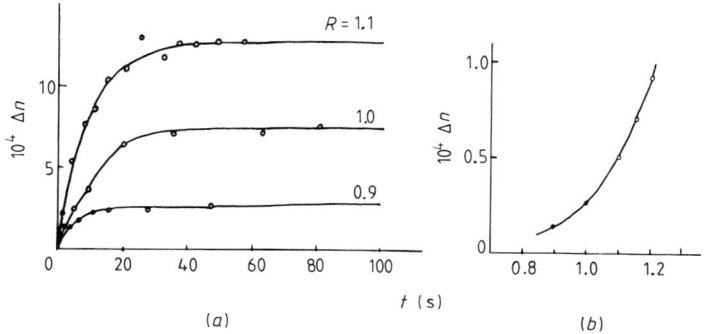

Figure 8.15 Time dependences of light-induced variation of birefringence Δn in $LiNbO_3$ crystals for a different value of $R = $ Li/Nb (*a*) and the dependence of Δn on $R = $ Li/Ta for a $LiTaO_3$ crystal (*b*) [68].

exceeds a month. It has been shown [13] that the storage time is equal to half the dielectric relaxation time $\frac{1}{2}\varepsilon\rho$, where ε is the dielectric permittivity and ρ is the specific resistance of the material.

In the course of reading, i.e. exposure, a hologram is erased. In undoped crystals of $LiNbO_3$, the curves of increase of diffraction efficiency and degradation during read-out are symmetric in time. But an application of a DC electric field $E = 20\ kV\ cm^{-1}$ breaks this symmetry. The energy of light necessary for recording decreases by an order of magnitude, whereas the erasure curve remains almost unchanged. A substantial symmetry of 'recording–erasure' curves is observed in doped crystals. The erasure energy in $LiNbO_3$:Fe crystals is approximately an order of magnitude higher than the recording energy, which enables several holograms to be recorded in the same region of the crystal. In doped crystals of $BaTiO_3$ and $KNbO_3$, the thermal erasure constant increases up to 300 s at room temperature [67].

A method of thermal fixation of holograms recorded in $LiNbO_3$ and in doped Ba_2NaNbO_{15} was proposed [14, 69, 70]. According to this method, a crystal with a recorded hologram is heated in the dark up to $120°C$ for 5 min and is then cooled to room temperature.

Under exposure to a homogeneous reading light, the diffraction efficiency increases from zero to a stationary value close to the one obtained in the recording. A fixed hologram is not erased during a long-time reading. So, during a 16 h exposure to a light of $0.5\ W\ cm^{-2}$, no signs of destruction were observed. The thermal fixation effect is explained by the fact that, upon heating up to a temperature sufficient for activating ion defect motion (~ 1.1 eV in $LiNbO_3$) but insufficient for exciting the electrons captured in traps, the charged ion defects are redistributed in the crystal and compensate for the field of the entrapped electrons. As a result, after cooling at the beginning of the reading process the diffraction efficiency is equal to zero. Upon reading by a homogeneous light, electrons are released from traps, and the grating due to localized electrons is 'cleaned out'. Thus, the information is recorded already in the grating of charged ion defects which is insensitive to light. The obtained 'ion' recording can be erased by heating up to $300°C$. The described fixation method is applicable to crystals in which ion conduction is observed at room temperature or slightly above.

In crystals with a relatively low Curie temperature and a small coercive field ($BaTiO_3$, $KNbO_3$, $Ba_{0.25}Sr_{0.75}Nb_2O_6$), fixation is realized by an electric field [62]. So, if a crystal of $BaTiO_3$ with 0.1 wt.% of Fe, in which a hologram is recorded, has an electric field pulse equal to $E = 0.9E_c$ (E_c is the coercive field) applied along the ferroelectric axis for 0.1 s, then under a reading light with an intensity of $50\ W\ cm^{-2}$ the storage time reaches five hours. Holograms may be fixed electrically [62, 66, 71–74]. The following model of the phenomenon is proposed. In a single-domain crystal, when an external field is zero, a hologram is recorded in the form of an electron density grating of the entrapped electrons. This affects the internal field $E(x)$. When an external field, close to a coercive one, is applied, it is composed of the field $E(x)$. As a result, in some regions of the crystal the field exceeds the coercive one, while in other regions it is insufficient for polarization reversal. This induces the appearance of a domain structure whose space distribution coincides with the density distribution of the entrapped photoelectrons.

This domain structure is light-resistant. In reading, electrons are photo-excited and escape from the hologram region, leaving non-compensated ion charges. The hologram is erased by an instantaneous application of an electric field that significantly exceeds the coercive field, after which the crystal becomes single domain. Figure 8.16 demonstrates electric fixation and erasure in $Ba_{0.25}Sr_{0.75}Nb_2O_6$ crystals. The hologram was recorded at the wavelength of an Ar laser with a radiation power density of $1\ W\ cm^{-2}$. The holographic diffraction efficiency was measured by a weak He–Ne laser beam. A crystal with 8 mm between the electrodes was primarily polarized by an applied voltage of 1700 V. After optical erasure during reading, the crystal was again polarized. For fixing the hologram, a voltage of -1000 V was applied for 0.5 s. After fixation, the diffraction efficiency value $\eta = 51\%$, reached during reading, remained unchanged under a continuous 10 h reading. For electric erasure, a pulse of 1700 V was applied for 1 s.

8.4.3 *Diffraction efficiency control*

In electro-optic PLZT ceramics (9% of La) with a Curie point below room temperature, the induced variation of the refractive index is based on the quadratic electro-optic effect. The linear effect is absent, since room-temperature ceramics are in the centrosymmetric phase. Figure 8.17 shows a ceramic specimen in which a grating is recorded. An electric field E_a is applied perpendicular to the lines of the grating. The photo-excited electrons

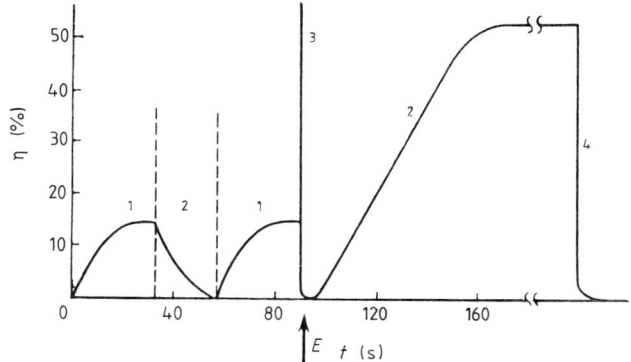

Figure 8.16 Diffraction efficiency variation in $Ba_{0.25}Sr_{0.75}Nb_2O_6$ crystals: (1) during recording; (2) during optical erasure (under reading); (3) during electric fixation; (4) during electric erasure [72].

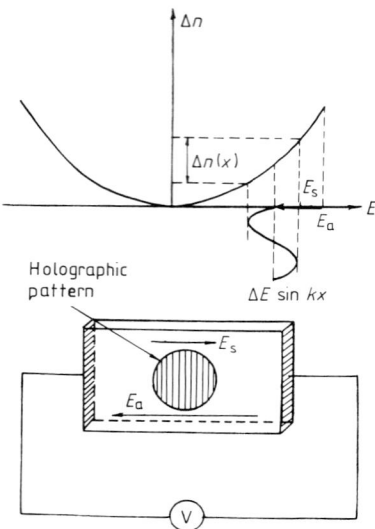

Figure 8.17   Photo-induced variation of the refractive index in a material with a quadratic electro-optic effect [71].

drift from an exposed to an unexposed region and create in the exposed region a mean space charge field E_s directed opposite to the applied field E_a. Along the direction of the applied field E_a, the sinusoidal light density distribution creates space modulation of $E(x)$:

$$E(x) = \Delta E \sin(Kx) \tag{8.40}$$

where K is the spatial frequency. In the hologram region, the spatial refractive index variation is given by the expression

$$\Delta n(x) = (A/2)[E_a - E_s + \Delta E \sin(Kx)]^2 = (A/2)(E_a - E_s)^2$$
$$+ (A/2)\Delta E^2 \sin^2(Kx) + (E_a - E_s)\Delta E \sin(Kx). \tag{8.41}$$

The first and second terms in equation (8.41) do not contribute to the diffraction efficiency. The term that does contribute is proportional to $(E_a - E_s)\Delta E$; in this term E_s and E_a increase with increasing light density. Equation (8.41) implies that $\Delta n(x)$, and therefore the diffraction efficiency, may be changed by an electric field. Figure 8.18 demonstrates the results of the field effect upon the diffraction efficiency of a hologram recorded on a plate made of electro-optic PLZT ceramics (9% of La). It is seen that a biasing field may transfer a hologram into a latent state for which $\eta = 0$ and may also increase the diffraction efficiency.

In $Ba_{0.25}Sr_{0.75}Nb_2O_6$ crystals, which have a smeared phase transition near 50°C, the room-temperature refractive index variation is substantially

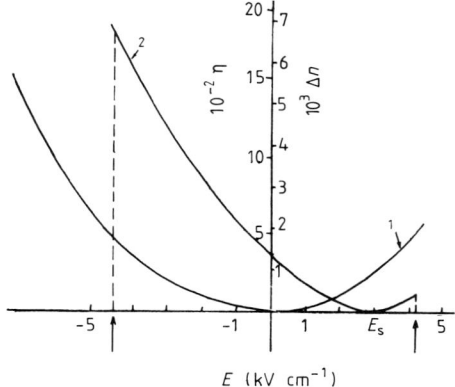

Figure 8.18 Field dependences of (1) induced variation of birefringence and (2) diffraction efficiency under recording η on a PLZT (9% of La) ceramic plate (350 μm thick) [71].

due, as well as to the linear electro-optic effect, also to the quadratic effect:

$$\Delta n = - (rn^3 E/2) - (Rn^3 E^2/2). \tag{8.42}$$

The dependence Δn has a minimum for non-zero E. The curve $\Delta n(E)$ plotted in figure 8.19 has a minimum at E_B' lying between E_0 and E_1, where $(E_0 - E_1)$ is the limiting modulation of the internal field in a crystal induced by charge separation due to photoexcitation. If such an external field E for which the mean field in the region with a recorded hologram is equal to E_B' is applied to a crystal, we have zero diffraction efficiency when we read out the recorded hologram. By reversing polarity of the applied field, i.e. by passing to

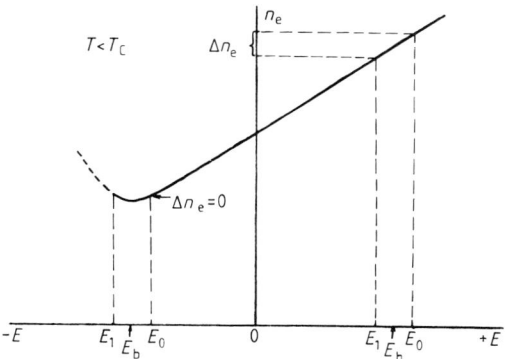

Figure 8.19 Schematic representation of the dependence of the refractive index n_e on the external field in a crystal at $T < T_C$ [62].

the point E_B during reading, we obtain a high diffraction efficiency. Figure 8.20 presents experimental results on diffraction efficiency control in $Ba_{0.25}Sr_{0.75}Nb_2O_6$ crystals.

The possibility described above of transferring a recorded hologram into a 'latent' state and bringing it back to a 'working' state with a high diffraction efficiency is used for creating layer memory. A model of the device for recording layer memory is shown in figure 8.21. The model uses $Ba_{0.25}Sr_{0.75}Nb_2O_6$ crystalline plates which are cut out parallel to the c axis and are so arranged that for one and the same position of a reference and object beams the recording can be made in any of the plates with voltage applied to its electrodes. By changing the voltage the recorded hologram is

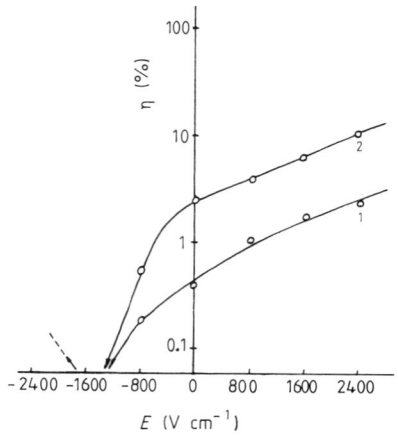

Figure 8.20 Diffraction efficiency of holograms as a function of the applied field: (1) under a non-fixed and (2) under a fixed recording [62].

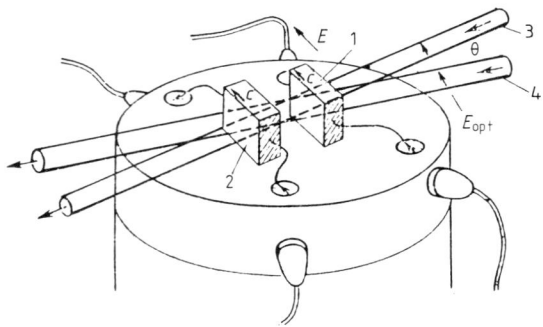

Figure 8.21 Model of layered memory [62]: (1, 2) crystal plates; (3, 4) reference and objective beams respectively.

transferred to a 'latent' state, and then the recording is made on the next plate. In reading by means of a field applied to the plate, the hologram is brought out into a 'working' state. The hologram can be read out either from each plate separately or simultaneously from all the plates. No mechanical displacements are used.

8.4.4 *Erasure of holograms*

At the present time, several hologram erasure methods are known, among which optical and thermal erasures are most widespread. Thermal erasure is made by a short-time heating of a $LiNbO_3$ crystal up to 300°C. At this temperature, the trapped electrons are thermally excited, and the charge relief disappears. It has been shown that optical erasure in $LiNbO_3$ is most effective in a crystal exposed to light with a wavelength of about 0.4 μm, which corresponds to the maximum sensitivity to recording [4]. We have already described the electric erasure procedure. Bordogna *et al* [76] described a random erasure of holograms recorded in one volume of a $LiNbO_3$:Fe (0.15%) crystal which proved to be highly resistant to optical erasure. The idea of the method is that the angle between the reference and object beams is the same for the recording and erasure, but to the reference beam one introduces a phase shift of π, which induces a compensating space modulation δn such that $\delta n' + \delta n = 0$, where $\delta n'$ is the modulation of the refractive index of the erased hologram. The other holograms recorded in the same volume remain unchanged, but the scattering losses of light somewhat increase.

8.5 Photoelectric and photorefractive properties of barium sodium niobate crystals

Undoped barium sodium niobate (BNN) crystals are light-resistant [94], which is why they are used for the generation of a second optical harmonic of a solid state laser with $\lambda = 1.06$ μm. But impurities and the crystal lattice defects occurring in the course of growing, and thermoelectric treatment of a crystal under polarization, lower significantly the laser radiation resistance of BNN. We have investigated the photoelectric properties of BNN crystals and the influence of various factors upon the photoelectric and photorefractive properties. We have studied the influence of thermoelectric treatment upon crystal polarization, the influence of Fe and Mo additions and also the replacement of part of the Na by K [36, 39–41, 50, 95].

8.5.1 *Polydomain crystals of barium sodium niobate*

The optical and photoelectric properties of polydomain BNN crystals not subjected to the action of an electric field and temperature are fairly reproducible.

The crystals used in the experiments had been grown mainly from the composition $Ba_{2.09}Na_{0.72}Nb_{5.02}O_{15}$ (such crystals have been discussed in detail in chapter 5). Their optical quality is higher than that of the crystals grown from a material of stoichiometric composition. They exhibit no pronounced striation. To measure the photoconduction, we used specimens aligned with the crystallographic axes. The electrode material did not affect greatly the results of measurements. The light absorption was investigated in crystals measuring from 50 μm to 5 mm in thickness.

The absorption in the visible and UV regions was measured by a spectrophotometer. The absorption coefficients were determined by the method of two thicknesses, with allowance made for reflection at the crystal–air boundary. For $\alpha l \ll 1$, allowance was made for multiple reflection. The transmission coefficient was calculated by the formula [96]

$$T = (1 - R)^2 e^{-\alpha l}(1 - 2Re^{\alpha l})^{-1} \qquad (8.43)$$

where T is the light transmission coefficient, R is the reflection coefficient

$$R = \left(\frac{n-1}{n+1}\right)^2 \qquad (8.44)$$

and the refractive indices are given in chapter 5. Figure 8.22 presents the absorption coefficients of a BNN crystal with the coordinates $\ln \alpha - \lambda$. The region of large α corresponds to band–band absorption; the flattened portion of the curve corresponds to light absorption by defects.

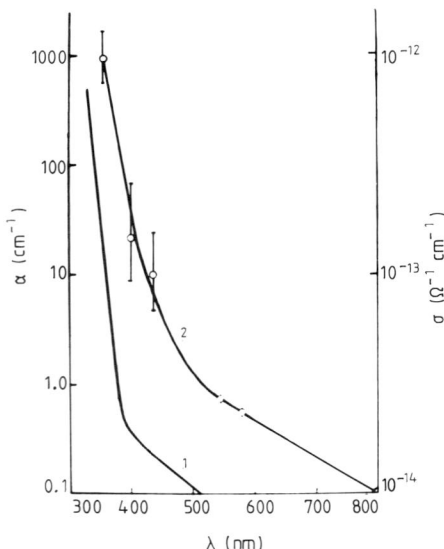

Figure 8.22 Spectra of absorption (1) and photoconductivity (2) of barium sodium niobate crystals.

Near the self-absorption edge, α obeys the law $\alpha \sim (hv - E_g)^r$ [96]. For indirect transitions, where photons participate, $r = 2$ and E_g^T corresponds to the thermal width of the forbidden band. For direct transitions, $r = \frac{1}{2}$ and E_g^O determines the optical width of the forbidden band. Below $\alpha \sim 300\,\text{cm}^{-1}$ the law $\alpha \sim (hv - E_g)^2$ holds, i.e. indirect transitions take place. For high values of α direct transitions dominate. For BNN crystals, $E_g^O = 3.65 \pm 0.02$ eV and $E_g^T = 3.47 \pm 0.05$ eV.

The device for measuring photoconduction is schematically shown in figure 8.23. A mercury lamp served as a radiation source. The radiation intensity was controlled by the choice of light filters, the operation regime and the width of the gaps in the monochromator. The samples were uniformly exposed; the electrodes were in a darkened region.

The spectra of photoconduction σ_{ph} of BNN crystals were plotted in figure 8.22, from which one can see that the photoconduction increases monotonically with decreasing wavelength. The values of σ_{ph} measured perpendicular to and along the ferroelectric axis were equal, within experimental error. The dependence of σ_{ph} on the radiation intensity in the indicated range of power remained linear except for the wavelength $\lambda = 365$ nm, where it was superlinear. The dark specific conductance of BNN crystals was $\sigma_d = 1.0 \times 10^{-14}\,\Omega^{-1}\,\text{cm}^{-1}$.

The photoconduction of crystals is determined by the expression [96]

$$\sigma_{ph} = \frac{1}{l} e\mu\beta\tau I_0 (1 - e^{-\alpha l}) \tag{8.45}$$

where e is the electron charge, μ is the mobility, β is the quantum yield, τ is the lifetime of carriers and I_0 is the incident light intensity. For small absorption coefficients $\alpha l \ll 1$ the equation $\sigma_{ph} = e\mu\beta\tau\alpha I_0$ holds. The similarity of the absorption and photoconduction spectra presented in figure 8.22

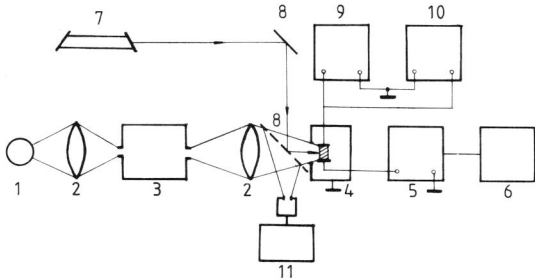

Figure 8.23 Set-up for measuring photoconductivity. (1) mercury lamp; (2) lens; (3) monochromator; (4) crystal in a screen; (5) electrometric amplifier; (6) two-coordinate recorder; (7) He–Ne laser with $\lambda = 628$ nm or He–Cd with $\lambda = 422$ nm; (8) rotating mirror; (9) constant-voltage stabilizer; (10) digital voltmeter; (11) optical power meter.

implies that the product $\mu\beta\tau$ is essentially independent of the wavelength of light. The data on the photoconduction and absorption spectra make it possible to calculate the product $\mu\beta\tau$ for the wavelengths $\lambda = 405$ nm and 436 nm. For these values of λ, the absorption coefficient α is sufficiently small, so that the absorption is uniform over the entire crystal thickness and can be measured with a satisfactory degree of accuracy. The numerical values of $\mu\beta\tau$ are presented in table 8.1.

8.5.2   Polarization effect

The effect of polarization upon the electric and optical properties of BNN has already been considered in chapter 5.

In photoconduction studies it has been discovered that in polarized samples the light-induced electric conduction does not vanish immediately after the light is turned off, but remains for a long time. For instance, after exposure to light with $\lambda = 442$ nm and $I = 1$ mW cm^{-2}, the current relaxes to the initial value for more than 24 h. Switching off the voltage does not affect the relaxation time. Figure 5.29 shows the photoconduction kinetics of BNN polydomain crystals and crystals polarized in different ways. The analysis of the kinetics of photoconduction relaxation has shown that the rate of its variation decreases with time by the hyperbolic law

$$\sigma = \sigma_0 (1 + at)^{-b}. \tag{8.46}$$

For polarized crystals $b = 0.33$ and for polydomain crystals $b = 1.67$.

The long-time conduction relaxation in polarized BNN crystals is not a unique phenomenon—it is observed in a number of semiconductors and oxides [97]. The phenomenon of residual conduction in homogeneous crystals cannot be explained only by a set of corresponding recombination and trapping levels. The majority of the models associate this residual conduction with the existence of macroscopic potential barriers caused by various inhomogeneities in crystals. In a polarized BNN crystal, inhomogeneities may be caused by the pinning of vacancies and impurities at block boundaries under electrolysis which accompanies polarization.

Table 8.1   The values of the products $\mu\beta\tau$ and $\alpha\mu\beta\tau$ for BNN crystals [36].

Crystal	Wavelength λ (nm)	$\mu\beta\tau$ (m^2 V^{-1})	$\alpha\mu\beta\tau$ (m V^{-1})
$Ba_{2.09}Na_{0.72}Nb_{5.02}O_{15}$	405	3.9×10^{-14}	1.2×10^{-16}
	436	3.6×10^{-14}	0.85×10^{-16}
$Ba_{0.25}Sr_{0.75}Nb_2O_6$	405	7.6×10^{-12}	1.2×10^{-13}
	436	8.4×10^{-12}	0.84×10^{-13}
$K(Nb, Ta)O_3$	633	—	6×10^{-15}
	488	—	2.5×10^{-12}

From figure 5.29 (curve 1) it is seen that, after the light is turned off, the current in a polydomain crystal falls gradually to a stationary value comparable to the dark current. This phenomenon is explained by electron redistribution over the levels inside the forbidden band; this redistribution is induced by thermal excitation to free bands and by a repeated trapping [8].

8.5.3 *Photorefractive phenomena*

The variation of birefringence $\Delta(n_e - n_o)$ induced by the light of a He–Cd laser with $\lambda = 441.6$ nm has been studied on polarized BNN crystals. The birefringence was measured by means of a modified phase method with a self-acting phase compensation [4].

For a 1 mm thick specimen, the accuracy of $\Delta(n_e - n_o)$ measurement was 6.7×10^{-6} and the maximum variation of birefringence was 9×10^{-4}. An increase in the specimen size proportionally increased the accuracy of measurement.

The photorefractive effect was measured in two regimes: with and without an applied electric field. In the former case, the effect was observed only under the action of focused radiation.

Figure 8.24 presents the time dependence of $\Delta(n_e - n_o)$ for a polarized BNN crystal. The power density of He–Cd laser radiation in the lens focus was 100 W cm^{-2}; the crystal was placed in the lens caustic. It is seen from the figure that, after the laser is turned on, $\Delta(n_e - n_o)$ increases rapidly, then the process decelerates and reaches saturation within a time of about 80 s. As soon as the laser is turned off, part of the light-induced $\Delta n \approx 7 \times 10^{-6}$ disappears within less than 3 s, the remaining part decreasing within ~ 200 s. On crystals polarized at a high temperature, the effect of optical inhomogeneity formation is weaker than on crystals polarized at a low temperature.

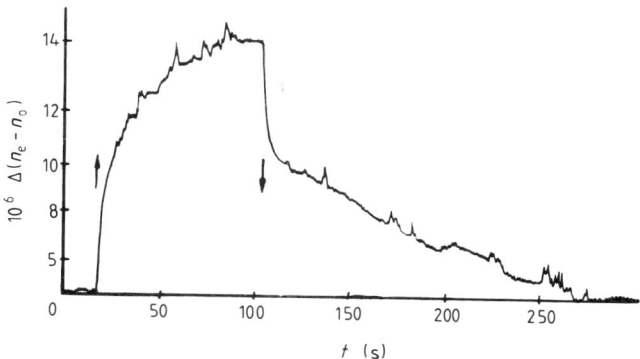

Figure 8.24 Time dependence of birefringence variation under the action of He–Cd laser radiation ($\lambda = 0.44$ μm, $I = 100$ W cm^{-2}): ↑ light on; ↓ light off.

Under exposure to light with $\lambda = 0.44$ μm, $\Delta(n_e - n_o)$ arises both due to photorefraction and due to crystal temperature variation upon laser radiation absorption. A comparison of the calculated value with the data of figure 8.24 shows that the slowly decreasing part of $\Delta(n_e - n_o)$ is associated with the radiation heating of the crystal, and the true photorefractive effect has the same value and disappears within less than 3 s.

The photo-induced birefringence variation of BNN crystals in an external electric field as a function of temperature is plotted in figure 8.25. A field $E = 2.7 \text{ kV cm}^{-1}$ is applied to a crystal, and the birefringence changes due to the electro-optic effect. The photorefraction is induced by unfocused He–Cd laser radiation ($I = 0.1 \text{ W cm}^{-2}$, beam diameter = 2 mm). About 100 s after the light is turned on, the quantity $\Delta(n_e - n_o)$ is saturated. After the light is switched off, at first the photo-induced birefringence gradient changes rapidly. There then comes the stage of slow variation. In an external electric field, a photo-induced birefringence inhomogeneity is observed over several hours. The effect disappears several seconds after the field is switched off. Repeated application of an external field causes the appearance of the effect. Thus the effect is transferred from the 'hidden' into the 'visible' state using an electric field. The storage time of a photo-induced refractive index gradient does not depend on whether it is in the 'hidden' or in the 'visible' state, and it is close to the time of long-time photoconduction relaxation.

The decrease of $\Delta(n_e - n_o)$ under exposure is connected with the field compensation in a crystal by the space charge field occurring upon drift of photo-excited carriers and their entrapment on the periphery of the exposed region. If the size of the exposed region is comparable with the crystal size in the direction of the field, then the applied field is not compensated completely.

The long-time storage of information and the effect of its transfer into the 'hidden' or 'visible' state is explained by the above mentioned long-time

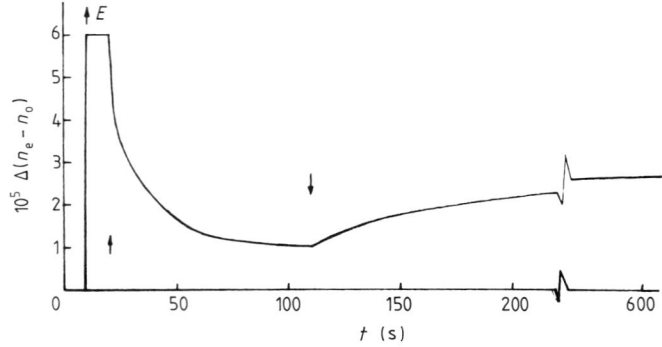

Figure 8.25 Photorefraction kinetics in a BNN crystal in an external electric field $E = 2.7 \text{ kV cm}^{-1}$. ↑$E$ field on; ↑ light on; ↓ light off.

photoconduction relaxation. Turning on the light does not lead to an instantaneous decrease in the concentration of free carriers. In the region that has been previously exposed, the electric conduction remains higher than that in the unexposed part. In an external electric field this leads to a non-uniform field distribution, and therefore to a non-uniform refractive index variation in the crystal. With the external field off, the space charges are cleaned out during the time $\tau \sim \varepsilon\varepsilon_0/\sigma$. When the field is on again, it is distributed in the crystal following the electric conduction distribution. The information storage time is determined by the time of long-time photoconduction relaxation. The estimation of the photorefractive sensitivity $S = \Delta n/W$ on recording in an external electric field $E = 2.7\,\mathrm{kV\,cm^{-1}}$ gives the value $S = 1.5 \times 10^{-4}\,\mathrm{J^{-1}\,cm^2}$, which is comparable with the LN:Fe crystal sensitivity [88]. It is of interest to compare the measured value S with the calculated one:

$$S = \alpha\beta\mu\tau E\left(\frac{n^3\ r_{33}}{2\ \varepsilon_0\varepsilon_{33}}\right)\frac{e}{h\nu}. \tag{8.47}$$

Substituting the known values of n, r_{33}, ε_{33} and the product $\alpha\beta\mu\tau$ from table 8.2, we obtain $S = 1.7 \times 10^{-4}\,\mathrm{J^{-1}\,cm^2}$, which is in close agreement with the value observed experimentally.

8.5.4 *Fe and Mo ion doped barium sodium niobate crystals*

An introduction of Fe and Mo to BNN crystals has been shown [69, 86] to lead to a substantial photorefractive effect, which makes these crystals suitable for recording volume phase holograms with a diffraction efficiency reaching 67%.

The investigations were carried out on both pure and 0.15 wt.% Fe- and Mo-doped BNN crystals. The effect was induced by focused He–Ne and He–Cd laser radiation. The maximum magnitude of the effect was $\Delta(n_e - n_o) \approx (1-2) \times 10^{-4}$. The variation of the refractive index of an extraordinary beam, Δn_e, calculated from the data of holographic measurements [69] at a spatial recording frequency of $10^4\,\mathrm{cm^{-1}}$ gave in saturation the value $\Delta n_e \approx 0.5 \times 10^{-4}$. These data suggest that the diffusion mechanism cannot explain the photorefractive effect in BNN Fe- and Mo-doped crystals, since the increase of the spatial frequency from zero upon light beam recording to $10^4\,\mathrm{cm^{-1}}$ upon hologram recording does not cause any variation of the refractive index in saturation.

The dark conduction σ_d, photoconduction σ_{ph} and photocurrent j were measured on single-domain crystals exposed to He–Ne and He–Cd laser radiation with $\lambda = 0.63\,\mu\mathrm{m}$ and $\lambda = 0.44\,\mu\mathrm{m}$ with the plane of electric vector oscillations perpendicular to the c axis of the crystal. In pure and doped BNN crystals a photogalvanic effect has been discovered, whose value is an order of magnitude less than that in LN crystals.

The kinetics of the photo-response in an Fe- and Mo-doped BNN single

Table 8.2 Photoelectric properties of BNN crystals [40, 41].

Material	σ_d ($10^{-15}\,\Omega^{-1}\,cm^{-1}$)	α (cm^{-1})	(σ_{ph}/I) ($10^{-14}\,W^{-1}\,\Omega^{-1}\,cm$)	(j_{phg}/I) ($10^{-11}\,A$ $cm^2\,W^{-1}$)	E_{sat} ($V\,cm^{-1}$)	K_G ($10^{-10}\,A\,cm\,W^{-1}$)	$\mu\beta\tau$ ($cm^2\,V^{-1}$)
BNN pure	12						
$\lambda = 0.63\,\mu m$		0.08	300	2.5	8.4	3.1	7.6×10^{-11}
$\lambda = 0.44\,\mu m$		0.3	1100	10	9.1	3.3	1.0×10^{-10}
BNN:Fe:Mo	0.8 ± 0.3						
$\lambda = 0.63\,\mu m$		0.5	4.3	18.8	4.3×10^3	3.7	1.7×10^{-13}
$\lambda = 0.44\,\mu m$		5.0	21	90	4.3×10^3	3.0	2.0×10^{-13}

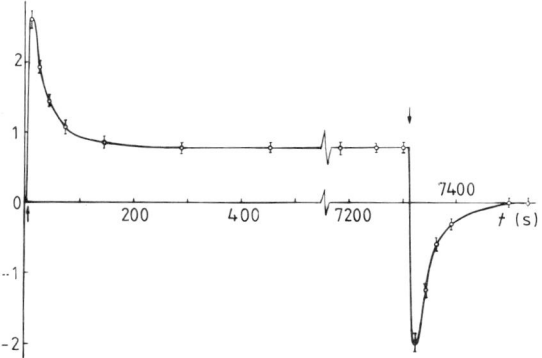

Figure 8.26 Photo-response kinetics in a BNN:Fe:Mo single crystal in a specimen exposed to light with $\lambda = 0.63$ μm, $I = 43$ mW cm^{-2}. $\uparrow$ light on; $\downarrow$ light off.

crystal for a uniform illumination of a specimen is presented in figure 8.26. Within the first 100 s one can observe a transition current induced by the pyroelectric effect. Then a stationary current is established due to the volume photogalvanic effect which occurs in ferroelectrics. The transition process caused by the pyroelectric effect is still observed when the light is turned off. The time dependence of the photo-response in an undoped crystal has the same form. The current density of the photogalvanic effect is proportional to the intensity of light incident on the crystal.

When an external voltage is applied to a crystal, the current through the crystal is a superposition of the current connected with photoconduction and dark conduction, and with the current of the photogalvanic effect [19]

$$j = (\sigma_d + \sigma_{ph})E - j_{phg}. \tag{8.48}$$

The value of the photogalvanic current is given by

$$j_{phg} = K_G \alpha I \tag{8.49}$$

where I is the intensity of light incident on the crystal, K_G is the Glass coefficient and α is the light absorption coefficient. If the electrodes of the exposed crystal are not close, then after the end of the transition process an electric field sets in this crystal, and the resultant current becomes $j = 0$. Equations (8.48) and (8.49) imply that

$$E_{sat} = \frac{K_G \alpha I}{\sigma_d + \sigma_{ph}} \tag{8.50}$$

from which one can calculate the value of the photo-induced birefringence variation

$$\Delta n = \tfrac{1}{2}(n_e^3 r_{33} - n_o^3 r_{13})E_{sat}. \tag{8.51}$$

The $V-I$ characteristics of the photoelectric current in a BNN:Fe:Mo crystal, measured under a uniform exposure of the whole crystal to He–Ne and He–Cd lasers with an equal power density of 60 mW cm^{-2}, are plotted in figure 8.27. The $V-I$ characteristics make it possible to find the parameters of the photoconduction and of the photogalvanic effect. Table 8.2 presents the experimental values of σ_d, σ_{ph}/I, j_{phg}/I and the Glass coefficients calculated by these parameters. Figure 8.28 shows the absorption spectrum of BNN:Fe:Mo crystals for different light polarizations. For large absorption coefficients $\alpha l \gg 1$, one should necessarily take into account the non-uniformity of light intensity distribution over the sample thickness. The expression (8.49) will then take the form

$$j_{phg} = \frac{K_G}{l}(1 - e^{-\alpha l}). \tag{8.52}$$

Table 8.2 also gives the values of the product $(\mu\beta\tau)$ of the photoconduction phenomenological parameters, which were estimated using relation (8.45).

The dark electric conduction of crystals on the linear portion of the $V-I$ characteristic of an unexposed crystal is $12 \times 10^{-15}\,\Omega^{-1}\,\text{cm}^{-1}$ for a pure crystal and $0.8 \times 10^{-15}\,\Omega^{-1}\,\text{cm}^{-1}$ for a doped one.

The photo-induced birefringence was found, using equation (8.51), to be $\Delta n = 1.2 \times 10^{-4}$. The calculated and experimental values coincide. This suggests that the photorefractive effect in BNN:Fe and BNN:Mo is determined by the photogalvanic effect.

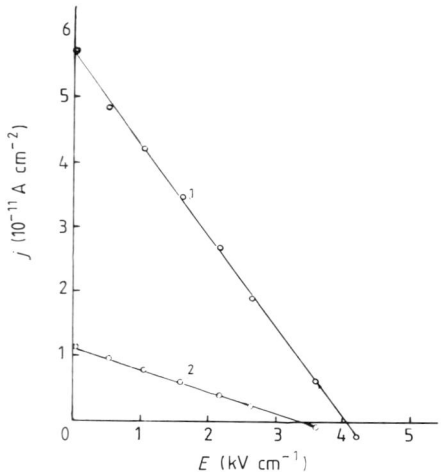

Figure 8.27   Voltage–current characteristics of a stationary photocurrent in a BNN:Fe:Mo crystal under a uniform exposure to He–Cd (1) and He–Ne (2) lasers with $I = 60$ mW cm^{-2}.

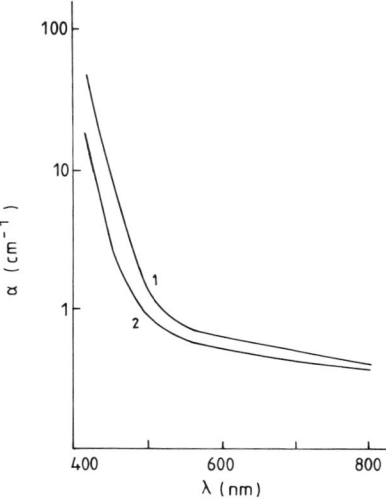

Figure 8.28 Absorption spectrum of a BNN:Fe:Mo crystal. Light polarization (1) $E \| c$ and (2) $E \perp c$.

The Glass coefficient $K_G \approx 3.5 \times 10^{-10}$ A cm W^{-1} does not depend essentially on the introduction of Fe and Mo. Similar results for LN:Fe crystals were reported in [98]. An impurity of Fe and Mo lowers substantially the dark conduction and photoconduction. This leads to increases in the absorption and in the proportional amplification of photogalvanic current, which in turn causes an increase in the photo-induced field and therefore an increase in photorefraction.

In pure BNN crystals the value $E_{sat} \approx 10$ V cm^{-1}, which may induce a birefringence of 3×10^{-7}. By holographic and polarization optical measurements, the value $\Delta n_e \approx 5 \times 10^{-6}$. Thus the photogalvanic effect cannot cause photorefraction in pure BNN.

8.5.5 *Crystals of barium sodium potassium niobate $Ba_2Na_{1-x}K_xNb_5O_{15}$ with compositions from $x = 0$ to $x = 0.6$*

The dark electric conduction, photoconduction and photogalvanic current were measured in crystals exposed to He–Ne ($\lambda = 0.63 \ \mu$m) and He–Cd ($\lambda = 0.44 \ \mu$m) lasers.

As the content of potassium increases, the electric conduction grows monotonically, and for the composition $x = 0.6$ it is two orders of magnitude higher than in BNN crystals. The experimental data on photoconduction and photogalvanic current are presented in figure 8.29(a,b). The value of the photogalvanic field at saturation decreases with increasing x, and its maximum value does not exceed 10 V cm^{-1}. As a result of the electro-optic

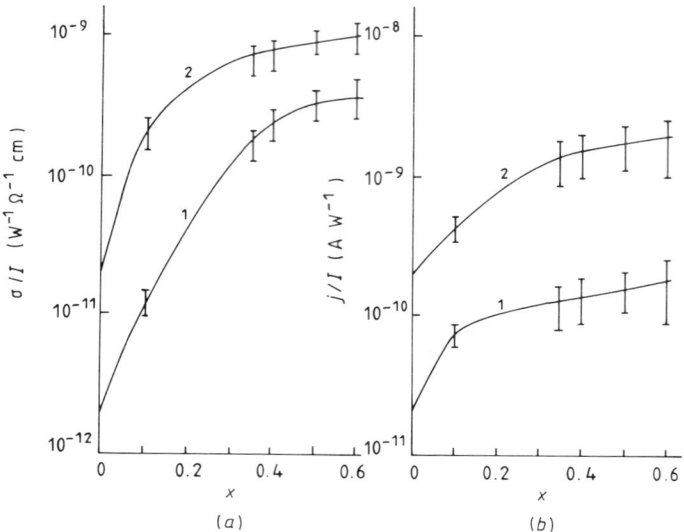

Figure 8.29 Photoconductivity (a) and photogalvanic current (b) of $Ba_2K_xNa_{1-x}Nb_5O_{15}$ crystals. (1) $\lambda = 0.63$ μm; (2) $\lambda = 0.44$ μm.

effect, this field induces a birefringence gradient $\Delta(n_e - n_o) \approx 3 \times 10^{-7}$. Under the action of focused He–Cd laser radiation, the photorefraction was measured to be $\Delta(n_e - n_o) < 1 \times 10^{-5}$, which is close to that observed in BNN crystals. Thus the replacement of part of Na by K does not cause an increase in photorefraction.

8.6   Photorefractive effects in barium strontium niobate crystals

Crystals of solid solutions of barium strontium niobate (BSN) $Ba_xSr_{1-x}Nb_2O_6$ are among the most sensitive photorefractive materials. A number of papers [62, 71, 77] are devoted to the study of holographic recording in essentially pure and doped BSN crystals of different compositions. BSN crystals seem to be a convenient object for studying the photorefraction mechanism because they possess a ferroelectric phase transition in an admissible temperature interval (see chapter 4) and a significant photoconduction [31]. Because of this fact, one can investigate the connection of the photorefractive processes with the ferroelectric and electric properties of these crystals, which, as is known, is practically impossible with LN crystals.

The photorefractive characteristics of BSN crystals were examined by holographic (figure 8.10) and polarization optical methods [33, 34, 37, 38]. Recording was made by a He–Cd laser ($\lambda = 0.44$ μm) whose vector E was perpendicular to the c axis of the crystal (ordinary polarization). Read-out was made by the same light or by a He–Ne laser beam ($\lambda = 0.63 \mu$m) whose vector E lay in the plane of incidence of the recording beams (extraordinary polarization). In several experiments, recording and reading were made by a He–Ne laser beam of extraordinary polarization. The photorefraction value was estimated by the diffraction efficiency of the hologram.

We have investigated the optical and photoelectric properties of BSN crystals [36]. The photoconduction spectrum is reproduced in figure 8.30 (curve 2). The photoconduction increases monotonically with decreasing wavelength. The photoconduction was almost absent for $\lambda > 800$ nm (1.5 eV), which coincides with the activation energy of the electric conduction. The dark conduction of BSN crystals is $\sigma_d = 7.5 \times 10^{-12} \, \Omega^{-1} \, \text{cm}^{-1}$.

Figure 8.30 (curve 1) presents the absorption spectrum of BSN crystals with $x = 0.25$. There are two rectilinear regions on the graph. The region of large α corresponds to band–band absorption, the flattened region to light absorption by defects. For a BSN crystal, the widths of the optical and thermal forbidden bands are respectively equal to $E_g^O = 3.44 \pm 0.02$ eV and $E_g^T = 3.25 \pm 0.02$ eV.

The product of the phenomenological parameters of photoconduction ($\mu\beta\tau$) for $\lambda = 405$ and 436 nm was estimated from the absorption and conduction spectra (table 8.3).

The room-temperature dependences of $\Delta(n_e - n_o)$ on the exposure time and on the 'damaging' light intensity for BSN 0.25 and BSN:Ce crystals without

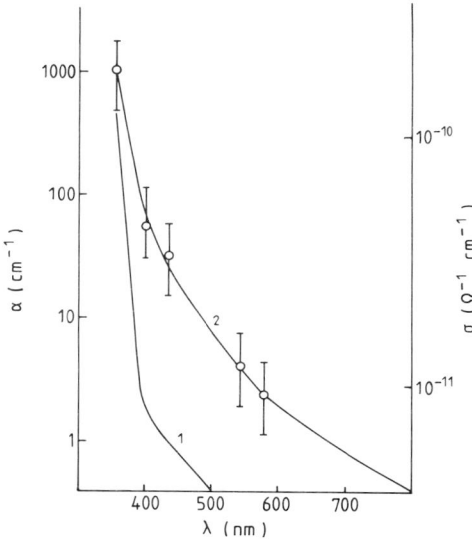

Figure 8.30 Spectra of (1) absorption and (2) photoconductivity of BSN crystals of the composition $x = 0.25$.

Table 8.3 Optical and photoelectric properties of BSN crystals.

Composition	Degree of CeO_2 doping (wt.%)	α ($\lambda = 4400$ Å) (cm^{-1})	σ_d (10^{-13} Ω^{-1} cm^{-1})	σ_{ph} (10^{-10} Ω^{-1} cm^{-1})	K_G (10^{-10} A W^{-1} cm)	$\mu\beta\tau$ (10^{-10} cm^2 V^{-1})	β (%)	$\mu\tau$ (10^{-10} cm^2 V^{-1})
BSN-39	pure	0.29	170	$2.53\,I^{0.5}$	3.1	140.38†	4.7	2900
	0.05	5.8				2.2	27	8
	0.10	11.5				2.3	41	5.6
BSN-25	0.05	8.6	3.5	$1.70\,I^{0.7}$	3.4	0.029	0.07	4.1

† Product $\mu\beta\tau$ is determined from photoelectrical measurements.

an external field are plotted in figure 8.31. The value of $d\Delta(n_e - n_o)/dt$ which characterizes the photorefractive sensitivity of the material is seen to increase sharply in a doped crystal. Figure 8.32 shows the variations of $\Delta(n_e - n_o)$ under the action of He–Ne laser radiation in a DC electric field.

The photorefractive effect in pure and doped BSN crystals is not connected with the appearance of an anomalous photogalvanic current, because for damaging light intensities of about 100 W cm^{-2} the values of the occurring photoelectric voltage measured by the compensation method do not exceed 0.1–0.2 V. The corresponding field strength is 10–20 V cm^{-1}, whereas the observed experimental photorefraction values $\Delta(n_e - n_o) \approx 10^{-4}$ for the same radiation intensities correspond to fields of about 300–500 V cm^{-1}.

The photorefraction in BSN might be associated with the screening by

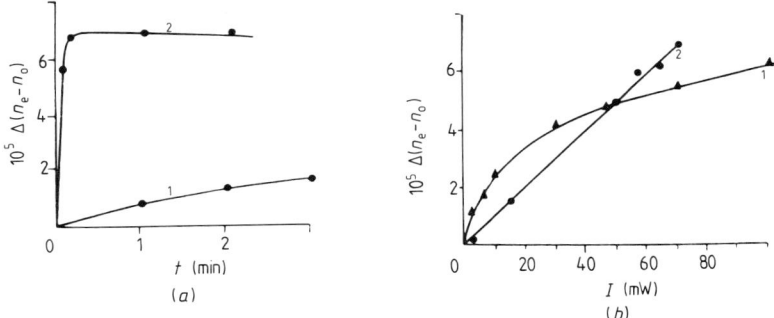

Figure 8.31 Dependences of photorefraction of BSN-0.25 (1) and BSN:Ce (2) crystals (a) on the time of exposure $I = 20$ W cm^{-2} and (b) on damaging light intensity for $\lambda = 0.488$ μm.

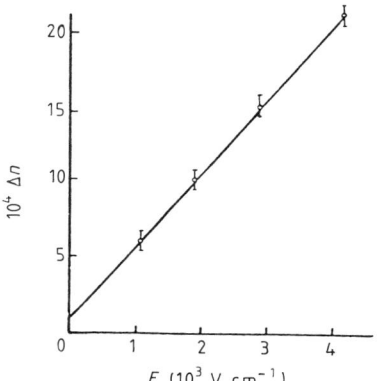

Figure 8.32 Photorefractive effect as a function of external electric field in BSN-0.25 crystals. $\lambda = 633$ nm, $I = 10$ W cm^{-2}.

photo-carriers of pyroelectric fields which occur upon absorption of an intense light [99, 100]. But in this case the dependence of the sensitivity $S(\lambda)$ should have correlated with the dependence $\alpha(\lambda)$, which is not the case, as shown below.

In BSN crystals not specially doped, photorefraction is not a 'proper' effect but is connected with the screening of the internal field in the volume of an inhomogeneous crystal.

Crystals of BSN solid solutions are known to possess a considerable structural inhomogeneity. This is due not only to the growth conditions, but also to non-stoichiometry of the composition. As shown in chapter 4, in the volume of BSN crystals there exist substantial composition fluctuations. Structural inhomogeneities are most clearly pronounced in crystals with a large amount of Sr: in particular in BSN 0.25 which has a smeared phase transition (see chapter 4).

Structural inhomogeneity of BSN crystals is confirmed by the long-time residual conduction observed in them. The phenomenon of residual conduction is now assumed to be due to the crystal structure inhomogeneity, i.e. to the existence of a randomly distributed potential relief which hampers the recombination of carriers.

Residual conduction is observed in all crystals of the BSN family. The characteristics of residual conduction depend, as do the photorefractive properties of BSN, essentially on crystal treatment conditions. In unpolarized room-temperature BSN 0.25 crystals, the ratio $\sigma_{res}/\sigma_d \sim 10$ and the time of decrease to the equilibrium dark value is 1–2 h (figure 8.33). In polarized crystals, $\sigma_{res}/\sigma_d \approx 10^2$, and the time of decrease rises up to 20–25 h. Residual conduction in BSN 0.25 is most clearly pronounced in the direction of the polar c axis.

Polarization of BSN 0.25 crystals by the methods described in chapter 4 is responsible, not only for an increase of residual conduction, but also for an increase of crystal unipolarity; the loops of dielectric and optical hysteresis of polarized crystals exhibit stable asymmetry. The photorefractive sensitivity increases with increasing degree of unipolarity and residual conduction. In

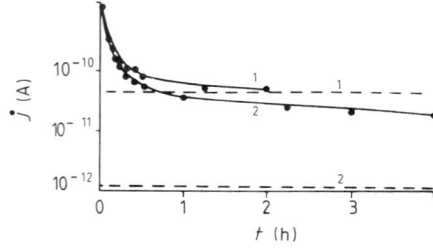

Figure 8.33 Decrease of remanent conduction in BSN-0.25 crystals. (1) Polydomain and (2) single-domain specimens, room temperature.

crystals with a low degree of unipolarity the photorefractive effect is not observed (within experimental error).

Thus a polarized BSN 0.25 crystal is a set of 'micro-grains' characterized by different values of spontaneous polarization P_s. A non-uniform P_s distribution in the volume (div $P_s \neq 0$) leads to a corresponding non-uniform space charge distribution

$$\rho(z) = e(N_d - N_a + p - n)$$

where N_d and N_a are respectively the concentrations of donor and acceptor impurities, and there appears a system of collective barriers responsible for the phenomenon of residual conduction. Such inhomogeneous structure is characterized by the existence in the crystal volume of inhomogeneity of the distributed field

$$E_i = (1/l)dF/dz$$

(F is the Fermi level) whose magnitude is determined by the degree of crystal unipolarity.

When a crystal is exposed locally, the photo-excited carriers undergo volume separation in the macroscopic field E_i; the generated space charge field E_{sch} is responsible for the birefringence variation in the exposed region.

The above consideration is qualitative, but the experimental facts listed below confirm the assumption about the existence of a macroscopic field in the volume of unipolar BSN 0.25 crystals.

(i) A preliminary uniform exposure of BSN 0.25 crystals to a photo-active light sharply lowers the photorefractive sensitivity S, which is apparently caused by the screening of the field E_i in the entire volume. In a number of cases, a subsequent illumination with IR light brings the crystal to the initial state, and the sensitivity S attains its initial value.

(ii) The highest photorefractive sensitivity is exhibited by strongly unipolar BSN 0.25 crystals with an asymmetric electro-optic hysteresis loop. The lowered asymmetry of the electro-optic 'butterfly' in the exposed region is obviously also due to the screening of the field E_i. As a result of thermal or optical photorefraction erasure, the 'butterfly' restores the initial symmetry (figure 8.34). The values of E_{sch} estimated independently by the values of photorefraction $\Delta(n_e - n_o)$ and of the 'butterfly' displacement are in good agreement.

The proposed model of the appearance of photorefraction in an inhomogeneous ferroelectric can explain the poor reproducibility of the values $\Delta(n_e - n_o)$ in BSN 0.25 as well as the dependence of photorefraction upon a preliminary crystal annealing and a substantial non-uniformity of its distribution in the crystal volume. The spatial distribution of photorefraction in BSN crystals has the form similar to the one reported by Chen [3] for LiNbO$_3$, which testifies in favour of the existence of a space charge field E_{sch} in the exposed region.

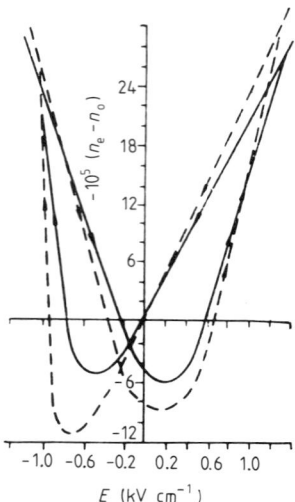

Figure 8.34 Change in the shape of electro-optic hysteresis loop as a result of the photorefractive effect. The broken curve represents the initial shape of an electro-optic 'butterfly' in a unipolar BSN-0.25 crystal. The full curve represents the shape of an electro-optic butterfly in a damaged region of the crystal.

In pure BSN crystals, the characteristic times (of recording, erasure and storage) do not correlate with the dielectric relaxation time $\tau_{dr} \sim \varepsilon/\sigma$. The storage time τ_{st} in BSN 0.25 is usually approximately an order of magnitude less than the σ_{dr} calculated from the data on the equilibrium dark value σ_d. This can be explained by the long-time residual conduction. Illumination of single-domain BSN 0.25 crystals by an intense photo-active light leads to an increase of the dark conduction up to a non-equilibrium value σ_{neq}, which is observed for 24 h. Accordingly, the times of photorefraction storage in the dark, τ_{st}, should be determined not by the equilibrium value σ_d but rather by the actual value σ_{neq} of the conduction excited by light of a given intensity. Indeed, the times τ_{st} can be described in the first approximation by the 'non-equilibrium' time of dielectric relaxation $\tau'_{dr} \sim \varepsilon/\sigma_{neq}$.

Crystals of BSN 0.25 exhibit the effect of field suppression of residual conduction which consists of a sharp σ_{neq} decrease caused by a large external field pulse. Since τ_{st} of photorefraction in BSN 0.25 is determined by the σ_{neq} value, the field suppression of residual conduction leads to an increase of τ_{st}. In the investigated single-domain BSN 0.25 crystals, $\sigma_{neq} \approx 10^{-12} \, \Omega^{-1} \, cm^{-1}$ and $\tau_{st} \approx 2$–3 h. An application of an external field pulse of $\pm(600$–$700)$ V cm^{-1} for 1 min led to a decrease of σ_{neq} by almost an order of magnitude and to a corresponding increase of τ_{st} up to 10–15 h.

Thus, the use of the field-induced suppression of residual conduction in BSN 0.25 crystals allows us somehow to vary the photorefraction storage times.

8.7 Photoelectric and photorefractive properties of cerium-doped crystals of barium strontium niobate

Cerium was introduced into BSN crystals $(Sr_x Ba_{1-x})_{1-y}(Nb_2 O_6)_y$ $(x = 0.61,$ $y = 0.4993)$. This is a congruently melting composition, according to [101]. In the grown crystals, striation was weakly pronounced. Crystals were doped by the addition of 0.05 and 0.10 wt.% of CeO_2 to the starting material and polarized by the application of an electric field $(5-10 \, kV \, cm^{-1})$ at a temperature of 150°C and a subsequent cooling down to room temperature below the field. The half-wave voltage of polarized crystals was 240 V for $\lambda = 0.63 \, \mu m$.

For optical measurements, plates were cut out whose ferroelectric c axis lay in the plane of the plate.

8.7.1 *Photoelectric properties*
Silver electrodes were deposited onto the faces perpendicular to the c axis. The resistance of the contacts was controlled by measuring the $V-I$ characteristics (VACs) and measuring the electric field along the c axis. VACs of the crystals were linear in the range from 1 to 200 V cm^{-1}. For higher E, the voltage dependence of the current was quadratic. Measurements of the electro-optic effect at a constant volume along the c axis of a sample has shown field inhomogeneity in the crystal to be within $\pm 10\%$ (figure 8.35).

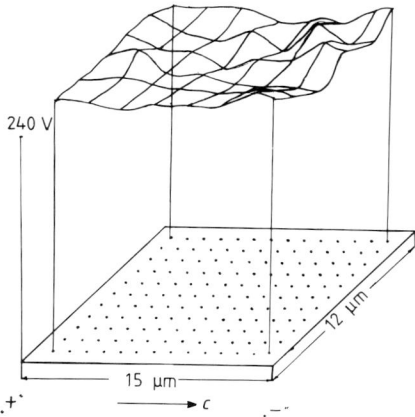

Figure 8.35 Distribution of half-wave voltage $V_{\lambda/2}$ over a BSN:Ce crystal plate $(x = 0.61)$. Light polarization $E \| c$.

Measurements of electric conduction, photoconduction and photoelectric currents were carried out in a direct current. Crystals were illuminated with light of ordinary polarization with $\lambda = 0.44\ \mu$m in the radiation intensity range from 4 to 100 mW cm^{-2}. The experimental dependence of photoconduction on the radiation intensity is plotted in figure 8.36. The photoconduction of a pure BSN crystal obeys the equation

$$\sigma_{ph} = 2.53 \times 10^{-10} I^{0.50}\ \Omega^{-1}\ cm^{-1} \tag{8.53}$$

and the photoconduction of a crystal doped with 0.10 w.% of CeO_2 obeys the equation

$$\sigma_{ph} = 1.70 \times 10^{-10} I^{0.70}\ \Omega^{-1}\ cm^{-1} \tag{8.54}$$

where I is expressed in W cm^{-2}.

For the wavelength $\lambda = 0.488\ \mu$m, another dependence was obtained:

$$\sigma_{ph} = 1.2 \times 10^{-10} I\ \Omega^{-1}\ cm^{-1}. \tag{8.55}$$

The dark conduction σ_d of BSN:Ce crystals is two orders of magnitude lower than that of undoped crystals. The electric and photoelectric properties of BSN and BSN:Ce crystals are listed in table 8.4.

The lux–ampere characteristic of BSN:Ce photoconduction for $\lambda = 0.488\ \mu$m is linear throughout the intensity range 10^{-1}–10^2 W cm^{-2} (figure 8.37). The corresponding dependence of $\tau \sim \varepsilon/\sigma_{ph}$ on the irradiating light intensity shown in figure 8.37 has been calculated on the basis of the obtained dependences of the photoconduction.

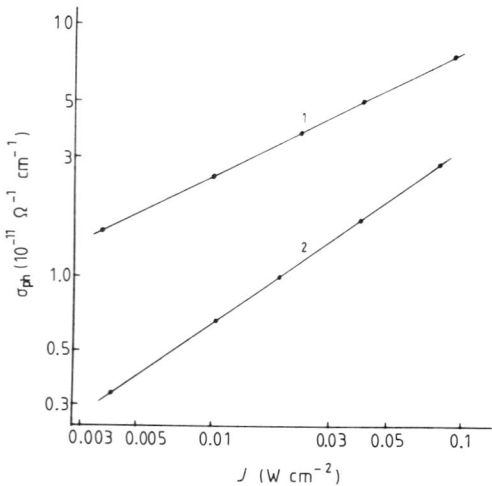

Figure 8.36   Photoconductivity of BSN crystal ($x = 0.61$) as a function of exposure to light with $\lambda = 0.44\ \mu$m. (1) a pure crystal of BSN; (2) BSN:Ce.

Table 8.4 Sensitivity of ferroelectric crystals to an optically induced variation of the refractive index.

Material	Impurity (wt.%)	λ (nm)	Sensitivity $S = \eta/W$ (10^{-2} J^{-1} cm^2)	Sensitivity $S = \Delta n/W$ (J^{-1} cm^2)	External field (kV cm^{-1})	References
LiNbO$_3$	—	488	6×10^{-3} (5)†	2×10^{-8}	—	[76]
	0.05Fe	488	22 (5)	7×10^{-5}	—	[76]
	0.03Fe	476	500 (1)	7.6×10^{-3}	—	[90]
	0.03Fe	488	400 (1)	6.2×10^{-3}	—	[90]
	0.03Fe	633	0.5 (3–6)	3.4×10^{-6}	—	[88, 89]
	0.03Fe	442	24 (3–6)	1.1×10^{-4}	—	[88, 89]
	0.05Cu	488	4.2 (5)	1.3×10^{-5}	—	[76]
	0.05Ni	488	0.1 (5)	3.1×10^{-7}	—	[76]
	0.05Mn	488	0.54 (5)	1.7×10^{-6}	—	[76]
	0.05Rh	488	12.6 (3)	3.9×10^{-5}	—	[93]
LiTaO$_3$	—	488	2×10^{-5} (5)	6×10^{-11}	—	[87]
Ba$_2$NaNb$_5$O$_{15}$	Fe	488	7.0 (5)	2×10^{-5}	—	[87]
Ba$_2$NaNb$_5$O$_{15}$	0.036Fe + 0.002Mo	488	5.2 (3, 2)	2.5×10^{-5}	—	[85, 86]
Ba$_{0.25}$Sr$_{0.75}$Nb$_2$O$_6$	—	488	330 (8)	6.4×10^{-4}	3	[62]
PLZT ceramics (9% La)	—	488	0.033 (0.350)	1.5×10^{-6}	4.4	[71]
K(Ta, Nb)O$_3$	—	488	—	1×10^{-6}	3	[92]
KNbO$_3$	—	488	—	1.6×10^{-7}	—	[67]
	Co	488	—	2.4×10^{-7}	—	[67]
	Fe	488	—	5.0×10^{-6}	—	[67]
	Fe + W	488	—	5.0×10^{-5}	—	[67]
BaTiO$_3$	—	488	10 (5)	—	—	[14]
5PbO·3GeO$_2$	—	633	—	1×10^{-7}	—	[91]
K(Ta, Nb)O$_3$	—	530	Two-photon recording	4×10^{-3}	—	[21]

† Sample thickness is given in mm in brackets.

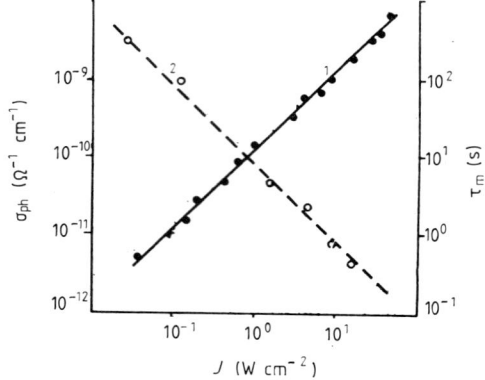

Figure 8.37 Intensity–current characteristic of photoconductivity (1) and dependence of dielectric relaxation time τ on light intensity ($\lambda = 0.488\ \mu$m) in a BSN:Ce crystal.

The absorption coefficients α for $\lambda = 0.44\ \mu$m under ordinary polarization of light were calculated from the transmission spectra of samples of different thickness (figure 8.38).

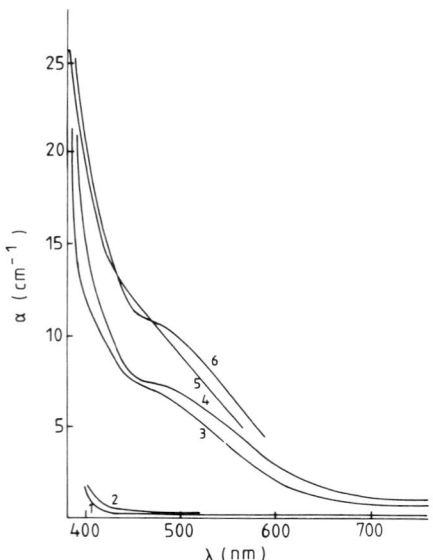

Figure 8.38 Absorption spectra of BSN single crystals. (1) BSN, ordinary beam; (2) BSN, extraordinary beam; (3) BSN:0.05 Ce, ordinary beam; (4) BSN:0.05 Ce, extraordinary beam; (5) BSN:0.10 Ce, ordinary beam; (6) BSN:0.10 Ce, extraordinary beam.

The addition of Ce to BSN crystals leads to the appearance of an impurity maximum on the absorption curve and the corresponding 'tail' of photo-conduction in the region of 0.5 μm.

The product of the phenomenological photoconduction parameters ($\mu\beta\tau$) was determined from the relation (8.45). Since σ_{ph} has a power-law dependence on the radiation intensity, the product ($\mu\beta\tau$) also depends on the radiation intensity, the ($\mu\beta\tau$) value decreasing with increasing I. Such a behaviour is explained by the nonlinear recombination whose rate is increased and whose lifetime is therefore shortened as the number of non-equilibrium carriers increases. Comparing the ($\mu\beta\tau$) values for a pure and a doped crystal, one may conclude that the lowering of ($\mu\beta\tau$) is connected with the lowering of τ, since the quantum yield β does not change greatly as a rule. Furthermore it is hardly probable that an introduction of a small amount of impurity will affect the mobility value μ, while the lifetime is determined by the recombination and entrapment and varies substantially with an introduction of impurity.

The current of the photogalvanic effect of Ce-doped and pure crystals depends linearly on the radiation intensity. The introduction of Ce increases the current of the photogalvanic effect in proportion to the absorption coefficient. The Glass coefficients for pure and doped crystals are also listed in table 8.4. The value of the Glass coefficient in BSN crystals is an order of magnitude less than in LN crystals and coincides with the K_G value for BNN.

8.7.2 Photorefractive properties of BSN:Ce

A typical picture of hologram recording and reading by a He–Cd laser beam is presented in figure 8.39 [78]. To achieve a 10% diffraction efficiency in LiNbO$_3$:Fe crystals, 400 mJ cm^{-2} is required, whereas under similar recording conditions in BSN:Ce crystals 60 mJ cm^{-2} is enough to achieve the same diffraction efficiency.

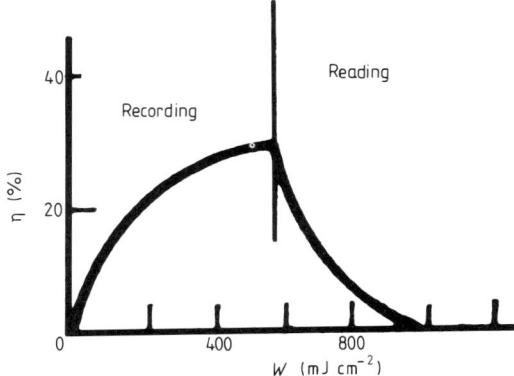

Figure 8.39 Recording–reading cycle for light with $\lambda = 0.44$ μm. Diffraction efficiency was determined by a weak light with $\lambda = 0.63$ μm.

The variation of the refractive index was calculated from the diffraction efficiency of a hologram by Kogelnik's formula [63] for the case when the vector E lies in the plane of incidence

$$\eta = \sin^2 \left(\frac{\pi \Delta n_e d \cos \theta}{\lambda \cos \theta/2} \right) \qquad (8.56)$$

where d is the crystal thickness and $\theta/2$ is the angle of refraction. The sensitivity was determined for an η level equal to 5%.

We have investigated the dependence of the sensitivity and maximum η_{max} on the spatial frequency and the intensity of recording beams. The spatial frequency K ($K = 2\pi/L$, where L is the period of the recording grating and $L = \lambda/2n \sin \theta/2$) was varied by varying the angle of incidence of the recording beams. The experimental dependence of the sensitivity S on the spatial frequency is presented in figure 8.40. The sensitivity increases linearly with grating frequency up to 700 lines mm^{-1}, reaches saturation and then falls. For recording by light with $\lambda = 0.44\ \mu m$ the maximum sensitivity is $S = 2 \times 10^{-3}\ J^{-1}\ cm^2$, and for recording by an extraordinary polarized light with $\lambda = 0.63\ \mu m$ the sensitivity $S = 2.2 \times 10^{-4}\ J^{-1}\ cm^2$. The dependence of the maximum η_{max} on the spatial frequency is plotted in figure 8.41. The recording light intensity does not affect the maximum η_{max} in the recording of a grating with a frequency of 700 lines mm^{-1} in the range 4 to 40 mW cm^{-2}.

As distinguished from hologram recording by an ordinarily polarized light, in the case of an extraordinarily polarized light the energy is pumped into a beam which has a negative projection of the wavevector onto the direction of spontaneous polarization P_s in the crystal. This shows that the photo-excited charge carriers have a negative sign, i.e. the electrons are excited [102]. The time dependence of the intensities of transmitted beams in a recording by a light of extraordinary polarization with $\lambda = 0.63\ \mu m$ is presented in figure 8.42. The effect observed can be explained as follows. In

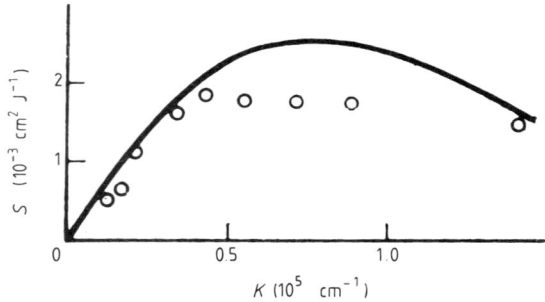

Figure 8.40 Sensitivity S of a BSN:Ce crystal to light with $\lambda = 0.44\ \mu m$ as a function of spatial frequency K.

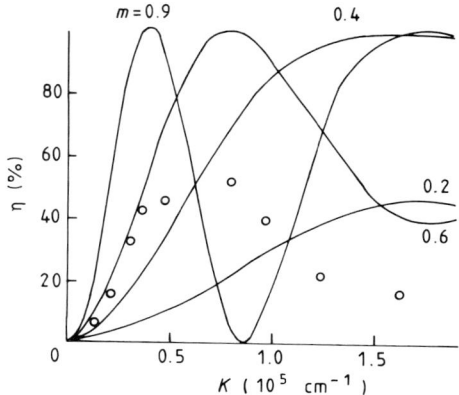

Figure 8.41 Dependence of maximum diffraction efficiency (η) on spatial frequency K for different values of modulation ratio m.

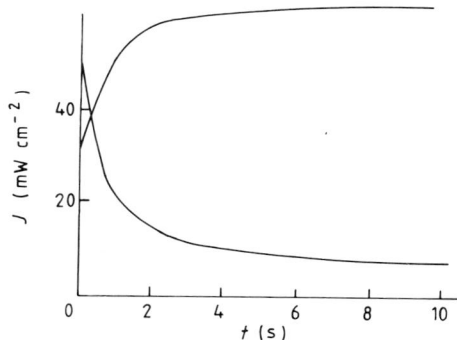

Figure 8.42 Redistribution of intensities of beams after they have passed through a crystal under hologram recording by light beams of extraordinary polarization with $\lambda = 0.63\ \mu$m.

a recording by light of extraordinary polarization, a dynamic hologram is formed which interacts with interfering beams and causes their redistribution [13]. As a result of redistribution, the contrast of the interference pattern reduces, which leads to a decrease of the diffraction efficiency. The interaction of the beams with the hologram is proportional to the value of the electro-optic coefficient. In BSN crystals, the electro-optic coefficient $r_{33} \gg r_{13}$, and therefore in a recording by light of ordinary polarization the intensity redistribution is much weaker, which is confirmed by experiment.

The kinetics of optical erasure of photorefraction in BSN:Ce for all exposures to an erasing light is described satisfactorily by the expression

$$\frac{\Delta(n_e - n_o)_t}{\Delta(n_e - n_o)} = \exp(C_{op\ er} W) \qquad (8.57)$$

where $\Delta(n_e - n_o)$ is the initial photorefraction value and $W = It$ is the erasing light energy. The absorption was taken into account using the intensity value $I = I_0(1 - e^{-\alpha d})/(\alpha d)$, and the optical erasure constant $C_{op\ er}$, which characterizes the sensitivity to erasure and depends only on the wavelength of light, as $C_{op\ er} = 1/\tau_{er} I \ s^{-1} \ W^{-1} \ cm^2$ (figure 8.43).

In the photo-active region 0.4–0.65 μm, the dependence $C_{op\ er}(\lambda)$ correlates with $\sigma_{ph}(\lambda)$. In the near IR there exists an additional maximum of optical photorefraction erasure for $\lambda \approx 1.0$ μm which is not accompanied by absorption or photoconductivity anomalies. The optical erasure constants for the wavelengths $\lambda = 0.488$ μm and 1.15 μm are respectively $C_{op\ er} = 1.2 \times 10^{-1}$ and $C_{op\ er} = 1.1 \times 10^{-2}$, that is, the efficiency of IR erasure is an order of magnitude lower than the efficiency of erasure by photo-active light.

The hologram storage time is one of the important parameters of a reversing recording medium. Holograms recorded in BSN:Ce and stored in darkness remain practically unchanged over 14 days.

A material's ability for re-recording and reversibility can be estimated using a continuously rated recording, since under real conditions of a continuously rated crystal exposure the recording and erasure of holograms proceed simultaneously. So a continuous hologram recording for 24 h with

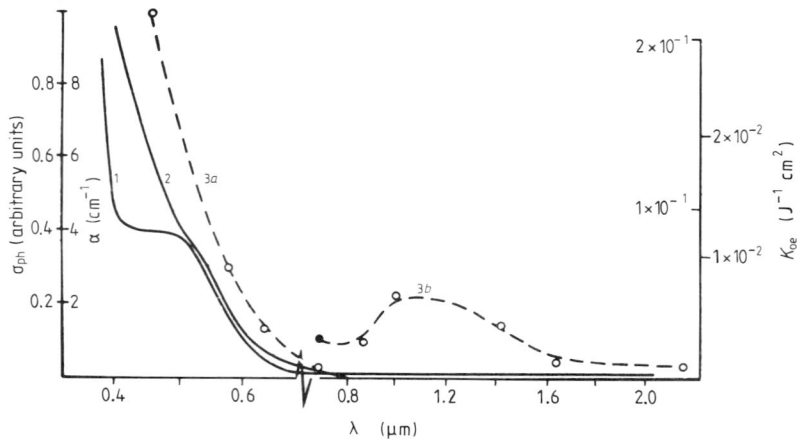

Figure 8.43 Spectral dependence of optical erasure constant of photo-refraction $K_{oe}(\lambda)$ (3a, b) in BSN:Ce crystals. Spectral dependences of the absorption coefficient $\alpha(\lambda)$ (1) and of photoconductivity $\sigma_{ph}(\lambda)$ (2) are given for comparison. In the IR region the K_{oe} is given on a smaller scale (3b).

a power density of $40 \, \mathrm{mW} \, \mathrm{cm}^{-2}$ produced no effect upon crystal characteristics. This dose of exposure is equivalent to 10^5 recording cycles.

8.7.3 *Photorefraction mechanism in* BSN:Ce *crystals*

A nonlinear dependence of σ_{ph} on I leads to the fact that the field strength of photogalvanic saturation is a function of light intensity and increases with the intensity $E_{\mathrm{sat}} = K_\mathrm{G} \alpha I / (\sigma_\mathrm{d} + \sigma_{\mathrm{ph}})$, whence one can calculate the refractive index variation Δn_e which is induced by the field caused by the photogalvanic effect. From the calculations it follows that the photogalvanic effect cannot be the cause of the variation $\Delta n_\mathrm{e}(I)$ in BSN:Ce crystals at intensities $I = 0.001\text{--}0.1 \, \mathrm{W} \, \mathrm{cm}^{-2}$ used in hologram recording. The mechanisms associated with charge drift in the internal electric field, or with the change of the electron state of impurity [103], cannot explain the experimental results either. In these models, neither the diffraction efficiency nor the sensitivity depend on the spatial frequency of recording, at least for low spatial frequencies. The most adequate mechanism which explains the experimental results is the diffusion mechanism [13].

In the diffusion mechanism, the charge redistribution results in the appearance of the spatial charge field $(E_{\mathrm{sp\,ch}})$ which at the initial region of hologram formation changes by the law [104]

$$E_{\mathrm{sp\,ch}} = \frac{eg_0 mtKD\tau}{(1 + K^2 D\tau)\varepsilon_0\varepsilon} \tag{8.58}$$

where g_0 is the free electron photo-generation rate, $g_0 = \alpha\beta I / h\nu$, m is the interference pattern contrast, D is the diffusion coefficient and K is the spatial frequency.

The variation of the refractive index is related to the spatial charge field as [105]

$$\Delta n = \tfrac{1}{2} n_\mathrm{e}^3 (r_{13} \sin^2 \theta / 2 + r_{33} \cos^2 \theta / 2) E_{\mathrm{sp\,ch}}. \tag{8.59}$$

The values of r_{33} and r_{13} are given in chapter 4.

The relation for the sensitivity to recording is obtained by substitution of equations (8.58) and (8.59) and the Einstein equation $D = kT\mu/e$ into the expression for the sensitivity $S = \Delta n / \Delta W$:

$$S = \frac{mK \cos^2(\theta/2)\lambda\alpha\mu\beta\tau}{2\varepsilon_0\varepsilon V_{\lambda/2}[1 + K^2(kT/e)\mu\tau]}. \tag{8.60}$$

Photoelectric measurements gave $(\mu\beta\tau) = 3.4 \times 10^{-10} \, \mathrm{V}^{-1} \, \mathrm{cm}^2$ for $I = 0.02 \, \mathrm{W} \, \mathrm{cm}^{-2}$; dielectric measurements gave $\varepsilon = 750$. Assuming $T = 300 \, \mathrm{K}$, $m = 0.9$ and $\beta = 0.1$, we obtain for the sensitivity

$$S = 0.6 \times 10^{-7} \frac{\cos^2(\theta/2)K}{1 + 8.7 \times 10^{-11} K^2}. \tag{8.61}$$

The calculated dependence $S(K)$ is presented in figure 8.40 by the full curve. The experimental values of S correspond qualitatively to the theoretical curve.

The dependences $\eta_{max}(K)$ calculated on the basis of photoelectric measurements for different m values are presented in figure 8.41. The best agreement between theory and experiment is observed for $m = 0.6$. The calculated value of η_{max} coincides with the experimental points up to frequencies of 600 lines mm^{-1}. The further difference between the theoretical and experimental curves can be explained by the fact that the holograms were recorded by beams with gaussian intensity distribution which, according to [106], strongly decreases η_{max}.

The beam intensity redistribution at the output from the hologram, which is shown in figure 8.42, is indicative of the existence of phase shift between the interference pattern and a periodic change of the refractive index close to $\pi/2$. Such a shift is typical of the diffusion mechanism of recording.

From the data of table 8.4 and photorefraction studies, it follows that the role of Ce reduces to an increase of absorption at a recording wavelength and to a substantial decrease of conductivity and photoconductivity.

From the material presented in the present section and the review in table 8.4, one can readily see that crystals of alkaline and alkaline earth niobates and tantalates are widely used in holography, particularly in elements of holographic memory. When enhanced hundreds of times by alloying with transition elements, the susceptibility of these crystals to laser radiation, which was earlier only an obstacle for employing these crystals as electro-optic modulators, now widens the possibilities of their device use.

Crystal ferroelectrics used for hologram recording have a number of advantages over other materials. There is first of all the possibility of recording volume phase holograms with a high recording density and a high diffraction efficiency, and the possibility of erasing this recording by heating, intense light pulses or by applying an electric field. The latter fact is especially promising for operative memory control. It should, however, be noted that there exist technological difficulties in obtaining optically homogeneous crystals.

9 Relation between Ferroelectric and Nonlinear Optical Properties of Oxygen Octahedral Ferroelectrics

To advance the technology and application of ferroelectrics, one should investigate the inter-relation between the ferroelectric characteristics (Curie temperature, spontaneous polarization, pyroelectric coefficients and others) and the optical and nonlinear properties. Several measured parameters will make it possible to judge approximately the other properties and the prospects for utilizing a given compound.

9.1 Curie temperature and spontaneous polarization

All the oxygen octahedral ferroelectrics considered above are of the displacive type. Simple experimental relationships between homopolar atomic displacement and macroscopic ferroelectric characteristics, such as Curie temperature T_C and spontaneous polarization P_s, which are valid for many materials, are established [1]. Physically, these experimenal relationships reflect the connection between lattice vibration energies and the formation of a ferroelectric state.

The properties of several displacive-type ferroelectrics are listed in table 9.1. The displacement of a homopolar atom of metal along the polar direction at a temperature $T \ll T_C$ is denoted Δz. The experimental Δz and T_C values for ten displacive-type ferroelectrics are presented in figure 9.1. The least-

Table 9.1 Properties of several displacive-type ferroelectrics [1].

Compound	T_c (K)	Space group transformation under transition to the paraelectric phase	Homopolar atom	Δz (Å)	P_s ($\mu C\ cm^{-2}$)
1 $NaNbO_3$	73 ± 10	monocl.–$Pbma$	Nb	0.060	11.7 ± 5
2 SbSI	296 ± 2	$Pna2_1$–$Pnam$	Sb	0.144 ± 0.040	25 ± 3
3 $Ba_{0.25}Sr_{0.75}Nb_2O_6$	348 ± 15	$P4bm$–$P4b2$	Nb	0.106 ± 0.022	31.3 ± 1
4 $Pb_{10}Fe_5Nb_5O_{30}$	388 ± 15	$R3m$–$R\bar{3}m$	Nb	0.091 ± 0.100	—
5 $BaTiO_3$	399 ± 5	$P4mm$–$Pm3m$	Ti	0.132 ± 0.009	25 ± 1
6 $Ba_6Ti_2Nb_8O_{30}$	505 ± 15	$P4bm$–$P4b2$	Ti, Nb	0.174 ± 0.100	22 ± 1
7 $KNbO_3$	708 ± 5	$Pmm2$–$Pm3m$	Nb	0.160 ± 0.014	30 ± 3
8 $PbTiO_3$	763 ± 15	$P4mm$–$Pm3m$	Ti	0.299 ± 0.040	—
9 $LiTaO_3$	891 ± 5	$R3C$–$R\bar{3}C$	Ta	0.197 ± 0.008	50 ± 2
10 $Bi_4Ti_3O_{12}$	949 ± 5	$Fmm2$–$Fmmm$	Ti	0.215	50 ± 10
11 $LiNbO_3$	1468 ± 15	$R3C$–$R\bar{3}C$	Nb	0.269 ± 0.006	71 ± 2
12 $Ba_{10}Cu_5W_5O_{30}$	1473 ± 15	$P4mm$–$Pm3m$	W	0.328 ± 0.100	—

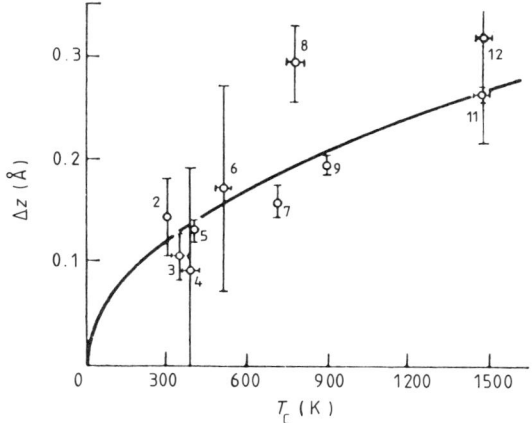

Figure 9.1 Experimental dependence of ferroelectric ion displacement Δz on Curie temperature T_C [1]. Numbers at the points indicate the number of the material presented in table 9.1. The full curve corresponds to the expression (9.1).

squares technique applied to experimental data gives the relation

$$T_C = (2.00 \pm 0.09) \times 10^4 (\Delta z)^2 \qquad (9.1)$$

where T_C is expressed in K and Δz in Å. The dependence (9.1) can also be expressed as

$$kT_C = \tfrac{1}{2} K (\Delta z)^2 \qquad (9.2)$$

where k is the Boltzmann constant, and $K = (5.52 \pm 0.25) \times 10^4$ dyn cm^{-1} can be estimated from either of the two latter equations.

The quantity K is comparable with the force constant of interatomic interaction in a crystal. Thus it may be interpreted as the mean force constant of the interaction of a homopolar atom with the oxygen atmosphere along the polar direction. The simple physical interpretation of equation (9.2) is as follows. At T_C, the amplitude of thermal oscillations of a homopolar atom turns out to be equal to Δz, i.e. the mean displacement of this atom becomes zero. Below T_C, when the amplitude of thermal oscillations drops rapidly, the mean displacement increases up to its asymptotic value Δz.

There exist only five displacive-type ferroelectrics for which P_s and Δz values are reliably measured (figure 9.2). The full line in this figure shows the experimental data processed using the least-squares technique, and corresponds to the relation

$$P_s = (258 \pm 9)\Delta z \qquad (9.3)$$

where P_s is expressed in μC cm^{-2} and Δz in Å. It should be noted that the

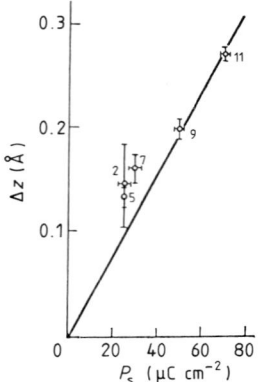

Figure 9.2   Experimental dependence of ferroelectric ion displacement Δz on spontaneous polarization P_s [1]. Numbers at points indicate the number of the material presented in table 9.1. The full line corresponds to expression (9.3).

value of the constant in the expression (9.3) must be actually dependent on the class of compound. The spontaneous polarization of the compounds listed in table 9.1 was determined for their Curie temperature on the basis of the above relations; the latter were found to be in close agreement with experiment.

We present another empirical relation which can be derived from the experimental data of table 9.1:

$$T_C = (0.303 \pm 0.018)P_s^2 \tag{9.4}$$

where T_C is expressed in K and P_s in $\mu C\ cm^{-2}$. The value of the constant estimated from equations (9.1) and (9.2) is 0.300 ± 0.020.

Thus, the above empirical relations hold for a wide range of displacive-type ferroelectrics and may appear to be useful in estimating the ferroelectricity of new compounds. The dependence of the Curie temperature on the normalized relation $\zeta = 10^{1/2}\ c/a$ for the crystal lattice parameters of tetragonal potassium tungsten bronzes was pointed out in [2] for the compounds obtained in $KNbO_3$–$LiNbO$–$BaNb_2O_6$, $KSr_2Nb_5O_{15}$–$KBa_2Nb_5O_{15}$ and $KBaNb_5O_{15}$–$KPb_2Nb_2O_{15}$ systems (figure 6.26).

9.2   Relation between the output of the second harmonic and the Curie temperature

The extensive experimental material obtained mainly on polycrystals, which permitted the establishment of correlation between the output of the second harmonic R_h and the Curie temperature of the compounds, was summarized in [3]. The second harmonic output with respect to the KDP crystal was found to be

$$R_h = I^{2\omega}/I^{2\omega}_{KDP}.$$

The available experimental data make it possible to choose material for frequency doubling on the basis of dielectric measurements, which is methodically simpler than measuring nonlinear optical characteristics. Besides, dielectric measurements can be made on powder and one need not grow single crystals to make preliminary estimations.

The relation for R_h was verified on a large number of compounds in which replacement ions were varied, and therefore the number of vacancies in the structure was varied too. Namely, the authors of [3] performed:

(i) variations of the number of vacancies in sites with coordinate numbers 15, 12 and 9 in the compounds $Ba_x Li_{5-2x} Nb_5 O_{15}$ ($2.02 \leqslant x \leqslant 2.20$) and $Ba_x Na_{5-2x} Nb_5 O_{15}$ ($1.93 \leqslant x \leqslant 2.22$);

(ii) cation replacements in sites with coordinate numbers 15 and 12 in $ABCNb_5 O_{15}$ type compounds, where A = Ca, Sr, Ba, Pb; B = Ca, Sr, Ba, Pb, and C = Na, K;

(iii) cation replacements in positions with coordinate numbers 6 and 8 in the compounds $Ba_{2.14} Li_{0.71} Nb_{5(1-x)} Ta_{5x} O_{15}$ ($0 \leqslant x \leqslant 1$) and $Ba_2 CNb_{5(x-1)} Ta_{5x} O_{15}$, where C = Na, K ($0 \leqslant x \leqslant 1$);

(iv) replacements in positions with coordinate numbers 15, 12 and 6 in the compound $Sr_{2-x} K_{1+x} Nb_{5-x} W_x O_{15}$ ($0 \leqslant x \leqslant 1$);

(v) replacements in cation and anion sublattices of the compounds $Ba_2 NaNb_{5-x} Ti_x O_{15-x} F_x$ ($0 \leqslant x \leqslant 0.50$) and $Ba_{2-x} Na_{1+x} Nb_5 O_{15-x} F_x$ ($0 \leqslant x \leqslant 1$).

Figure 9.3 shows that the output of the second harmonic R_h increases with Curie temperature. The experimental points are bounded by two broken lines.

Some of the investigated phases which may be of interest for second harmonic generation are listed in table 9.2.

When choosing compounds for parametric transformation, one should take into account the possibility of phase matching between the primary radiation and the second harmonic, as well as the possibility of obtaining a single-crystal compound.

It should be noted that the composition of a grown BSN single crystal differs from the melt composition. Besides, the congruent compositions recommended for growing have, as a rule, excess Ba and deficient Na. Table 9.2 demonstrates that, as soon as the quantity x in $Ba_x Na_{5-2x} Nb_5 O_{15}$ varies from 2.06 to 2.11, the output of the second harmonic decreases by a factor of 12. This implies that an effective transformation will take place in a composition close to stoichiometry.

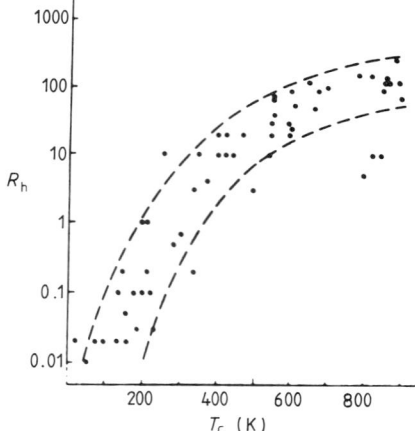

Figure 9.3 Output of second harmonic ($\lambda = 1.06\ \mu$m) as a function of Curie temperature for various ferroelectrics with tungsten bronze-type structure [3].

9.3 The theory of electro-optic and nonlinear optic effects

Several simplifying assumptions are made in the analysis of the features of oxide ferroelectrics. The first and the most important assumption suggests that the lattices of all single crystals are composed of BO_6 octahedra with sides equal to about 4 Å. The octahedra are arranged along the principal crystallographic symmetry axes. By disregarding slight distortions or slight orientational changes in the BO_6 octahedron structure of a real crystal, we disregard a slight crystal biaxiality, i.e. all polarized single crystals are treated as uniaxial. This assumption is confirmed experimentally. The basic structural element—the BO_6 octahedron—is represented in figure 9.4. In the absence of polarization the structure is centrosymmetric with symmetry O_h. When polarization occurs along the four-fold symmetry axis, the symmetry becomes C_{4v}; when along the three-fold symmetry axis the symmetry is C_{3v}; and when along the two-fold symmetry axis, the symmetry becomes C_{2v}. All the considerations in the sequel refer to perovskite-type BO_6 octahedra (see figure 1.1). To reduce other structures to the perovskite type, one should introduce the packing density η, which for an ideal perovskite is equal to unity and for other crystalline structures is defined as the ratio of the amount of B ions per unit volume of a given crystal to their amount per unit volume of a perovskite-type crystal [6]:

$$\eta = \frac{\rho_x M_p}{\rho_p M_x} \tag{9.5}$$

where $\rho_{x,p}$ is the density (g cm^{-3}) and $M_{x,p}$ the molecular weight of the

Table 9.2 Output of second harmonic with $\lambda = 1.06\ \mu$m as a function of the Curie temperature for several ferroelectrics [4, 5].

Composition	T_C (K)	R_h	Composition	T_C (K)	R_h
$Ba_xNa_{5-2x}Nb_5O_{15}$			$Ba_xLi_{5-2x}Nb_5O_{15}$		
$x = 1.93$	858	140	$x = 2.02$	899	70
2.00	853	120	2.08	896	120
2.06	853	120	2.14	883	250
2.11	843	10	2.18	863	120
2.17	818	10	2.20	852	90
2.22	798	5			
$Ba_2NaNb_{5(1-x)}Ta_{5x}O_{15}$			$Ba_{2.14}Li_{0.71}Nb_{5(1-x)}Ta_{5x}O_{15}$		
$x = 0.05$	786	150	$x = 0.05$	820	150
0.15	707	100	0.15	672	90
0.35	550	75	0.25	590	30
0.50	400	20	0.50	346	10
0.70	278	0.5	0.75	209	1
0.80	200	0.1	0.85	197	1
0.85	154	0.05	1.00	185	0.03
$Ba_2KNb_{5(1-x)}Ta_{5x}O_{15}$			Composition		
$x = 0.05$	600	25	$Sr_2NaNb_5O_{15}$	539	20
0.15	463	20	$Ba_2NaNb_5O_{15}$	853	120
0.25	421	10	$BaSrNaNb_5O_{15}$	547	40
0.35	230	0.03	$Sr_2KNb_5O_{15}$	437	10
0.50	158	0.02	$Ba_2KNb_5O_{15}$	668	50
0.60	136	0.02	$BaSrKNb_5O_{15}$	551	70
0.70	96	0.02	$BaCaNaNb_5O_{15}$	598	90
0.85	72	0.02	$SrCaKNb_5O_{15}$	546	30
1.00	60	0.01	$BaCaKNb_5O_{15}$	539	10
			$Pb_2KNb_5O_{15}$	647	120

formula unit; the indices x and p refer to the complex oxide and the perovskite crystal respectively. Within such a description, in the tungsten bronze-type structure one does not distinguish between octahedra containing BI and BII positions, so that the formula unit of these compounds can be written as ABO_3, where $A = (AI)_{1/5}(AII)_{2/5}C_{2/5}$ and $B = (BI)_{1/5}(BII)_{4/5}$.

9.3.1 *Sellmeier formula*

From the dispersion theory, the refractive index n of a dielectric medium in the weak absorption region is known to be defined [7] as

$$n^2(\omega) - 1 = \sum [f_i/(\omega_i^2 - \omega^2)]. \qquad (9.6)$$

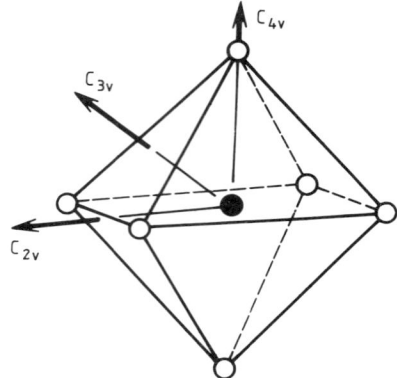

Figure 9.4 Octahedron BO_6 (the basic structural element) and its symmetry axes [6].

This equation takes into account intra-band optical transitions (occurring in a maximum density of states) of an individual dipole oscillator with frequency ω and force f_i. Equation (9.6) can be rewritten in the more convenient form

$$n^2(\lambda) - 1 = \sum S_i \lambda_i^2 [1 - (\lambda_i/\lambda)^2]^{-1} \qquad (9.7)$$

where S_i is the force factor. Equation (9.7) is known as the Sellmeier dispersion formula, utilized for the various approximations in the long-wavelength region of the spectrum. One oscillator is normally assumed to dominate, and then the action of other oscillators is denoted by a constant.

Approximation of equation (9.7) to the long-wavelength region, where oscillator parameters keep their values, leads to the Sellmeier relation for one term:

$$n^2(\lambda) - 1 = \frac{S_0 \lambda_0^2}{1 - (\lambda_0/\lambda)^2} \qquad (9.8)$$

where λ_0 is the mean wavelength and S_0 is the mean oscillator force, both differing in the general case from the parameters of individual oscillators in equation (9.7).

The parameters S_0 and λ_0 can be obtained from experiment by plotting $1/(n^2 - 1)$ versus $1/\lambda^2$. The slope of this straight line specifies the ratio $1/S_0$, and its intersection with the $1/(n^2 - 1)$ axis gives the value $1/S_0\lambda_0^2$. For the uniaxial crystals one has to establish two series of parameters S_a, λ_a and S_c, λ_c, whose subscripts indicate perpendicular (a) and parallel (c) directions of the axis, i.e.

$$n_a^2 - 1 = \frac{S_a \lambda_a^2}{1 - (\lambda_a/\lambda)^2} \qquad n_c^2 - 1 = \frac{S_c \lambda_c^2}{1 - (\lambda_c/\lambda)^2}. \qquad (9.9)$$

It should be noted that in a uniaxial crystal the oscillator forces and frequency dependences of individual optical transitions for light polarized along the a and c axes differ in the general case. The Sellmeier parameters S_0 and $\mathscr{E}_0 = hc/e\lambda_0$ (c is the velocity of light, h the Planck constant, e the electron charge) for several compounds are listed in table 9.3.

We should note that, for the compounds listed, the ratios $\mathscr{E}_0/S_0$ in the expressions for the electro-optic and nonlinear optic coefficients remain approximately constant. The mean value $\mathscr{E}_0/S_0 = (6 \pm 0.5) \times 10^{-14}$ eV m^2. One can thus arrive at the conclusion that the refractive indices of the various materials containing BO_6 octahedra possess similar dispersion properties.

The general formula for the birefringence dispersion can be derived from the system of equations (9.9). For a small birefringence $\Delta n = n_a - n_c$; in the first approximation the following expression holds:

$$n_0 \Delta n/(n_0^2 - 1) = (1/S_0\lambda_0^2)(\Delta\lambda_0/\lambda_0)\{1 + K[1 - (\lambda_0/\lambda)^2]\} \quad (9.10)$$

where

$$K = \frac{1}{2}\frac{\Delta S_0/S_0}{\Delta\lambda_0/\lambda_0}$$

$$\Delta S_0 = S_a - S_c; \quad \Delta\lambda_0 = \lambda_a - \lambda_c; \quad S_0 \approx S_a \approx S_c; \quad \lambda_0 \approx \lambda_a \approx \lambda_c \quad (9.11)$$

because $n_0 \approx n_a \approx n_c$. Equation (9.10) makes it possible to calculate the birefringence dispersion, provided that equations (9.9) hold for each light polarization direction.

Birefringence of a crystal can be caused either by its structure or by the field. In the former case, ΔS_0 and $\Delta\lambda_0$ are due to anisotropy of S_0 and λ_0

Table 9.3 Sellmeier parameters for several ferroelectrics (room temperature) [6].

Material	S_0 (10^{14} m^3)	$\mathscr{E}_0$ (eV)	$\mathscr{E}_0/S_0$ (10^{-14} eV m^2)	References
$BaTiO_3$	0.91	5.88	6.4	[8]
$KNbO_3$	0.9812 ± 0.005	6.68 ± 0.01	6.81 ± 0.05	[9]
$KTaO_3$	0.89	5.74	6.5	[6]
$KTa_{0.65}Nb_{0.35}O_3$	0.94	6.17	6.6	[6]
$LiNbO_3$	1.16; 1.15†	7.08; 6.74	6.1; 5.9	[6]
$LiTaO_3$	1.27	7.49	5.9	[6]
$Ba_2NaNb_5O_{15}$	1.29; 1.09	7.33; 6.42	5.7; 5.9	[6]
$Ba_2LiNb_5O_{15}$	1.15; 1.02	7.04; 6.90	6.12; 6.76	[10]
$Pb_3MgNb_2O_9$	0.97	5.71	5.88	[8]
$Ba_6Ti_2Nb_8O_{30}$	1.32; 1.36	6.42; 6.17	4.86; 4.54	[11]
$Pb_3ZnNb_2O_9$	1.089	5.88	5.40	[12]

† From here on the first value corresponds to polarization along the c axis and the second along the a axis.

along the *a* and *c* axes of the crystal. The anisotropy of ΔS_0 and $\Delta \lambda_0$ induced by the field results from the bias-stimulated forces and states of the oscillators along two orthogonal crystallographic axes. Comparing equation (9.10) with the experimental dependence of birefringence on the wavelength, one can find the birefringence dispersion constant K. The parameters $\mathscr{E}_0/S_0$ and K permit a detailed description of the behaviour of optical crystal dispersion, the ratio $\mathscr{E}_0/S_0$ depending on the characteristics of intra-band transitions in concrete crystals.

9.3.2 *Relation between Sellmeier parameters and band structure*

Oxygen electron levels O(2p) form an occupied valence band and the lowest unoccupied levels of the conduction band. The remaining portion of the conduction band is formed by d orbitals of transition metals. Kahn and Leyendecker [13] calculated the structure of the energy band of $SrTiO_3$. The picture obtained is schematically shown in figure 9.5. One can see that the d band splits into sub-bands d_ε and d_y. The order of sub-band succession can be established by considering the electrostatic energy of the cation localized in the centre of regular anion octahedra. In this configuration, d orbitals are directed between anions and are overlapped with oxygen p orbitals, mostly through the pdπ interaction. As a result of this interaction, the d_y band is assumed to lie on a higher energy level than the d_ε band. As a consequence of a large pd overlapping, the d band has a large width of $1-3$ eV. The width of the d_y band is mainly specified by the integral of pdσ

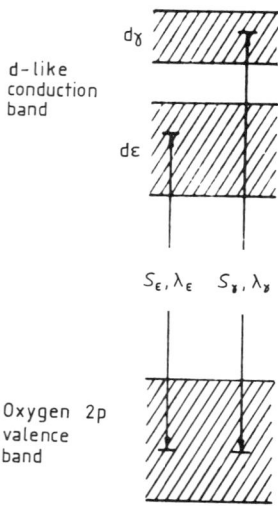

Figure 9.5   The structure of ferroelectric bands of $SrTiO_3$ [13].

overlapping. The upper valence band is relatively flat because of a small overlap of $pp\pi$ and $pp\sigma$ orbitals.

Although the energy bands were calculated by Kahn and Leyendecker [13] for $SrTiO_3$, Wemple and Di Domenico [6] proposed to extend these results, at least qualitatively, to a wide class of oxygen octahedral ferroelectrics. This conclusion was based on the experiments carried out by Kurtz [14], who had analysed the UV reflection spectra and shown that the band picture of figure 9.5 is common for all oxygen octahedral ferroelectrics. High-lying energy levels of the conduction band are strongly affected by the oxide composition, since these levels depend critically on the nature of metallic ions in the A position. Figure 9.6 presents the spectrum of the imaginary part of the optical dielectric permittivity, which was obtained by Kurtz [14] and Cardona [15] from the analysis of UV reflection. The figure indicates possible intra-band S_ε, λ_ε and S_γ, λ_γ transitions (see figure 9.5). Each of the two peaks, which fall on 5 ± 0.5 and 9 ± 1 eV, evidently involves more than one critical point or critical line in the Brillouin zone. The intra-band transition S, for example, may involve $X_{5'}(\text{upper}) \rightarrow X_3$, $X_{5'}(\text{upper}) \rightarrow X_5$ and $X_{5'}(\text{lower}) \rightarrow X_3$ transitions. Kahn and Leyendecker [13], as well as Zook and Casselman [16], singled out the $X_{5'}(\text{lower}) \rightarrow X_5$ transition as being mainly responsible for the transition S_ε. The energy of this transition, calculated by Kahn and Leyendecker [13], is in good agreement with the experimental measurements reported by Cardona [15].

According to the theoretical calculations carried out by Brews [17] for perovskite-type crystals, the crystal lattice polarization due to displacement

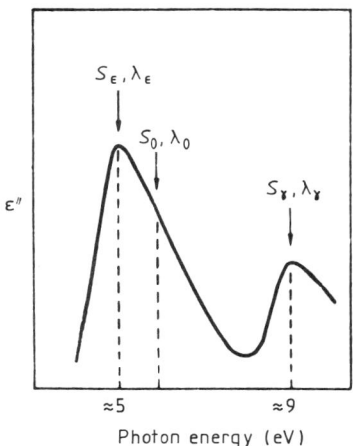

Figure 9.6 Absorption spectrum of $SrTiO_3$ [14, 15]. S_ε, λ_ε and S_γ, λ_γ are intra-band transitions; S_0, λ_0 are averaged parameters of the Sellmeier oscillator.

of B cations relative to O anions in the direction of the octahedral n-fold symmetry axis shifts the level of the d_ε conduction band towards higher energies. This shift is caused by variation of the pdπ overlapping integral with varying distance B–O. The shifts in the d_γ band are substantially smaller since pdσ overlappings are insensitive to B–O variations. For a quantitative description of the energy band shift due to polarization, one uses the polarization potential tensor introduced by Zook and Casselman [16]:

$$\Delta \mathscr{E}^n(\bar{k}) = \sum_{ij} \sigma_{ij}^n(\bar{k}) \bar{P}_i \bar{P}_j \qquad (9.12)$$

where $\Delta \mathscr{E}(\bar{k})$ is the shift of n-band energy levels at the point K of the Brillouin zone, $\sigma_{ij}^n(\bar{k})$ is the corresponding polarization potential tensor, and $\bar{P}_i$, $\bar{P}_j$ the crystal lattice polarization components. The energy level shifts are all assumed to be measured relative to the energy bands of the crystal in the centrosymmetric phase.

Since there exist several close transitions contributing to the transition S_ε, and since a detailed band structure and the intra-band selection rules for oxygen octahedral ferroelectrics are not reliably known, it is convenient to introduce an effective polarization potential β, which is the potential σ_{ij}^n averaged by corresponding selection rules for each polarization direction of light. The polarization potential β is supposed to range from 1 to 10 eV m^4 C^{-2}. This supposition rests upon theoretical calculations and experimental data. The level shifts of the energy $\mathscr{E}_\varepsilon = hc/e\lambda$ of the band d_ε are thus given by a simple expression $\Delta \mathscr{E}_\varepsilon = \beta P^2$. In a detailed analysis one finds that β depends on the light and crystal polarization directions relative to the BO$_6$ octahedron axes. For the polarization of light parallel ($\parallel$) and perpendicular ($\perp$) to the crystal polarization axis, we have

$$\Delta \mathscr{E}_\varepsilon(\parallel) = \beta_{11} P^2 \qquad \Delta \mathscr{E}_\varepsilon(\perp) = \beta_{12} P^2$$
$$\Delta \mathscr{E}_\varepsilon(\parallel) - \Delta \mathscr{E}_\varepsilon(\perp) = \beta_{44} P^2. \qquad (9.13)$$

9.3.3 Parameters of the Sellmeier oscillator

The relationships between the Sellmeier parameters (S_0, λ_0) and the band-structure parameters $(S_\varepsilon, S_y, \lambda_\varepsilon, \lambda_y)$ will be established using the two-band model (figure 9.5).

The displacements of the oscillator states S_ε and S_y and variations of the oscillator forces λ_ε and λ_y due to polarization R of the crystal are specified by the relations

$$\lambda_\varepsilon(\bar{P}) = \lambda_\varepsilon + \Delta \lambda_\varepsilon(\bar{P}) \qquad (9.14)$$

$$\lambda_y(\bar{P}) = \lambda_y + \Delta \lambda_y(\bar{P}) \qquad (9.15)$$

$$S_\varepsilon(\bar{P}) = S_\varepsilon + \Delta S_\varepsilon(\bar{P}) \qquad (9.16)$$

$$S_y(\bar{P}) = S_y + \Delta S_y(\bar{P}). \qquad (9.17)$$

To simplify the model, one assumes that $\Delta S_\gamma/S_\gamma = \Delta S_\varepsilon/S_\varepsilon = \Delta\lambda_\varepsilon/\lambda_\varepsilon = \Delta\lambda_\gamma/\lambda_\gamma = 0$. This is demonstrated in figure 9.7, in which the energy levels of oscillators are shown in the ground state and shifted. Since in the case of interest $\lambda_\gamma \ll \lambda_\varepsilon$ and $S_\gamma \lesssim S_\varepsilon$, then the inequality $R\lambda_\gamma^2/\lambda_\varepsilon^2 \ll 1$ holds, where $R = S_\gamma\lambda_\gamma^2/S_\varepsilon\lambda_\varepsilon^2$. Omitting the calculations, we present the final result

$$\frac{\Delta S_0}{S_0} \approx -\frac{\Delta\lambda_\varepsilon}{\lambda_\varepsilon}\frac{4R}{(1+R)} \qquad (9.18)$$

$$\frac{\Delta\lambda_0}{\lambda_0} \approx \frac{\Delta\lambda_\varepsilon(1+2R)}{\lambda_\varepsilon(1+R)}. \qquad (9.19)$$

Substituting equations (9.18) and (9.19) into (9.11), we obtain the expression for the birefringence dispersion constant K:

$$K \approx -2R/(1+2R) \qquad (9.20)$$

according to which this constant depends on R only. It is readily seen that the constant K is non-zero even for $\Delta S_\varepsilon = \Delta S_\gamma = 0$. From experiments on UV reflection in perovskite-like materials, it follows that $R \lesssim 0.5$. Substituting this value into equation (9.20), we find that $K \gtrsim -0.5$, which is in satisfactory agreement with experiment.

Thus the two-oscillator model (high-frequency and low-frequency oscillators) describes fairly well the dispersion of birefringence.

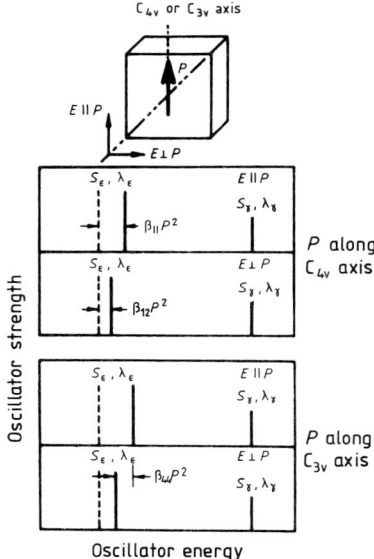

Figure 9.7 Schematic representation of oscillator shifts induced by polarization, $\beta_{ij}P^2$.

9.3.4 *The electro-optic effect*

Knowing the energy band structure and the Stark level shift induced by polarization, one can calculate the quadratic electro-optic coefficients g. In a zero approximation of the two-oscillator model, the coefficients of a clamped crystal can be computed. The clamping effect (piezoelectric effect or electro-striction) is of great importance for low frequencies. Mechanical strain of the crystal starts playing a significant part. In the high-frequency range, when mechanical strains developed in a crystal are small, the electro-optic effect is only due to crystal lattice poling and electron level shift in the energy band. The quadratic electro-optic coefficients of a clamped perovskite-like ferroelectric crystal in the centrosymmetric phase are determined by the equation

$$(1/n^2)_{ij} = \sum g_{ijkl} P_k P_l. \tag{9.21}$$

This relation was extended in [6] to the other oxide ferroelectrics isostructural to perovskites.

By combining equations (9.10) and (9.19), one derives the expression

$$\frac{n_0 \Delta n}{(n_0^2 - 1)^2} = \frac{e}{hc} \frac{\mathscr{E}_0}{S_0} \frac{1 + 2R}{(1 + R)^{1/2}} \frac{\Delta \lambda_\varepsilon}{\lambda_\varepsilon^2} \left[1 + K \left(1 - \frac{\lambda_0^2}{\lambda^2} \right) \right] \tag{9.22}$$

where e is the electron charge, which is introduced in order that all the other quantities can be represented in MKS units, and $\mathscr{E}_0/S_0$ has the dimensions of eV m^2.

The change in the refractive index Δn due to the oscillator shift $\Delta\lambda_\varepsilon/\lambda_\varepsilon^2 = (e/hc)\Delta\mathscr{E}_\varepsilon$ can be related to the quadratic electro-optic coefficients g and polarization potential β by means of equations (9.13) and the equations

$$\Delta n = \tfrac{1}{2} n_0^3 g_{ij} \bar{P}^2. \tag{9.23}$$

The subscripts i, j correspond to the subscripts at the polarization potential in equations (9.13).

Substituting equations (9.13) and (9.23) into (9.22), we obtain

$$g_{ij} = 2 \left(\frac{e}{hc} \right)^2 \frac{\mathscr{E}_0}{S_0} \beta_{ij} N_0 \frac{1 + 2R}{(1 + R)^{1/2}} \left[1 + k \left(1 - \frac{\lambda_0^2}{\lambda^2} \right) \right] \tag{9.24}$$

where

$$N_0 = (1 - 1/n_0^2)^2. \tag{9.25}$$

For IR waves, equation (9.24) along with equation (9.20) gives

$$g_{ij} \approx \frac{2}{(1 + R)^{1/2}} \left(\frac{e}{hc} \right)^2 \frac{\mathscr{E}_0}{S_0} \beta_{ij} N_0. \tag{9.26}$$

For perovskite-type structures, for which $R \lesssim 0.5$, $N_0 \approx 0.65$ and $\mathscr{E}_0/S_0 \approx$

$6.5 \times 10^{-14} \, \text{eV m}^2$, equation (9.26) leads to

$$g_{ij} \approx \beta_{ij}/20 \tag{9.27}$$

where g_{ij} and β_{ij} have respectively the units $\text{m}^4 \, \text{C}^{-2}$ and $\text{eV m}^4 \text{C}^{-2}$.

The expression (9.27) implies that the quadratic electro-optic coefficients g are directly related to the components of the polarization potential tensor. For perovskites [18–20] $g_{11} \approx 0.17$, $g_{12} \approx 0.04$ and $g_{44} \approx 0.12 \, \text{m}^4 \, \text{C}^{-2}$. On substitution into equation (8.27), these values give $\beta_{11} = 3.4$, $\beta_{12} \approx 0.8$ and $\beta_{44} \approx 2.4 \, \text{eV m}^4 \text{C}^{-2}$. These estimates are in good agreement with the theoretical calculations [17], with the experimental results of electro-reflection measurements [21–22] and with forbidden bandwidth measurements [23]. This allowed Di Domenico and Wemple to conclude that the electro-optic effect is caused by variation of pdπ overlapping of electron orbits induced by B cation displacement in an oxygen octahedral cell. Oxygen octahedral ferroelectrics have similar band structures, and therefore almost identical polarization potentials and identical quadratic electro-optic coefficients.

To extend the above perovskite analysis to other complex oxide ferroelectrics, it is necessary, first, to introduce a correction to the packing density of BO_6 octahedra; second, to transform the tensor of rank 4, g, for polarising along other principal symmetry axes of a unit cubic cell (figure 9.4); and finally, to associate the electro-optic coefficients with the corresponding coefficients of perovskites.

If each octahedron is assumed to have a dipole moment μ induced by an applied field or spontaneous polarization, then the polarization P_x occurring in a complex oxide and normalized to the ideal perovskite polarization P_p gives a packing density determined by the equation (9.5)

$$P_x/P_p = \eta.$$

Omitting tensor notation, one can write the expressions for the quadratic electro-optic effect as follows:

$$\Delta(1/n_p^2) = g_p P_p^2 \qquad \Delta(1/n_x^2) = g_x P_x^2 \tag{9.28}$$

where the subscript p refers to the perovskite and x to the complex oxide. Using expressions (9.28) and the relation for η, we obtain

$$\frac{g_p}{g_x} = \eta^2 \frac{n_x^4 \, \Delta n_p^2}{n_p^4 \, \Delta n_x^2}. \tag{9.29}$$

The linear dielectric susceptibility $\chi \approx n^2 - 1$ is directly related to η as follows:

$$n^2 - 1 = A\eta.$$

For oxygen octahedral ferroelectrics, $A \approx 4$. Substituting the dielectric susceptibility into (9.29), we obtain

$$\frac{g_\mathrm{p}}{g_x} = \eta \frac{(1 + A\eta)^2}{(1 + A)^2}. \tag{9.30}$$

Since for oxygen octahedral ferroelectrics $\eta \leqslant 1.2$ and $A \approx 4$, the expression (9.3) can be approximated to the form

$$g_\mathrm{p}/g_x \approx \eta^3. \tag{9.31}$$

In the calculations of the linear electro-optic coefficients, the linear electro-optic effect was assumed to be essentially quadratic due to spontaneous polarization.

From the structural point of view, there is a difference between the spontaneous polarization direction along the four-fold symmetry axis of an elementary octahedron (figure 9.4), as in perovskites and tungsten bronzes, and along the three-fold symmetry axis, as in lithium niobate.

As mentioned above, the polar structures of tungsten bronzes have symmetry C_{4v}. A LiNbO$_3$ type rhombohedral cell can be formed of two perovskite cells which are located along the three-fold octahedral symmetry axis and have an almost reversed orientation. Thus the LiNbO$_3$ crystal may be regarded as having symmetry C_{6v}.

These non-centrosymmetric structures possess the linear electro-optic effect described by the expression

$$\Delta\left(\frac{1}{n^2}\right)_{ij} = \sum f_{ijk} P_k \tag{9.32}$$

where for both the point groups C_{4v} and C_{6v} the tensor components f_{33}, $f_{13} = f_{23}$ and $f_{51} = f_{42}$ are non-zero. The relation between the coefficients of the tensor of the linear electro-optic effect described by the expression (9.32) and the tensor of the quadratic electro-optic effect was specified by Wemple and Di Domenico [6]. For crystals belonging to the point symmetry group C_{4v} we can write the equalities

$$f_{13} = 2g_{12} P_\mathrm{s} \qquad f_{33} = 2g_{11} P_\mathrm{s} \qquad f_{51} = g_{44} P_\mathrm{s}. \tag{9.33}$$

The total birefringence Δn_t in the general case is the sum of the contribution induced by polarization and the temperature-dependent structural contribution:

$$\Delta n_\mathrm{t} = -\tfrac{1}{2} n_0^3 (g_{11} - g_{12}) P_\mathrm{s} + \Delta n_0(T). \tag{9.34}$$

For structures with symmetry C_{6v}, the corresponding equations will have the form

$$\begin{aligned} f_{13} &= \tfrac{2}{3}(g_{11} + 2g_{12} - g_{44})P_\mathrm{s} \\ f_{33} &= \tfrac{2}{3}(g_{11} + 2g_{12} + 2g_{44})P_\mathrm{s} \end{aligned} \tag{9.35}$$

$$\begin{aligned} f_{51} &= \tfrac{2}{3}(g_{11} - g_{12} + \tfrac{1}{2}g_{44})P_\mathrm{s} \\ \Delta n_\mathrm{t} &= -\tfrac{1}{2} n_0^3 g_{44} P_\mathrm{s}^2 + \Delta n_0(T). \end{aligned} \tag{9.36}$$

To find the coefficients g contained in equations (9.33)–(9.36), we make use of the measured electro-optic coefficients defined by the relation

$$\Delta(1/n^2)_{ij} = \sum r_{ijk} E_k \tag{9.37}$$

where E_k is an applied field. We then employ the known relations

$$P_k = \varepsilon_0(\varepsilon_k - 1)E_k \qquad f_{ijk} = \frac{r_{ijk}}{\varepsilon_0(\varepsilon_k - 1)} \tag{9.38}$$

where ε_k is the dielectric permittivity along the k-fold symmetry axis and ε_0 is the electrical permittivity of a vacuum.

Combining equations (9.38) with (9.31), (9.33) and (9.35), we obtain for the perovskites the following relations between the measured linear (r) and quadratic (g_p) coefficients, for symmetry C_{4v}:

$$r_{13} = 2(g_{12})_p \varepsilon_0 (\varepsilon_c - 1) P_s / \eta^3$$

$$r_{33} = 2(g_{11})_p \varepsilon_0 (\varepsilon_c - 1) P_s / \eta^3 \tag{9.39}$$

$$r_{51} = 2(g_{44})_p \varepsilon_0 (\varepsilon_a - 1) P_s / \eta^3$$

and for the symmetry C_{6v}:

$$r_{13} = \tfrac{2}{3}[(g_{11})_p + 2(g_{12})_p - (g_{44})_p] \varepsilon_0 (\varepsilon_c - 1) P_s / \eta^3$$

$$r_{33} = \tfrac{2}{3}[(g_{11})_p + 2(g_{12})_p + 2(g_{44})_p] \varepsilon_0 (\varepsilon_c - 1) P_s / \eta^3 \tag{9.40}$$

$$r_{51} = \tfrac{2}{3}[(g_{11})_p - (g_{12})_p + \tfrac{1}{2}(g_{44})_p] \varepsilon_0 (\varepsilon_a - 1) P_s / \eta^3.$$

In these relations ε_c and ε_a are dielectric permittivities along the c and a axes. Substituting the experimental values of the linear coefficients r_{ij}, the spontaneous polarization P_s and the dielectric permittivities into equations (9.39) and (9.40), one can obtain quadratic electro-optic coefficients for some ferroelectric materials (table 9.4). Note that the values of these coefficients are close to those for oxygen octahedral ferroelectrics. This implies that the polarization tensor coefficients are also identical for these classes of compounds.

The similarity of polarization potentials β in similar crystalline and band structures is thus experimentally confirmed.

9.3.5 The nonlinear optic effect

Making use of the polarization potential, one can estimate the coefficients of the optical nonlinear susceptibility tensor. The optical nonlinearity is regarded as occurring due to Stark energy level shifts induced by electron polarization, by analogy with the polarization shift responsible for the electro-optic effect. We restrict our consideration to the electron processes only, i.e. nuclei are assumed to be fixed. It should be noted, however, that a combination of low-frequency (ion) and high-frequency (electron) polarization

Table 9.4 Dielectric constants and quadratic electro-optic coefficients for several ferroelectrics at room temperature (in MKS system) [6].

Material	T_C (°C)	η	P_s (C m^{-2})	ε_a	ε_c	$g_{11}-g_{12}$	(g_{44})	(g_{11})	(g_{12})
LiNbO$_3$	1195	1.2	0.71	30	84	0.12	0.11	0.16	0.043
LiTaO$_3$	610	1.2	0.50	45	51	0.14	0.12	0.17	0.03
Ba$_2$NaNb$_5$O$_{15}$	560	1.03	0.40	51	242	0.12	0.12	0.17	0.048
Ba$_{0.5}$Sr$_{0.5}$Nb$_2$O$_6$	115	1.06	0.25	300	—	0.13	—	—	—
BaTiO$_3$	120	1.0	0.25	160	3000	0.14	—	—	—

potentials can in principle describe the majority of fundamental nonlinear optical problems.

To calculate the optical polarization potential, we assume that the main contribution to the electron polarization is given by 2p orbitals of oxygen ions. The electron polarization P_e can be considered as a result of shift and distortion of 2p orbitals. The effect of the optical field upon the pdπ energy overlapping integral is similar to the static effect induced by the shift of the B cation with respect to the O anion.

To describe nonlinear optical interactions in acentric crystals, one usually assumes the presence of a specific nonlinear polarization 'source' p_i^α, which is defined by the expression

$$p_i^\alpha = \sum_{j,k} d_{ijk}^{\alpha\beta\gamma} E_j^\beta E_k^\gamma \qquad (9.41)$$

where E_j^β and E_k^γ are electric fields, i, j, k are the principal axes in the crystal, and α, β, γ are the indices of the radiation frequencies $\omega_\alpha, \omega_\beta, \omega_\gamma$ which satisfy the energy conservation law $\omega_\alpha = \omega_\beta \pm \omega_\gamma$.

Miller has shown [24] that it is physically more meaningful to introduce a nonlinear field source e_i by means of the expression

$$e_i^\alpha = \sum_{j,k} \delta_{ijk} P_j^\beta P_k^\gamma \qquad (9.42)$$

where P_j^β and P_k^γ are crystal polarizations at frequencies ω_β and ω_γ, and δ_{ijk} are experimentally established material-independent nonlinear coefficients.

The tensor δ_{ijk}, as shown below, is directly connected with the polarization potential tensor β_{ij}^0 which describes the energy band shift by a light wave.

The polarization source p_j^α and the field source e_i^α were ascertained by Di Domenico and Wemple to be related through linear susceptibility

$$p_i^\alpha = \varepsilon_0 \chi_{ii}^\alpha e_i^\alpha. \qquad (9.43)$$

Substituting (9.43) into (9.42) and comparing with (9.41), we obtain [24–27]

$$d_{ijk}^{\alpha\beta\gamma} = \varepsilon_0^3 \chi_{ii}^\alpha \chi_{jj}^\beta \chi_{kk}^\gamma \delta_{ijk}. \qquad (9.44)$$

Since the susceptibilities in equation (9.44) possess dispersion, Kleinman's symmetry conditions for the coefficients δ_{ijk} are fulfilled approximately. In experiments on SHG, where $\omega_\alpha = 2\omega$ and $\omega_\beta = \omega_\gamma = \omega$, according to (9.44) we can write the relations

$$d_{15}^{2\omega} = \chi_1^{2\omega} \chi_1^\omega \chi_3^\omega \delta_{15} \qquad d_{13}^{2\omega} = \chi_3^{2\omega} \chi_1^\omega \chi_1^\omega \delta_{31}.$$

For crystals possessing either C_{4v} or C_{6v} symmetry, the condition $\delta_{31} = \delta_{15}$ holds, from which one can obtain

$$d_{15}^{2\omega}/d_{31}^{2\omega} = (\chi_1^{2\omega}/\chi_1^\omega)(\chi_3^\omega/\chi_3^{2\omega}).$$

It should be noted that according to [6] the equality $d_{15}^{2\omega} = d_{31}^{2\omega}$ holds within

5% accuracy. Thus, in the visible region, where the dispersion of the refractive indices is not too high, Kleinman's conditions are approximately fulfilled.

We are now in a position to establish the relation between the nonlinear susceptibility tensor δ_{ijk} and the polarization potential tensor β_{ijk} in acentric crystals by assuming β_{ijk} to remain constant both in low-frequency and in high-frequency regions. Equations (9.41) and (9.44) give

$$p_i^\alpha = \varepsilon_0 \sum_{j,k} \chi_{ii}^\alpha \delta_{ijk} P_j^\beta P_k^\gamma. \tag{9.45}$$

In electro-optics, $\omega_\gamma = 0$ and $\omega_\alpha = \omega_\beta = \omega$, so that along the principal crystallographic directions equation (9.45) is transformed to become

$$p_i^\omega = 2\varepsilon_0 \sum_{j,k} \chi_{ii}^\omega \delta_{ijk} P_i^\omega P_k^0 \tag{9.46}$$

where P_k^0 is the applied static polarization, and the coefficient 2 results from bilateral constraint.

The coefficients p_i^ω can be expressed in terms of the linear electro-optic coefficients of the polarizing crystal. Along the principal axes

$$p_i^\omega = \varepsilon_0 \Delta\chi_{ii}^\omega E_i^\omega$$

where $\Delta\chi_{ii}^\omega$ is an induced change in linear susceptibility. It has been shown [6] that along principal crystallographic axes

$$\Delta\chi_{ii}^\omega = -n_{ii}^4 \Delta(1/n^2)_{ii} \tag{9.47}$$

where $\Delta(1/n^2)_{ii}$ is connected with the clamped crystal effect by means of the relation (9.32). The relation (9.32) leads to the expression

$$\Delta\chi_{ii}^\omega = -n_{ii}^4 \sum_k f_{iik} P_k^0. \tag{9.48}$$

Since

$$p_i^\omega = \varepsilon_0 \Delta\chi_{ii}^\omega E_i^\omega = \frac{\Delta\chi_{ii}^\omega}{\chi_{ii}^\omega} P_i^\omega$$

from equation (9.48) we obtain

$$\delta_{iik} = -\left[2\varepsilon_0 \left(1 - \frac{1}{n_{ii}^2} \right)^2 \right]^{-1} f_{iik}. \tag{9.49}$$

For crystals possessing C_{4v} and C_{6v} symmetries, we find particular expressions for nonlinear coefficients. For the symmetry C_{4v}

$$\delta_{15} = g_{12}^c P_s / \varepsilon_0 N_a \qquad \delta_{33} = g_{11}^c P_s / \varepsilon_0 N_c$$

and for the symmetry C_{6v}

$$\delta_{15} = (g^c_{11} + 2g^c_{12} - g^c_{44})\frac{P_s}{3\varepsilon_0 N_a}$$

$$\delta_{33} = (g^c_{11} + 2g^c_{12} + 2g^c_{44})\frac{P_s}{3\varepsilon_0 N_c}$$

(9.50)

where $N_a = (1 - 1/n_a^2)^2$, $N_c = (1 - 1/n_c^2)^2$. Here n_a and n_c are refractive indices in directions respectively perpendicular and parallel to the c axis. The coefficients g^c are quadratic electro-optic coefficients of the clamped crystal.

From the symmetry permutation relations for δ_{ijk} it was found that $\delta_{31} = \delta_{15}$. Thus, structures with C_{4v} or C_{6v} symmetry have only two independent nonlinear susceptibility constants δ_{33} and δ_{15}.

Substituting in equations (9.50) the equations (9.26) and (9.31), we derive the following important relations: for symmetry group C_{4v}

$$\delta_{15} = \frac{D\beta_{12}P_s}{\eta^3\varepsilon_0} \qquad \delta_{33} = \frac{D\beta_{11}P_s}{\eta^3\varepsilon_0}$$

(9.51)

and for symmetry group C_{6v}

$$\delta_{15} = D(\beta_{11} + 2\beta_{12} - \beta_{44})P_s/3\eta^3\varepsilon_0$$

$$\delta_{33} = D(\beta_{11} + 2\beta_{12} + 2\beta_{44})P_s/3\eta^3\varepsilon_0$$

where

$$D = [2/(1 + R)^{1/2}](e/hc)^2(\mathscr{E}_0/S_0)$$

(9.52)

in the assumption that $N_a \approx N_c \approx N_0$. In oxygen octahedral ferroelectrics $R \lesssim 0.5$ and $\mathscr{E}_0/S_0 \approx 6 \times 10^{-14}$ eV m^2, so that from equation (9.52) it follows that $D \approx 0.07$ eV^{-1}. Employing the numerical values of the static coefficients β, we determine δ_{15} and δ_{33} (J m^3 C^{-3}) for both symmetry groups C_{4v} and C_{6v}:

$$\delta_{15} \approx 0.7 \times 10^{10} P_s/\eta^3 \qquad \delta_{33} \approx 2.8 \times 10^{10} P_s/\eta^3.$$

(9.53)

Comparison of (9.53) with experiment can only be made for a few materials, because there are very few data on SHG. Table 9.5 gives a list of experimental and theoretical values of these quantities for several crystals possessing C_{4v} and C_{6v} symmetry. The experimental values were referred to the quartz standard, δ^Q_{11}. The absolute values can be obtained by assuming for quartz $\delta^Q_{11} = (0.25 \pm 0.05) \times 10^{10}$ J m^3 C^{-3}. Table 9.5 exhibits a rather close agreement between the measured and calculated coefficients δ_{ijk}. It can be concluded that the polarization potential β^0 for the optical frequency does not differ substantially from the static potential which describes the electro-optic effect, but these two potentials are essentially distinct.

It is of interest that crystals with the symmetry group C_{6v} (LiNbO$_3$) are less suitable in their nature for SHG than crystals with the symmetry group

Table 9.5 Coefficients of nonlinear optical susceptibility tensor for several ferro-electrics [6].

Material	η	P_s (C m^{-2})	δ_{33} (10^{10} J m^3 C^{-3}) expt.	calc.	δ_{15} (10^{10} J m^3 C^{-3}) expt.	calc.
$K_3Li_2Nb_5O_{15}$	1	0.25	0.5 ± 0.15	0.7	0.3 ± 0.12	0.18
$BaTiO_3$	1.0	0.25	0.14 ± 0.06	0.7	0.20 ± 0.08	0.18
$Ba_2NaNb_5O_{15}$	1.03	0.40	0.6 ± 0.18	1.0	0.35 ± 0.1	0.25
$LiTaO_3$	1.2	0.50	0.8 ± 0.25	0.8	0.05 ± 0.02	0.21
$LiNbO_3$	1.2	0.71	1.1 ± 0.3	1.2	0.18 ± 0.06	0.29

$C_{4v}(Ba_2NaNb_5O_{15})$: the ratio of the nonlinear coefficients $\delta_{15}(Ba_2NaNb_5O_{15})/\delta_{15}(LiNbO_3)$ is approximately 3.

Making use of equation (9.44) and Kleinman's conditions, one can calculate the nonlinear susceptibility constant d. From equation (9.8) for $\lambda \gg \lambda_0$, we have

$$\chi = n^2 - 1 \approx S_0\lambda_0^2 = (hc/e)^2 S_0/\mathscr{E}_0^2. \tag{9.54}$$

Substituting expression (9.54) into (9.44), we obtain

$$d_{31} \approx d_{15} \approx [(hc/e)^2\varepsilon_0(S_0/\mathscr{E}_0)]^3(\delta_{31}/\mathscr{E}_0^3) \tag{9.55}$$

$$d_{33} \approx [(hc/e)^2\varepsilon_0(S_0/\mathscr{E}_0)]^3(\delta_{33}/\mathscr{E}_0^3). \tag{9.56}$$

Since $S_0/\mathscr{E}_0$ is a constant quantity and $\delta \sim \beta P_s$, taking into account equation (9.51) we have

$$d/P_s \sim \beta/\eta^3\mathscr{E}_0^3. \tag{9.57}$$

From the last relation one can draw an instructive practical conclusion. As has been shown above, $\mathscr{E}_0$ decreases with decreasing forbidden band width of the crystal. Consequently, d/P_s must be larger in materials with narrower bands. The forbidden band increases with the principal quantum number of the B cation: i.e. in the Ti–Nb–Ta ion series, the titanates with perovskite and tungsten bronze-type structures should therefore be expected to possess optimal nonlinear optical properties.

Thus, from the energy band picture, one can see that the coefficients of electro-optic and nonlinear optic tensors are similar in all oxygen octahedral ferroelectrics. This is the result of BO_6 octahedral structure of these crystals, which leads to similar energy band structures and identical Stark splittings induced by polarization.

9.4 Optical properties of ferroelectrics with a smeared phase transition

The effect of disorder upon the behaviour of ferroelectrics was considered in [28–31]. Ferroelectrics with a relatively large degree of disorder (or a large impurity concentration) constitute a special group in which the refractive index varies gradually between high-temperature and low-temperature phases, in contrast with the sharp change in 'pure' ferroelectrics. Burns [32] called such materials 'dirty' displacive-type ferroelectrics. This term refers to crystals in which each unit cell is slightly different from another one, but the mean translational symmetry of the whole crystal is nonetheless conserved. Smolensky defined such materials as ferroelectrics with a diffuse phase transition [33]. The temperature dependence of the dielectric permittivity of such ferroelectrics has a broad peak, and the diffuse phase transition occurs as a result of the macroscopic variation of composition, which is responsible for the distinct Curie temperatures in separate macroscopic regions. In such crystals, the peak position of the permittivity–temperature curve is conditionally taken for the Curie temperature. Typical representatives of crystals with smeared phase transition may be $Pb_3MgNb_2O_9$, $Pb_3ZnNb_2O_9$ and $Ba_xSr_{1-x}Nb_2O_6$.

The shape of the permittivity–temperature curve is a good indicator of crystal structure deviation from the order. Materials with a smaller degree of structure ordering have a broader peak of the dependence $\varepsilon(T)$ [34–36]. This effect is demonstrated by the dependences $\varepsilon(T)$ obtained in [35] and depicted in figure 6.15.

Burns and colleagues [28–30, 32] believe that the optical properties of the materials are more strongly affected by lack of identity of unit cells than by a macroscopical variation of the composition. The temperature dependences of the refractive indices, $n(T)$, of ferroelectric $K_{2-2x}Sr_{4+x}Nb_{10}O_{30}$ tungsten bronzes are presented in [32]. For $x = 0$, the compound has the stoichiometric composition $K_2Sr_4Nb_{10}O_{30}$ in which all unit cells are identical. For $x > 0$, the degree of disorder in a crystal may in principle change. Experiments show that with an increase of crystal lattice disorder the behaviour of the dependence $n(T)$ becomes different from the expected scheme depicted in figure 9.8. Figure 9.9 presents the dependences $n(T)$ of the crystals $K_{1.43}Sr_{3.76}Nb_{10}O_{29.47}$ (curve 1) and $K_{0.74}Sr_{3.88}Nb_{10}O_{29.26}$ (curve 2). The refractive indices were measured using the method of minimum deflection of a He–Ne laser beam ($\lambda = 632.8$ nm) by the crystal prism. Samples 1 show a slight temperature hysteresis of $n(T)$, which coincides with the hysteresis of $\varepsilon_c(T)$. The presence of hysteresis of $\varepsilon_c(T)$ and $n(T)$ characterizes a first-order phase transition. The more disordered composition (curve 2) exhibits no temperature hysteresis.

The changes in the refractive indices of the indicated crystals over the entire temperature range make up $\Delta n_a = 0.002$ and $\Delta n_c = 0.07$. The observed anisotropy of Δn is much higher than for other known tungsten bronze-type

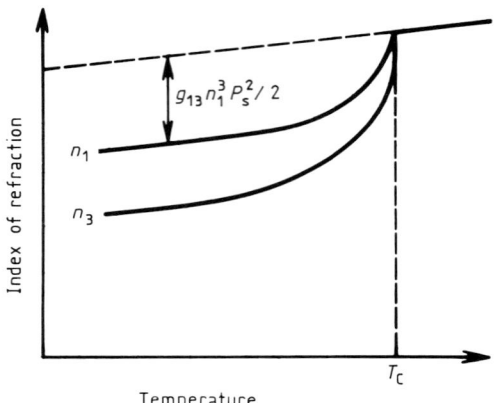

Figure 9.8 Schematic representation of temperature dependence of refractive indices n_1 and n_2 for 'pure' ferroelectrics [29]. The deviations from the rectilinear dependence at $T < T_C$ are due to spontaneous polarization of crystals.

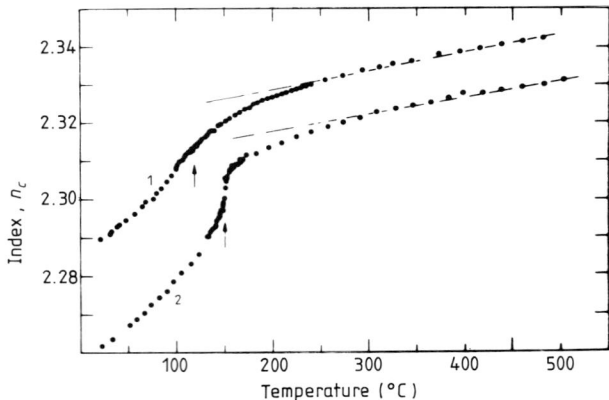

Figure 9.9 Temperature dependences of refractive indices of (1) $K_{1.43}Sr_{3.76}Nb_{10}O_{29.47}$ and (2) $K_{0.74}Sr_{3.88}Nb_{10}O_{29.26}$ crystals under light polarization along the ferroelectric axis c ($\lambda = 632.8$ nm) [32]. Arrows correspond to maxima of temperature dependences of dielectric permittivity of crystals.

ferroelectrics. As has already been mentioned, the form of the dependence $n_c(T)$ differs qualitatively from what is expected for ordered ferroelectrics (figure 9.8). We shall not attempt a discussion of this phenomenon at this stage, since this forms the subject of a more detailed treatment given below. The main concepts employed in interpretation of the dependence $n(t)$ are associated with the quadratic electro-optic effect (9.21) which occurs in all crystals.

As is well known, the spontaneous electro-optic effect in the high-temperature symmetric ferroelectric phase is quadratic and is determined by the expression (9.21)

$$\frac{1}{n_{ij}^2} = \sum g_{ijkl} P_k P_l.$$

We represent the polarization as the sum of two parts P_s and P_m, the former being the spontaneous polarization and the latter the polarization due to an applied field.

For the indicated oxygen octahedral ferroelectrics, the dielectric permittivity along the axis of spontaneous polarization is equal to $\varepsilon_3 \gg 1$, and for the induced polarization P_m one can write

$$P_m = \frac{\varepsilon_3}{4\pi} E_3. \tag{9.58}$$

Using the conventional brief form of notation, we obtain from (9.21) and (9.58) the equation

$$\Delta(1/n^2)_j = g_{j3} P_s^2 + (g_{j3} P_s \varepsilon_3 / 2\pi) E_3. \tag{9.59}$$

The large value of spontaneous polarization P_s near the Curie point leads to distortion of the centrosymmetric crystal lattice and to a linear electro-optic effect induced by an applied field E_3. In this sense, the linear electro-optic effect will be the continuation of the quadratic effect. If it is written in the conventional form, as the product $r \times E$, then the linear electro-optic coefficient can be expressed in terms of the polarization P_s:

$$r = g P_s \varepsilon_3 / 2\pi.$$

The quadratic electro-optic coefficient g above T_C and the linear electro-optic coefficient r below T_C were measured for the $K_2 Sr_4 Nb_{10} O_{30}$ type structure with known P_s and ε [32]. The observed similarity of these two quantities (g and r) confirmed the fact that the lattice distortion is responsible for the occurrence of the linear electro-optic effect. Thus the simple idea of lattice distortion caused by spontaneous polarization, which explains the nature of the linear electro-optic effect, was supported by experimental evidence. Moreover, it was ascertained in [37–39] that the temperature dependence of g is weak.

The change in the value of the refractive index Δn can be obtained by differentiating equation (9.59). The second term on the right-hand side of this equation can be neglected, and one obtains

$$\Delta n_j = -\tfrac{1}{2}(n_j^0)^3 g_{j3} P_3^2. \tag{9.60}$$

The expression implies that as the polarization varies from 0 to P, the refractive index changes by Δn_j.

It should be emphasized that Δn is proportional to the polarization squared. If polarizations of separate crystal regions are antiparallel and irreversible, the refractive index of the non-polar phase will still change. This can be explained by assuming a crystal in the centrosymmetric phase to have local polarization. Using equation (9.60) and the data of figure 9.9, one can determine the mean local polarization of crystals. To this end one should know the refractive index of unpolarized crystal n_0 and the squared electro-optic coefficient g.

For $K_2Sr_4Nb_{10}O_{30}$, direct measurements above T_C have shown that $g_{33} - (n_1/n_3)^3 g_{13} = 0.1 \text{ m}^4 \text{C}^{-2}$ [37]. For this crystal, one can neglect the quantity g_{13} and assume $g_{33} = 0.1 \text{ m}^4 \text{C}^{-2}$. The n_3^0 value is determined by $n(T)$ extrapolation from the high-temperature region (broken line in figure 9.9). The lowering of $n(T)$ at $T \ll T_C$ is presumably due to the higher-order terms in the expressions (9.21) and (9.60). The full lines in figure 9.10(*a*), (*b*) depict the calculated polarizations for both

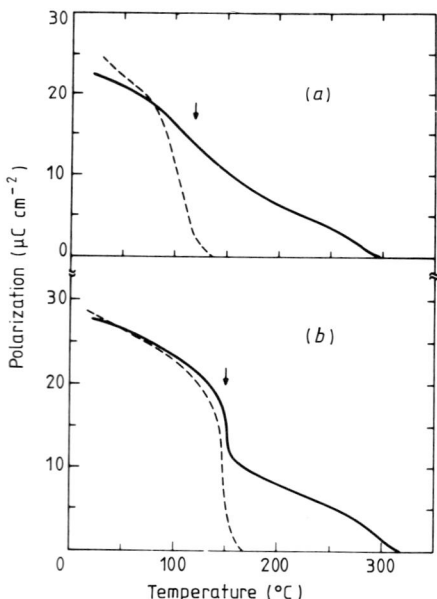

Figure 9.10 Temperature dependences of complete polarization P of (*a*) $K_{1.43}Sr_{3.76}Nb_{10}O_{29.47}$ and (*b*) $K_{0.74}Sr_{3.88}Nb_{10}O_{29.26}$ crystals calculated from temperature dependences of refractive indices [32]. Temperature dependences of reversible spontaneous polarization of these specimens (broken curves) are taken from [23].

compositions (figure 9.9). It is readily seen that for the less ordered sample $(K_{0.74}Sr_{3.88}Nb_{10}O_{29.26}$, figure 9.10($b$)) the polarization, or more precisely $|P|$, is over half the value at room temperature. In the more ordered sample $(K_{1.43}Sr_{3.76}Nb_{10}O_{29.47}$, figure 9.10($a$)) the polarization at T_C is lower, but still has a noticeable value. The reversible spontaneous polarization measured from the dielectric hysteresis loops is plotted in figure 9.10(a), (b) (broken lines). The consistency of both curves below T_C may be regarded as more or less satisfactory if one notes that the full curves were obtained from the expression (9.60) without any fitting parameters. A similar picture is observed for $Pb_3MgNb_2O_9$ and $Pb_3ZnNb_2O_9$ crystals [29].

Let us now discuss the existence of appreciable polarization largely above T_C. As follows from the expression (9.60), the change in the refractive index is determined by the square of polarization projection onto the c axis. This implies that refractive indices must be sensitive, for example, to the local polarization which can be caused by a broken charge electroneutrality in the crystal volume. Clearly, a crystal region with excess or deficiency of charge may exhibit an irreversible spontaneous polarization. Consequently, the polarization calculated from the dependence $n(T)$ will be higher than the reversible spontaneous polarization P_s, since a crystal always comprises regions with irreversible polarization. The local fields in these volumes cause local disordering of atoms and determine the magnitude and direction of polarization. This polarization is irreversible because no reversible polarization can be observed above T_C.

9.5 Microscopic parameters

The nonlinear coefficients above have been considered from a phenomenological point of view and their relationships with macroscopical quantities have been established. We are now in a position to carry out a microscopical analysis and to make an attempt to associate the nonlinear coefficients d_{ijk}, responsible for SHG, with the crystalline structure of octahedral ferroelectrics.

Robinson [40] was the first to suggest that the nonlinear optic effects may be related to the nonlinear polarization of individual bonds which are described by additive tensor quantities. Jeggo and Boyd [41] examined the individual bonds Nb–O in niobates and used the empirical approach to associate the nonlinear optic coefficients of the material with polarization of these bonds.

Polarization of the Nb–O bond exhibits a cylindrical symmetry relative to the ferroelectric axis and therefore can be described by two independent tensor coefficients $\beta_\parallel = \beta_{zzz}$ and $\beta_\perp = \beta_{xxz} = \beta_{yyz}$. The calculations carried out in [42] and the model representations of charge transfer [43] have shown that polarization of the Nb–O bond has a high anisotropy, i.e. $\beta_\parallel \gg \beta_\perp$, so that the tensor coefficient $\beta_\perp$ can be neglected.

Let us introduce the notation $\beta_\| = \beta_i$ and the direction cosines l_i, m_i, n_i. The SHG coefficients d_{ijk} of a crystal can then be represented by the expressions

$$d_{311} = v^{-1}\sum n_i l_i^2 \beta_i \qquad d_{333} = v^{-1}\sum n_i^3 \beta_i \qquad d_{222} = v^{-1}\sum m_i^3 \beta_i \tag{9.61}$$

where sums are taken over all bonds in a unit cell of volume v.

It should be noted that the nonlinear coefficients of SHG are sensitive functions of the refractive indices, which are in turn determined by the number of bonds in a unit cell. Besides, there is experimental evidence that the high electron polarizability of ions in the lattice composition is responsible for the high coefficients of linear and nonlinear optic effects. Therefore the tensor coefficients β characterize the microscopical properties of compounds.

Following Jeggo and Boyd, consider the relation between β and the nonlinear coefficients on an example of $LiNbO_3$, since the crystalline structure of this compound is known and d_{ijk} are measured.

Lithium niobate (with point symmetry group 3m) has three independent nonlinear coefficients, $d_{311} \approx d_{131}$, d_{333} and d_{222}, given by the lattice symmetry and by Kleinman's symmetry conditions [44]. The rhombohedral cell of $LiNbO_3$ [45] contains 12 Nb–O bonds, six of which are 1.89 Å and six 2.11 Å long. In niobates of alkali and alkaline earth metals, the nonlinear optic properties are due to distortion of NbO_6 octahedra. Consequently, the nonlinear optic polarization can be caused by Nb–O bonds only. Lithium–oxygen bonds can be neglected because they have an ionic nature and are weakly polarized. Ascribing to 1.89 and 2.11 Å Nb–O bonds the polarizations β_1 and β_2 respectively, and using equations (8.61), we obtain expressions for the nonlinear coefficients d_{ijk} [41]:

$$d_{311} = v^{-1}(1.104\beta_1 - 1.106\beta_2)$$

$$d_{222} = v^{-1}(0.396\beta_1 - 0.195\beta_2) \tag{9.62}$$

$$d_{333} = v^{-1}(0.643\beta_1 - 1.796\beta_2).$$

The coefficients β_{ij} and $d_{333} = (70 \pm 9)d_{312}^{KDP}$ were calculated from the experimental values of d_{311} and d_{222}, which are consistent with the experimental value $d_{333} = (86 \pm 22)d_{312}^{KDP}$. It should be noted that the relative signs of d remain unknown. The β values, which exhibit the closest correspondence for all three equations for d_{ijk}, are as follows: $\beta_1 = (-51 \pm 7) \times 10^{-28} d_{312}^{KDP}$ and $\beta_2 = (-64 \pm 8) \times 10^{-28} d_{321}^{KDP} m^3$. The nonlinear coefficients for crystals of $LiNbO_3$ are listed in table 9.6. The close agreement between the measured values and those calculated by formula (9.62) confirm the participation of bonded electrons in SHG.

Before proceeding to other niobates of alkaline earth metals, one should examine the coefficient β versus the Nb–O bond length r. In [41] it is indicated that $\beta \sim r^n$, where n is the degree of bond ionicity which decreases with increasing ionicity. For RX type compounds $n = 7$. The exact value of

Table 9.6 Calculated and measured nonlinear coefficients d_{ijk} for several niobates [41]†.

Material	d_{ijk}	Calculation	Experiment
LiNbO$_3$	d_{311}	$+13.9 \pm 3.9$	$+12.3 \pm 1.8$
	d_{333}	$+78 \pm 17$	86 ± 22
	d_{222}	-7.2 ± 1.4	-6.5 ± 0.6
LiTaO$_3$	d_{311}	$+3.1 \pm 1.0$	2.7 ± 0.5
	d_{333}	$+36 \pm 7$	41 ± 5
	d_{222}	-4.9 ± 1.6	4.4 ± 0.5
K$_3$Li$_2$Nb$_5$O$_{15}$	d_{311}	$+13.5 \pm 1.7$	14.8 ± 3.1
	d_{333}	$+21 \pm 7$	27 ± 4
Ba$_2$NaNb$_5$O$_{15}$	d_{311}	$+16.0 \pm 2.8$	27 ± 5
	d_{322}	$+18.6 \pm 2.5$	32 ± 3
	d_{333}	$+36 \pm 10$	32 ± 2
Ba$_{0.27}$Sr$_{0.75}$Nb$_2$O$_6$	d_{311}	$+5.0 \pm 3.0$	—
	d_{333}	$+12.8 \pm 3.6$	—

† The quantities d_{ijk} are determined relative to the quantity d_{312} (KDP) and the signs relative to the sign of d_{311} (LiNbO$_3$).

$\beta(r)$ is not necessary because r varies within a very small range. The lengths and the direction cosines of Nb–O bonds for the known Ba$_2$NaNb$_5$O$_{15}$ and K$_3$Li$_2$Nb$_5$O$_{15}$ structures were estimated. The lithium niobate polarizabilities $\beta = (-39.7 - 0.13r^7) \times 10^{-28} d_{312}^{KDP}$ m^3 were used to evaluate the nonlinear coefficients d_{ijk}. And as before, the Nb–O bonds alone were taken into account. The results obtained, as well as the experimental values, are listed in table 9.6. The calculations show a satisfactory agreement with experiment for K$_3$Li$_2$Nb$_5$O$_{15}$ and an appreciable disagreement for Ba$_2$NaNb$_5$O$_{15}$, due to a disregard of the other bonds. The presumed signs of the coefficients of each material relative to the sign of the coefficient d_{311} (LiNbO$_3$) were later confirmed by Miller and Nordland [46].

Modelling of nonlinear coefficients, which exploits the nonlinearity of bonds, takes into account only electronic processes, which in their nature satisfy Kleinman's symmetry conditions [44]. This model neglects the lattice contributions to nonlinear optic effects and is therefore inapplicable to electro-optic effects. Indeed, Miller and Nordland revealed that all electro-optic coefficients of LiNbO$_3$ and LiTaO$_3$ have identical signs, which confirms the importance of ionic effects.

Dispersions of the quadratic susceptibilities χ_{ij} of LiNbO$_3$ crystals, barium sodium niobate of congruent composition Ba$_{2.09}$Na$_{0.711}$Nb$_5$O$_{15}$, as well as barium strontium niobate (BSN), were measured experimentally in [47–49]. The susceptibility dispersion of these compounds at room temperature was

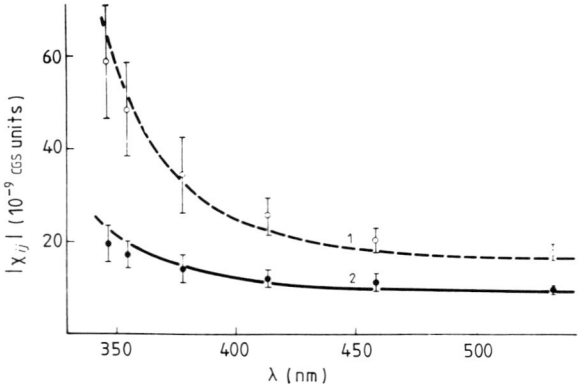

Figure 9.11   Dispersion dependences: (1) χ_{31} and (2) χ_{33} of BSN crystals near the edge of the absorption band [48].

calculated theoretically within the model of the perfectly anharmonic oscillator by analogy with the computations made by Jeggo and Boyd [41] (figure 9.11).

9.6   Second harmonic generation by ferroelectric crystals

Under second harmonic generation by nonlinear crystals, a plain optical wave with amplitude E_1, wavevector k and frequency ω in a piezoelectric medium gives rise to a polarization quadratic with respect to the field, having a frequency 2ω and described by the equation

$$P_2 = \chi_2^{2\omega} E_1^2$$

where χ_{ij} is the quadratic susceptibility of the crystal.

The second harmonic field strength induced by these sources consists of two parts [44, 50]:

$$E_2 = \frac{1}{2}\left[\frac{2\pi\chi_2^{2\omega}\varepsilon_1^2}{n_2^2 - n_1^2} \exp\{2i(k_1 x - \omega t)\} - \frac{n_1 + 1}{n_2 + 1} \exp\{i(k_2 x - 2\omega t)\} \right]$$

where k_1 and k_2 are the wavevectors and n_1 and n_2 are the refractive indices of the fundamental wave and the second harmonic respectively. The second term, called a free wave, is the solution of the wave equation in the form of a continuous plane wave propagating with phase velocity $v_2 = c/n_2$. The first term is a particular solution of the wave equation and is referred to as an intensity wave, which propagates with a phase velocity $2\omega/2|k_1| = c/n_1$, i.e.

is spatially associated with the fundamental wave. Usually $n_1 \neq n_2$, and two waves propagate with different phase velocities, and therefore interference takes place. The intensity I_2 of the second harmonic generated by the fundamental wave of intensity I_1 in a crystal of length l is determined by the expression [25]

$$I_2 = \frac{2^{12}\pi^5(\chi_2^{2\omega})^2 I_1^2 n_2 l^2}{c\lambda_2^2(n_2+1)^3(n_1+1)^3(n_2+n_1)} \frac{\sin^2(\pi l/2l_c)}{(\pi l/2l_c)^2} = K l^2 \frac{\sin^2(\pi l/2l_c)}{(\pi l/2l_c)^2} \tag{9.63}$$

where I_1 and I_2 are intensities measured beyond the nonlinear crystal and l_c is the coherence length for phase matching of the ordinary and extraordinary waves in the nonlinear crystal. The wavelength λ_2 of the second harmonic in free space and the coherence length l_c are related as

$$l_c = \frac{\lambda_2}{2|n_2 - n_1|} = \frac{\pi}{|k_2 - 2k_1|} \tag{9.64}$$

where k_1 and k_2 are the wavevectors of the fundamental wave and the second harmonic respectively. For $l_c \gg l$, the second harmonic intensity is proportional to the squared crystal length: $I_2 = K l^2$. When $l_c \ll l$, the intensity I_2 oscillates between 0 and $4K l_c^2/\pi^2$. The condition under which $n_1 = n_2$ is termed phase matching.

Consider the effect produced by ferroelectric domain structure upon SHG for light beams normal to both directions of spontaneous polarization, i.e. a ferroelectric is divided into 180° domains. On a 180° wall, the intensity wave phase changes by π, then induces the change of the free wave amplitude by a quantity equal in magnitude but opposite in direction to the intensity wave. If the domain walls have a length l_c, we observe the second harmonic amplification which is equivalent to the change in the effective harmonic generation length from l_c to the crystal length l, i.e. we have arrived at phase matching.

For N randomly arranged 180° walls, part of the second harmonic power, which is on the average non-zero, increases with the coordinate x as $2N$. SHG amplification due to randomly arranged domain walls was observed in several ferroelectric crystals. If $l_c > l$, walls separate domains of opposite signs and act in a contrary manner. A more general case, when a laser beam makes an angle with the domain walls, was treated by Freund [51], who revealed that in this case nonlinear diffraction presumably occurs under which the principal wave and its harmonic leave the crystal in different directions.

Now the phase matching conditions will be considered for the example of LiNbO$_3$. An optically uniaxial negative crystal of LiNbO$_3$ belongs to the point symmetry group 3m. Second-order polarization P_2 with frequency 2ω and optical electric light wave fields varying with frequency ω are related through the system of equations

$$P_x = 2\chi_{15} E_x E_z - 2\chi_{22} E_x E_y$$
$$P_y = -\chi_{22} E_x^2 + \chi_{22} E_y^2 + 2\chi_{15} E_y E_z \qquad (9.65)$$
$$P_z = \chi_{31} E_x^2 + \chi_{31} E_y^2 + \chi_{33} E_z^2.$$

Assume that the fundamental light wave E_y is ordinary, and all waves propagate along the x axis. Then the system of equations gives rise to a second harmonic polarized along a c axis. To achieve phase matching, the following equality should hold:

$$n_e^{2\omega} = n_o^\omega$$

where $n_e^{2\omega}$ is the refractive index of an extraordinary second harmonic wave and n is the refractive index of a fundamental ordinary wave. The dispersion of the refractive indices of $LiNbO_3$ [52] shows that, for He–Ne laser radiation with $\lambda = 1.15$ μm, the phase matching condition is fulfilled for a beam propagating at an angle $\theta_{phm} = 67°$ to the optic axis. In uniaxial crystals, phase matching holds both for a negative ($n_e < n_o$) and a positive ($n_e > n_o$) birefringence [53].

For an effective transformation, the characteristic radiation in the volume of a nonlinear crystal should be focused. But focusing causes an amplification of the undesirable phenomenon called the aperture effect. Suppose phase matching is achieved, i.e. the wavevectors $2k_1$ and k_2 are equal and collinear. The two Poynting vectors determining the energy propagation directions normal to the indicatrices of refractive indices are not parallel to each other in the general case. An extraordinary second harmonic wave practically has a tendency to move aside from the fundamental wave, which decreases the efficiency of transformation of the fundamental wave into a second harmonic. This effect can be eliminated by achieving phase matching for $\theta_{phm} = 90°$, i.e. when a fundamental wave and a second harmonic propagate along the principal axis of the ellipsoid of the refractive indices.

Such phase matching can be achieved for several ferroelectric crystals due to the temperature dependence of the refractive indices. Second harmonic generation for $\theta_{phm} = 90°$ is called non-critical phase matching, which was first attained in $LiNbO_3$ under the temperature matching of refractive indices. The non-critical character for $\theta_{phm} = 90°$ refers to the angular divergence of the principal wave and to crystal position [54]. The classes of crystals for which non-critical phase matching holds belong to ten polar classes and also include classes 4 and $\bar{4}2m$. Very few crystals in which a 90° phase matching holds are ferroelectrics. Two of them, $LiNbO_3$ and $Ba_2NaNb_5O_{15}$, are of appreciable practical value.

The coefficients of second-order polarization, which describe nonlinear interactions between laser beams and ferroelectric crystals, differ by several orders of magnitude for different materials.

9.6.1 *Miller's coefficients*

Miller [24] approaches second harmonic generation from the energetic viewpoint, which differs from the preceding approach in that the energy is expressed in terms of the corresponding electric polarization and not in terms of the electric field strength. The portion of free energy responsible for SHG can be represented in the form

$$F(\boldsymbol{P}) = -\tfrac{1}{3}\delta P^3 \tag{9.66}$$

where P is the resultant amplitude of corresponding polarization waves, i.e. $P = P^\omega + P^{2\omega}$. We then have

$$-\frac{\partial F_i(\boldsymbol{P})}{\partial P_i^{2\omega}} = e_i^{2\omega} = \sum \delta_{ijk} P_j P_k. \tag{9.67}$$

From this expression Miller obtained the relations which were presented in §9.3 (see (9.42)–(9.44)). Nonlinear susceptibilities were estimated for a large number of materials. The mean value $\hat{\delta}$ for 120 ferroelectric compounds is 1.46×10^{-6} CGS units with a standard deviation of 0.88×10^{-6} [25].

It was shown in [26] that the approximate constancy of the nonlinear susceptibility δ for different ferroelectrics follows from the simple anharmonic oscillator model. The anharmonic oscillator approximation applicable for ferroelectric crystals, which in the paraelectric phase are not piezoelectrics, leads to the relation $\delta \approx 2 \times 10^{-8} P_s$ CGS units, where spontaneous polarization P_s is expressed in μC cm^{-2} [27]. This may cause a temperature dependence of δ similar to that of P_s, which is supported by experimental evidence for BaTiO$_3$ and LiTaO$_3$. In the case of LiNbO$_3$, however, the temperature dependences of δ and P_s did not show perfect agreement.

Figure 9.12 is a plot of δ versus P_s for several ferroelectric crystals at a temperature of 300 K. Line 1 is calculated in the anharmonic oscillator approximation. Line 2 corresponds to experimental data processes using the least-squares technique, where all points have equal statistical weights. Line 3 shows the dependence of the mean value of $\hat{\delta}$ having a standard deviation $\pm\sigma$. The figure also presents experimental δ values for several compounds, most of which lie within the interval $\pm\sigma$.

Figure 9.12 demonstrates that δ variations for different materials are smaller in the general case than the variations of corresponding d_{ijk} (table 9.6). This suggests that, to obtain more fundamental dependences, the nonlinear phenomena should be described in terms of dielectric polarization rather than the electric field strength. Such assumptions were also made in the attempts to consider the quadratic electro-optic effect in several perovskites.

On introducing the nonlinear susceptibility δ, it became clear that a crystal possessing higher linear susceptibilities will exhibit larger nonlinear effects for corresponding frequencies. The dependence of d_{ijk} on linear dielectric properties of the medium established by Armstrong *et al* [50] differs from a similar dependence reported by Miller [24].

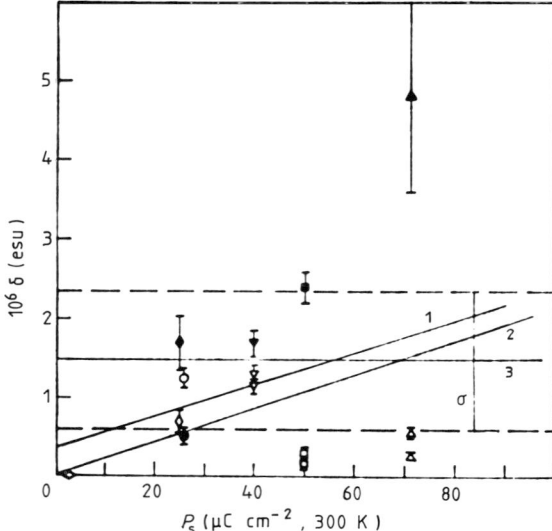

Figure 9.12   (1) Calculated and (2) experimental dependences of nonlinear optical susceptibility δ on spontaneous polarization; (3) the line shows the mean value $\hat{\delta}$ obtained for 120 ferroelectric materials; σ is the standard deviation from the mean value. Experimental points correspond to individual materials involved in the figure [25]. $\triangle$ LiNbO$_3$; $\square$ LiTaO$_3$; ∇ Ba$_2$NaNb$_5$O$_{15}$; $\bigcirc$ BaTiO$_3$; $\diamond$ K$_{0.6}$Li$_{0.4}$NbO$_3$; $\diamond$ TGS average. δ_{33} is shown by full shapes.

9.6.2   Relative signs of nonlinear optical coefficients

We have already mentioned relative signs of nonlinear optical coefficients, which are responsible for the transformation of laser radiation frequency. The signs of these coefficients affect the output of the second harmonic.

Miller and Nordland (chapter 6, [35]) proposed the method for determining relative signs of nonlinear optical coefficients which is based on the use of plane parallel single-domain plates of a crystal cut out along XZ or YZ planes. The plate thickness should suffice to observe the peaks of Maker's second harmonic (chapter 6, [33]), which occur when the plate is rotated around the vertical c axis which lies in the cut plane and is normal to the principal laser wave. For the characteristic radiation source one commonly uses a YAG: Nd laser with the radiation wavelength $\lambda = 1.064$ μm. The second harmonic as a function of the angle of rotation φ is recorded using the standard methods mentioned in chapter 5.

A laser beam initially polarized at 45° to the vertical c axis passes through the polarizer, which transmits radiation polarized at an angle β to the vertical. Installed behind the crystal plate is a second harmonic analyser, polarization being directed at an angle α to the vertical.

If a Y-cut plate is rotated around the c axis, the second harmonic power $P_2(\alpha, \beta)$ can be written in the form

$$P_2(0, \pi/2) = K[d_{31}l_{31}E_1^2/(n_1^o + 1)^2]^2 \sin^2(\pi l/2l_{31})$$

$$\equiv P_2(0, \pi/2)_0 \sin^2(\pi l/2l_{31})$$

$$P_2(0, 0) = K[d_{33}l_{33}E_1^2/(n_1^e + 1)^2]^2 \sin^2(\pi l/2l_{31})$$

$$\equiv P_2(0, 0)_0 \sin^2(\pi l/2l_{33}). \tag{9.68}$$

In these expressions K is a constant for a given crystal, $l = l(\varphi)$ is the path length in the crystal, E_1 is the amplitude of the fundamental wave incident on the polarizer, and l_{ij} is the coherent length corresponding to d_{ij} and specified by the relation

$$l_{ij} = \lambda/4(n_2^{2\omega} - n_1^{\omega})$$

where λ is the fundamental wavelength in a vacuum, $n_2^{2\omega}$ and n_1^{ω} are the refractive indices at harmonic and fundamental frequencies respectively.

Upon crystal rotation, the second harmonic power oscillates between zero and $P_2(\alpha, \beta)$. If the polarizer is oriented at an arbitrary angle β to the vertical, the second harmonic power resulting from the mutual action of the coefficients d_{31} and d_{33} is described by the expression

$$P_2(0, \beta) = 4K[a_{31}(a_{31} + a_{33}) \sin^2(\pi l/2l_{31}) + a_{33}(a_{31} + a_{33}) \sin^2(\pi l/2l_{33})$$

$$- a_{31}a_{33} \sin^2 \pi l/2(l_{31} - l_{33})]. \tag{9.69}$$

The quantities a_{31} and a_{33} are determined by the relations

$$(n_1^o + 1)^2 a_{31} = d_{31}l_{31}E_1^2 \cos^2(\beta - \pi/4) \sin^2 \beta$$

$$(n_1^e + 1)^2 a_{33} = d_{33}l_{33}E_1^2 \cos^2(\beta - \pi/4) \cos^2 \beta.$$

If we match β to β_0, for which $|a_{31}| = |a_{33}| \equiv a$, then $\tan^4 \beta_0 = P_2(0, 0)_0/P_2(0, \pi/2)_0$. Next, for $a_{31} = -a_{33}$ we obtain

$$P_2(0, \beta_0) = 4Ka^2 \sin^2[\pi l/2(l_{31} - l_{33})]. \tag{9.70}$$

Thus in this case the second harmonic is determined by the simple oscillating function l, and therefore by the angle of rotation φ. The interference coherence length is determined by the expression

$$l_{\beta_0} = l_{31}l_{33}/(l_{33} - l_{31}).$$

Note that each l_{ij} has a sign. Under the conditions $a_{31} = +a_{33}$, equations (9.68) give

$$P_2(0, \beta_0) = 4Ka^2\{2[\sin^2(\pi l/2l_{31}) + \sin^2(\pi l/2l_{33})] - \sin^2[\pi l/2(l_{31} - l_{33})]\}. \tag{9.71}$$

In this case $P_2(0, \beta_0)$ must be a composite function of φ.

It can thus be seen that, depending on φ, the function $P_2(0, \beta_0)$ gives directly the relative signs of a_{31} and a_{33}. Such a definition does not require a detailed knowledge of l_{ij} or d_{ij}. The intensity ratio of the second harmonic power maxima can also be used to establish the sign of a_{ij}. For example, for identical signs of a_{31} and a_{33} the maximum $P_2(0, \beta_0)$ is four times the one for opposite signs of the same coefficients.

As an example, figure 9.13 gives the dependences of second harmonic powers $P_2(0, \pi/2)$, $P_2(0, 0)$ and $P_2(0, \beta_0)$, determined respectively by the coefficients d_{31}, d_{33} and $d_{31} + d_{33}$, on the angle φ for a constant $\beta_0 = 31°$ in $LiNbO_3$. These graphs imply opposite signs of a_{31} and a_{33}. The fact that the interference coherence length l_{β_0} is smaller than l_{33} is an indication of a negative l_{31}, since l_{33} cannot be negative [55]. The nonlinear coefficients d_{33} and d_{31} in $LiNbO_3$ have, accordingly, negative signs; this is also the case with $LiTaO_3$. A simple modification of the above technique made it clear that the coefficients d_{22} and d_{31} have opposite signs in $LiTaO_3$ as well as in $LiNbO_3$. This conclusion was drawn from observations of $P_2(\pi/2, \beta_0)$ and $P_2(\pi/2, \pi/2)$ in a rotation of the X-cut of a $LiTaO_3$ plate around the c axis. On the assumption that Kleinman's condition $d_{15} = d_{31}$ holds, we have

$$\tan \beta_0 = |2d_{15}l_{15}(n_1^o + 1)/d_{22}l_{22}(n_1^e + 1)|.$$

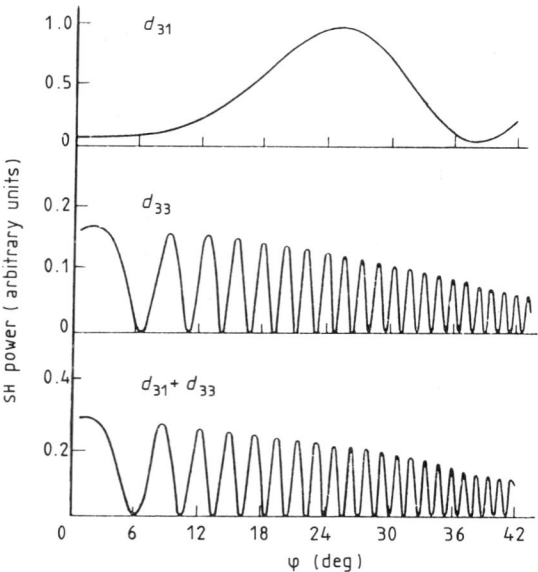

Figure 9.13 Dependences of second harmonic powers $P_2(0, \pi/2)$, $P_2(0, 0)$ and $P_2(0, \beta_0)$ generated by a $LiNbO_3$ crystal plate on the angle of rotation φ about the ferroelectric axis c for nonlinear coefficients d_{31}, d_{33} and $d_{31} + d_{33}$ respectively (chapter 8, [35]).

9.7 Relationships between pyroelectric and ferroelectric parameters

According to the Devonshire–Ginzburg phenomenological theory of ferro-electricity, the free energy $F(P)$ of ferroelectrics, to which an electric field is applied along a polar axis, can be expanded in terms of polarization P [56, 57]:

$$F(P) = \frac{1}{4\pi\varepsilon_0}\left[\frac{1}{2}\frac{4\pi}{C}(T - T_C)P^2 + \frac{1}{4}a_1 P^4 + \frac{1}{6}a_2 P^6 + \ldots\right] \quad (9.72)$$

where T_C and C are Curie temperature and Curie constant respectively. The calculations were made in MKS units, and a_1 and a_2 have dimensionality of P^{-2} and P^{-4} respectively. Conversion to the CGS electrostatic system requires fulfilment of the condition $4\pi\varepsilon_0 = 1$.

Spontaneous polarization P_s can be determined by equating the electric field $E(P)$ to zero:

$$E(P_s) = \left(\frac{\partial F}{\partial P}\right)_{P = P_s} = \frac{1}{4\pi\varepsilon_0}\left[\frac{4\pi}{C}(T - T_C)P_s + a_1 P_s^3 + a_2 P_s^5 + \ldots\right] = 0.$$
$$(9.73)$$

The dielectric susceptibility χ is found from the expression

$$\frac{1}{\chi\varepsilon_0} = \frac{\partial^2 F}{\partial P^2}\bigg|_{P = P_s} = \frac{1}{4\pi\varepsilon_0}\left[\frac{4\pi}{C}(T - T_C) + 3a_1 P_s^2 + 5a_2 P_s^4 + \ldots\right]. \quad (9.74)$$

Differentiating equation (9.73) with respect to T and making use of equation (9.74), we obtain

$$\frac{dP_s}{dT}\left(\frac{4\pi}{\chi}\right) + \frac{4\pi}{C}P_s = 0$$

which implies that

$$\gamma/\chi = P_s/C \quad (9.75)$$

where $\gamma = -dP_s/dT$ is a pyroelectric coefficient. This ratio specifies the temperature dependence of the pyroelectric effect. Restricting the expansion (9.74) to the term P^4, we find the solution for P^2. For a second-order transition $(a_1, a_2 > 0)$

$$P_s^2 = -\frac{a_1}{2a_2} + \frac{a_1}{2a_2}\left[1 + 4\left(\frac{4\pi}{C}\right)\frac{a_2}{a_1^2}(T_C - T)\right]^{1/2} \quad (9.76)$$

and for a first-order transition $(a_1 < 0; a_2 > 0)$

$$P_s^2 = \frac{|a_1|}{2a_2} + \frac{|a_1|}{2a_2}\left[1 + 4\left(\frac{4\pi}{C}\right)\frac{a_2}{a_1^2}(T_C - T)\right]^{1/2}. \quad (9.77)$$

Differentiating with respect to T, we derive

$$2P_s\gamma = \frac{4\pi}{C}\frac{1}{|a_1|}\left[1 + 4\left(\frac{4\pi}{C}\right)\frac{a_2}{a_1^2}(T_C - T)\right]^{-1/2} \tag{9.78}$$

for both types of phase transitions.

Next, multiplying the expressions (9.78) by (9.75), we derive

$$\frac{\gamma^2}{\chi} = P_0^2 \bigg/ 2CT_C\left[1 + 4\left(P_0^2\frac{a_2}{|a_1|}\right)\left(1 - \frac{T}{T_C}\right)\right]^{1/2} \tag{9.79}$$

where

$$P_0^2 \equiv T_C\left(\frac{\partial P_s^2}{\partial T}\right)_{T \le T_c} = \frac{4\pi T_C}{C|a_1|}$$

which follows from equations (9.76) and (9.77). Near T_C we have

$$\gamma^2/\chi \approx P_0^2/2CT_C \tag{9.80}$$

or

$$\gamma/\sqrt{\chi} \approx P_0/(2CT_C)^{1/2}. \tag{9.80}$$

The ratio (9.80) establishes the relation between the pyroelectric parameters γ, χ and the ferroelectric parameters P_0, T_C and C. In the general case P_0 depends on the temperature interval in which equations (9.76) and (9.77) hold. Presumably, P_0 is comparable with P_s at a temperature at which P_s reaches its maximum value P_s^{max}. On the other hand, P_s^{max} is rather simple to determine experimentally since polarization has a tendency towards saturation at low temperatures.

It would not be out of place to note here that Abrahams *et al* [1] reported a constancy of the ratio $(P_s^{max})^2/T_C$ over a wide temperature range for displacive-type ferroelectrics. All the materials which they observed belong to the class of ferroelectrics with a high Curie constant. The measurements thus show that $\gamma/\sqrt{\chi}$ is a constant. Combining this quantity with Abrahams *et al*'s ratio $(P_s^{max})^2/T_C$, we obtain

$$\gamma/\sqrt{\chi} \approx P_s^{max}/\sqrt{CT_C} \approx \text{const.} \tag{9.81}$$

Figure 9.14 shows a logarithmic dependence of γ on χ for a wide range of pyroelectric materials at room temperature. This dependence is most convenient for practical needs. The straight lines have a slope equal to 0.5; the central line corresponds to the expression

$$\gamma/\sqrt{\chi} = \sqrt{10} \times 10^{-9}\, \text{C cm}^{-2}\, \text{K}^{-1}.$$

The two broken lines differ from the central value by $\pm\gamma$, which is the standard deviation. In solid solutions of tungsten bronzes, the quantity $\gamma/\sqrt{\chi}$ can be expected to vary within a narrower range than in figure 9.14.

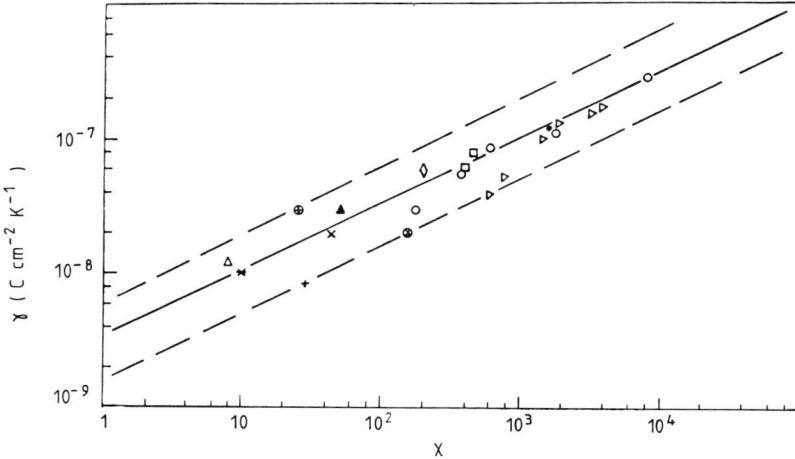

Figure 9.14 Dependence of pyroelectric coefficient γ on dielectric susceptibility χ for ferroelectric materials at room temperature. Broken lines indicate the standard deviation from the mean value γ. Experimental points correspond to individual materials included in the figure [51]. ○ SBN [9]; □ SBN [20]; ▷ PLZT ceramics [20]; × LiTaO$_3$ [8]; + LiNbO$_3$ [8]; ◇ PbTiO$_3$ [21]; ● 2% La–SBN [7]; * LiSO$_4$.H$_2$O [1]; △ NaNO$_2$ [1]; ⊗ BaTiO$_3$ [1]; ⊕ ATGS [22]; ▲ TGS [19].

The expressions (9.79)–(9.81) obtained from the phenomenological theory can be verified most thoroughly on materials for which there are enough experimental data. Several such materials are listed in table 9.7, which comprises their parameters χ T_C, C as well as the quantities $\gamma/\sqrt{\chi}$ and $P_s^{max}/\sqrt{CT_C}$. The latter two quantities are relatively constant for the velocity of ferroelectrics listed in table 9.7, which embraces ferroelectrics with first- and second-order transitions, displacive and order–disorder type ferroelectrics, as well as those with low and high Curie constants.

The expression (9.81) thus proves more universal than was expected from the Devonshire–Ginzburg theory. In this respect the relation is similar to Miller's rule [24] for nonlinear susceptibilities in asymmetric materials.

In the molecular field approximation described by Liu and Zook [56], there exists a universal constant P_{02} which is related to the crystal lattice volume v per unit dipole as

$$P_{02} = [\varepsilon_0 kC/v]^{1/2} \tag{9.82}$$

where k is the Boltzmann constant.

The expression (9.82) is approximately fulfilled for all types of internal field potentials of ferroelectrics. The parameters in the expression (9.82) are

Table 9.7 Parameters of several ferroelectric crystals [57, 58].

Crystal	T_C (K)	Unit cell volume (Å)³	Lattice volume per dipole v (Å)³	C (10⁵ K)	P_{02} (μC cm⁻²)†	P_s^{max} (μC cm⁻²)	$P_{\gamma\epsilon}$ (μC cm⁻²)‡	$P_s^{max}/(CT_C)^{1/2}$ (10⁻³ μC cm⁻² K⁻¹)	$\gamma\chi^{1/2}$ (10⁻³ μC cm⁻² K⁻¹)
Ba$_{0.5}$Sr$_{0.5}$Nb$_2$O$_6$	392	614.21	61.42	1.76	59.2	34	25.0	2.89	3.0
Ba$_{0.5}$Sr$_{0.5}$Nb$_2$O$_6$ +1% La$_2$O$_3$	335	613.23	61.32	2.66	72.8	32	28.3	2.83	2.93
Ba$_2$NaNb$_5$O$_{15}$	856	1238.69	61.94	2.6	71.6	40	—	2.68	—
LiTaO$_3$	891	105.7	52.85	1.2	52.7	50	27.0	4.18	2.68
LiNbO$_3$	1468	106.05	53.03	2.5	75.9	71	30.0	3.70	1.57
BaTiO$_3$	383	64.11	64.11	1.7	56.9	26	12.7	3.22	1.58
KNbO$_3$	633	64.84	64.84	2.4	67.5	30	—	2.29	—
PbTiO$_3$	722	62.78	62.78	4.1	89.3	57	39.0	5.42	4.24

† $P_{02} = (\epsilon_0 kC/v)^{1/2}$.
‡ $P_{\gamma\epsilon} = (CT_C)^{1/2}\gamma/\chi^{1/2}$.

presented in table 9.7. The ratio P_s^{max}/P_{02} varies between 1 and 0.25 for a wide range of ferroelectrics. The volume v can be smaller than the unit cell volume. For perovskite-type ferroelectrics, v corresponds to a unit cell and comprises one BO_6 octahedron. The choice of v in more complex crystals (e.g. BSN) is based on the assumption that the correlation effects in such crystals should be similar to those in ferroelectrics with one BO_6 octahedron per unit cell.

It has already been mentioned that the ratio $(P_s^{max})/\sqrt{CT_C}$ is almost completely material-independent and coincides with $\gamma/\sqrt{\chi}$. Table 9.7 also presents the values of $P_{\gamma\varepsilon} = \sqrt{CT_C}\gamma/\sqrt{\chi}$ which determine the polarization at room temperature. The ratio $P_{\gamma\varepsilon}/P_s^{max}$ ranges from 0.5 to 1.0 for different ferroelectrics. This type of ratio was proposed by Liu and Zook [56] on the basis of the two-level ferroelectric model.

Thus, the above dimensionless ratios of several ferroelectric parameters are almost independent of the form of interatomic potential. This fact confirms the applicability of the effective field model for the various ferroelectrics. The established ratios can serve for testing new materials for which complete data are not yet available. These ratios can also predict the applicability of some materials as pyroelectric transducers [59], and provide a deeper insight into the relationships of ferroelectric properties. These ratios imply that materials whose parameter $\gamma/\sqrt{\chi}$ would be appreciably higher than in the now known ferroelectrics are unlikely to be found, because the volume v per unit dipole cannot be noticeably diminished (see table 9.7).

9.8 Figure of merit for electro-optic applications

The figure of merit for electro-optic deflectors and modulators may be the ratio r^2/χ. As shown in §9.3, the electro-optic coefficient r is proportional to the polarization: $r \sim \chi P$. The figures of merit of the electro-optic and pyroelectric properties of one ferroelectric crystal can be related by the equality (9.75)

$$r^2/\chi \sim \chi P^2 = (C\gamma/\sqrt{\chi})^2.$$

Thus the figure of merit for electro-optic applications is proportional to the squared figure for pyroelectric applications, both figures showing a weak temperature dependence.

The angle of laser beam deflection by an electro-optic deflector is given by the expression [60]

$$\phi_m = 4(\Delta n)_p (L/W) \tag{9.83}$$

where $(\Delta n)_p$ is the maximal (peak) change of the refractive index induced by the field, L is the optical path length in the deflector and W is the beam width in the deflection plane at the output. The value $(\Delta n)_p$ is determined by the maximal electric field applied to the crystal:

$$(\Delta n)_p = \tfrac{1}{2}(n^3 r E_p). \tag{9.84}$$

Another important deflector characteristic is resolution, which is defined as the ratio of the maximal beam deflection angle to the diffraction limit:

$$N_0 = \phi_m W / e\lambda \tag{9.85}$$

where λ is the radiation wavelength, and e is a quantity of the order of unity depending on uniformity of deflector lighting and on the resolution criterion.
From (9.83) and (9.85), N_0 can be written in the form

$$N_0 = 2n^3 r E_p L / e\lambda. \tag{9.86}$$

If radiation focusing in the crystal is taken into account, the deflector resolution is specified by the expression

$$N_R = N_0(1 - N_0/4N_g) \tag{9.87}$$

where $N_g = W^2/ne\lambda L$ is the Fresnel number for a deflector.
The ratio of deflector resolution to the energy τP_a of a beam declined by the deflector is proportional to the factor

$$\frac{N_R^{5/2}}{\tau P_a} \sim \frac{n^7 r^2}{\varepsilon(e\lambda)^3}$$

which is a combination of material constants. It determines the speed power (transmission) of a given electro-optic material and may serve the criterion of its electro-optic performance. This criterion can also be applied to electro-optic modulators [61]. The maximal value of N_R depends on the crystal size and on the maximal change of the refractive index induced by the field. The crystal size imposes a stronger limit upon the quantity W than upon L, since the optical path L can always be extended by using a prism or a repeated beam passage through the deflector. Table 9.8 gives a list of material parameters of DKDP, lithium niobate and BSN crystals. High-quality DKDP crystals can be of considerable size, so that a crystal may serve as a deflector with $W = 7$ cm. Optically satisfactory crystals of lithium niobate now reach a size of 3 cm. Due to a strong optical inhomogeneity, the expected maximal size of a BSN crystal may reach 1.5 cm. The $(\Delta n)_p$ tabulated above are calculated on the basis of the known electro-optic coefficients and the electric fields which make up $\tfrac{2}{3}$ of the breakdown voltages.
Three materials presented in table 9.8 belong to three different ferroelectric classes. The parameters of these materials demonstrate typical problems encountered in manufacturing deflectors. The DKDP crystal possesses high optical qualities but low electro-optic coefficients. The high-quality lithium niobate crystals now available have a small size. The high-quality crystals

Table 9.8 Parameters of electro-optic materials and of deflectors made of these materials (for $e = 1.6$ and $e\lambda = 10^{-4}$ cm) [60].

Crystal	n	$r \ (10^{-9} \ \text{cm V}^{-1})$	$(e\lambda/2n^3 r)$ (V)	$\varepsilon_j/\varepsilon_0$	r_{ij}/ε_j $(\text{m}^2\,\text{K}^{-1})$	n^7
KD$_2$PO$_4$	1.50	2.4	6170	49	0.055	17.1
LiNbO$_3$	2.20	3.10	1520	26.4	0.133	250
Ba$_{0.25}$Sr$_{0.75}$Nb$_2$O$_6$	2.30	134	31	6500	0.0233	340

Crystal	$n^7 r^2/\varepsilon(e\lambda)^3$ $(10^7 \ \text{J}^{-1})$	$\Delta n_p \ (10^{-4})$	W_{max} (cm)	L/W	N_R
KD$_2$PO$_4$	2.26	0.8	7	55.5	1050
LiNbO$_3$	103	10	2	19.1	1280
Ba$_{0.25}$Sr$_{0.75}$Nb$_2$O$_6$	1061	30	1	11.3	1130

of BSN with large electro-optic coefficients and dielectric constant are no longer accessible for extensive use.

The maximal resolution N_R, as shown in table 9.8, is approximately equal for all three crystals. This implies that the electro-optic effect can be balanced by increasing the crystal size, which leads however to undesirable effects: an increased optical aperture, a low beam deflection rate, a low acoustic resonance frequency, and difficulties in removing heat due to dielectric losses.

The performance criterion of a crystal as an electro-optic deflector is inversely proportional to the dielectric constant, but the electro-optic coefficient r is proportional to ε. To estimate the crystal quality, it is thus useful to employ the polarization factor $f = r/\varepsilon$, also presented in table 9.8.

The quantity f is seen from table 9.8 to vary only slightly from material to material. It was found experimentally [62] and theoretically [63] that the electro-optic effect is essentially the polarization effect rather than the one induced by the field. With the use of the polarization factor f, the performance criterion is proportional to εf^2. A high dielectric constant, say of BSN, is thus responsible for a higher beam deflection rate as compared with materials with lower values of ε, e.g. lithium niobate.

The problems encountered in using ferroelectric materials as deflectors near the Curie temperature, where ε is high, are discussed in [62].

References

Introduction and Chapter 1

[1] Chen F S and Geusic J E 1966 *J. Appl. Phys.* **37** 388
[2] Ashkin A, Boyd G D, Dziegzic J M *et al* 1966 *Appl. Phys. Lett.* **9** 72
[3] Jona F and Shirane G 1962 *Ferroelectric Crystals* (Oxford: Pergamon)
[4] Smolensky G A, Bokov, V A, Isupov V A, Krainik N N, Pasynkov R E and Shur A S 1971 *Ferroelectrics and Antiferroelectrics* (Leningrad: Nauka)
[5] Fesenko E G 1972 *Perovskite Family and Ferroelectricity* (Moscow: Atomizdat)
[6] Bursian E V 1974 *Nonlinear Crystal of Barium Titanate* (Moscow: Nauka)
[7] Karasev M D 1959 *Usp. Fiz. Nauk* **69** 217
[8] Mirlin D N 1958 *Semiconductors in Science and Engineering* ed A F Ioffe (Moscow: Izdatel'stvo Akad. Nauk SSSR) p 516
[9] Johnson K M 1962 *J. Appl. Phys.* **33** 2826
[10] Rez I S 1969 *Usp. Fiz. Nauk* **93** 633
[11] Berezhnoi A A, Bukhman V N, Kudinova L T *et al* 1968 *Fiz. Tverd. Tela* **10** 255
[12] Smolensky G A, Krainik N N and Berezhnoi A A 1968 *Fiz. Tverd. Tela* **10** 417
[13] Smolensky G A *et al* 1960 *Physics of Dielectrics: Proc. II All-Union Conf. on Physics of Dielectrics* ed G I Skanavi (Moscow: Izdatel'stvo Akad. Nauk SSSR) p 339
[14] Smolensky G A and Agranovskaya A I 1959 *Fiz. Tverd. Tela* **1** 1562
[15] Fesenko E G, Filip'ev V S and Kupriyanov M F 1968 *Ferroelectrics* ed E G Fesenko (Rostov-na-Donu: Izdatel'stvo Rostovskogo Universiteta) p 63
[16] Belyaev I N 1958 *Izv. Akad. Nauk SSSR: Ser. Phys.* **22** 1436
[17] Venevtsev Yu N, Zhdanov G S *et al* 1958 *Kristallografiya* **3** 373
[18] Venevtsev Yu N, Soloviev S P *et al* 1959 *Kristallografiya* **4** 855
[19] Matthias B P 1951 *Science* **113** 591
[20] Smolensky G A and Kozhevnikova N K 1951 *Dokl. Akad. Nauk SSSR* **6** 519
[21] Isupov V A 1954 *Izv. Akad. Nauk SSSR: Ser. Phys.* **28** 653
[22] Huchua N P, Bova V F and Myl'nikova I E 1968 *Fiz. Tverd. Tela* **10** 257
[23] Matthias B T, Wood E A and Holden A N 1949 *Phys. Rev.* **76** 175; Matthias B T 1949 *Phys. Rev.* **75** 1771
[24] Shirane G, Danner H, Pavlovic A *et al* 1954 *Phys. Rev.* **93** 672
[25] Hurst J J and Linz A 1971 *Mater. Res. Bull.* **6** 163

[26] Tomashpol'sky Yu Ya and Venevtsev Yu N 1968 *Kristallografiya* **13** 791
[27] Wood E A 1951 *Acta Crystallogr.* **4** 353
[28] Katz L and Megaw H D 1967 *Acta Crystallogr.* **22** 639
[29] Vousden P 1951 *Acta Crystallogr.* **4** 373
[30] Shirane G, Newnham R and Pepinsky R 1954 *Phys. Rev.* **96** 581
[31] Wainer E and Wentworth C 1952 *J. Am. Ceram. Soc.* **35** 207
[32] Matthias B T 1950 *Helv. Phys. Acta* **23** 167
[33] Venevtsev Yu N and Zhdanov G S 1957 *Izv. Akad. Nauk SSSR: Ser. Phys.* **21** 275
[34] Smolensky G A 1950 *Zh. Tekh. Fiz.* **20** 137; *Dokl. Akad. Nauk SSSR* **70** 405
[35] Megaw H D 1952 *Acta Crystallogr.* **5** 739; 1954 *Acta Crystallogr.* **7** 186
[36] Vousden P 1951 *Acta Crystallogr.* **4** 545
[37] Blokhin M A 1954 *Dokl. Akad. Nauk SSSR* **95** 965, 1165
[38] Granovsky V G 1962 *Kristallografiya* **7** 604
[39] Lambert M and Comes R 1969 *Solid State Commun.* **7** 305
[40] Hewat A W 1973 *J. Phys. C: Solid State Phys.* **6** 1074
[41] Hewat A W 1973 *J. Phys. C: Solid State Phys.* **6** 2559
[42] Matthias B T and Remeika J P 1951 *Phys. Rev.* **82** 727
[43] Nunes A C, Axe J D and Shirane G 1971 *Ferroelectrics* **2** 291
[44] Kaifu Y, Komatsu T, Hirano T *et al* 1967 *J. Phys. Soc. Japan* **23** 903
[45] Cotts R M and Knight W D 1954 *Phys. Rev.* **96** 1285
[46] Volger J 1952 *Philips Res. Rep.* **7** 21
[47] Blattner H, Kanzig W and Merz W 1949 *Helv. Phys. Acta* **22** 35
[48] Shirane G and Takeda A 1952 *J. Phys. Soc. Japan* **7** 1
[49] Todd S S and Lorenson R E 1952 *J. Am. Chem. Soc.* **74** 2043
[50] Triebwasser S and Halpern J 1955 *Phys. Rev.* **98** 1962
[51] Triebwasser S 1956 *Phys. Rev.* **101** 993
[52] Hall J J and Triebwasser S 1959 *J. Am. Chem. Soc.* **81** 6394
[53] Merz W 1949 *Phys. Rev.* **16** 1221
[54] Fukuda T, Hirano H and Uematsu Y 1974 *Jap. J. Appl. Phys.* **13** 1021
[55] Uematsu Y and Fukuda T 1971 *Jap. J. Appl. Phys.* **10** 507
[56] Timofeeva V A and Popova A S 1962 *Kristallografiya* **7** 300
[57] Deshmukh K G and Ingle S G 1969 *Proc. Nucl. Phys.: Solid State Phys. Symp. 14th* p 362
[58] Last J T 1957 *Phys. Rev.* **105** 1740
[59] Wiesendanger E 1973 *Czech. J. Phys.* **23** 91
[60] Deshmukh K G and Ingle S G 1971 *J. Phys. D: Appl. Phys.* **4** 1633
[61] Deshmukh K G and Ingle S G 1972 *Ind. J. Pure Appl. Phys.* **10** 29
[62] Deshmukh K G and Ingle S G 1971 *J. Phys. D: Appl. Phys.* **4** 124
[63] Deshmukh K G and Ingle S G 1970 *Curr. Sci.* **39** 368
[64] Deshmukh K G and Ingle S G 1972 *Ind. J. Pure Appl. Phys.* **10** 353
[65] Tanaka M 1969 *J. Phys. Soc. Japan Suppl.* **28** 386
[66] Kulkarni R H, Chaudari R M and Ingle S G 1973 *J. Phys. D: Appl. Phys.* **6** 1816
[67] Mohamed Abdel Aziz Gaffor 1975 *Growth and Study of Sodium Niobate Crystals: Cand. Dissertation* (Moscow: Moscow State University)
[68] Fukuda T and Uematsu Y 1977 *Izv. Akad. Nauk SSSR: Ser. Phys.* **41** 548
[69] Wiesendanger E 1974 *Ferroelectrics* **6** 264
[70] Kurtz S K and Perry T T 1968 *J. Appl. Phys.* **39** 3798
[71] Uematsu Y 1974 *Jap. J. Appl. Phys.* **13** 1362

[72] Wiesendanger E 1970 *Ferroelectrics* **1** 141
[73] Di Domenico M and Wemple S H 1969 *J. Appl. Phys.* **40** 720
[74] Gunter P 1974 *Opt. Commun.* **11** 285
[75] Fukuda T and Uematsu Y 1972 *Jap. J. Appl. Phys.* **11** 163
[76] Miller R C and Nordland W A 1970 *Appl. Phys. Lett.* **16** 174
[77] Hobden M V 1967 *J. Appl. Phys.* **38** 4365
[78] Uematsu Y and Fukuda T 1973 *Jap. J. Appl. Phys.* **12** 841
[79] Stein A and Kaplan R A 1970 *Appl. Phys. Lett.* **16** 338
[80] Smith R G 1970 *IEEE J. Quantum Electron.* **QE-6** 215
[81] Geusic J E, Levinstein H J, Singh S *et al* 1968 *Appl. Phys. Lett.* **12** 306
[82] Midwinter J E 1967 *Appl. Phys. Lett.* **11** 128
[83] Reisman A and Holtzberg F 1955 *J. Am. Chem. Soc.* **77** 2115
[84] Joly A 1877 *Ann. Sci. de l' École Normale Superieur, Paris* **6** 164
[85] Reisman A, Holtzberg F, Triebwasser S *et al* 1956 *J. Am. Chem. Soc.* **78** 719
[86] Timofeeva V A and Bychkov V Z 1964 *Crystal Growth* ed A V Shubnikov (Moscow: Nauka) p 140
[87] Miller C E 1958 *J. Appl. Phys.* **29** 233
[88] Fukuda T, Uematsu Y and Ito T 1974 *J. Cryst. Growth* **24** 450
[89] Fedulov S A, Shapiro Z I and Ladyzhinsky P B 1965 *Kristallografiya* **10** 268
[90] Wilcox W R and Fullmer L D 1966 *J. Am. Ceram. Soc.* **49** 415
[91] Shtenberg A A, Lapsker Ya E and Kuznetsov V A 1967 *Kristallografiya* **12** 957

Chapter 2

[1] Devonshire A F 1949 *Phil. Mag.* **40** 1040
[2] Boner W A, Dearborn E F and Van Uitert L G 1965 *J. Am. Ceram. Soc.* **44** 9
[3] Fay H 1967 *Mater. Res. Bull.* **2** 679
[4] Triebwasser S 1959 *Phys. Rev.* **114** 63
[5] Fay H 1967 *Rev. Sci. Instrum.* **88** 197
[6] Von Raalte A 1967 *J. Opt. Soc. Am.* **57** 671
[7] Nye J F 1964 *Physical Properties of Crystals* (Oxford: Clarendon)
[8] Geusic J E, Kurtz S K, Van Uitert L G *et al* 1964 *Appl. Phys. Lett.* **4** 141
[9] Geusic J E, Kurtz S K, Nelson T J *et al* 1963 *Appl. Phys. Lett.* **2** 185
[10] Spitzer W G, Miller R G, Kleinman D A *et al* 1962 *Phys. Rev.* **126** 1710
[11] Reisman A, Holtzberg F and Berry M 1956 *J. Am. Chem. Soc.* **78** 4514
[12] Kirgintsev A I and Avvakumov E G 1965 *Izv. Akad. Nauk SSSR: Ser. Inorg. Mater.* **1** 638
[13] Reisman A, Triebwasser S and Holtzberg F 1955 *J. Am. Chem. Soc.* **77** 4228; 1958 *J. Am. Chem. Soc.* **80** 1877
[14] Ivleva L I, Sal'nikov V D and Kuz'minov Yu S 1970 *Chemical Reagents and Compounds* (Moscow: VNII "IREA") vol 30 p 299
[15] Bonner W A, Dearborn E F and Van Uitert 1968 *Growth of K(Ta, Nb)O_3 Single Crystals for Optical and Semiconductor Studies: Preprint* (Murray Hill, New Jersey: Bell Telephone Laboratories)

Chapter 3

[1] Smolensky G A and Agranovskaya A I 1958 *Zh. Tekhn. Fiz.* **28** 1491
[2] Smolensky G A, Isupov V A and Agranovskaya A I 1959 *Fiz. Tverd. Tela* **1** 170
[3] Smolensky G A, Agranovskaya A I, Popov S N *et al* 1958 *Zh. Teor. Fiz.* **28** 2152
[4] Bokov V A and Myl'nikova I E 1960 *Fiz. Tverd. Tela* **2** 2728
[5] Yokomiro Y and Nomura S 1970 *J. Phys. Soc. Japan* **28** 150
[6] Krainik N N, Gokhberg L S and Myl'nikova I E 1970 *Fiz. Tverd. Tela* **12** 2360
[7] Megerhoffer D 1958 *Phys. Rev.* **112** 413
[8] Smolensky G A, Krainik N N, Berezhnoi A A *et al* 1968 *Fiz. Tverd. Tela* **10** 2675
[9] Berezhnoi A A 1969 *Author's abstract, Cand. Dissertation* (Leningrad: Izdatel'stvo Akad. Nauk SSSR)
[10] Miller R C 1964 *Appl. Phys. Lett.* **5** 17
[11] Berezhnoi A A 1967 *Izv. Akad. Nauk SSSR: Ser. Phys.* **31** 1154
[12] Myl'nikova I E and Bokov V A 1961 *Crystal Growth* ed A V Shubnikov (Moscow: Nauka) **3** p 438
[13] Smolensky G A, Agranovskaya A I and Popov S N 1959 *Fiz. Tverd. Tela* **1** 167
[14] Smolensky G A, Krainik N N, Berezhnoi A A *et al* 1968 *Fiz. Tverd. Tela* **10** 467
[15] Kirillov V V and Isupov V A 1969 *Izv. Akad. Nauk SSSR: Ser. Phys.* **32** 313
[16] Isupov V A 1956 *Zh. Tekh. Fiz.* **26** 1912
[17] Rolov B N 1964 *Fiz. Tverd. Tela* **6** 2128
[18] Skanavi G I 1949 *Physics of Dielectrics* (Moscow: Gostekhizdat)
[19] Sal'nikov V D, Kuz'minov Yu S and Venevtsev Yu N 1971 *Izv. Akad. Nauk SSSR: Ser. Inorg. Mater.* **7** 1277
[20] Smolensky G A, Berezhnoi A A, Krainik N N *et al* 1969 *Izv. Akad. Nauk SSSR: Ser. Phys.* **33** 282
[21] Krainik N N, Trepakov V A, Kamzina L S *et al* 1975 *Fiz. Tverd. Tela* **17** 208
[22] Kamzina L S, Krainik N N and Berezhnoi A A 1973 *Fiz. Tverd. Tela* **15** 3011
[23] Kurtz S K and Robinson F N 1967 *Appl. Phys. Lett* **10** 62
[24] Smolensky G A, Berezhnoi A A, Pisarev P V *et al* 1969 *Fiz. Tverd. Tela* **11** 1120
[25] Kamzina L S, Krainik N N and Nesterova N N 1972 *Fiz. Tverd. Tela* **14** 2147
[26] Berezhnoi A A 1972 *Fiz. Tverd. Tela* **14** 2035
[27] Kamzina L S, Krainik N N, Gene V V *et al* 1971 *Izv. Akad. Nauk SSSR: Ser. Phys.* **35** 1862
[28] Sonin A S and Lomova L G 1967 *Izv. Akad. Nauk SSSR: Ser. Phys.* **31** 1145
[29] Smolensky G A, Kamzina L S and Krainik N N 1975 *Izv. Akad. Nauk SSSR: Ser. Phys.* **39** 805
[30] Martynova S V and Petrov V M 1970 *Izv. Akad. Nauk SSSR* **34** 2594
[31] Smolensky G A, Krainik N N, Trepakov V A *et al* 1975 *Izv. Akad. Nauk SSSR: Ser. Phys.* **39** 791
[32] Adrianova I I, Berezhnoi A A, Nefedova E V *et al* 1973 *Opt. Spektrosk.* **35** 888
[33] Adrianova I I, Popov Yu V, Sokolova R S *et al* 1968 *Opt. Mekh. Promyshlennost'* **10** 40
[34] Berezhnoi A A 1972 *Fiz. Tverd. Tela* **14** 2576
[35] Agranovskaya I I 1960 *Izv. Akad. Nauk SSSR: Ser. Phys.* **2** 1271
[36] Bokov V A, Myl'nikova I E and Smolensky G A 1962 *Zh. Exp. Teor. Fiz.* **42** 643
[37] Bokov V A and Myl'nikova I E 1961 *Fiz. Tverd. Tela* **3** 841
[38] Myl'nikova I E and Bokov V A 1959 *Kristallografiya* **4** 433

[39] Myl'nikova I E and Bokov V A 1961 *Crystal Growth* ed A V Shubnikov (Moscow: Nauka) p 438
[40] Sholokhovich M L and Fesenko E G 1960 *Izv. Akad. Nauk SSSR: Ser. Phys.* **24** 1242
[41] Bonner W A and Van Uitert Z G 1965 *Am. Ceram. Soc. Bull.* **44** 9

Chapter 4

[1] Lenzo P V, Spenser E G and Ballman A A 1967 *Appl. Phys. Lett.* **11** 23
[2] Glass A M 1964 *J. Appl. Phys.* **40** 4699
[3] Venturini E L, Spenser E G and Ballman A A 1969 *J. Appl. Phys.* **40** 1622
[4] Abrahams S C, Jamieson P B and Bernstein J L 1971 *J. Chem. Phys.* **54** 2355
[5] Micheron F, Mayeux C and Trotier J C 1974 *Appl. Opt.* **13** 913
[6] Jamieson P B, Abrahams S C and Bernstein J L 1968 *J. Chem. Phys.* **48** 5048
[7] Megumi K, Negatsuma K, Kashiwada Y *et al* 1976 *J. Mater. Sci.* **11** 1583
[8] Displanches G, Barrand J L and Lazennec Y 1974 *J. Cryst. Growth* **23** 149
[9] Whipps P W 1972 *J. Solid State Chem.* **4** 281
[10] Boniort J, Brehm C, Displanches G *et al* 1975 *J. Cryst. Growth* **30** 357
[11] Carruthers J R and Crasso M 1970 *Solid State Sci.* **117** 1426
[12] Peterson G E, Bridenbaugh P M and Green P 1967 *J. Chem. Phys.* **46** 4009
[13] Sakamoto S and Yazaki T 1973 *Appl. Phys. Lett.* **22** 429
[14] Rolov B N 1969 *Izv. Akad. Nauk SSSR: Ser. Phys.* **33** 227
[15] Fox A J 1973 *J. Appl. Phys.* **44** 254
[16] Maciolek R B and Liu S T 1975 *J. Electron. Mater.* **4** 517
[17] Liu S T and Maciolek R B 1975 *J. Electron. Mater.* **4** 91
[18] Kuz'minov Yu S 1975 *Lithium Niobate and Tantalate as Materials for Nonlinear Optics* (Moscow: Nauka)
[19] Cockayne B, Chesswas M, Plant J G *et al* 1969 *J. Mater. Sci.* **4** 565
[20] Dubovik M F, Drogaitsev E A and Teplitskaya T S 1975 *Monokristally i Tekhnika* (Khar'kov: All-Union Research Institute for Single Crystals) vol 5 p 23
[21] Elinson M I, Zakharov L Yu, Morozov N A *et al* 1974 *Mikroelektronika* **3** 459
[22] Kopylov Yu L, Kravchenko V B and Dudnik O F 1978 *Piezo- and Ferro-electric Materials and Their Applications* (Moscow) p 86
[23] Kopylov Yu L, Kravchenko V B, Kucha V V *et al* 1978 *Mikroelektronika* **7** 412
[24] Kopylov Yu L, Kravchenko V B and Moiseev V P 1979 *Kristall und Technik* **14** 697
[25] Bush A A, Chechkin V V, Leichenko *et al* 1977 *Izv. Akad. Nauk SSSR: Ser. Inorg. Mater.* **13** 2214
[26] Elsen E, Grabmaier J and Graf P 1971 *Optics and Laser Technology* **3** 218
[27] Furuhata Y 1977 *Izv. Akad. Nauk SSSR: Ser. Phys.* **41** 573
[28] Voronov V V, Desyatkova S M, Ivleva L I, Kuz'minov Yu S *et al* 1973 *Fiz. Tverd. Tela* **15** 2198
[29] Zharikov E V, Ivleva L I, Kuz'minov Yu S *et al* 1975 *Izv. Akad. Nauk SSSR: Ser. Inorg. Mater.* **11** 875
[30] Vaivod P A, Voronov V V, Ivleva L I, Kuz'minov Yu S *et al* 1977 *Fiz. Tverd. Tela* **19** 3163

[31] Voronov V V, Desyatkova S M, Ivleva L I, Kuz'minov Yu S and Osiko V V 1974 *Kristallografiya* **19** 401

[32] Kuz'minov Yu S and Osiko V V 1974 *Priroda* **12** 28

[33] Chang T S, Amzallag E and Rokni M 1971 *Ferroelectrics* **3** 57

[34] Abrahams S C, Kurtz S K and Jamieson P B 1968 *Phys. Rev.* **172** 555

[35] Tada K, Murai T, Aoki M *et al* 1972 *Jap. J. Appl. Phys.* **11** 1622

[36] Aleksyuk V E, Kovtun R N, Kushnir P I *et al* 1975 *Izv. Akad. Nauk SSSR: Ser. Inorg. Mater.* **11** 2098

[37] Pearson C L and Bardeen J 1949 *Phys. Rev.* **75** 865

[38] Brice J C, Hill O F, Whiffen *et al* 1971 *J. Cryst. Growth* **10** 133

[39] Venturini E L, Spenser E G, Lenzo P V *et al* 1968 *J. Appl. Phys.* **39** 343

[40] Wemple S H and Di Domenico M 1969 *Phys. Rev. Lett.* **23** 1156

[41] Andreichuk A E, Dorozhkin L M, Kuz'minov Yu S *et al* 1984 *Kristallografiya* **20** 1094–101

[42] Wemple S H and Di Domenico M 1969 *J. Appl. Phys.* **40** 720

[43] Rice R R 1971 *IEEE J. Quantum Electron.* **QE-7** 284

[44] Nomura S, Kojima H, Hattori Y and Kotsuka H 1974 *Jap. J. Appl. Phys.* **13** 1185

[45] Dudnik O F, Kopylov Yu L and Kravchenko V B 1973 *Pis'ma v ZhETF* **18** 407

[46] Hulme K F 1971 *IEEE J. Quantum Electron.* **QE-7** 236

[47] Hulme K F, Davies P H and Counda V M 1969 *J. Phys. C: Solid State Phys.* **2** 855

[48] Lutber-Davies B, Davies P H, Cound V M *et al* 1970 *J. Phys. C: Solid State Phys.* **3** L106

[49] Smolensky G A 1970 *J. Phys. Soc. Japan Suppl.* **28** 26

[50] Ballman A A and Brown H 1967 *J. Cryst. Growth* **1** 311

[51] Dudnik O F, Kravchenko V B, Gromov A K and Kopylov Yu L 1972 *Crystal Growth* ed N N Sheftal' and E I Givargizov (Moscow: Nauka) vol 9 p 130

[52] Zharikov E V, Kuz'minov Yu S and Osiko V V 1974 *Izv. Akad. Nauk SSSR: Ser. Inorg. Mater.* **10** 1081

[53] Losev N F 1969 *Quantitative X-ray Diffraction Fluorescent Analysis* (Moscow: Nauka)

[54] Abramov N A, Ivleva L I, Kuz'minov Yu S *et al* 1977 *Kristall und Technik* **12** 1157

[55] Byer R L, Young J F and Feigelson R S 1970 *J. Appl. Phys.* **41** 2320

[56] Barns R L and Carruthers J R 1971 *Lithium Tantalate Single Crystal Stoichiometry: Preprint* (Murray Hill, New Jersey: Bell Telephone Laboratories)

[57] Voronov V V, Zharikov E V, Ivleva L I, Kuz'minov Yu S *et al* 1976 *Kristall und Technik.* **11** 1113

[58] Maciolek R B and Liu S T 1973 *J. Electron. Mater.* **2** 191

[59] Voronov V V, Ivleva L I, Kuz'minov Yu S *et al* 1977 *Kristallografiya* **22** 552

[60] Scott B A and Burns G 1972 *J. Am. Ceram. Soc.* **55** 225

[61] Zheludev I S 1968 *Physics of Crystal Dielectrics* (Moscow: Nauka)

[62] Distler G I *et al* 1976 *Decoration of Solid Surfaces* (Moscow: Nauka)

[63] Ito Y and Furuhata Y 1974 *Phys. Status Solidi* (a) **23** 147

[64] Ito Y, Kozuka H, Kashiwada Y and Furuhata Y 1975 *Jap. J. Appl. Phys.* **14** 1443

[65] Revkevich G P, Ivleva L I, Katsnel'son A A and Kuz'minov Yu S 1979 *Kristallografiya* **24** 1079

[66] Guinier A 1959 Heterogeneities in solid solutions *Solid State Physics* vol 9, ed F Seitz and D Turnbull (New York: Academic Press)

[67] Francombe M 1960 *Acta Crystallogr.* **13** 131
[68] Boniort J, Brehm C, Displanches G *et al* 1973 *J. Cryst. Growth* **30** 357
[69] Gorislavsky B I, Kisel' N G, Nikiforov L G *et al* 1975 *Izv. Akad. Nauk SSSR: Ser. Inorg. Mater.* **11** 1049
[70] Carruthers J R, Blank S L, Grasso M *et al* 1974 *J. Cryst. Growth* **23** 195
[71] Megumi K, Kozuka H, Nagatuma K and Kashiwada Y 1974 *Collected Abstracts, Int. Conf. on Crystal Growth, ICCG-4, Tokyo* p 466
[72] Abrahams S C, Reddy J M and Bernstein J L 1966 *J. Phys. Chem. Solids* **27** 999
[73] Kuz'minov Yu S, Novikova E M, Turkina K Ya *et al* 1969 *Izv. Akad. Nauk SSSR: Ser. Inorg. Mater.* **5** 1982
[74] Dudnik O F, Gromov A K, Kravchenko V B *et al* 1970 *Kristallografiya* **15** 386
[75] Kazakov Yu V, Kuz'minov Yu S, Osiko V V and Polozkov N M 1984 *Kristallografiya* **29** 576–8
[76] Dudnik O F, Kopylov Yu L and Kravchenko V B 1975 *Kristallografiya* **20** 1013
[77] Kim K M, Witt A F and Gatos H G 1972 *J. Electrochem. Soc.* **119** 1218
[78] Van Uitert L G, Rubin J J and Bonner W A 1968 *IEEE J. Quantum Electron.* **QE-4** 622
[79] Wiffin P A C and Brice J C 1971 *J. Cryst. Growth* **10** 91
[80] Brehm C, Boniort J and Margotin P 1973 *J. Cryst. Growth* **18** 191
[81] Cockane B, Chesswas M, Born R J and Filby J D 1969 *J. Mater. Sci.* **4** 236
[82] Park K C, Berbenblit M, Herrell D J *et al* 1972 *J. Electron. Mater.* **2** 201
[83] Dubovik M F *et al* 1972 *Monokristally i Tekhnika* ed E F Chaikivsky (Khar'kov: VNII Monokristallov) p 130
[84] Ivleva L I 1979 *Candidate Dissertation: The Study of Physico-Chemical Conditions of Obtaining and Properties of Single Crystals of BSN Solid Solutions* (Moscow: Lebedev Physical Institute, USSR Acad. Sci.)
[85] Voronov V V, Ivleva L I, Kuz'min G P, Kuz'minov Yu S *et al* 1976 *Preprint No 191* (Moscow: Lebedev Physical Institute, USSR Acad. Sci.)
[86] Voronov V V, Karlov N V, Kuz'min G P, Kuz'minov Yu S *et al* 1977 *Kvant. Electron.* **4** 1903
[87] Furuhata Y 1976 *J. Jap. Assoc. Cryst. Growth* **3** 1
[88] Savitsky V G, Aleksyuk V E and Protsakh P F 1975 *Fiz. Khim. Tverd. Tela* vol 6, ed Yu N Venevtsev (Moscow: Nauka) p 137

Chapter 5

[1] Geusic J E, Levinstein H J, Rubin J J *et al* 1967 *Appl. Phys. Lett.* **11** 269
[2] Rubin J J, Van Uitert L G and Levinstein H J 1967 *J. Cryst. Growth* **1** 315
[3] Sonin A S and Vasilevskaya A S 1971 *Electro-optic Crystals* (Moscow: Atomizdat)
[4] Van Uitert L G, Singh S, Levinstein H J *et al* 1967 *Appl. Phys. Lett.* **11** 161
[5] *Bell Lab. Rec.* 1967 **45** No 9 p 304
[6] Geusic J E, Marcos H M and Van Uitert L G 1964 *Appl. Phys. Lett.* **4** 182
[7] Smith R G, Nassau K and Galvin M K 1965 *Appl. Phys. Lett.* **7** 256

[8] Geusic J E, Levinstein H J, Singh S *et al* 1968 *Appl. Phys. Lett.* **12** 306

[9] Van Uitert L G, Levinstein H J, Rubin J J *et al* 1968 *Mater. Res. Bull.* **3** 47

[10] Tsuya H, Fujino Y and Sugibuchi K 1970 *J. Appl. Phys.* **41** 2557

[11] Ryabtsev N G 1972 *Materials for Quantum Electronics* (Moscow: Sovetskoye Radio)

[12] Smith R G, Geusic J E, Levinstein H J *et al* 1968 *IEEE J. Quantum Electron.* **QE-4** 354

[13] Nash F R, Turner E H, Bridenbaugh P M *et al* 1972 *J. Appl. Phys.* **43** 1

[14] Warner A W, Coquin G A, Meitzler A H and Fink J I 1969 *Appl. Phys. Lett.* **14** 34

[15] Warner A W, Coquin G A and Fink J L 1969 *J. Appl. Phys.* **40** 4353

[16] Spenser E G and Van Uitert L G 1968 *Phys. Lett.* **27A** 626

[17] Bonner W A, Grodkiewicz W H and Van Uitert L G 1967 *J. Cryst. Growth* **1** 318

[18] Jamieson P B, Abrahams S C and Bernstein S L 1969 *J. Chem. Phys.* **50** 4352

[19] Giess E A, Scott B A, Burns G *et al* 1969 *J. Am. Ceram. Soc.* **52** 276

[20] Abrahams S C and Keve E T 1971 *Ferroelectrics* **2** 129

[21] Abell J C, Barraclough K G, Harris I R *et al* 1971 *J. Mater. Sci.* **6** 1084

[22] Ballman A A, Kurtz S K and Brown H 1971 *J. Cryst. Growth* **10** 185

[23] Vere A W, Plant J G, Cockayne B *et al* 1969 *J. Mater. Sci.* **4** 1075

[24] Ballman A A, Carruthers J R and O'Bryan 1970 *J. Cryst. Growth* **6** 184

[25] Barns R L 1968 *J. Appl. Crystallogr.* **1** 290

[26] Scott B A, Giess E A and O'Kane D F 1969 *Mater. Res. Bull.* **4** 107

[27] Carruthers J R and Grasso M 1969 *Mater. Res. Bull.* **4** 413

[28] Aizu K 1967 *J. Phys. Soc. Japan* **27** 387

[29] Zharikov E V, Kalabukhova V F, Kuz'minov Yu S and Osiko V V 1977 *Crystal Growth* (Erevan: Izdatel'stvo Erevanskogo Univers.) vol 12, p 255

[30] Lidiard A B 1957 *Ionic conductivity: Handbuch der Physik* vol XX, Teil 2 (Berlin: Springer) p 246

[31] Burns G, Giess E A and O'Kane D F 1970 *Proc. 2nd Int. Mtg on Ferroelectrics, Kyoto, Japan* 1969 pp 153–8

[32] Bonner W A, Carruthers J R and O'Bryan H M 1970 *J. Mater. Res. Bull.* **5** 243

[33] Rice R R, Fay H, Dess H M and Alford W J 1969 *J. Electrochem. Soc.* **116** 839

[34] Singh S, Draegert D A and Geusic J E 1970 *Phys. Rev.* B **2** 2709

[35] Singh S, Levinstein H J and Van Uitert L G 1970 *Appl. Phys. Lett.* **16** 176

[36] Voronov V V, Zharikov E V, Kuz'minov Yu S *et al* 1974 *Fiz. Tverd. Tela* **16** 162

[37] Van Uitert L G, Rubin J J, Grodkiewicz W H and Bonner W A 1969 *Mater. Res. Bull.* **4** 63

[38] Zharikov E V, Ivleva L I, Kuz'minov Yu S and Osiko V V 1975 *Izv. Akad. Nauk SSSR: Ser. Inorg. Mater.* **11** 875

[39] Singh S, Draegert D A, Geusic J E *et al* 1968 *IEEE J. Quantum Electron.* **QE-4** 352

[40] Hooton J A and Merz W J 1955 *Phys. Rev.* **98** 409

[41] Zharikov E V, Kuz'minov Yu S, Osiko V V *et al* 1977 *Crystal Growth* ed A A Chernov (Erevan: Izdatel'stvo Erevanskogo Univers.) vol 12 p 219

[42] Aleksandrovskii A L and Nagaev A I 1983 *Phys. Status Solidi* (a) **78** 431

[43] Klyuev V P, Safonov A I, Baryshev S A and Angert N B 1969 *Elektron. Tekh. Ser. 14: Mater.* **3** 95

[44] Voronov V V and Kuz'minov Yu S 1978 *Fiz. Tverd. Tela* **20** 389

[45] Arsenev P A, Linda J G, Sarafanova N N *et al* 1973 *Kristall und Technik* **8** 615

[46] Smith R G, Fraser D B, Denton R T and Rich T C 1968 *J. Appl. Phys.* **39** 4600
[47] Rez I S 1969 *Izv. Akad. Nauk SSSR: Ser. Phys.* **33** 289
[48] Allen L and Jones D G C 1967 *Principles of Gas Lasers* (London: Butterworth)
[49] Aleksandrovskii A L 1981 *Vestnik Mosk. Univers.: Ser. Phys., Astron.* **22** 51
[50] Dolino G 1972 *Phys. Rev.* B **6** 4025–35
[51] Barraclough K G, Harris I R, Cockayne B *et al* 1970 *J. Mater. Sci.* **5** 389
[52] Giess E A, Scott B A and Olson B L 1970 *J. Am. Ceram. Soc.* **53** 14
[53] Linares R C and Sigsway R L 1968 *Mater. Res. Bull.* **3** 825
[54] Cockayne B, Plant J G and Vere A W 1972 *J. Cryst. Growth* **15** 167
[55] Carruthers J R, Blank S L, Grasso M and Biolsi W A 1974 *J. Cryst. Growth* **23** 195
[56] Bonner W A 1970 *J. Electrochem. Soc.* **117** 849
[57] O'Kane D F, Burns G, Giess E A *et al* 1969 *J. Electrochem. Soc.* **116** 1555
[58] Giess E A, Scott B A and O'Kane D F 1969 *Mater. Res. Bull.* **4** 741
[59] Ainger F W and Bickley W P 1968 *J. Cryst. Growth* **3/4** 278
[60] Ballman A A, Kurtz S K and Brown H 1971 *J. Cryst. Growth* **10** 185
[61] Zupp R R, Nielsen J W and Vittorio P V 1969 *J. Cryst. Growth* **5** 269
[62] Van Uitert L G 1970 *Platinum Metals Rev.* **14** 118
[63] Brice J C and Whiffin P A C 1973 *Platinum Metals Rev.* **17** 46
[64] Brice J C, Hill O F and Whiffin P A C 1970 *J. Cryst. Growth* **6** 297
[65] Ballman A A and Carruthers J R *Patent, France No 70 18841*
[66] Valyashko E G, Kukina A M and Rashkovich L N 1973 *Spektroskopiya Kristallov. Proc. 3rd Symp. on Crystal Spectroscopy* ed P P Feofilov (Leningrad: Nauka) p 187
[67] Maskayev Yu A, Rashkovich L N and Koptsik V A 1975 *Kristallografiya* **20** 622
[68] Maskayev Yu A and Rashkovich L N 1976 *Kristallografiya* **21** 1043
[69] Bursian E V 1974 *Nonlinear Crystal of Barium Titanate* (Moscow: Nauka)
[70] Sheinkman M K and Shik A Ya 1976 *Fiz. Tekh. Poluprov.* **10** 209
[71] Nash F R, Boyd G D, Sargent M and Bridenbaugh P M 1970 *J. Appl. Phys.* **41** 2564
[72] Nash F R, Bergman J G, Boyd G D and Turner E H 1969 *J. Appl. Phys.* **40** 5201
[73] Bobb L C, Lefkowitz I and Muldawur L 1969 *J. Appl. Crystallogr.* **2** 189
[74] Cockayne B and Gates M P 1967 *J. Mater. Sci.* **2** 118
[75] Wilcox W R and Fullmer L D 1965 *J. Appl. Phys.* **36** 2201
[76] Linares R C and Gurski T R 1969 *Mater. Res. Bull.* **4** 663
[77] Nassau K, Levinstein H J and Loiacono G M 1966 *J. Phys. Chem. Solids* **27** 983
[78] Camlibel I 1969 *J. Appl. Phys.* **40** 1690
[79] Schneck J, Primot J and Raver J 1977 *Solid State Commun.* **21** 207
[80] Van Uitert L G, Rubin J J and Bonner W A 1968 *IEEE J. Quantum Electron.* **QE-4** 622
[81] Barry J D and Kennedy C J 1975 *IEEE J. Quantum Electron.* **QE-11** 575
[82] Rubin J J and Capio C D 1968 *Mater. Res. Bull.* **3** 47
[83] Burns G and O'Kane D F 1969 *Phys. Lett.* **28** A 776
[84] Cockane B, Chesswas M, Plant J G and Vere A W 1969 *J. Mater. Sci.* **4** 565
[85] Toledano J C 1975 *Phys. Rev.* B **12** 943
[86] Wemple S H and Di Domenico M D 1968 *J. Appl. Phys. Lett.* **12** 209
[87] Quandt H D and Rothrock L R 1977 *J. Cryst. Growth* **42** 435
[88] Yamanaka T, Iwasaki H and Nizeki N 1970 *J. Appl. Phys.* **41** 4141

Chapter 6

[1] Scott B A, Giess E A, O'Kane D F and Burns G 1970 *J. Am. Ceram. Soc.* **53** 106
[2] Smith A W, Burns G and O'Kane D F 1971 *J. Appl. Phys.* **42** 250
[3] Voronov V V, Kuz'minov Yu S, Osiko V V, Prokhorov A M *et al* 1974 *Dokl. Akad. Nauk SSSR* **218** 1317
[4] Voronov V V, Kuz'minov Yu S, Osiko V V and Prokhorov A M 1975 *Kvant. Elektron.* **2** 525
[5] Burns G and Smith A W 1968 *IEEE J. Quantum Electron.* **QE-4** 584
[6] Wemple S, Di Domenico M and Camlibel I 1968 *Appl. Phys. Lett.* **12** 209
[7] Geusic J E, Kurtz S K, Van Uitert L G and Wemple S H 1964 *Appl. Phys. Lett.* **4** 141
[8] Byer R L, Harris S E, Kuizenga D J and Young J F 1969 *J. Appl. Phys.* **40** 444
[9] Ivleva L I, Kuz'minov Yu S and Shumskaya L S 1972 *Fiz. Tverd. Tela* **14** 3137
[10] Hirano H, Takei H and Koide S 1970 *Jap. J. Appl. Phys.* **9** 580
[11] Born M and Wolf E 1964 *Principles of Optics* (Oxford: Pergamon)
[12] Bernecker O, Mattnes H and Marshall A 1973 *Phys. Status Solidi* (a) **17** 453
[13] Mattnes H 1972 *J. Cryst. Growth* **15** 157
[14] Giess E A, Burns G, O'Kane D F and Smith A W 1967 *Appl. Phys. Lett.* **11** 293
[15] Clarke R and Ainger F W 1974 *Ferroelectrics* **7** 101
[16] Ainger F W, Beswick J A, Porter S G and Clarke R 1973 *Ferroelectrics* **3** 321
[17] Van Uitert L G 1970 *Appl. Phys. Lett.* **16** 84
[18] Scott B A, Burns G, Giess E A and O'Kane D F 1969 *Bull. Am. Phys. Soc.* **14** 60
[19] Kitahiro I, Yano T and Watanabe A 1969 *Jap. J. Appl. Phys.* **8** 807
[20] Watanabe A, Sato Y, Yano T and Kitahiro I 1970 *J. Phys. Soc. Japan Suppl.* **28** 93
[21] Yano T, Ohta T and Watanabe A 1970 *Jap. J. Appl. Phys.* **9** 1008
[22] Ohta T, Watanabe A and Tamada M 1970 *Jap. J. Appl. Phys.* **9** 721
[23] Spencer E G, Lenzo P V and Ballman A A 1967 *Proc. Inst. Electron. Engnrs* **55** 2074
[24] O'Kane D F, Burns G, Scott B A and Giess E A 1968 *J. Electrochem. Soc.* **115** 1081
[25] Burns G and O'Kane D F 1969 *Phys. Lett.* **28** A 776
[26] Burns G, O'Kane D F, Giess E A and Scott B A 1968 *Solid State Commun.* **6** 223
[27] Scott B A, Giess E A, Burns G and O'Kane D F 1968 *Mater. Res. Bull.* **3** 831
[28] Krainik N N, Isupov V A, Bryzhina M F and Agranovskaya A I 1964 *Sov. Phys.–Crystallogr.* **9** 352
[29] Yoichiro M and Masanobu W 1973 *Jap. J. Appl. Phys.* **12** 1101
[30] Itok Y and Iwasahi H 1973 *J. Phys. Chem. Solids* **34** 1639
[31] Fang P H, Roth R S and Forrat F 1961 *C. R. Acad. Sci., Paris* **253** 1039
[32] Stephenson N C 1965 *Acta Crystallogr.* **18** 496
[33] Maker P D, Terhuner R W, Nisenaff M and Savage C M 1962 *Phys. Rev. Lett.* **8** 21
[34] Jerphagnon J and Kurtz S K 1970 *J. Appl. Phys.* **41** 1667
[35] Miller R C and Nordland W A 1970 *Appl. Phys. Lett.* **16** 174
[36] Jerphagnon J 1970 *Phys. Rev. B* **2** 1091
[37] Ikeda T, Kato J and Fujimura I 1977 *Izv. Akad. Nauk SSSR: Ser. Phys.* **41** 567
[38] Ikeda T and Fujimura I 1974 *Jap. J. Appl. Phys.* **13** 57
[39] Ikeda T and Kato J 1974 *Jap. J. Appl. Phys.* **13** 1792

[40] Tatsuya N and Ikeda T 1973 *Jap. J. Appl. Phys.* **12** 199
[41] Smith A W, Burns G, Scott B A and Edmonds H D 1971 *J. Appl. Phys.* **42** 684
[42] Kleinman D A 1963 *Phys. Rev.* **131** 95
[43] Miller R C and Savage A 1966 *Appl. Phys. Lett.* **9** 169
[44] Scott B A, Giess E A, Olson B L *et al* 1970 *Mater. Res. Bull.* **5** 47
[45] Reisman A and Holtzberg F 1958 *J. Am. Chem. Soc.* **80** 6503
[46] Fukuda T 1969 *Jap. J. Appl. Phys.* **8** 122
[47] Grodkiewicz W H, Dearborn E F and Van Uitert L G 1967 *J. Cryst. Growth* **1** 441
[48] Ainger F W, Beswick J A, Bickley W P *et al* 1971 *Ferroelectrics* **2** 183
[49] Fukuda T, Hirano H and Koide S 1970 *J. Cryst. Growth* **6** 293

Chapter 7

[1] Willand C S and Albrecht A C 1986 *Opt. Commun.* **57** 146
[2] Chen C, Bochang Wu, Aidong J and Guiming J 1985 *Scientia Sinica Ser. B* **28** 235
[3] Chen C, Fan Y X, Eckardt R C and Byer R L 1986 '*Cleo*' (*Proc. Int. Conf. Lasers and Electro-optics*) *California* p 3322
[4] *NBS (US) Monogr.* 1965 **25** §4
[5] Hubner K H 1969 *Neues Jahrb. Mineral, Monatsh.* 335–42
[6] Lu Shao-Fang, Ho Mei-Jun and Huang Lin-Ling 1982 *Acta Phys. Sinica* **31**
[7] Ivleva L I, Gordadze I G, Kuz'minov Yu S, Voronov V V and Ivanovskaya V M 1989 *Izv. Akad. Nauk SSSR: Ser. Inorg. Mater.* **25** 804–7
[8] Ivleva L I, Kiselev D T, Kuz'minov Yu S and Polozkov N M 1988 *Izv. Akad. Nauk SSSR: Ser. Inorg. Mater.* **24** 1153–7
[9] Chen C, Wu B, Yiang A and You G 1985 *Scientia Sinica* (b) **28** 235
[10] Chen C 1985 *Preprint, Fujian Institute of Research on the Structure of Matter: Academia Sinica*
[11] Adhav R S, Adhav S R and Pelaprat J M 1987 *Laser Focus Electro-optics* September 88–100
[12] Kurtz S K, Jerphagnon J and Choy M M 1979 *Nonlinear Dielectric Susceptibilities, I: Landolt-Bornstein, Numerical Data and Functional Relationships in Science and Technology* vol 11, ed K H Hellwege and A M Hellwege (Berlin: Springer)
[13] Boyd G D and Kleinman D A 1968 *J. Appl. Phys.* **39** 3597–639
[14] Armstrong J A, Bloemberg N, Ducuing J and Pershan P S 1962 *Phys. Rev.* **127** 1918–39
[15] Matveev Y A, Hikoggosyan D N, Kabelka V and Piskarkas A 1978 *Sov. J. Quantum Electron.* **QE-8** 386–8
[16] Kato K 1986 *IEEE J. Quantum Electron.* **QE-22** 1013–14
[17] Dewey H J 1976 *IEEE J. Quantum Electron.* **QE-12** 303
[18] Fan Y X, Eckardt R C, Byer R L, Chen C and Ying A D 1986 June 9–13 Post deadline paper Th T4-1, *Conf. on Lasers and Electro-Optics, San Francisco, June 1986*
[19] Vanherzeele H, Dupont de Nemours E I and Chen C 1985 *Preprint*
[20] Zumsteg F C, Bierlein J D and Gier T E 1976 *J. Appl. Phys.* **47** 4980
[21] Yao J Q and Fahlen Th 1984 *J. Appl. Phys.* **55** 65
[22] Fendt A, Kranitzky W, Lauberau A and Kaiser W 1979 *Opt. Commun.* **28** 142
[23] Sellmeier A and Kaiser W 1980 *Appl. Phys.* **23** 113

[24] Kung A H 1974 *Appl. Phys. Lett.* **25** 653
[25] Sukhorukov A P and Schednova A K 1971 *ZhETP* **60** 1251
[26] Schröder W A 1984 *Opt. Commun.* **49** 75
[27] Zumsteg F C, Bierlein J D and Gier T E 1976 *J. Appl. Phys.* **47** 4980–5
[28] Tordjman I, Masse R and Guitel J C 1974 *Z. Kristallogr.* **139** 103–15
[29] Leonov A P, Voronkova V I, Stefanovich S Yu and Yanovskii V K 1988 *Sov. Tech. Lett.* **11** 34
[30] Voronkova V I, Stefanovich S Yu and Yanovskii V K 1988 *Kvant. Elektron.* **15** 752–6
[31] Yanovskii V K, Voronkova V I, Leonov A P and Stefanovich S Yu 1985 *Fiz. Tverd. Tela* **27** 2516
[32] Yanovskii V K and Voronkova V I 1986 *Phys. Status Solidi* (a) **93** 665
[33] Kalesnikas V A, Pavlova N I, Rez I S and Grichas I P 1982 *Litovskii Fizicheskii Sbornik* **22** 87–92
[34] Yanovskii V K and Voronkova V I 1985 *Fiz. Tverd. Tela* **27** 2183
[35] Voronkova V I, Leonov A P, Stefanovich S Yu and Yanovskii V K 1985 *Pis'ma v ZhETF* **11** 531
[36] Bierlein J D and Arweiler C B 1986 *Appl. Phys. Lett.* **49** 917–19
[37] Bierlein J D and Ahmed F 1987 *Appl. Phys. Lett.* **51** 1322–3
[38] Yariv Y 1984 *Optical Waves in Crystals* (New York: Wiley)
[39] Stolzenberger R A 1988 *Appl. Opt.* **27** 3883–6
[40] Miller R C 1964 *Phys. Rev.* **134** A1313
[41] Liu Yung S, Dentz D and Belt R 1984 *Opt. Lett.* **9** 76
[42] Smith R G 1970 *IEEE J. Quantum Electron.* **QE-6** 215
[43] Murray J E and Harris S E 1970 *Appl. Phys.* **41** 609
[44] Bordui P F, Jacco J C and Loiacano G M 1987 *J. Cryst. Growth* **84** 403–8
[45] Kurtz S K and Perry T T 1968 *J. Appl. Phys.* **39** 3798
[46] Andreichuk A E, Dorozhkin L M, Kuz'minov Yu S *et al* 1984 *Kristallografiya* **29** 1094–101
[47] Lin J T and Chen C 1987 *Laser Optronics* November
[48] Lin J T 1987 *Proc. Int. Conf. on Lasers* (VA: STS Press) p 262

Chapter 8

[1] Ashkin A, Boyd G D, Driedric J H *et al* 1966 *Appl. Phys. Lett.* **9** 72
[2] Chen F S, Geusic J E, Kurtz S K *et al* 1966 *J. Appl. Phys.* **37** 388
[3] Chen F S 1969 *J. Appl. Phys.* **40** 3389
[4] Serreze H B and Goldner R B 1973 *Appl. Phys. Lett.* **22** 626
[5] Turner E H 1966 *Appl. Phys. Lett.* **8** 303
[6] Lenze P V, Turner E H, Spencer E G *et al* 1966 *Appl. Phys. Lett.* **8** 81
[7] Chen F S, Le Macchina J T and Fraser D B 1968 *Appl. Phys. Lett.* **13** 233
[8] Bube R H 1960 *Photoconductivity in Solids* (New York: Wiley)
[9] Johnston W D 1970 *J. Appl. Phys.* **41** 3279
[10] Miller R C and Savage A 1959 *J. Appl. Phys.* **30** 808
[11] Landau L D and Lifshitz E M 1957 *Electrodynamics of Continuous Media* (Moscow: Gostekhizdat)

[12] Nassau K and Lines M E 1970 *J. Appl. Phys.* **41** 533

[13] Amodei J J and Staebler D L 1972 *RCA Rev.* **33** 71

[14] Staebler D L and Amodei J J 1972 *Ferroelectrics* **3** 107

[15] Amodei J J 1971 *Appl. Phys. Lett.* **18** 22

[16] Amodei J J 1971 *RCA Rev.* **32** 185

[17] Auston D H, Glass A M and Ballman A A 1972 *Phys. Rev. Lett.* **28** 897

[18] Glass A M and Auston D H 1972 *Opt. Commun.* **5** 45

[19] Glass A M, von der Linde D and Negran T 1974 *J. Appl. Phys. Lett.* **25** 233

[20] Glass A M and Auston D H 1974 *Ferroelectrics* **7** 187

[21] Von der Linde D, Glass A M and Rodgers K 1974 *Appl. Phys. Lett.* **25** 155

[22] Von der Linde D, Auston D N, Glass A M *et al* 1974 *Solid State Commun.* **14** 137

[23] Di Domenico M J and Wemple S H 1969 *J. Appl. Phys.* **40** 720

[24] Belincher V I, Malinovsky V K and Struman B I 1977 *ZhETF* **73** 692

[25] Tsuya H 1977 *Izv. Akad. Nauk SSSR: Ser. Phys.* **41** 740

[26] Von Baltz R 1978 *Phys. Status Solidi* (b) **89** 419

[27] Drivotin O I, Korovin L I and Kudinov E K 1978 *Ferroelectrics* **22** 667

[28] Fridkin V M 1977 *Appl. Phys.* **13** 357

[29] Voronov V V, Ionov P V, Kuz'minov Yu S *et al* 1976 *Fiz. Tverd. Tela* **18** 286

[30] Ionov P V, Mikhailina K A, Ivleva L I *et al* 1973 *Kratkie Soobshcheniya po Fizike* **10** 24

[31] Volk T R, Kochev K D and Kuz'minov Yu S 1975 *Kristallografiya* **20** 583

[32] Fridkin V M, Kochev K D, Kuz'minov Yu S *et al* 1976 *Phys. Status Solidi* (a) **33** K137

[33] Ginzberg A V, Kochev K D, Kuz'minov Yu S *et al* 1978 *Ferroelectrics* **18** 71

[34] Volk T R, Kovalevich V I and Kuz'minov Yu S 1978 *Ferroelectrics* **22** 659

[35] Volk T R, Kovalevich V I, Polozkov N M *et al* 1979 *Fiz. Tverd. Tela* **21** 2591

[36] Voronov V V, Kuz'minov Yu S and Lukina I G 1976 *Fiz. Tverd. Tela* **18** 1047

[37] Ginzberg A V, Kochev K D, Kuz'minov Yu S *et al* 1975 *Phys. Status Solidi* (a) **29** 309

[38] Volk T R, Kovalevich V I, Kuz'minov Yu S *et al* 1980 *Avtometriya* **1** 46

[39] Voronov V V, Kuz'minov Yu S and Osiko V V 1976 *Kvant. Elektron.* **3** 2101

[40] Voronov V V, Kuz'minov Yu S, Osiko V V and Prokhorov A M 1980 *Kristallografiya* **25** 1208

[41] Voronov V V, Kuz'minov Yu S and Osiko V V 1979 *Fiz. Tverd. Tela* **21** 3061

[42] Levanyuk A P and Osipov V V 1975 *Izv. Akad. Nauk SSSR: Ser. Phys.* **39** 686

[43] Levanyuk A P and Osipov V V 1975 *Fiz. Tverd. Tela* **17** 3595

[44] Phillips W, Amodei J J and Staebler D L 1972 *RCA Rev.* **33** 94

[45] Shvarz K K 1977 *Izv. Akad. Nauk SSSR: Ser. Phys.* **41** 788

[46] Volk T R, Ginzberg A V, Kovalevich V I *et al* 1977 *Izv. Akad. Nauk SSSR: Ser. Phys.* **41** 783

[47] Kovalevich V I, Shuvalov L A and Volk T R 1978 *Phys. Status Solidi* (a) **45** 249

[48] Abramov N A, Voronov V V and Kuz'minov Yu S 1978 *Ferroelectrics* **22** 649

[49] Abramov N A and Voronov V V 1979 *Fiz. Tverd. Tela* **21** 1234

[50] Voronov V V 1980 *Candidate Dissertation 'Photoelectric and Photorefractive Properties of Ferroelectric Niobate Crystals'* (Moscow: Lebedev Physical Institute, USSR Acad. Sci.)

[51] Ohnishi N 1977 *Jap. J. Appl. Phys.* **16** 1451

[52] Ohnishi N 1978 *1st Meeting on Ferroelectrics, Kyoto, Japan* p 145

[53] Sugh K, Iwasaki H, Miyazawa S *et al* 1973 *J. Cryst. Growth* **18** 159

[54] Krausslish J and Mohring H 1974 *Kristall und Technik.* **9** 811

[55] Mustel' E P and Parygin V N 1970 *Methods of Modulation and Scanning of Light* (Moscow: Nauka)

[56] Fridkin V M 1979 *Photoferroelectrics* (Moscow: Nauka)

[57] Krivoglaz M A 1967 *The Theory of Scattering of X-rays and Thermal Neutrons by Real Crystals* (Moscow: Nauka)

[58] Ohnishi N and Jizuka T 1975 *J. Appl. Phys.* **46** 1063

[59] Ohnishi N and Jizuka T 1974 *Ferroelectrics* **7** 269

[60] Fridkin V M, Grekov A A, Ionov P V *et al* 1974 *Ferroelectrics* **8** 433

[61] Evlanova N F 1978 *Candidate Dissertation 'Domain Structure of Czochralski Grown Lithium Metaniobate Single Crystals* (Moscow: Moscow State University)

[62] Thaxter J B and Kestigian M 1974 *Appl. Opt.* **13** 913

[63] Kogelnik H W 1969 *Bell Syst. Tech. J.* **48** 2909

[64] Tufte O N and Chen D 1973 *IEEE Spectrum* **10** 26

[65] Staebler D L and Phillips W 1974 *Appl. Opt.* **13** 788

[66] Micheron F and Besmuth G 1972 *J. Phys. Suppl.* **33** C2 149

[67] Trotier J C and Micheron F 1974 *J. Phys. Suppl.* **35** C3 119

[68] Angert N B, Pashkov V A and Solovieva N M 1972 *ZhETF* **62** 1666

[69] Amodei J J, Staebler D L and Stephens W 1971 *Appl. Phys. Lett.* **18** 507

[70] Amodei J J and Staebler D L 1971 *Appl. Phys. Lett.* **18** 540

[71] Micheron F, Mayeir C and Trotier J C 1974 *Appl. Opt.* **13** 784

[72] Micheron F and Besmuth G 1973 *Appl. Phys. Lett.* **23** 71

[73] Micheron F 1973 *Rev. Tech. Thomson CSF* **5** 263

[74] Huignard J P, Herran J P and Micheron F 1975 *Appl. Phys. Lett.* **26** 256

[75] Ohmori Y, Yasojima Y, Addachi U *et al* 1973 *Technol. Rep. Osaka University* **23** No 1166 p 105

[76] Bordogna J, Keneman S A and Amodei J J 1974 *RCA Rev.* **33** 227

[77] Megumi K, Kozuca H, Kobayashi K *et al* 1977 *Appl. Phys. Lett.* **36** 631

[78] Voronov V V, Gulanyan E X, Dorosh I R *et al* 1979 *Kvant. Elektron.* **6** 1993

[79] Megumi K, Nogatsuma N, Kashiwada Y *et al* 1976 *J. Mater. Sci.* **11** 1583

[80] Lenzo P V, Spencer E G and Ballman A A 1967 *Appl. Phys. Lett.* **11** 23

[81] Voronov V V, Dorosh I R, Kuz'minov Yu S *et al* 1980 *Kvant. Elektron.* **7** 2313

[82] Dorosh I R, Kuz'minov Yu S, Osiko V V *et al* 1981 *Fiz. Tverd. Tela* **23** 609

[83] Dorosh I R, Kuz'minov Yu S and Tkachenko N V 1981 *Avtometriya* **5** 27

[84] Dorosh I R, Kuz'minov Yu S and Polozkov N M 1981 *Phys. Status Solidi* (a) **65** 513

[85] Amodei J J, Staebler D L and Stephens A W 1971 *Appl. Phys. Lett.* **18** 507

[86] Amodei J J, Phillips W and Staebler D L 1971 *IEEE J. Quantum Electron.* **QE-7** 321

[87] Spinhirne J M and Estle T L 1974 *Appl. Phys. Lett.* **25** 38

[88] Bobrinev V I, Vasiliev Z G, Gulanyan E H *et al* 1973 *Pis'ma v ZhETF* **18** 267

[89] Gulanyan E H, Vasiliev Z G, Gabrielyan V T *et al* 1974 *Voprosy Radioelektroniki: Ser. Gen. Phys.* **1** 40

[90] Pradeen Shan, Rabson T A, Tittel F K and Gaylord T K 1970 *Appl. Phys.* **24** 130

[91] Ionov V P, Voronov V V and Gabrielyan V T 1975 *Fiz. Tverd. Tela* **17** 1144

[92] King S R, Hartmick T S and Chase A V 1972 *Appl. Phys. Lett.* **21** 312

[93] Ishida A, Mikami O, Miyazawa S *et al* 1972 *Appl. Phys. Lett.* **21** 192

[94] Van Uitert L G, Levinstein H J, Rubin J J, Capio C D *et al* 1968 *Mater. Res. Bull.* **3** 47–57

[95] Voronov V V and Kuz'minov Yu S 1978 *Fiz. Tverd. Tela* **20** 389–94

[96] Rivkin S M 1963 *Photoelectric Phenomena in Semiconductors* (Moscow: Fizmatgiz)

[97] Bursian E V 1974 *Nonlinear Crystal of Barium Titanate* (Moscow: Nauka) p 295

[98] Belobaev K G, Markov V B and Odulov S G 1978 *Fiz. Tverd. Tela* **20** 2520–2

[99] Belincher V I, Kopaev I T, Malinivsky V K and Struman B I 1976 *Fiz. Tverd. Tela* **18** 798

[100] Schein L B, Cressman P I and Gross L E 1978 *J. Appl. Phys.* **49** 798

[101] Furuhata Y 1976 *J. Jap. Assoc. Cryst. Growth* **3** 1–47

[102] Staebler D L and Amodei J J 1972 *J. Appl. Phys.* **43** 1042–9

[103] Levanyuk A P and Osipov V V 1977 *Izv. Akad. Nauk SSSR: Ser. Phys.* **41** 752

[104] Young L, Wong W K Y, Thewalt M L W and Corhish W D 1974 *Appl. Phys. Lett.* **24** 264–5

[105] Dorosh I R 1979 *Voprosy Radioelektroniki* **3** 119–25

[106] Solymar L and Jordan M P 1977 *Opt. Quantum Electron.* **9 m** 437–44

Chapter 9

[1] Abrahams S C, Kurtz S K and Jamieson P B 1968 *Phys. Rev.* **172** 555

[2] Giess E A, Scott B A, Burns G, Smith A W, Olson B L and O'Kane D F 1970 *Mater. Res. Bull.* **5** 109

[3] Ravez J, Perron A, Hagenmuller P and Rivoallan L 1975 *Mater. Res. Bull.* **10** 201

[4] Ravez J, Perron A, Chaminade J P, Rivoallan L and Hagenmuller P 1974 *J. Solid State Chem.* **10** 274

[5] Ravez J and Budin J P 1972 *C. R. Acad. Sci., Paris* **274** 635

[6] Wemple S H and Di Domenico M 1969 *J. Appl. Phys.* **40** 720

[7] Born M and Wolf E 1964 *Principles of Optics* (Oxford: Pergamon)

[8] Kamzina L S, Krainik N N and Berezhnoi A A 1973 *Fiz. Tverd. Tela* **15** 3011

[9] Yataka U 1974 *Jap. J. Appl. Phys.* **13** 1362

[10] Bernecker O, Mattnes H and Marshall A 1973 *Phys. Status Solidi* (a) **17** 453

[11] Itoh Y and Iwasaki H 1973 *J. Phys. Chem. Solids* **34** 1639

[12] Kojima F, Kuwata J and Nomura S 1977 *Proc. 1st Mtg. on Ferroelectric Materials and Their Applications* (*Kyoto, Japan*)

[13] Kahn A H and Leyendecker A J 1964 *Phys. Rev.* **135** A1321

[14] Kurtz S K 1966 *Proc. Int. Mtg. on Ferroelectricity: Czechoslovak Academy of Sciences* vol 1 (Prague) p 413

[15] Cardona M 1965 *Phys. Rev.* **140** A651

[16] Zook J D and Casselman T N 1966 *Phys. Rev. Lett.* **17** 960

[17] Brews J R 1967 *Phys. Rev. Lett.* **18** 662

[18] Wemple S H, Di Domenico M and Gamlibel I 1968 *Appl. Phys. Lett.* **12** 209

[19] Geusic J E, Kurtz S K, Van Uitert L G and Wemple S H 1964 *Appl. Phys. Lett.* **4** 141

[20] Wemple S H and Di Domenico M 1969 *J. Appl. Phys.* **40** 735

[21] Frova A and Bodoy B 1967 *Phys. Rev.* **153** 606

[22] Zook J D 1968 *Phys. Rev. Lett.* **20** 848

[23] Di Domenico M and Wemple S H 1968 *Phys. Rev.* **166** 565

[24] Miller R C 1964 *Appl. Phys. Lett.* **5** 17

[25] Miller R C 1970 *J. Phys. Soc. Japan Suppl.* **28** 15

[26] Garrett C G B and Robinson F N H 1966 *IEEE J. Quantum Electron.* **QE-2** 328

[27] Kurtz S K and Robinson F N H 1967 *Appl. Phys. Lett.* **10** 62

[28] Burns G and Scott B A 1973 *Solid State Commun.* **13** 417

[29] Burns G and Scott B A 1973 *Solid State Commun.* **13** 423

[30] Burns G and Burstein E 1974 *Ferroelectrics* **7** 297

[31] Yacoby Y and Just D 1974 *Solid State Commun.* **15** 715

[32] Burns G 1976 *Phys. Rev.* B **13** 215

[33] Smolensky G A 1970 *J. Phys. Soc. Japan Suppl.* **28** 26

[34] Scott B A, Giess E A, Olson B L, Burns G, Smith A W and O'Kane D F 1970 *Mater. Res. Bull.* **5** 47

[35] Ainger F W, Beswick J A, Porter S G and Clarke R 1972 *Ferroelectrics* **3** 321

[36] Clarke R and Burfoot J C 1974 *Ferroelectrics* **8** 505

[37] Burns G and Smith A W 1968 *IEEE J. Quantum Electron.* **QE-4** 584

[38] Smith A W, Burns G and O'Kane D F 1971 *J. Appl. Phys.* **42** 250

[39] Smith A W, Burns G, Scott B A and Edmonds H D 1971 *J. Appl. Phys.* **42** 684

[40] Robinson F N H 1967 *Bell Syst. Tech. J.* **46** 913

[41] Jeggo C R and Boyd G D 1971 *J. Appl. Phys.* **41** 2741

[42] Flytrains C and Ducuing J 1969 *Phys. Rev.* **178** 1218

[43] Phillips J C and Van Vechten J A 1969 *Phys. Rev.* **183** 709

[44] Kleinman D A 1962 *Phys. Rev.* **128** 1761

[45] Kuz'minov Yu S 1975 *Lithium Niobate and Tantalate as Materials for Nonlinear Optics* (Moscow: Nauka)

[46] Miller R C and Nordland W A 1970 *Appl. Phys. Lett.* **16** 174

[47] Dorozhkin L M, Kisel' V A, Shigorin V D and Shipulo G P 1976 *Pis'ma v ZhETF* **24** 363

[48] Dorozhkin L M, Kisel' V A, Kuz'minov Yu S *et al* 1977 *Kvant. Elektron.* **4** 2266

[49] Sakudo T and Fuji E 1977 *Izv. Akad. Nauk SSSR: Ser. Phys.* **41** 517

[50] Armstrong J A, Bloembergen N, Ducuing J and Pershan P S 1962 *Phys. Rev.* **127** 1918

[51] Freund I 1968 *Phys. Rev. Lett.* **21** 1404

[52] Boyd G D *et al* 1964 *Appl. Phys. Lett.* **5** 234

[53] Midwinter J E and Warner J 1965 *Br. J. Appl. Phys.* **16** 1135

[54] Miller R C, Boyd G D and Savage A 1965 *Appl. Phys. Lett.* **6** 77

[55] Bergman J G 1968 *Appl. Phys. Lett.* **12** 92

[56] Liu S T and Zook J D 1974 *Ferroelectrics* **7** 171

[57] Liu S T, Zook J D and Long D 1975 *Ferroelectrics* **9** 39

[58] Zook J D and Liu S T 1976 *Ferroelectrics* **11** 371

[59] Voronov V V, Karlov N V, Kuz'min G P, Kuz'minov Yu S, Kuritsin B A, Nikiforov S N, Osiko V V and Prokhorov A M 1977 *Kvant. Elektron.* **4** 1903

[60] Zook J D 1974 *Appl. Opt.* **13** 875

[61] Chen F S 1970 *Proc. IEEE* **58** 1440

[62] Wemple S H and Di Domenico M 1972 *Applied Solid State Science* (New York: Academic Press) vol 3, p 264

[63] Zook J D 1973 *Proc. Int. Conf. on Modulation Spectroscopy, Surf. Sci.* **37** 244

Index